MW00760066

Sonic and Photonic Crystals

Sonic and Photonic Crystals

Editors

Lien-Wen Chen
Jia-Yi Yeh

MDPI • Basel • Beijing • Wuhan • Barcelona • Belgrade • Manchester • Tokyo • Cluj • Tianjin

Editors
Lien-Wen Chen
National Cheng Kung University
Taiwan

Jia-Yi Yeh
Chung Hwa University of Medical Technology
Taiwan

Editorial Office
MDPI
St. Alban-Anlage 66
4052 Basel, Switzerland

This is a reprint of articles from the Special Issue published online in the open access journal *Crystals* (ISSN 2073-4352) (available at: https://www.mdpi.com/journal/crystals/special_issues/sonic_photonic_crystals).

For citation purposes, cite each article independently as indicated on the article page online and as indicated below:

LastName, A.A.; LastName, B.B.; LastName, C.C. Article Title. *Journal Name* **Year**, *Article Number*, Page Range.

ISBN 978-3-03936-660-6 (Hbk)
ISBN 978-3-03936-661-3 (PDF)

Contents

About the Editors

Lien-Wen Chen obtained his Ph.D. in Mechanical Engineering, Rensselaer Polytechnic Institute, USA. Then, he joined the member of department of Mechanical Engineering of National Cheng Kung University, Taiwan. He is expert in photonic and phononic crystal system, plate, shell and composite structure dynamics and he is the author of more than 165 papers about applications of those investigation topics.

Jia-Yi Yeh obtained the Ph.D. in department of Mechanical Engineering of National Cheng Kung University, Taiwan. He works in Dept. of Information Management, Chung Hwa University of Medical Technology, Tainan, Taiwan and act as associate professor and vice president for Library and Information Services. He majored in structure vibration and control and also published in sonic crystal applications. He is the author of more than 40 research papers and 60 conferences paper.

Preface to "Sonic and Photonic Crystals"

This Special Issue on "Sonic and Photonic Crystals" is focused on broad applications of the results involving characterizations of the sonic and photonic crystal properties.

Various applications of photonic crystals are presented. The gradient cavity, waveguide, switch, and spatial beam filtering with autocloned photonic crystals are studied and discussed [Huang et al., Wang et al., Jao et al., Francis et al., Azizpour et al., Ren et al.]. Studies of the metamaterial of crystal structures can be found in [Luan, Nguyen et al., Li et al., Chang et al.], in which the negative effects of star-shaped structures were investigated and also discussed with respect to application.

Elastic wave propagations in phononic crystals is also an interesting topic. Elastic wave propagations in metamaterials, elastic piezoelectric phononic crystals, and energy harvesting phononic devices are also covered in [Le et al., Lv et al., Liang et al., Deng et al.].

Additional studies examining various photonic crystal applications also form part of this collection. The deterministic insertion of KTP nanoparticles into polymeric structures, flexible photonic nanojets with cylindrical graded-index lens, and photonic crystal fiber investigations are presented [Nguyen and Lai, Liu, Zhang et al., Yang et al.]. The design of a polarization splitter and converter based on square lattice photonic crystal fiber is investigated in [17, 18] and includes a discussion of the polarization characteristics of photonic crystals. Finally, a review paper of the recent advances in colloidal photonic crystal-based anticounterfeiting materials is included [Ren et al.].

This Special Issue presents and discusses the work of scientists studying a wide range of sonic/photonic crystal applications toward advancing this field.

Lien-Wen Chen, Jia-Yi Yeh

Editors

Article

Trapping and Optomechanical Sensing of Particles with a Nanobeam Photonic Crystal Cavity

Lin Ren *, Yunpeng Li, Na Li and Chao Chen

School of Aviation Operations and Services, Aviation University of air force, Changchun 130022, China; ew_radar@163.com (Y.L.); lina0629@163.com (N.L.); chenchao.19830823@163.com (C.C.)

* Correspondence: renlin_ok@163.com; Tel.: +86-13504329261

Received: 5 December 2018; Accepted: 17 January 2019; Published: 22 January 2019

Abstract: Particle trapping and sensing serve as important tools for non-invasive studies of individual molecule or cell in bio-photonics. For such applications, it is required that the optical power to trap and detect particles is as low as possible, since large optical power would have side effects on biological particles. In this work, we proposed to deploy a nanobeam photonic crystal cavity for particle trapping and opto-mechanical sensing. For particles captured at 300 K, the input optical power was predicted to be as low as 48.8 μW by calculating the optical force and potential of a polystyrene particle with a radius of 150 nm when the trapping cavity was set in an aqueous environment. Moreover, both the optical and mechanical frequency shifts for particles with different sizes were calculated, which can be detected and distinguished by the optomechanical coupling between the particle and the designed cavity. The relative variation of the mechanical frequency achieved approximately 400%, which indicated better particle sensing compared with the variation of the optical frequency (±0.06%). Therefore, our proposed cavity shows promising potential as functional components in future particle trapping and manipulating applications in lab-on-chip.

Keywords: optical force; photonic crystal cavity; particle trapping; optomechanical sensing

---

1. Introduction

Optical force is regarded as an ideal tool for trapping and manipulating vulnerable particles due to its contactless and nondestructive properties [1]. A photonic crystal cavity, formed by introducing a defect into a periodic arrangement of different materials, is able to build an optical potential around the defects and traps of the manipulated particles with the advantage of the cavity-enhanced optical force [2–4]. Therefore, the enlarged optical force via a high-quality factor photonic crystal cavity is promising for trapping and manipulating various vulnerable particles. Compared with two-dimensional photonic crystal structures [5–8], one-dimensional nanobeam photonic crystal cavities have attracted considerable attention due to their compact nanobeam structures [9], ease of integration of coupling waveguides [10], high quality factors [11] and small mode volumes [12]. The relationship of the optical gradient force between the quality factor, mode volume and transmission has been systematically analyzed in the nanobeam photonic crystal cavities [13]. Among these works on cavity enhanced optical forces, the primary target is to trap and manipulate particles with optical power as low as possible, since the enhanced optical absorption by the high-quality-factor cavities would have side effects on biological particles such as bacteria, proteins, and viruses due to a severe increase in temperature [14]. X. Serey et al. have compared several photonic crystal cavities and showed that polystyrene particles with different diameters could be trapped with the input optical power of 10 mW [15]. N. Descharmes et al. have presented 500-nm dielectric particles with low intracavity powers of only 120 μW for long-time optical trapping [16]. Despite these advances, optical trapping power still needs to be further lowered to reduce the influence on biomolecules. Furthermore, the optomechanical coupling between the cavity and the particle due to the self-induced back-action optical trapping is expected to isolate, analyze, or sort different particles depending on their mechanical performances [1,16–18].

In this work, we proposed to utilize a larger-center-hole nanobeam photonic crystal cavity [19,20] in silicon for trapping and opto-mechanical sensing particles. The radius of air holes increases in the center of the defect region in the designed cavity so that large optical force can be obtained. For the calculation of optical force, a polystyrene particle was introduced and located at the center above the cavity, while the trapping cavity was set in an aqueous environment. The optical force and potential on the introduced particle with radius of 150 nm were calculated at a temperature of 300 K. For particles captured at this condition, the input optical power is required to be as low as 48.8 μW to generate an optical potential of 4.14×10^{-20} J (10 k_BT). We additionally calculate the optical and mechanical frequency shift for the particles with radius ranging from 100 nm to 500 nm. Accordingly, the change of the mechanical frequency shift can be detected and analyzed based on the optomechanical coupling between the particle and the designed cavity.

2. Results

2.1. Design of a Photonic Crystal Cavity

We designed a larger-center-hole nanobeam photonic crystal cavity fabricated on silicon-on-insulator for particle trapping and sensing. The structure of the cavity is shown in Figure 1a. The radius of air holes, r_n, linearly increases from the mirror region to the center of the cavity defect region, where $r_n = r_0 + 4 \times n$ (nm). Here, r_0 is fixed at a radius of 100 nm in the whole mirror region, and the largest hole in the center of the cavity defect region is set with a radius of r_7 = 128 nm. For the entire structure, the lattice constant a, the distance between two adjacent holes, is kept as 365 nm. The width and the thickness of the nanobeam are 480 nm and 220 nm, respectively.

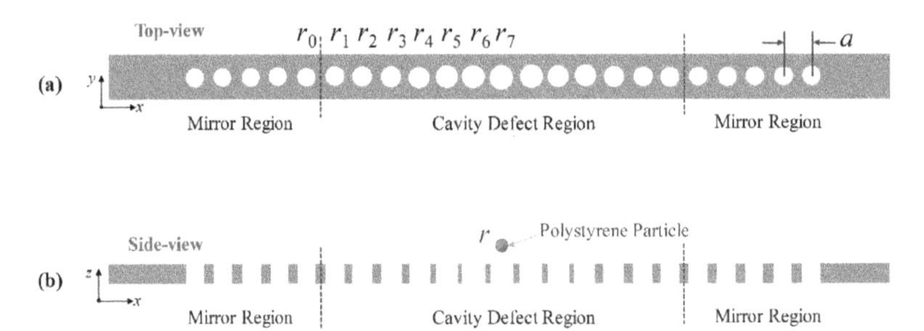

Figure 1. (**a**) Top-view and (**b**) Side-view of the structure of the larger-center-hole nanobeam photonic crystal cavity, above which a polystyrene particle is introduced. The position of the polystyrene particle deviates from the center of the largest hole of the cavity in the x direction is denoted as Δx. The distance between the center of the polystyrene particle and the top surface of the nanobeam cavity in the z direction is denoted as Δz.

For particle trapping and sensing, the cavity was simulated in an aqueous environment with a polystyrene particle located above the center of the cavity, as shown in Figure 1b. The position of the polystyrene particle deviated from the center of the largest hole of the cavity in the x direction is denoted as Δx. The distance between the center of the polystyrene particle and the top surface of the nanobeam cavity in the z direction is denoted as Δz. The optical trapping force and the optical potential on the manipulated polystyrene particle were analyzed by varying its position based on the designed larger-center-hole nanobeam cavity. In the simulation, the material of the designed cavity and the particle are silicon and polystyrene, respectively, with the corresponding refractive index of 3.42 and

1.59. The whole cavity is set in aqueous environment, whose refractive index is 1.33. The Young's modulus of the polystyrene is 3300 MPa.

The optical frequency of the defect mode in the photonic band structure of the designed cavity is presented in Figure 2a. The photonic band structure is calculated by plane wave expansion (PWE) method. It can be seen that the optical resonant mode is confined in the photonic band gap of the mirror structure. And the optical wavelength of the resonant cavity mode is calculated to be 1.5477 μm without the polystyrene particle. The optical transmission spectra of the designed cavity are simulated by finite-difference time-domain (FDTD) method and presented in Figure 2b. The optical quality factor (Q) of the cavity can be calculated from the transmission spectra and the Q factor for the designed cavity is 3968.

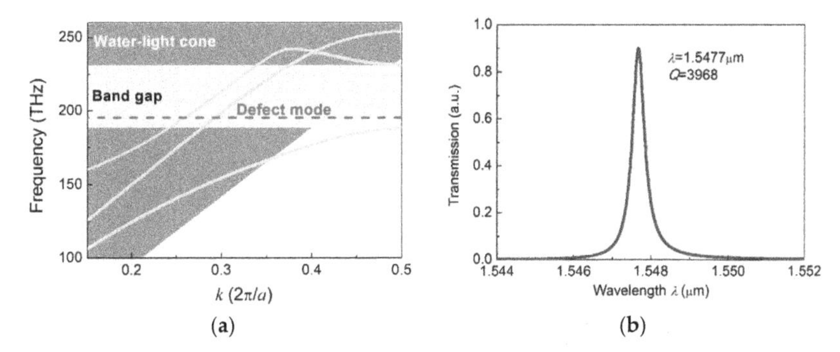

Figure 2. (**a**) The photonic band structure of the larger-center-hole nanobeam photonic crystal cavity. The optical frequency of the defect mode of the designed cavity is also presented in the photonic band gap with dash line. (**b**) The optical transmission spectra of the designed cavity calculated by finite-difference time-domain (FDTD) simulation with resonant wavelength of 1.5477 μm and Q factor of 3968.

To study the optical trapping force and the optical potential on the polystyrene particle, the optical fields in x and z direction of the designed cavity with the polystyrene particle located at different positions were simulated and given in Figure 3. Here, the location of the polystyrene particle is set as (a) $\Delta x = 0$ nm, $\Delta z = 5$ nm and (b) $\Delta x = 190$ nm, $\Delta z = 5$ nm, respectively. It can be seen that the particle is located at the position of the largest optical field with $\Delta x = 190$ nm, $\Delta z = 5$ nm in Figure 3b. Accordingly a strong optical trapping can be achieved for the polystyrene particle at this position. This result can also be obtained by the calculation for the optical force and potential being applied on the polystyrene particle in the following, where a strong optical trapping potential for the polystyrene particle is located at $\Delta x = 190$ nm, $\Delta z = 5$.

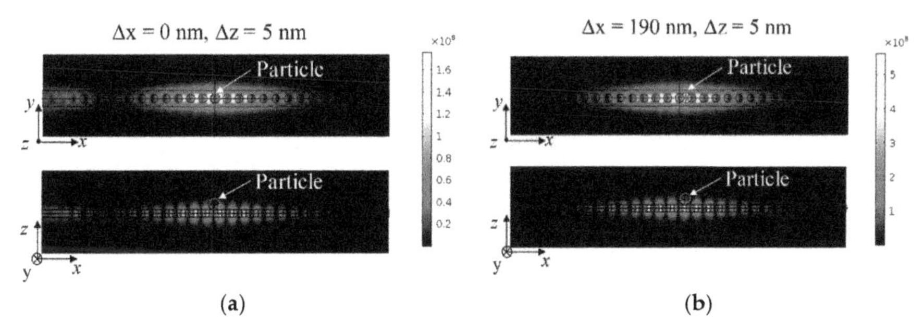

Figure 3. The optical fields in x and z direction of the designed cavity with polystyrene particle located at different position (**a**) $\Delta x = 0$ nm, $\Delta z = 5$ nm, and (**b**) $\Delta x = 190$ nm, $\Delta z = 5$ nm.

2.2. Optical Force and Potential on a Particle

The optical force on the particle was calculated by the distribution of the electric field and magnetic field with Maxwell Stress Tensor [21]. In our calculation, the optical field of the designed cavity with the introduced polystyrene particle was first simulated by finite element analysis. Then, the optical force (f) on the particle was calculated by integrating the Maxwell stress tensor over the surfaces of the particle:

$$f = \int_s T \cdot d\vec{n}, \tag{1}$$

Where S is the surface of the particle, $d\vec{n}$ is the surface normal and T denotes the Maxwell stress tensor given as:

$$T_{ij} \equiv \varepsilon\left(E_i E_j - \frac{1}{2}\delta_{ij}E^2\right) + \frac{1}{\mu_0}\left(B_i B_j - \frac{1}{2}\delta_{ij}B^2\right), \tag{2}$$

Where ε is the electric constant and μ_0 is the magnetic constant, E is the electric field, B is the magnetic field, and δ_{ij} is Kronecker's delta function.

It should be noted that the potential is affected by the presence of the particle, since the refractive index of the polystyrene particle is larger than that of water. Thus, the optical force and potential were calculated with the particle located above the cavity. In the simulation, the light is input from one port of the nanobeam waveguide. Its frequency is set as the optical resonant frequency of the optical cavity with the particle located 5 nm above the cavity for the x direction (Δx is changed with $\Delta z = 5$). The calculated results are presented in Figure 4a. The calculation of optical force and potential on the particle in the z direction, in which the particle is located at a fixed $\Delta x = 190$ nm and a varied Δz, is shown in Figure 4b. Here, the negative and positive optical force refer to the direction of the force on the particle. For example, in the x direction, the negative or positive optical force refers to the force direction opposite or along the x axis, respectively. For the z direction, the negative optical forces indicate that the direction of the force is opposite the z axis, so that the particle will be trapped just above the cavity.

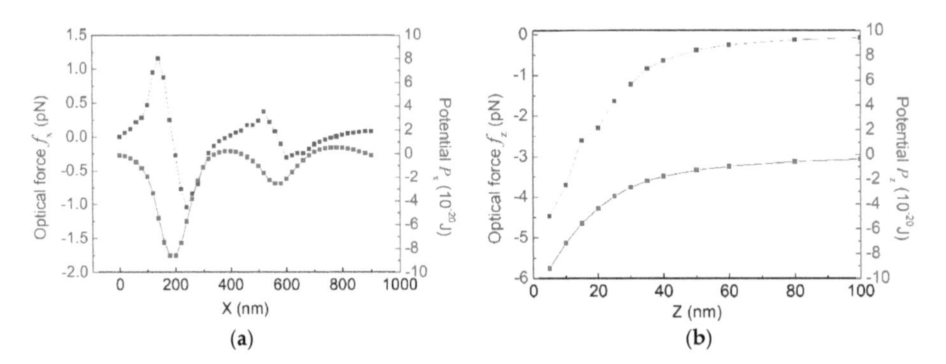

Figure 4. The optical force and potential on the particle (**a**) in the x direction and (**b**) in the z direction with the input light power of 100 μW.

Since dielectric particles are attracted in the region of strongest electric field, a strong optical trapping can be realized in an optical cavity. When the light is input into the cavity, an optical potential in the x direction and the z direction around the cavity defect can be built for trapping the polystyrene particle with radius of 150 nm. For particle capturing, the minimum potential of 4.14×10^{-20} J (10 k_BT) is required at temperature of 300 K. For x direction, the potential well depth is 8.48×10^{-20} J when the input power is 100 μW, as shown in Figure 4a. Since that the potential well depth increases linearly with the input light power, the input power is required to be 48.8 μW when the well depth reaches 4.14×10^{-20} J (10 k_BT). In addition, we also calculated the optical force and potential in the z direction while the particle is located at the strongest trapping of the potential well with Δx = 190 nm. As shown in Figure 4b, when the input power is 100 μW, the potential well depth is 9.17×10^{-20} J. Thus, when the well depth reaches 10 k_BT, the input power is required to be 45.1 μW. Taking the required input power for both x and z direction into consideration, it can be concluded that the input optical power is predicted as low as 48.8 μW for particles captured at 300 K. It should be mentioned that the coupling loss between the input light and the trapping cavity has been considered in our simulation. Thus, the input power for particle trapping using this proposed cavity is much lower compared with other reported work (120 μW–10 mW) [15,16], which benefits to mitigate the side effect on biological particles. In addition, it is possible to attract more than one particle from the water dispersion. It can be seen from Figure 4a that there is more than one potential well in the x direction. If the input optical power goes up, more potentials can be built for trapping multiple particles by the nanobeam photonic crystal cavity.

2.3. Optomechanical Sensing

Optomechanical coupling between cavity and particle inherently exists due to the mechanism of self-induced back-action optical trapping [16]. Based on optomechanical coupling, when we distinguish different particles, both the optical and mechanical frequency shift can be read out by the light output from the cavity. Here, the optical and mechanical frequency shift for the polystyrene particles with different radiuses ranging from 100 nm to 500 nm are calculated and shown in Figure 5a and 5b, respectively. The inset of Figure 5a shows the configuration for the detected nanobeam cavity with the particle located above the cavity.

When the particles with different radiuses, ranging from 100 nm to 500 nm, appear near the cavity, the surrounding refractive indices of the designed cavity change with a small variation, accordingly. Therefore, the optical resonant frequency of the cavity will shift dependent of the particles with different sizes. The changing percent of the optical resonant frequency of the cavity, $\Delta f / f_o$, was also calculated and presented in Figure 5a. Here, the polystyrene particle is located at Δx = 190 nm, Δz = 5 nm. It can be seen that there is only a relative variation of ±0.06% approximatively for the optical frequency when the particle enlarged with radius from 100 nm to 500 nm.

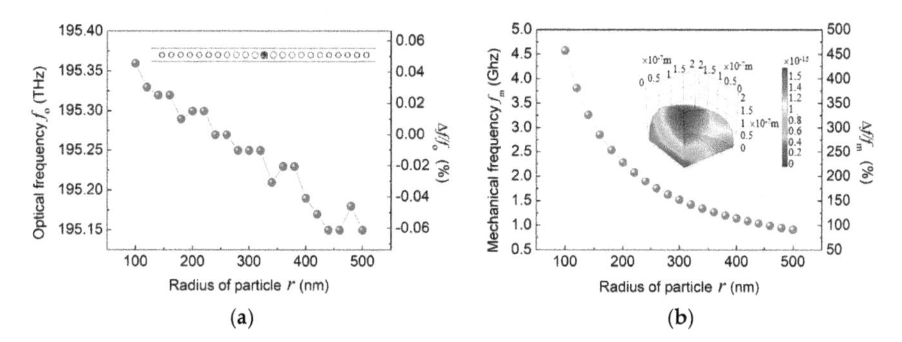

Figure 5. (**a**) The optical frequency shift for the particle with different radius, ranging from 100 nm to 500 nm. The relative changing percent of the optical resonant frequency of the cavity, $\Delta f / f_o$, was also calculated. Here, the polystyrene particle is located at Δx = 190 nm, Δz = 5 nm. The inset shows the configuration for the detected nanobeam cavity with the particle above the cavity. (**b**) The mechanical frequency shift and the relative changing percent of the mechanical frequency of the cavity, $\Delta f / f_m$ for the particle with a different radius. The inset shows the strain for the vibrated particle with a radius of 200 nm.

On the other hand, the mechanical frequency of the particle decreases from 4.6 GHz to 0.9 GHz with the increasing size of the particles from 100 nm to 500 nm in radius. Here, the mechanical mode of the particle is the lowest order mechanical mode with the symmetric strain in the x, y, and z axes, which can be detected by the optomechanical coupling between the cavity and the particle due to the self-induced back-action optical trapping. The corresponding changing percent of the mechanical frequency of the cavity, $\Delta f / f_{m}$, achieves approximately 400%. This is because the particle size significantly affects its mechanical frequency. Here, the strain for the vibrated particle with a radius of 200 nm is also given in the inset of Figure 5b. Therefore, compared with the optical frequency shift of ±0.06%, the mechanical frequency shift shows a better performance for particle sensing, which can be used to distinguish or analyze different particles with great potential.

3. Discussion

In this work, we calculated optical force and potential on an introduced polystyrene particle with radius of 150 nm generated by a larger-center-hole nanobeam photonic crystal cavity. The introduced polystyrene particle located over the cavity was simulated in an aqueous environment. For particles captured at a temperature of 300 K, the input optical power is required to be as low as 48.8 μW in the x direction and 45.1 μW in the z direction within the designed cavity. We also studied the sensing performance of the polystyrene particle dependent on different size by the resonant frequency shift of the optical cavity mode and mechanical mode. These results show a great potential of the designed nanobeam cavity for future lab-on-chip trapping and sensing applications.

Author Contributions: Conceptualization, L.R.; Data curation, L.R.; Formal analysis, L.R. and N.L.; Funding acquisition, Y.L.; Investigation, L.R., Y.L., N.L. and C.C.; Methodology, L.R. and Y.L.; Project administration, N.L.; Resources, C.C.; Software, L.R., Y.L. and N.L.; Supervision, N.L.; Validation, L.R., Y.L., N.L. and C.C.; Visualization, L.R.; Writing—original draft, L.R.; Writing—review & editing, L.R., Y.L., N.L. and C.C.

Funding: This research was funded by the National Natural Science Foundation of China (No. 61571462).

Acknowledgments: The authors would like to thank Chengzhi Yang for his invaluable discussion.

Conflicts of Interest: The authors declare no conflict of interest. The funders had no role in the design of the study; in the collection, analyses, or interpretation of data; in the writing of the manuscript, or in the decision to publish the results.

References

1. Van Thourhout, D.; Roels, J. Optomechanical Device Actuation through the Optical Gradient Force. *Nat. Photonics* **2010**, *4*, 211–217. [CrossRef]
2. Van Leest, T.; Caro, J. Cavity-Enhanced Optical Trapping of Bacteria Using a Silicon Photonic Crystal. *Lab. Chip* **2013**, *13*, 4358–4365. [CrossRef] [PubMed]
3. Quan, Q.; Floyd, D.L.; Burgess, I.B.; Deotare, P.B.; Frank, I.W.; Tang, S.K.Y.; Ilic, R.; Loncar, M. Single Particle Detection in CMOS Compatible Photonic Crystal Nanobeam Cavities. *Opt. Express* **2013**, *21*, 32225–32233. [CrossRef] [PubMed]
4. Renaut, C.; Cluzel, B.; Dellinger, J.; Lalouat, L.; Picard, E.; Peyrade, D.; Hadji, E.; de Fornel, F. On Chip Shapeable Optical Tweezers. *Sci. Rep.-UK* **2013**, *3*, 2290. [CrossRef] [PubMed]
5. Akahane, Y.; Asano, T.; Song, B.S.; Noda, S. High-Q Photonic Nanocavity in a Two-Dimensional Photonic Crystal. *Nature* **2003**, *425*, 944–947. [CrossRef]
6. Zhao, Q.; Cui, K.; Feng, X.; Liu, F.; Zhang, W.; Huang, Y. Low Loss Sharp Photonic Crystal Waveguide Bends. *Opt. Commun.* **2015**, *355*, 209–212. [CrossRef]
7. Zhao, Q.; Cui, K.; Huang, Z.; Feng, X.; Zhang, D.; Liu, F.; Zhang, W.; Huang, Y. Compact Thermo-Optic Switch Based on Tapered W1 Photonic Crystal Waveguide. *IEEE Photonics J.* **2013**, *5*. [CrossRef]
8. Zhao, Q.; Cui, K.; Feng, X.; Liu, F.; Zhang, W.; Huang, Y. Variable Optical Attenuator Based on Photonic Crystal Waveguide with Low-Group-Index Tapers. *Appl. Opt.* **2013**, *52*, 6245–6249. [CrossRef]
9. Deotare, P.B.; McCutcheon, M.W.; Frank, I.W.; Khan, M.; Loncar, M. High Quality Factor Photonic Crystal Nanobeam Cavities. *Appl. Phys. Lett.* **2009**, *94*, 121106. [CrossRef]
10. Quan, Q.; Deotare, P.B.; Loncar, M. Photonic Crystal Nanobeam Cavity Strongly Coupled to the Feeding Waveguide. *Appl. Phys. Lett.* **2010**, *96*, 203102. [CrossRef]
11. Quan, Q.; Loncar, M. Deterministic Design of Wavelength Scale, Ultra-High Q Photonic Crystal Nanobeam Cavities. *Opt. Express* **2011**, *19*, 18529–18542. [CrossRef] [PubMed]

12. Seidler, P.; Lister, K.; Drechsler, U.; Hofrichter, J.; Stoeferle, T. Slotted Photonic Crystal Nanobeam Cavity with an Ultrahigh Quality Factor-To-Mode Volume Ratio. *Opt. Express* **2013**, *21*, 32468–32483. [CrossRef] [PubMed]
13. Han, S.; Shi, Y. Systematic Analysis of Optical Gradient Force in Photonic Crystal Nanobeam Cavities. *Opt. Express* **2016**, *24*, 452–458. [CrossRef]
14. Chen, Y.; Serey, X.; Sarkar, R.; Chen, P.; Erickson, D. Controlled Photonic Manipulation of Proteins and Other Nanomaterials. *Nano Lett.* **2012**, *12*, 1633–1637. [CrossRef]
15. Serey, X.; Mandal, S.; Erickson, D. Comparison of Silicon Photonic Crystal Resonator Designs for Optical Trapping of Nanomaterials. *Nanotechnology* **2010**, *21*, 305202. [CrossRef] [PubMed]
16. Descharmes, N.; Dharanipathy, U.P.; Diao, Z.; Tonin, M.; Houdre, R. Observation of Backaction and Self-Induced Trapping in a Planar Hollow Photonic Crystal Cavity. *Phys. Rev. Lett.* **2013**, *110*, 123601. [CrossRef]
17. Huang, Z.; Cui, K.; Li, Y.; Feng, X.; Liu, F.; Zhang, W.; Huang, Y. Strong Optomechanical Coupling in Nanobeam Cavities Based on Hetero Optomechanical Crystals. *Sci. Rep.-UK* **2015**, *5*, 15964. [CrossRef]
18. Pan, F.; Cui, K.; Bai, G.; Feng, X.; Liu, F.; Zhang, W.; Huang, Y. Radiation-Pressure-Antidamping Enhanced Optomechanical Spring Sensing. *ACS Photonics* **2018**, *5*, 4164–4169. [CrossRef]
19. Li, Y.; Cui, K.; Feng, X.; Huang, Y.; Huang, Z.; Liu, F.; Zhang, W. Optomechanical Crystal Nanobeam Cavity with High Optomechanical Coupling Rate. *J. Opt.-UK* **2015**, *17*, 045001. [CrossRef]
20. Huang, Z.; Cui, K.; Bai, G.; Feng, X.; Liu, F.; Zhang, W.; Huang, Y. High-Mechanical-Frequency Characteristics of Optomechanical Crystal Cavity with Coupling Waveguide. *Sci. Rep.-UK* **2016**, *6*, 34160. [CrossRef]
21. Barton, J.P.; Alexander, D.R.; Schaub, S.A. Theoretical Determination of Net-Radiation Force and Torque for a Spherical-Particle Illuminated by a Focused Laser-Beam. *J. Appl. Phys.* **1989**, *66*, 4594–4602. [CrossRef]

Article

Polarization Converter Based on Square Lattice Photonic Crystal Fiber with Double-Hole Units

Zejun Zhang [1,*,†], Yasuhide Tsuji [2], Masashi Eguchi [3] and Chun-ping Chen [1]

[1] Department of Electrical, Electronics and Information Engineering, Kanagawa University, Yokohama 221-8686, Japan; chen@kanagawa-u.ac.jp
[2] Division of Information and Electronic Engineering, Muroran Institute of Technology, Muroran 050-8585, Japan; y-tsuji@mmm.muroran-it.ac.jp
[3] Department of Opt-Electronic System Engineering, Chitose Institute of Science and Technology, Chitose 066-8655, Japan; megu@ieee.org
* Correspondence: zhang-zj17@kanagawa-u.ac.jp; Tel.: +86-45-481-5661
† Current address: 3-27-1 Rokkakubashi, Kanagawa-ku, Yokohama 221-8686, Japan.

Received: 17 December 2018; Accepted: 21 January 2019; Published: 22 January 2019

Abstract: In this study, a novel polarization converter (PC) based on square lattice photonic crystal fiber (PCF) is proposed and analyzed. For each square unit in the cladding, two identical circular air holes are arranged symmetrically along the $y = x$ axis. With the simple configuration structure, numerical simulations using the FDTD analysis demonstrate that the PC has a strong polarization conversion efficiency (PCE) of 99.4% with a device length of 53 μm, and the extinction ratio is −21.8 dB. Considering the current PCF fabrication technology, the structural tolerances of circular hole size and hole position have been discussed in detail. Moreover, it is expected that over the 1.2∼1.7 μm wavelength range, the PCE can be designed to be better than 99% and the corresponding extinction ratio is better than −20 dB.

Keywords: polarization converter; photonic crystal fiber; square lattice

---

1. Introduction

With the rapid development of Internet, high frequency communication systems have been widely studied to realize the next-generation advanced communication system. Among several means of communication, the optical communication system attracts a lot of attention since it can achieve a high-speed and large-capacity data transmission. In recent years, studies on special optical fibers and high-performance optical devices promote the performance of optical communication system. Photonic crystal fiber (PCF) [1,2] is a newly emerging optical fiber with a periodic arrangement of microscopic air holes running along the fiber axis in the cladding region. For this kind of fiber, the refractive index of cladding can be easily controlled by adjusting the geometry and distribution of the holes. In comparison to conventional optical fiber, PCF shows basic properties like birefringence, nonlinearity and single-polarization transmission that can be tailored to achieve extraordinary outputs [3–5].

Polarization converters (PCs) [6–20] plays an important role in the modern communication systems and photonic integrated circuits, such as polarization division multiplexing systems [21], polarization diversity systems [22], and polarization switches [23]. Until now, several approches have been used to design PCs, which can be classified into two types. The first type is using the mode interference. The waveguide has a high birefringent core by using asymmetrical cross section, liquid crystal or with hybrid plasmonic. Two orthogonal guided modes are excited and beat together rhythmically with the light propagation. Polarization conversion can be achieved when the two guided modes accumulate a π-phase shift after undergoing each beat length. The second type of approach is on the basis of mode coupling. In this approach, at least two waveguides are used. Utilzing the birefringence of the input waveguide, the specified polarization mode is coupled and converted into the adjacent output waveguide by satisfying the phase matching condition. In contrast, another polarization mode in the input waveguide will keep propagating along because of the significant effective index difference between the two waveguides, resulting in very weak coupling. According to these two design mechanisms, plenty of PCs have been proposed based on waveguides or fibers. So far, several kinds of PCF based fiber type PCs have been designed by using a liquid crystal core [11–14], asymmetric core [15,16], plasmonic core [18] and elliptical hole core [19], etc. M. Hameed et al. proposed two kinds of PCs using a liquid crystal core [12–14] and an asymmetric core [15,16], respectively. Although these PCs can achieve a high polarization conversion efficiency (PCE) and a low extinction ratio (ER), the temperature dependence of liquid crystals and the non-Gaussian field distribution caused by the asymmetric structures limit the application of these PCs. L. Chen et al. designed a PC based on hybrid plasmonic PCF in 2014. The PC offers a 93% PCE with a device length of 163 μm. However, the PC suffers from a quite high insertion loss due to the use of metal copper. After then, in 2016, Z. Zhang et al. proposed a cross-talk free PC with a symmetric distribution using several tilted elliptical air holes in the core region [19]. Almost 100% PCE is achieved with a compact device length of only 31.7 μm. Whereas, it suffering a low structural tolerance due to the difficulty of producing elliptical holes in the PCF. Consequently, the use of different air hole shapes or multiple materials increases the manufacturing difficulties and leads to a low structural tolerance. Therefore, in designing a fiber type PC, in addition to achieve a high PCE and a low ER, a Gaussian-like field distribution and a large structural tolerance are also essential.

In this study, a novel PC element based on a square lattice PCF with a simple configuration structure is reported and analyzed. Since only one type of circular air hole is adopted to consist the PC, it can be easily fabricated with the method of stack and draw [24]. The setting of geometric parameters and the light propagation behavior are illustrated in the next section. After that, considering the current PCF fabrication technology, the structural tolerances of the circular air hole size and air hole position have been discussed in detail. Moreover, the wavelength dependence of the proposed PC has also given in Section 3. In this study, the full-vectorial finite-element method (FV-FEM) and finite-difference time-domain (FDTD) method have been used to estimate the modal effective index and the light propagation, respectively.

2. Square Lattice PC with Double-Hole Unit

2.1. Design of Square Lattice PC

In this section, a PC with double-hole unit cells in the cladding is illustrated. Figure 1 shows the cross section of our proposed PC based on a square lattice PCF. The light blue region and white circular area represent the background material SiO_2 and air holes, respectively. The square lattice pitch refers to Λ. For each unit cell, which is represented with red wire frame, two air holes (with the diameter d_c) are arranged symmetrically along the diagonal. The distance between the center of two air holes is represented by ΔD. The angle θ is 45 degrees. In the periodic lattice component, one defect is introduced to confine the light as the core region. Due to the symmetrical distribution along $y = x$, the polarization conversion can be achieved using two excited hybrid modes that are polarized in the $\pm 45^\circ$ directions with respect to the x axis.

According to the phase matching condition, the conversion length L_π is defined as

$$L_\pi = \frac{0.5\lambda}{n_{\text{eff},1} - n_{\text{eff},2}}, \tag{1}$$

where λ is the operation wavelength and $n_{\text{eff},1}$ and $n_{\text{eff},2}$ are the effective indices of fundamental and 1st-higher-order modes, respectively. As an incident light travels to L along the z direction, the PCE of a PC, which is an important performance indicator, is defined as follows [10]:

$$\mathrm{PCE} = \sin^2(2\varphi)\sin^2\left(\frac{\pi L}{2L_\pi}\right) \times 100\%, \tag{2}$$

where the rotation angle φ is defined by the fundamental mode field:

$$\tan\varphi = R = \frac{\iint n^2(x,y)H_x^2(x,y)dxdy}{\iint n^2(x,y)H_y^2(x,y)dxdy}. \tag{3}$$

Here, $n(x,y)$ is the refractive-index distribution, and $H_x(x,y)$ and $H_y(x,y)$ are the non-domain and domain magnetic fields of the fundamental mode, respectively. The modal hybrid, which is represented by R in Equation (3), is a quantity between 0 and 1. Therefore, $R = 1$ results in a PCE of 100% at $L = L_\pi$.

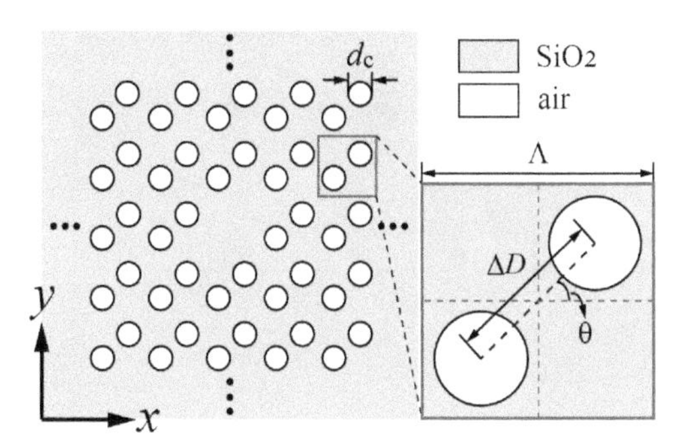

Figure 1. Cross-section view of square lattice PC with double-hole unit in the cladding.

The variation of conversion length with different geometric parameters is investigated at a wavelength of 1.55 μm, as illustrated in Figure 2. Here, the refractive indices of SiO_2 and air are 1.45 and 1, respectively, the lattice pitch is set to $\Lambda = 1$ μm. Simulation results using FV-FEM demonstrate that under these conditions, only two modes (the fundamental mode and the 1st-higher-order mode) are excited to dominate the polarization behavior. The parameter ΔS represents the horizontal distance between the center of air hole and the lattice for each unit cell. It is revealed that the PC has a shorter conversion length with a larger cladding hole size. Additionally, the variation of parameter ΔS has a small effect on the PC conversion length with a large hole size. In particularly, for $d_c/\Lambda = 0.6$, the conversion length is almost unaffected by the ΔS. In this case, considering that adjacent air holes on the same diagonal direction cannot overlap with each other, therefore, the parameter ΔS between 0.22 to 0.28Λ has been investigated to make the PC manufacturable. In this paper, the ΔS is fixed to 0.25Λ, i.e., the hole gap $\Delta D = \sqrt{2}/2\Lambda$. Moreover, the modal hybrids of fundamental and 1st-higher-order modes with different hole sizes have also been discussed. From Figure 3, as increasing the d_c/Λ from 0.4 to 0.6, the conversion length decreases from 384 μm to 53 μm, while the fundamental and 1st-higher-order modes have almost the same modal hybrid which are increasing from 0.914 to almost 1.000. It is evident from this figure that the values of R are better than 0.999 for $d_c/\Lambda > 0.48$.

Figure 4a–d show the magnetic field distributions of H_x and H_y components for fundamental and 1st-higher-order modes with $d_c/\Lambda = 0.6$ at a wavelength of 1.55 μm. It is apparent that each mode has almost the same magnetic field distributions, and the numerical results reveal that the modal hybrid is 0.999995 for the fundamental mode and 0.999996 for the 1st-higher-order mode. The light propagation behavior through the PC with $L_\pi = 53$ μm is shown in Figure 5. The 3D FDTD method with a grid size of $\Delta x = \Delta y = 0.015$ μm and $\Delta z = 0.05$ μm has been adopted in this investigation. The obtained normalized power against the propagation length is given in Figure 6. It is obvious that with a TM mode launched into the PC, it can be completely converted into the TE mode at $z = 53$ μm. In addition, the calculated PCE is better than 99.4% and the ER is better than −21.1 dB.

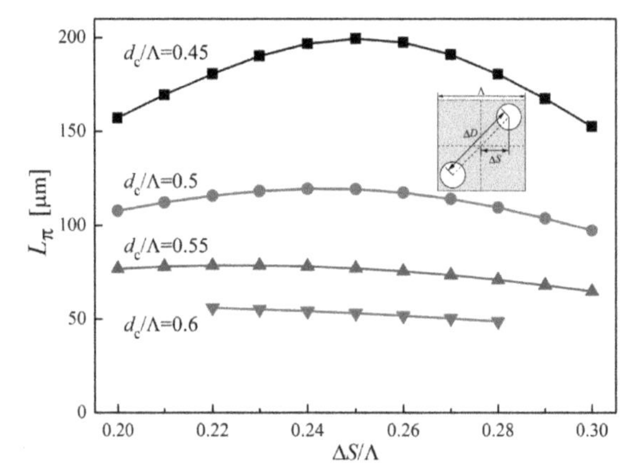

Figure 2. Conversion length as a function of ΔS with different cladding hole diameters at $\lambda = 1.55$ μm.

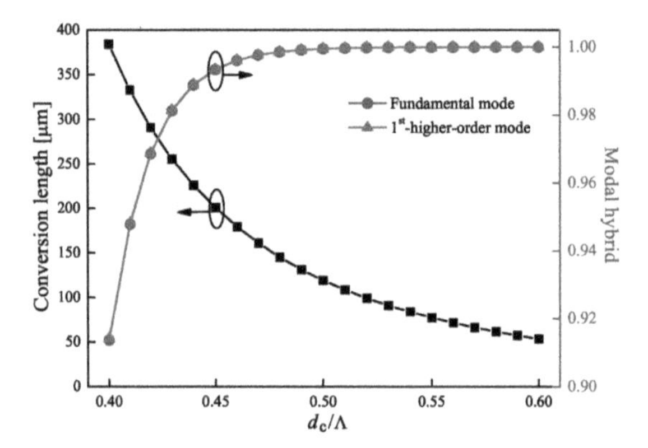

Figure 3. Conversion length as a function of d_c and the corresponding modal hybrids of fundamental and 1st-higher-order modes at $\lambda = 1.55$ μm.

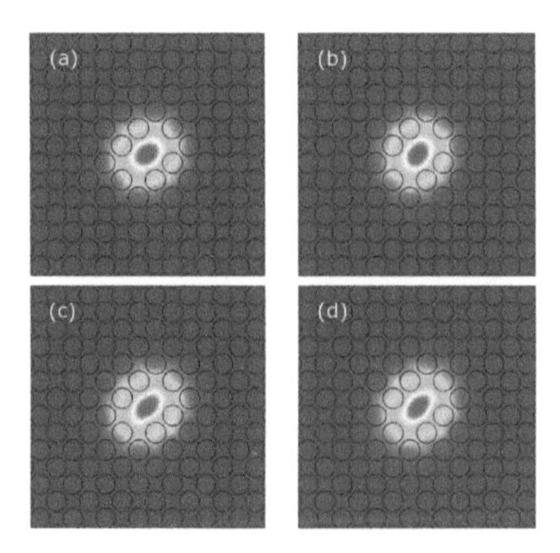

Figure 4. Magnetic field distributions of (**a**) H_x and (**b**) H_y components for the fundamental mode, and (**c**) H_x and (**d**) H_y components for the 1st-higher-order mode with $d_c/\Lambda = 0.6$ and $\Delta D/\Lambda = \sqrt{2}/2$.

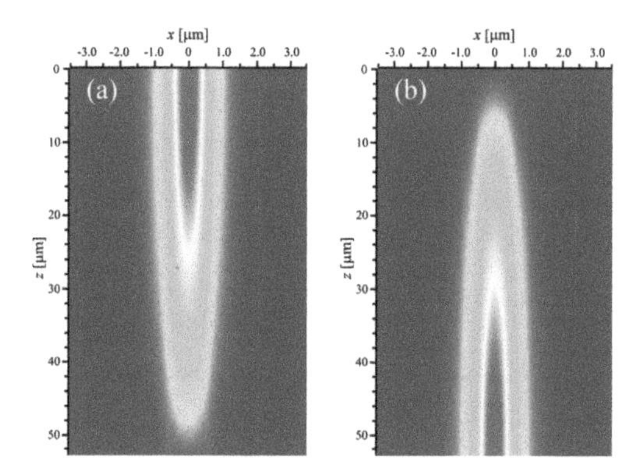

Figure 5. Propagation behavior of a TM mode incident light in the PC along z direction at $\lambda = 1.55$ μm. (**a**) H_x (**b**) H_y components.

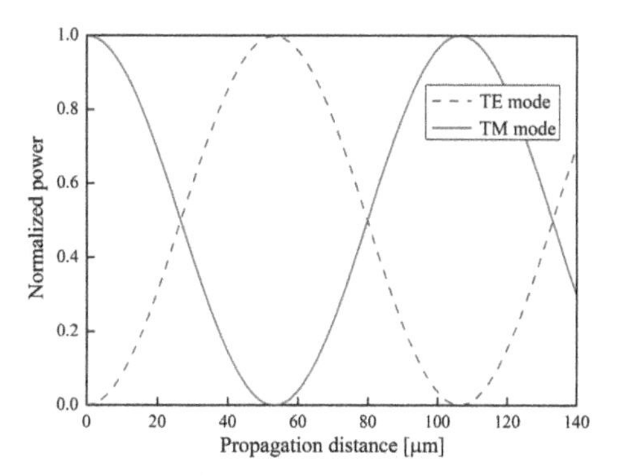

Figure 6. Normalized power variation of the TM mode incident light against the propagation distance.

2.2. Gaussian-Like Field Distribution of the PC

Considering the connection between a PC element and a conventional single-mode fiber (SMF), a Gaussian-like field distribution of a PC is necessary to suppress the insertion loss. Unlike previous studies using the asymmetric core distributions, our proposed PC can easily offer a Gaussian-like electromagnetic field distribution. In this study, we measure the mode matching ratio (MMR) between the fundamental mode of the PC and a Gaussian field distribution using the following overlap integral:

$$\text{MMR} = \frac{\left|\int \phi(x,y) g(x,y) dxdy\right|^2}{\int |\phi(x,y)|^2 dxdy \int |g(x,y)|^2 dxdy}, \tag{4}$$

where $\phi(x,y)$ is the field distribution of fundamental mode. The Gaussian field distribution is represented by the $g(x,y)$ as follows,

$$g(x,y) = \exp(-(x^2+y^2)/\delta^2), \tag{5}$$

where δ is the standard deviation of $g(x,y)$. Moreover, the spot size of a Gaussian field distribution, which is illustrated by w, is the diameter at which the light intensity $|g|^2$ drops to $1/e$ of its maximum value. The spot size is calculated by $w = 2\sqrt{2}\delta$. The maximum value of MMR between an excited fundamental mode of the PC and an appropriate Gaussian field distribution can be obtained by adjusting the value of δ. Table 1 illustrates the maximum MMR of each fundamental mode with an appropriate Gaussian field distribution (the corresponding spot size w/Λ is illustrated under each MMR). According to the calculated results, the excited modes of the proposed PC show quite a great agreement with Gaussian field distributions. Therefore, we believe our proposed PC has a great potential to be used with a low insertion loss.

Table 1. MMR of Each Fundamental Mode with an Appropriate Gaussian Field Distribution (w represents spot size for each Gaussian distribution).

d_c/Λ	0.45	0.5	0.55	0.6
MMR (w/Λ)	95.52% (3.05)	96.43% (2.63)	97.03% (2.35)	97.45% (2.15)

3. Structural Tolerance and Wavelength Dependence

So far, the propagation property of the PC with double-hole unit has been investigated. Compared to the previously proposed PCF based PCs [11–20], the optical device in this study has almost the same level of a high PCE and a low ER. However, the PC element with only one kind of circular air holes is the most prominent feature of this study. The simple structure distribution reduces the manufacturing difficulty and enlarges the structural tolerance simultaneously. Considering the current PCF fabrication technology, the structural tolerance should be discussed. In this study, we discuss the tolerances of hole size and hole position by investigating the variation of PCE, respectively. Additionally, the wavelength dependence of the PC has also been discussed. According to the calculation formula of the PCE in Equation (2), for a designed PC element, the modal hybrid (R) and the difference between the fixed device length (L) and the conversion length (L_π) are two primery factors affecting the final PCE. Therefore, the variation of parameters R and $L - L_\pi$ with different conditions (the deviation of hole size, change of hole position or variation of operation wavelength) should be taken into account.

3.1. Tolerance of Cladding Hole Size

Firstly, the structural tolerance of hole size in the PC has been investigated. The variations of parameters $L - L_\pi$ and R against the hole diameter with different deviation levels have been discussed and illustrated in Figure 7. Here, we claim that the fixed device length for each case is the completely conversion length of the PC with designed air hole sizes, i.e., $L = 199$ µm for $d_c/\Lambda = 0.45$, $L = 119$ µm for $d_c/\Lambda = 0.5$, $L = 77$ µm for $d_c/\Lambda = 0.55$, and $L = 53$ µm for $d_c/\Lambda = 0.6$. It is revealed from Figure 7 that with the deviation of d_c from -1.5% to 1.5%, the modal hybrids for d_c/Λ =0.5, 0.55, and 0.6 remain almost 1.000. Therefore, the final PCEs are mainly dependent on the value of $L - L_\pi$. On the contrary, for the $d_c/\Lambda = 0.45$, the parameter R increases from 0.991 to 0.996, and the variation of difference between the fixed device length and the conversion length is also relatively large. Consequently, the PCE is effected by the modal hybrid and the difference between L and L_π simultaneously.

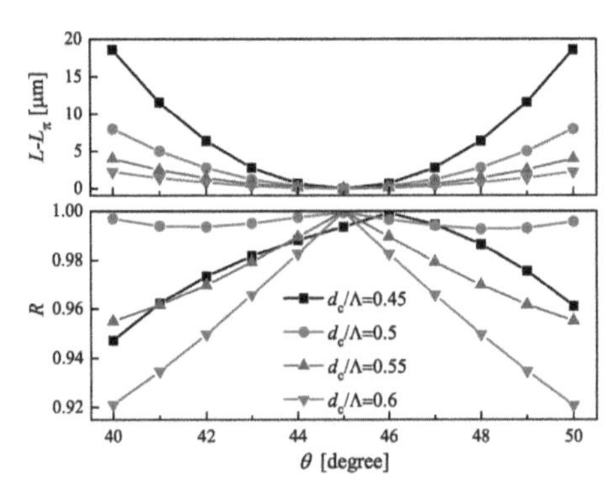

Figure 7. The variation of difference between a fixed device length and the corresponding conversion length ($L - L_\pi$) and the modal hybrid (R) against the hole diameter with different deviation levels at $\lambda = 1.55$ µm.

Figure 8 is the contour map of the PCE as a function of the deviation of hole diameter with different hole sizes from 0.4Λ to 0.6Λ. The contour line with a PCE of 99% is represented by a dash-dotted line. Simulation results reveal that within the deviation range of ±1.5%, the PC with a larger cladding hole size has a wider tolerance range to achieve the PCE better than 99%. In particularly, for d_c/Λ =0.6, the corresponding tolerance range is ±1.5% (±9.0 nm), which is 7 times as larger as that of the PC in [19]. Therefore, we believe that a simple structure PC with a relative large tolerance is achieved.

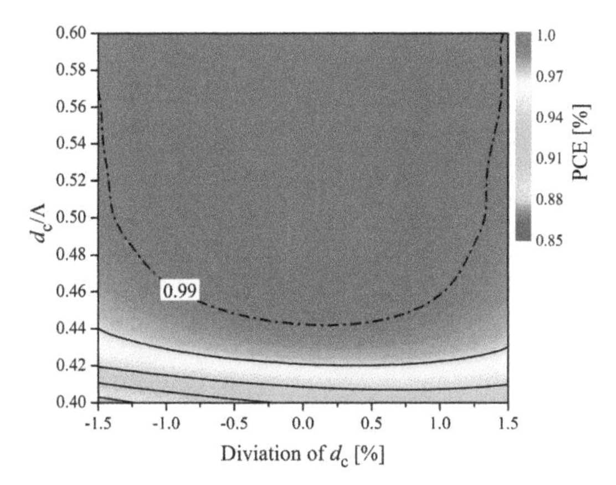

Figure 8. The variation of the PCE as a function of the deviation of hole diameter with different hole sizes.

3.2. Tolerance of Cladding Hole Position

Then, the tolerance of circular hole position in each unit cell is investigated. The square lattice PC with a symmetric structure along $y = x$ axis makes the angle between the modal optical axis and the horizontal axis to be 45 degrees. In this study, two circular air holes are arranged symmetrically along the diagonal for each unit cell to achieve the highest PCE. However, the positions of the circular holes will be slightly varied during the actual production process. The slight asymmetrical distribution of cladding holes results in a decrease of the modal hybrid. Here, we investigated the variation of PCE against the parameter θ with a constant hole gap of $\Delta D = \sqrt{2}/2\Lambda$. The variations of the two major factors have been investigated with different rotation angles from 40 to 50 degrees, as illustrated in Figure 9. The fixed device length for each hole size is the same as previously. For the variation of $L - L_\pi$, the differences increase as the angle θ deviates from 45 degrees. Additionally, for the variation of modal hybrid, the parameter R reaches the maximum value at $\theta = 45^\circ$ for d_c/Λ =0.5, 0.55 and 0.6, and the value gradually decreases as the angle varied. It is worth noting that the PC with a larger hole size has a stronger angle dependence. This is because the light is mainly confined in the core, and a larger cladding hole size leads to a stronger light confinement. Therefore, the light confinement region in the core has a strong dependence on the positions of first layer air holes in the cladding. A slight angle variation of the first layer holes may lead to a change in modal hybrid of excited modes. On the other hand, for a PC with d_c/Λ =0.45, the light confinement is relatively weak. Moreover, as the result shown in Figure 3, the corresponding modal hybrid is lower than 1.00 at $\theta = 45^\circ$. While for the simulation result in Figure 9, it is noted that the maximum value of modal hybrid exists at $\theta = 46^\circ$. Therefore, for PC with a small hole sizes, the maximum value of the PCE will shift in the direction of θ larger than 45 degrees.

Considering the two major factors, the variation of PCE against the parameter θ has been discussed and illustrated in Figure 10. The contour line with a PCE of 99% is represented by a dash-dotted line. It is evident that the PCE of a PC with a cladding hole size around 0.5Λ is better than 99% for θ from 40 to 49 degrees. For the PC hole size larger than 0.5Λ, the corresponding angle range decreases, and the maximum value of PCE remains at 45 degrees. While for the $d_c/\Lambda < 0.5$, the maximum value of PCE gradually shifts to a large rotation angle, which is in agreement with the aforementioned analysis.

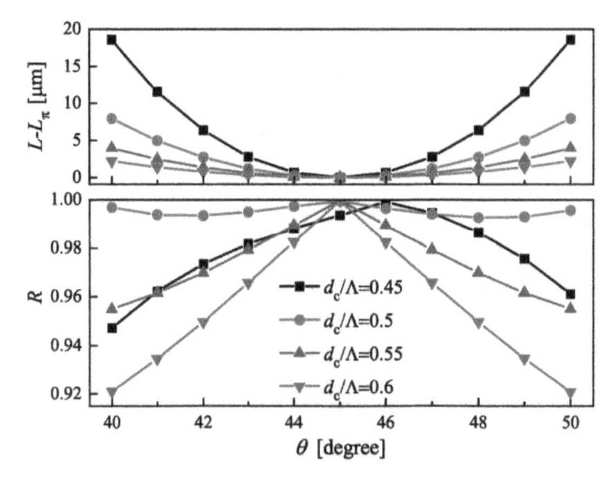

Figure 9. The variation of difference between a fixed device length and the corresponding conversion length ($L - L_\pi$) and the modal hybrid (R) against the angle θ with different values at $\lambda = 1.55$ μm.

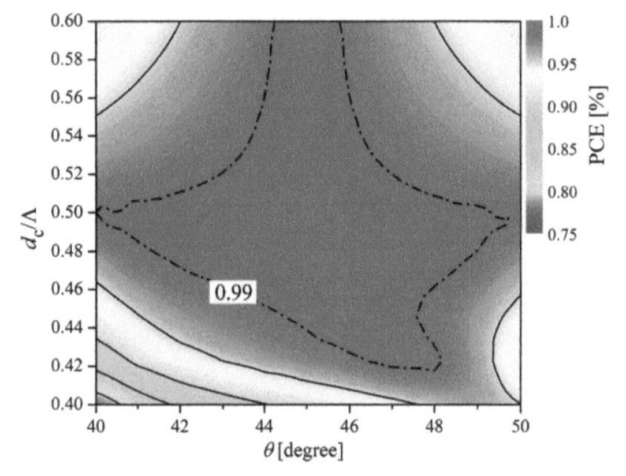

Figure 10. The variation of the PCE as a function of the deviation of angle θ with different hole sizes.

3.3. *Wavelength Dependence*

In order to confirm the wide-band transmission, the wavelength dependence of our proposed PC has also been discussed in detail. It is worth to note that more than two modes are excited for the PC with a short operation wavelength. Although other higher-order modes appear at a short wavelength (such as a wavelength range of 1.2~1.45 μm for the PC with $d_c = 0.6\Lambda$), the incident light of the FDTD simulation is only using the fundamental and the 1st-higher-order modes. Simulation results show that the other higher-order modes are not excited during the light propagation. Therefore, the effect of other modes on the polarization behavior is negligible. The variations of the difference between device length and conversion length, and the modal hybrid with different operation wavelengths are respectively illustrated in Figure 11. According to the simulation results, since the conversion length of the PC increases with wavelengths, the value of $L - L_\pi$ varies from positive to negative as the wavelength changes from 1.2 to 1.8 μm. Moreover, the modal hybrid of each case decreases as the wavelength increases. Consequently, the variation of PCE against the wavelength for different hole sizes is shown in Figure 12. It can be seen that the PCE is better than 99% for $d_c/\Lambda = 0.5$ within a wavelength range of 1.45 to 1.62 μm (about 170 nm). For $d_c/\Lambda = 0.6$, the corresponding wavelength bandwidth reaches 500 nm (from 1.2 to 1.7 μm), which is covering the O-band to the U-band. This is a great advantage for the wide application in the future.

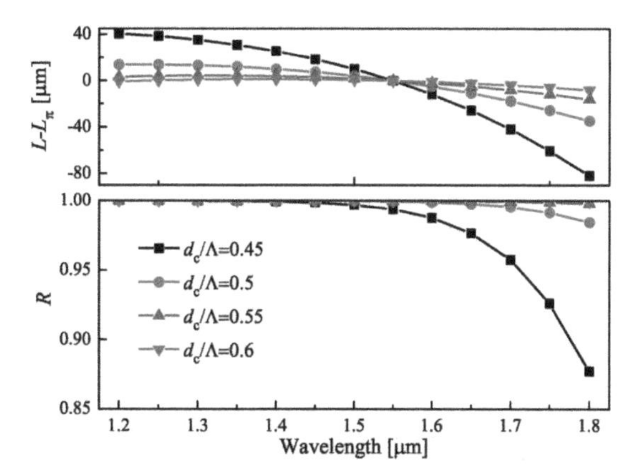

Figure 11. The variation of difference between a fixed device length and the corresponding conversion length ($L - L_\pi$) and the modal hybrid against different operation wavelengths.

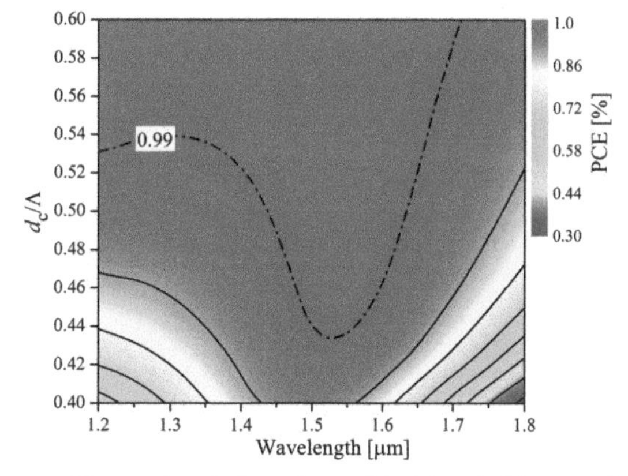

Figure 12. The wavelength dependence of the PC with different hole sizes.

4. Conclusions

In this paper, we focused on designing a novel square lattice PCF based PC element which has a simple construction and a large structural tolerance. Considering the fabrication technology of the PCF, only one type of circular air hole is adopted to consist the PC. In the periodic square lattice, one unit cell is defected to form the core, which allows the PC to achieve a Gaussian-like field distribution. FDTD simulation results reveal that a high PCE of 99.4% is achieved with a short device length of 53 μm, and the corresponding ER is lower than −21.8 dB. Large structural tolerances and low wavelength dependence have been demonstrated through our proposed PC. Therefore, we believe that this kind of PC has a great practical potential for optical communication systems in the future.

Author Contributions: Z.Z. conceived and designed the polarization converter; Y.T. gave the guideline for the paper; M.E. and C.-p.C. gave the suggestions and modifications for the paper.

Funding: This research received no external funding.

Conflicts of Interest: The authors declare no conflict of interest.

References

1. Knight, J.C.; Birks, T.A.; Russell, P.S.J.; Atkin, D.M. All-silica single-mode optical fiber with photonic crystal cladding. *Opt. Lett.* **1996**, *21*, 1547–1549. [CrossRef] [PubMed]
2. Russell, P.S.J. Photonic-crystal fibers. *J. Lightw. Technol.* **2006**, *24*, 4729–4749. [CrossRef]
3. Wang, L.; Yang, D. Highly birefringent elliptical-hole rectangular- lattice photonic crystal fibers with modified air holes near the core. *Opt. Express* **2007**, *15*, 8892–8897. [CrossRef] [PubMed]
4. Cerqueira, A., Jr.; Cordeiro, C.M.B.; Biancalana, F.; Roberts, P.J.; Hernandez-Figueroa, H.E.; Cruz, C.H.B. Nonlinear interaction between two different photonic bandgaps of a hybrid photonic crystal fiber. *Opt. Lett.* **2008**, *33*, 2080–2082. [CrossRef]
5. Wu, Z.; Shi, Z.; Xia, H.; Zhou, X.; Deng, Q.; Huang, J.; Jiang, X.; Wu, W. Design of highly birefringent and low-loss oligoporous-core thz photonic crystal fiber with single circular air-hole unit. *IEEE Photon. J.* **2016**, *8*, 4502711. [CrossRef]
6. Obayya, S.S.A.; Rahman, B.M.A.; El-Mikati, H.A. Vector beam propagation analysis of polarization conversion in periodically loaded waveguides. *IEEE Photon. Technol. Lett.* **2000**, *12*, 1346–1348. [CrossRef]
7. Watts, M.R.; Haus, H.A. Integrated mode-evolution-based polarization rotators. *Opt. Lett.* **2005**, *30*, 138–140. [CrossRef]
8. Fukuda, H.; Yamada, K.; Tsuchizawa, T.; Watanabe, T.; Shinojima, H.; Itabashi, S. Polarization rotator based on silicon wire waveguides. *Opt. Express* **2008**, *16*, 2628–2635. [CrossRef]
9. Zhang, J.; Yu, M.; Lo, G.; Kwong, D. Silicon-waveguide-based mode evolution polarization rotator. *IEEE J. Sel. Top. Quantum Electron.* **2010**, *16*, 53–60. [CrossRef]
10. Hsu, C.; Lin, H.; Chen, J.; Cheng, Y. Ultracompact polarization rotator in an asymmetric single dielectric loaded rib waveguide. *Appl. Opt.* **2016**, *55*, 1395–1400. [CrossRef]
11. Scolari, L.; Alkeskjold, T.; Riishede, J.; Bjarklev, A.; Hermann, D.; Anawati, A.; Nielsen, M.; Bassi, P. Continuously tunable devices based on electrical control of dual-frequency liquid crystal filled photonic bandgap fibers. *Opt. Express* **2005**, *13*, 7483–7496. [CrossRef] [PubMed]
12. Hameed, M.F.O.; Obayya, S.S.A. Analysis of polarization rotator based on nematic liquid crystal photonic crystal fiber. *J. Lightw. Technol.* **2010**, *28*, 806–815. [CrossRef]
13. Hameed, M.F.O.; Obayya, S.S.A. Polarization rotator based on soft glass photonic crystal fiber with liquid crystal core. *J. Lightw. Technol.* **2011**, *29*, 2725–2731. [CrossRef]
14. Hameed, M.F.O.; Obayya, S.S.A. Ultra short slica liquid crystal photonic crystal fiber polarization rotator. *Opt. Lett.* **2014**, *39*, 1077–1080. [CrossRef] [PubMed]
15. Hameed, M.F.O.; Obayya, S.S.A.; El-Mikati, H.A. Passive polarization converters based on photonic crystal fiber with L-shaped core region. *J. Ligthw. Technol.* **2012**, *30*, 283–289. [CrossRef]
16. Hameed, M.F.O.; Obayya, S.S.A. Design of passive polarization rotator based on silica photonic crystal fiber. *Opt. Lett.* **2011**, *36*, 3133–3135. [CrossRef] [PubMed]
17. Hameed, M.F.O.; Heikal, A.M.; Obayya, S.S.A. Novel passive polarization rotator based on spiral photonic crystal fiber. *J. Ligthw. Technol.* **2013**, *25*, 1578–1581. [CrossRef]
18. Chen, L.; Zhang, W.; Zhou, Q.; Liu, Y.; Sieg, J.; Zhang, L.; Wang, L.; Wang, B.; Yan, T. Polarization Rotator Based on Hybrid Plasmonic Photonic Crystal Fiber. *IEEE Photon. Technol. Lett.* **2014**, *26*, 2291–2293. [CrossRef]
19. Zhang, Z.; Tsuji, Y.; Eguchi, M. Design of cross-talk-free polarization converter based on square-lattice elliptical-hole core circular-hole holey fibers. *J. Opt. Soc. Am. B* **2016**, *33*, 1808–1814. [CrossRef]
20. Zhang, Z.; Tsuji, Y.; Eguchi, M.; Chen, C. Design of polarization converter based on photonic crystal fiber with anisotropic lattice core consisting of circular holes. *J. Opt. Soc. Am. B* **2017**, *34*, 2227–2232. [CrossRef]
21. Jansen, S.L.; Morita, I.; Schenk, T.C.W.; Tanaka, H. Long-haul transmission of 16 × 52.5 Gbits/s polarization-division multiplexed OFDM enabled by MIMO processing. *J. Opt. Netw.* **2008**, *7*, 173–182. [CrossRef]
22. Fukuda, H.; Yamada, K.; Tsuchizawa, T.; Watanabe, T.; Shinojima, H.; Itabashi, S. Silicon photonic circuit with polarization diversity. *Opt. Express* **2008**, *16*, 4872–4880. [CrossRef] [PubMed]

23. Sjödin, M.; Johannisson, P.; Wymeersch, H.; Andrekson, P.A.; Karlsson, M. Comparison of polarization-switched QPSK and polarization-multiplexed QPSK at 30 Gbit/s. *Opt. Express* **2011**, *19*, 7839–7846. [CrossRef] [PubMed]
24. Pysz, D.; Kujiwa, I.; Stepień, R.; Klimczak, M.; Filipkowski, A.; Franczyk, M.; Kociszewski, L.; Buźniak, J.; Haraśny, K.; Buczyński, R. Stack and draw fabrication of soft glass microstructured fiber optics. *Bull. Pol. Acad. Sci. Tech. Sci.* **2014**, *62*, 667–682. [CrossRef]

Article

Design of Polarization Splitter via Liquid and Ti Infiltrated Photonic Crystal Fiber

Qiang Xu [1,2,*,†], Wanli Luo [1,†], Kang Li [2,*], Nigel Copner [2] and Shebao Lin [1]

[1] College of Physics and Optoelectronic Technology, Baoji University of Arts and Sciences, Baoji 721016, China; 17789276164@163.com (W.L.); linshebao@163.com (S.L.)
[2] School of Engineering, University of South Wales, Cardiff CF37 1DL, UK; nigel.copner@southwales.ac.uk
* Correspondence: xuqiang@snnu.edu.cn (Q.X.); kang.li@southwales.ac.uk (K.L.); Tel.: +86-917-3364258 (Q.X.); +44-1443-482817 (K.L.)
† These authors contributed equally to this work.

Received: 31 January 2019; Accepted: 15 February 2019; Published: 18 February 2019

Abstract: We propose a new polarization splitter (PS) based on Ti and liquid infiltrated photonic crystal fiber (PCF) with high birefringence. Impacts of parameters such as shape and size of the air holes in the cladding and filling material are investigated by using a vector beam propagation method. The results indicate that the PS offers an ultra-short length of 83.9 μm, a high extinction ratio of −44.05 dB, and a coupling loss of 0.0068 dB and at 1.55 μm. Moreover, an extinction ratio higher than −10 dB is achieved a bandwidth of 32.1 nm.

Keywords: extinction ratio; polarization splitter; dual-core photonic crystal fiber; coupling characteristics

1. Introduction

Photonic crystal fiber (PCF) consists of a solid core and holes arranged in the cladding region non-periodically or periodically along the axis [1]. According to the mechanism of light transmission, PCF can be divided into refractive index guided PCF and photonic bandgap PCF. The refractive index guided PCF is similar to total internal reflection in the mechanism of light transmission. At present, refractive index guided PCF is the most mature and widely used optical fiber. PCF technology has made great progress in pharmaceutical drug testing, astronomy, communication, and biomedical engineering and sensing [2–8]. In recent years, the PCFs filled with materials have attracted great interests, because PCFs could provide excellent properties by filling different functional materials into the air holes [9–14]. Metal wire was filled into the air holes of PCFs for polarization splitting (PS) by Sun et al. and Fan et al. [10,11]. PCFs present high-quality channels that can be controllably filled with ultra-small volumes of analytes (femtoliter to subnanoliter), such as water [12], alcohol [13], and nematic liquid crystal [14].

The dual-core PCF is constructed by introducing two defect states in the periodic arrangement of air holes. When a polarized light beam is projected into the dual-core PCF with high birefringence, the coupling strength of two vertical polarization modes is weakened by the high birefringence [15]. Therefore, high birefringence could increase the difference in coupling length of the x-polarized mode and y-polarized mode of PCF, which is also beneficial to the miniaturization of PS. In general, high-birefringence fiber can be gained by breaking the symmetry implementing asymmetric defect structures, such as dissimilar air holes and elliptical holes along the two orthogonal axes, and asymmetric core design [16–18]. Another kind of high-birefringence fiber can also be manipulated by filling liquid into the air holes or hollow core [19]. At present, the dual-core PCF has a very mature application in polarization beam splitters.

The polarization handling devices based on PCFs filled with material, such as polarization splitters (PS) [20–22] and polarization rotators (PR) [23,24], have important applications in optical fiber sensing [25,26]. The PS could be divided into two fundamental modes (HE_{11}^{x} and HE_{11}^{y}) and propagate them in different directions. Its characteristics allow for significant applications in optical sensing, storage systems, communication systems, and integrated circuit systems [23,27]. In recent years, various PS structures based on PCFs have been reported in the literature [28–35]. These PS structures show good performance (see Table 1), such as an ultra-short length [31–33], a high extinction ratio [32], a low coupling loss [30,34], and an ultra-broad bandwidth [20,24,34], but these PS structures do not present these excellent properties at the same time. It is critical to design high-performance PS with ultra-short length, high extinction ratio, and low coupling loss [23]. In order to obtain a high-performance PS, we decided to use PCF filled with functional materials. It is well known that Ti demonstrates outstanding physical and chemical properties, such as light weight, anti-corrosion, biocompatibility, high melting point, durability, and high strength in extreme environments [36–38].

Table 1. Comparison of our proposed PS with earlier works.

References	Length/mm	Bandwidth/dB	Coupling loss/dB
[28]	1.7	40(<−11 dB)	not mentioned
[29]	4.72	190(<−20 dB)	not mentioned
[30]	8.7983	20(<−20 dB)	0.02
[31]	0.249	17(<−20 dB)	not mentioned
[32]	0.401	140(<−20 dB)	not mentioned
[33]	0.1191	249(<−20 dB)	not mentioned
[15]	14.662	13(<−10 dB)	not mentioned
[34]	4.036	430(<−20 dB)	0.011
[35]	0.775	32(<−20 dB)	not mentioned
Our work	0.0839	32.1(<−10 dB)	0.0068

In this paper, a novel ultra-short PS with low coupling loss and high birefringence is proposed based on the idea of material filled PCFs by using a vector beam propagation method (BMP) [39–41]. The numerical results present a 0.0839 mm-long PS with a coupling loss of 0.0068 dB and a high extinction ratio of −44.05 dB at the wavelength of 1.55 μm. Moreover, an extinction ratio higher than −10 dB is achieved at a bandwidth of 32.1 nm.

2. Physical Modeling

The physical modeling of the proposed PS is shown in Figure 1. The BPM-based commercially state-of-the-art software RSoft (Synopsys Inc., Mountain View, CA, USA) can be used to design and analyze optical telecommunication devices, optical systems and networks, optical components used in semiconductor manufacturing, and nano-scale optical structures. Figure 1 shows the structure of PS, where d, d_1, d_2, and d_3 represent the diameters of various air holes, respectively; a and b are the major and minor axes of the elliptical air hole; Λ represents distance of hole and hole (period); the ellipticity is expressed as $\eta = b/a$; and the air-filling ratio is d/Λ. The refractive index of background material is set as 1.45. Ti is filled into the two yellow air holes, and liquid (ethanol) is filled into the six blue holes. For ethanol (C_2H_5OH), variation of refractive index as a function of wavelength at a temperature of 20 °C is given by Reference [42].

$$n^2 - 1 = \frac{0.0165\lambda^2}{\lambda^2 - 9.08} + \frac{0.8268\lambda^2}{\lambda^2 - 0.01039} \quad (1)$$

where λ represents wavelength of the propagating light. The refractive index of ethanol is set as 1.35 at 1.55 μm. Figure 2 shows the refractive index function of Ti versus wavelength [43].

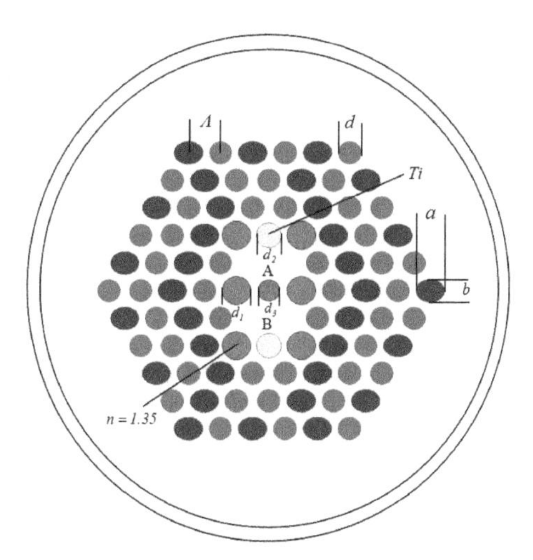

Figure 1. The cross-section of the proposed dual-core photonic crystal fiber (PCF).

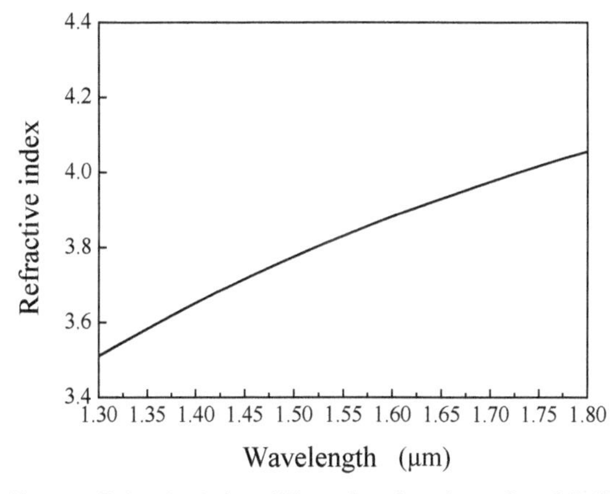

Figure 2. Refractive index of Ti as a function of wavelength [43].

The vector wave equation, which is the basis of BPM [39–41], can be expressed by

$$\nabla^2 \mathbf{E} + k^2 \mathbf{E} = 0 \tag{2}$$

$$\nabla^2 \mathbf{H} + k^2 \mathbf{H} = 0 \tag{3}$$

where $k \equiv \omega\sqrt{\mu\varepsilon}$. These two equations are known as the Helmholtz equations.

The electric field $E(x, y, z)$ can be separated into two parts: the fast change term of exp($-ikn_0z$) and the envelope term of $\phi(x, y, z)$ of slow change in the axial direction; n_0 is a refractive index in the cladding. Then, $E(x, y, z)$ is stated as

$$E(x, y, z) = \phi(x, y, z)\exp(ikn_0 z) \tag{4}$$

Substituting Equation (4) in Equation (1) results in

$$\nabla^2\phi - 2ikn_0\frac{\partial\phi}{\partial z} + k^2(n^2 - n_0^2) = 0 \tag{5}$$

Assuming the weakly guiding condition, we can approximate $n^2 - n_0^2 \cong 2n_0(n - n_0)$. Then Equation (5) can be rewritten as

$$\frac{\partial \phi}{\partial z} = -i\frac{1}{2kn_0}\nabla^2\phi + jk(n - n_0)\phi \tag{6}$$

A similar expression can be written for **H**. We find that $n \neq n_0$ if the fields vary in the transverse direction to propagation. Light propagation in various kinds of waveguides can be analyzed by the above method.

There are four modes of dual-core PCF on the basis of the principle of coupling mode, namely, even-mode of x-polarization, odd-mode of x-polarization, even-mode of y-polarization, and odd-mode of y-polarization. The coupling length has been defined by Reference [44] as

$$L_c^{x,y} = \frac{\lambda}{2\left|n_{even,\lambda}^{x,y} - n_{odd,\lambda}^{x,y}\right|} \tag{7}$$

where $n_{even}^{x,y}$, $n_{odd}^{x,y}$ denote the effective indexes of even-mode of x-polarization, odd-mode of x-polarization, even-mode of y-polarization, and odd-mode of y-polarization, respectively. When the coupling length of dual-core PCF satisfies $L = m\ L_c^x = n\ L_c^y$, the x-polarization and y-polarization launched into core A or B can be divided [33]. Hence, the coupling ratio (CR) can be defined as

$$CR = \frac{L_c^x}{L_c^y} = \frac{n}{m} \tag{8}$$

Assuming that the incident power is coupled into a certain core, the output power of x- or y-polarized light in the core can be expressed by the following equation [45]:

$$P_{out}^{x,y} = P_{in}^{x,y}\cos^2\left(\frac{\pi}{2}\cdot\frac{z}{L_c^{x,y}}\right) \tag{9}$$

where the transmission distance is denoted by z.

The extinction ratio is an important index to evaluate the performance of polarization splitter, which is expressed as

$$ER = 10\log_{10}\left(\frac{P_{out}^x}{P_{out}^y}\right) \tag{10}$$

where P_{out}^x, P_{out}^y represent the output energy of x-polarization and y-polarization, respectively [16,46].

The coupling loss of the PS can be described by

$$Loss = -10\log_{10}\left(\frac{P_{out}}{P_{in}}\right) \tag{11}$$

where P_{in} is the fundamental mode power at the input core [30].

Birefringence can be expressed as

$$B = \left|n_x - n_y\right| \tag{12}$$

where n_x and n_y are the effective refractive index of x-polarized and y-polarized fundamental modes [29].

3. Results and Discussion

First, the L_c and CR are examined for different period Λ, where d_1 = 0.8 μm, d/Λ = 0.7, d_2 = 0.7 μm, η = 0.8, and d_3 = 0.6 μm. The results are shown in Figure 3a, in which it is observed that the L_c is decreased when wavelength is increased for a constant period Λ. We also noticed that the L_c decreases with decreasing period Λ. Moreover, the coupling length of y-polarization is longer than the coupling length of x-polarization. As the period increases, the coupling between the cores becomes difficult. Hence, the L_c increases with the increase of the period. Interestingly, from Figure 3b, we noticed that when $\Lambda \leq 1.1$ μm, the size of the CR is higher for higher period Λ; when $\Lambda \geq 1.1$ μm, the size of the CR is higher for lower period Λ. According to Equation (8), when the CR is 3/4, the effective separation of the two orthogonal polarized lights can be achieved, so we choose the period value of 0.9 μm.

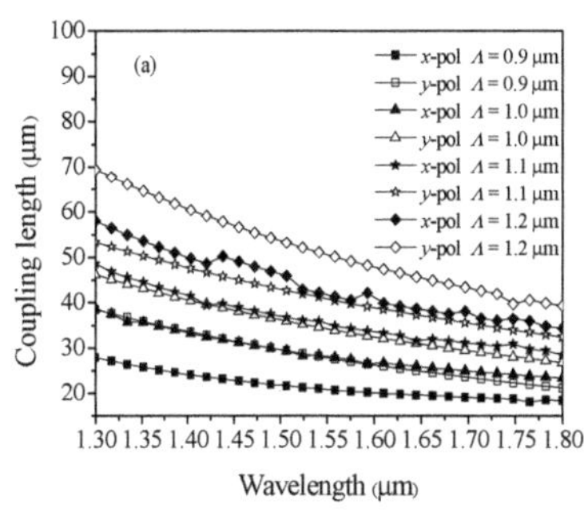

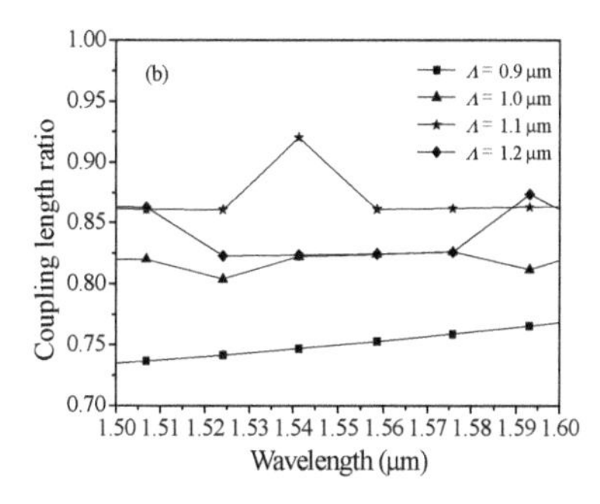

Figure 3. Coupling length (**a**) and coupling length ratio (**b**) as a function of wavelength for different period Λ.

Next, we analyze the L_c and CR as a function of wavelength for different air-filling ratio d/Λ, when η = 0.8, d_2 = 0.7 μm, Λ = 0.9, d_3 = 0.6 μm, and d_1 = 0.8 μm. From Figure 4a, it is observed that L_c is decreased when wavelength is decreased for the same value of air-filling ratio d/Λ. Moreover, we can find that the L_c decreases as the value of air-filling ratio increases, when air-filling ratio $d/\Lambda \leq 0.6$. This is owing to the restriction of the outer cladding to the light wave being enhanced as air-filling ratio increases. However, when $d/\Lambda \geq 0.6$, the result is opposite to the above. Meanwhile, the coupling length of *y*-polarization is longer than the coupling length of *x*-polarization. According to Figure 4b, it is found that when air-filling ratio $d/\Lambda \leq 0.6$, the size of the CR is higher for higher air-filling ratio d/Λ; when air-filling ratio $d/\Lambda \geq 0.6$, the size of the CR is higher for lower air-filling ratio d/Λ. When we choose an air-filling ratio d/Λ of 0.7, the CR is approximately 3/4 at 1.55 μm.

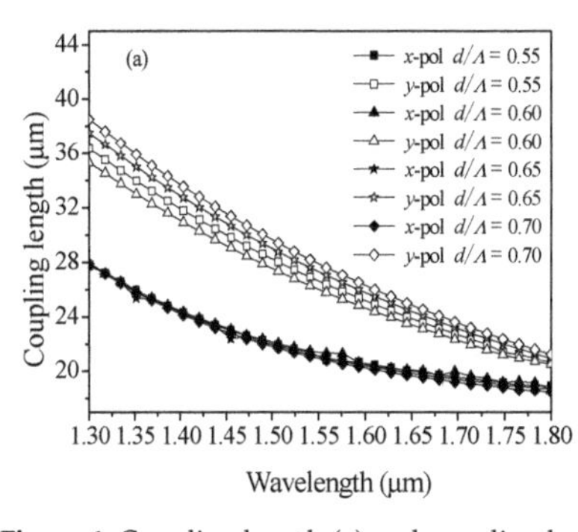

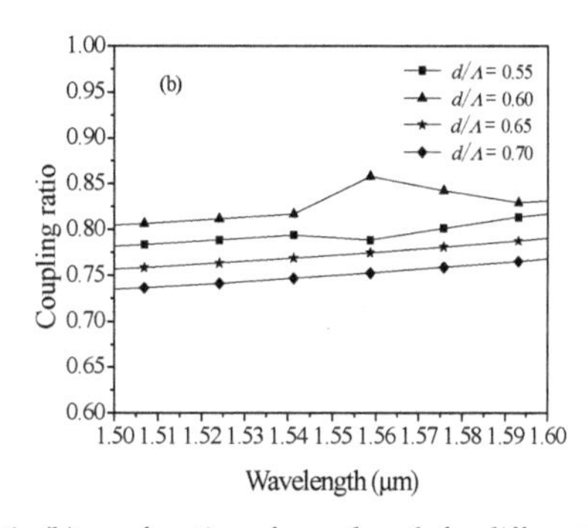

Figure 4. Coupling length (**a**) and coupling length ratio (**b**) as a function of wavelength for different air-filling ratio d/Λ.

Additionally, from Figure 5a, L_c is shown as a function of d_1, in which it is observed that the coupling length is increased if d_1 is increased. This phenomenon can be interpreted as the following: as the value of d_1 increases, the cores of PS can be compressed in the vertical direction, and fundamental modes in the horizontal direction will expand. Figure 5a also indicates that *x*-polarized coupling length is lower than *y*-polarized coupling length. As seen in Figure 5b, the size of the CR increases with increasing wavelength. According to the above discussed results, we determine that d_1 is 0.8 μm.

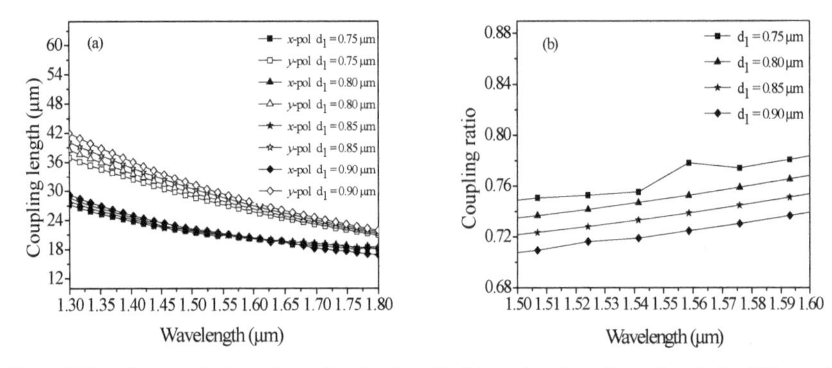

Figure 5. Coupling length (**a**) and coupling length ratio (**b**) as a function of wavelength for different d_1.

Figure 6 shows d_2 dependence on the L_c (Figure 6a) and CR (Figure 6b). From Figure 6a, it is evident that the L_c^x decreases with increasing the value of d_2. This result can be attributed to the following process: as the value of d_2 increases, the cores of PS can be compressed in the vertical direction, resulting in the increase of coupling length of *y*-direction. The L_c^y is shown as a function of d_2 in Figure 6a, in which it is observed that for $d_2 \leq 0.5$ μm, the value of CR is higher for higher d_2; for $d_2 \geq 0.5$ μm, the value of the CR is higher for lower d_2. This phenomenon is probably related to the ratio of compression. According to Figure 6b, we can clearly see that the size of the CR increases with a decrease of d_2. When the CR is 3/4, the effective separation of the two orthogonal polarized lights can be achieved, so we choose the d_2 value of 0.7 μm.

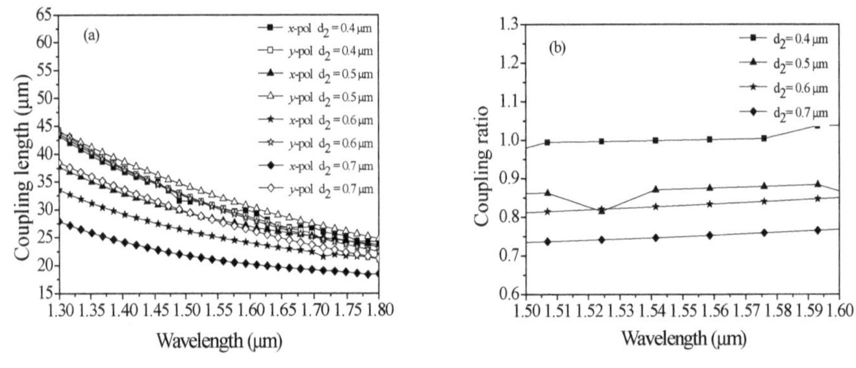

Figure 6. Coupling length (**a**) and coupling length ratio (**b**) as a function of wavelength for different d_2.

The L_c and CR with the variation of wavelength are demonstrated in Figure 7, when η = 0.65, 0.7, 0.75, and 0.8. It can be found that the L_c reduces with respect to wavelength. Figure 7a indicates that L_c^y is higher than L_c^x. It can also clearly be seen that the four *y*-polarized curves are extremely close to each other. This result can be explained that as the ellipticity η increases, the cores of PS can be compressed vertically, and the fundamental mode in the horizontal direction will expand, which makes the coupling of two cores easier. According to Figure 7b, the size of the CR increases with decreasing ellipticity η. According to the above discussed results, we determine that η is 0.8.

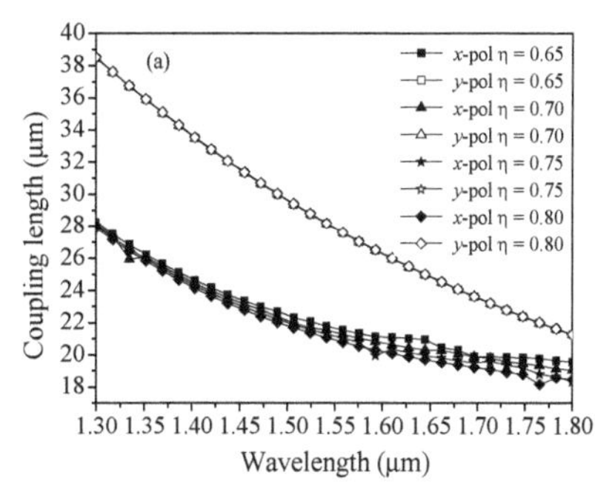

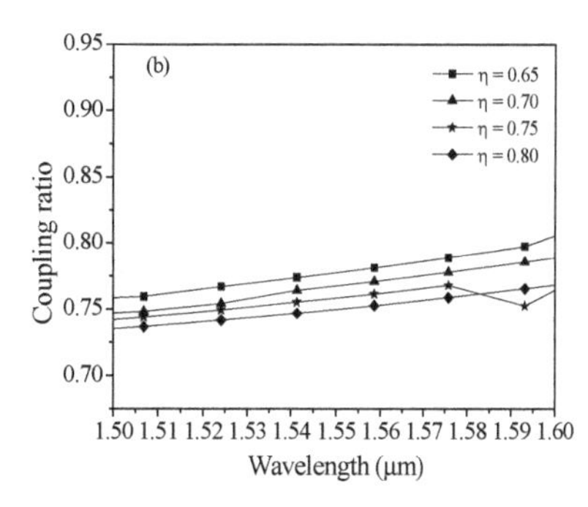

Figure 7. Coupling length (**a**) and coupling length ratio (**b**) as a function of wavelength for different η.

Finally, we found that the L_c and CR can both be impacted by d_1, d_2, d_3, d/Λ, Λ, and η. Hence, there exist optimized structural parameters, namely, d_1, d_2, d_3, Λ, d/Λ, and η is 0.8 μm, 0.7 μm, 0.6 μm, 0.9 μm, 0.7 and 0.8, respectively. Although the PS has many parameters, it can be easily influenced by the bulk polymerization process of polymers [47]. The optimized coupling length of x- and y-polarized direction are L^x = 20.91 μm and L^y = 27.96 μm at 1.55 μm, respectively. Figure 8 shows coupling characteristics of the PS. We observed that the separation of x- and y-polarized mode is achieved at the distance of 83.9 μm at 1.55 μm. Figure 9 shows the relationship between the birefringence and filling material for the optimized structural parameters. It is observed that birefringence of PS filled with liquid and Ti is higher than birefringence of PS filled with Ti. Meanwhile, the birefringence of PS filled with liquid and Ti can attain the order 10^{-2} at the wavelength of 1.55 μm, and the value of birefringence is about two orders of magnitude higher than that in References [29,48].

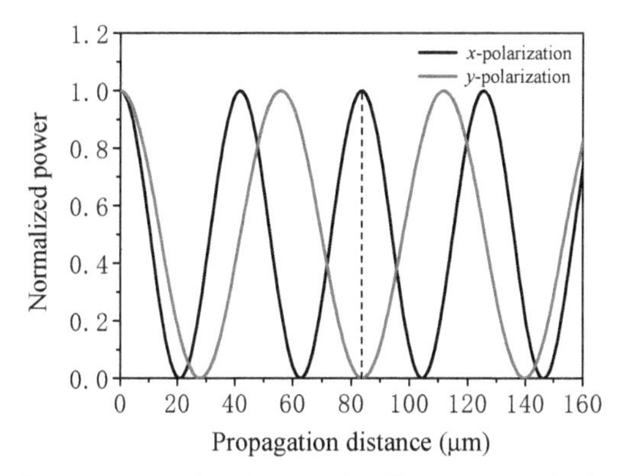

Figure 8. Normalized power as a function of propagation distance at optimized structural parameters.

Figure 10 shows the variation of coupling length with filling material for the optimized structural parameters. It can be seen that the coupling length of the PS with filled Ti is higher the coupling length of the PS with filled liquid and Ti. For the PS with filled Ti, coupling length of x- and y-polarized direction are L^x = 29.25 μm and L^y = 48.28 μm at 1.55 μm, respectively. According to Equation (8), the length of the PS with filled Ti is about 234 μm, which is much longer than the PS with filled liquid and Ti. Therefore, the PS filled with liquid and Ti has a shorter length and higher birefringence than the PS filled with Ti.

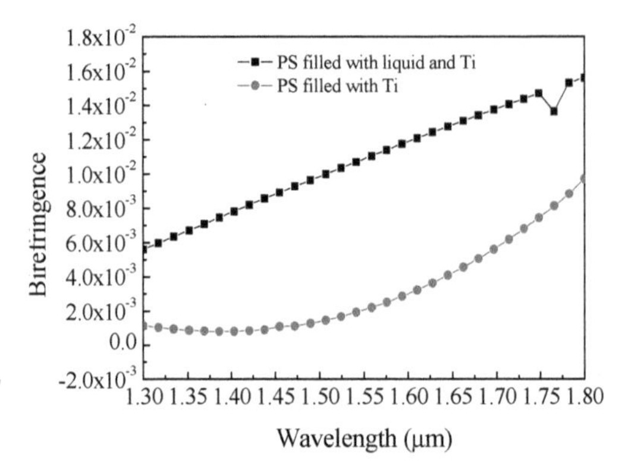

Figure 9. Birefringence as a function of wavelength for different filling materials.

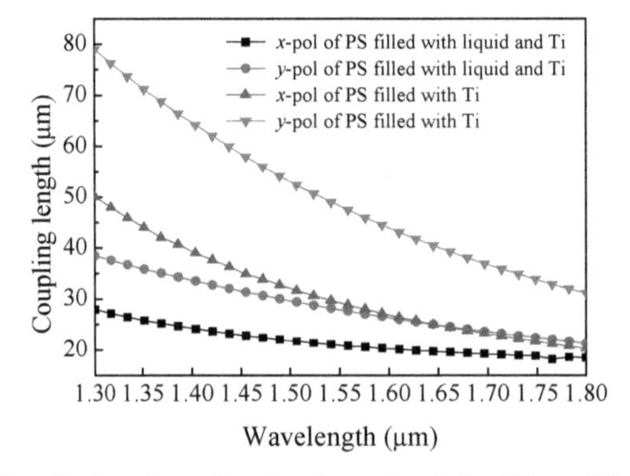

Figure 10. Coupling length as a function of wavelength for different filling materials.

Figure 11 shows the extinction ratio of PS with respect to wavelength at optimized structural parameters. The extinction ratio is measured in dB according to Equation (10). The PS has an extinction ratio of −44.05 dB at 1.55 μm, an extinction ratio better than −10 dB, and a bandwidth of 32.1 nm from 1567 nm to 1535 nm. Figure 12 shows the coupling loss of PS as function of wavelength. We observe that the PS has a coupling loss of 0.0068 dB at 1.55 μm. Performances in factors such as the length, coupling loss, and bandwidth are better than or at the same order of magnitude as those of the early works mentioned above (see Table 1).

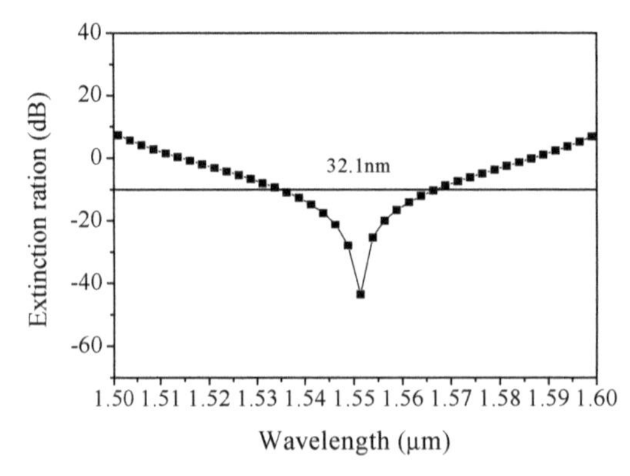

Figure 11. ER of PS as a function of wavelength at optimized structural parameters.

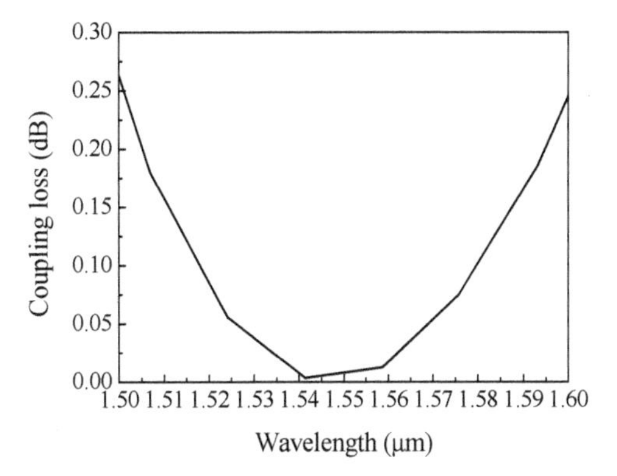

Figure 12. Coupling loss of PS as a function of wavelength at optimized structural parameters.

Figure 13 shows the mode field distribution of odd- and even-mode in x- and y-polarization direction. When a PS is incident upon core A or core B, both the odd- and even-mode of that polarization can be generated [49].

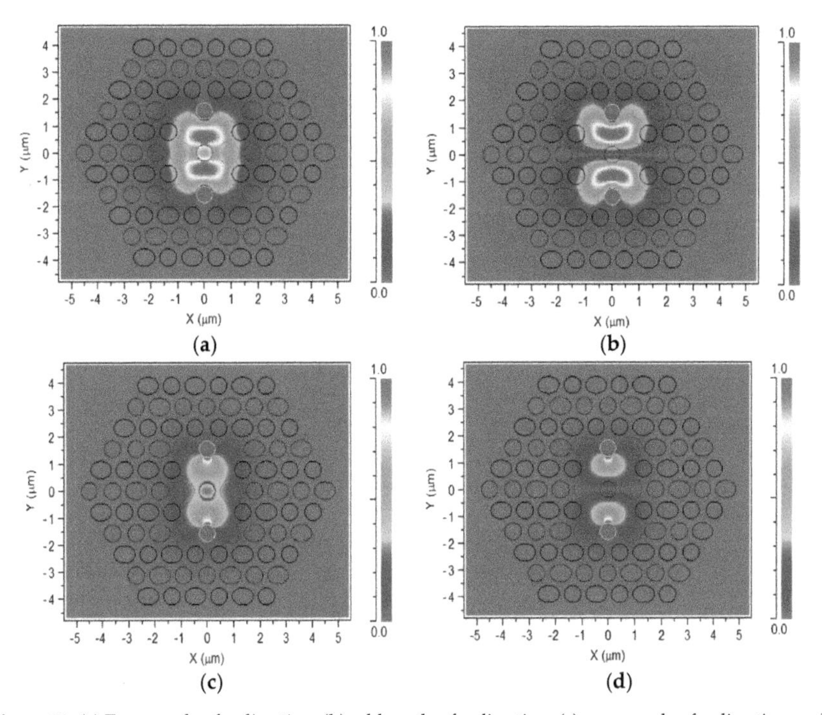

Figure 13. (**a**) Even-mode of x-direction, (**b**) odd-mode of x-direction, (**c**) even-mode of y-direction, and (**d**) odd-mode of y-direction for PS.

4. Conclusions

In conclusion, a novel ultra-short PS based on Ti and liquid infiltrated PCF with high birefringence have been demonstrated by using a vector beam propagation method. The designed PS shows an ultra-short length of 83.9 μm, a coupling loss of 0.0068 dB, a high extinction ratio of −44.05 dB, and a bandwidth of 32.1 nm at a wavelength of 1.55 μm. In addition, the birefringence of PS can attain the order 10^{-2} at the wavelength of 1.55 μm. The ultra-short PS with highly birefringent and low coupling loss properties is suitable for optical sensing, communication systems, storage systems, and integrated circuit systems.

Author Contributions: Analysis and writing, Q.X.; data curation, W.L.; project administration, S.L.; methodology, K.L. and N.C.

Funding: This work is supported by the National Natural Science Foundation of China (Grant No. 11647008) and the Scientific Research Program funded by Shaanxi Provincial Education Department (Grant No. 18JK0042).

Conflicts of Interest: The authors declare no conflicts of interest.

References

1. Hu, D.J.J.; Pui, H. Recent advances in plasmonic photonic crystal fiber: Design, fabrication and applications. *Adv. Opt. Photonics* **2017**, *9*, 259–314. [CrossRef]
2. Knight, J.C. Photonic crystal fiber. *Nature* **2003**, *424*, 847–851. [CrossRef]
3. Russell, P. Photonic crystal fiber. *Science* **2003**, *299*, 358–362. [CrossRef] [PubMed]
4. Kurokawa, K. Optical Fiber for High-Power Optical Communication. *Crystals* **2012**, *2*, 1382–1392. [CrossRef]
5. Chiang, J. Analysis of Leaky Modes in Photonic Crystal Fibers Using the Surface Integral Equation Method. *Crystals* **2018**, *8*, 177. [CrossRef]
6. Yu, Y.; Sun, B. Ultra-Wide-Bandwidth Tunable Magnetic Fluid-Filled Hybrid Connected Dual-Core Photonic Crystal Fiber Mode Converter. *Crystals* **2018**, *8*, 95.
7. Zhang, H.; Zhang, X.; Li, H.; Deng, Y.; Xi, L.; Tang, X.; Zhang, W. The Orbital Angular Momentum Modes Supporting Fibers Based on the Photonic Crystal Fiber Structure. *Crystals* **2017**, *7*, 286. [CrossRef]
8. Islam, M.S.; Sultana, J.; Dinovitser, A.; Faisal, M.; Islam, M.R.; Ng, B.W.; Abbott, D. Zeonex-based asymmetrical terahertz photonic crystal fiber for multichannel communication and polarization maintaining applications. *Appl. Opt.* **2018**, *4*, 666–672. [CrossRef]
9. Liu, Q.; Li, S.; Shi, M. Fiber Sagnac interferometer based on a liquid-filled photonic crystal fiber for temperature sensing. *Opt. Commun.* **2016**, *381*, 1–6. [CrossRef]
10. Sun, B.; Chen, M.; Zhang, Y.; Zhou, J. Polarization-dependent coupling characteristics of metal-wire filled dual-core photonic crystal fiber. *Opt. Quantum Electron.* **2015**, *47*, 441–451. [CrossRef]
11. Fan, Z.; Li, S.; Liu, Q.; Chen, H.; Wang, X. Plasmonic broadband polarization splitter based on dual-core photonic crystal fiber with elliptical metallic nanowires. *Plasmonics* **2016**, *11*, 1565–1572. [CrossRef]
12. Huang, Y.; Xu, Y.; Yariv, A. Fabrication of functional microstructured optical fibers through a selective-filling technique. *Appl. Phys. Lett.* **2004**, *85*, 5182–5184. [CrossRef]
13. Sultana, J.; Islam, M.S.; Ahmed, K.; Dinovitser, A. Terahertz detection of alcohol using a photonic crystal fiber sensor. *Appl. Opt.* **2018**, *57*, 2426–2431. [CrossRef]
14. Haakestad, M.W.; Alkeskjold, T.T.; Nielsen, M.D.; Scolari, L.; Riishede, J.; Engan, H.E.; Bjarklev, A. Electrically tunable photonic bandgap guidance in a liquid-crystal-filled photonic crystal fiber. *IEEE Photonics Technol. Lett.* **2005**, *17*, 819–821. [CrossRef]
15. Fan, Z.; Li, S.; Zhang, W.; An, G.; Bao, Y. Analysis of the polarization beam splitter in two communication bands based on ultrahigh birefringence dual-core tellurite glass photonic crystal fiber. *Opt. Commun.* **2014**, *333*, 26–31. [CrossRef]
16. Chen, M.; Yu, R.; Zhao, A. Highly birefringence rectangular lattice photonic crystal fiber. *J. Opt. A Pure Appl. Opt.* **2004**, *6*, 997–1000. [CrossRef]
17. Kim, S.; Kee, C.; Lee, C.G. Modified rectangular lattice photonic crystal fibers with high birefringence and negative dispersion. *Opt. Express* **2009**, *17*, 7952–7957. [CrossRef]
18. Yang, T.; Wang, E.; Jiang, H.; Hu, Z.; Xie, K. High birefringence photonic crystal fiber with high nonlinearity and low confinement loss. *Opt. Express* **2009**, *23*, 8329–8337. [CrossRef] [PubMed]
19. Jiang, L.; Zheng, Y.; Yang, J.; Hou, L.; Li, Z.; Zhao, X. An ultra-broadband single polarization filter based on plasmonic photonic crystal fiber with a liquid crystal core. *Plasmonics* **2017**, *12*, 411–417. [CrossRef]
20. Chiang, J.; Sun, N.; Lin, S.; Liu, W. Analysis of an ultrashort PCF-based polarization splitter. *J. Lightw. Technol.* **2010**, *28*, 707–713. [CrossRef]
21. Wu, H.; Tan, Y.; Dai, D. Ultra-broadband high-performance polarizing beam splitter on silicon. *Opt. Express* **2017**, *25*, 6069–6075. [CrossRef] [PubMed]

22. Jun, D.; Zhang, Z.; Zheng, H.; Sun, M. Recent progress on plasmon-enhanced fluorescence. *Nanophotonics* **2015**, *4*, 472–490.
23. Zhang, Y.; He, Y.; Wu, J.; Jiang, X.; Liu, R.; Qiu, C.; Jiang, X.; Yang, J.; Tremblay, C.; Su, Y. High-extinction-ratio silicon polarization beam splitter with tolerance to waveguide width and coupling length variations. *Opt. Express* **2016**, *24*, 6586–6593. [CrossRef] [PubMed]
24. Tanizawa, K.; Suzuki, K.; Ikeda, K.; Namiki, S.; Kawashima, H. Non-duplicate polarization-diversity 8×8 Si-wire PILOSS switch integrated with polarization splitter-rotators. *Opt. Express* **2017**, *25*, 10885–10892. [CrossRef]
25. González-Vila, Á.; Kinet, D.; Mégret, P.; Caucheteur, C. Narrowband interrogation of plasmonic optical fiber biosensors based on spectral combs. *Opt. Laser Technol.* **2017**, *96*, 141–146. [CrossRef]
26. Lu, X.; Soto, M.A.; Thévenaz, L. Temperature-strain discrimination in distributed optical fiber sensing using phase-sensitive optical time-domain reflectometry. *Opt. Express* **2017**, *25*, 16059–16071. [CrossRef]
27. Huang, Z.; Yang, X.; Wang, Y.; Meng, X.; Fan, R.; Wang, L. Ultrahigh extinction ratio of polarization beam splitter based on hybrid photonic crystal waveguide structures. *Opt. Commun.* **2015**, *354*, 9–13. [CrossRef]
28. Zhang, L.; Yang, C. Polarization splitter based on photonic crystal fibers. *Opt. Express* **2003**, *9*, 1015–1020. [CrossRef]
29. Li, J.; Wang, J.; Wang, R.; Liu, Y. A novel polarization splitter based on dual-core hybrid photonic crystal fibers. *Opt. Laser Technol.* **2011**, *43*, 795–800. [CrossRef]
30. Liu, S.; Li, S.; Yin, G.; Feng, R.; Wang, X. A novel polarization splitter in ZnTe tellurite glass three-core photonic crystal fiber. *Opt. Commun.* **2012**, *285*, 1097–1102. [CrossRef]
31. Zi, J.; Li, S.; An, G.; Fan, Z. Short-length polarization splitter based on dual-core photonic crystal fiber with hexagonal lattice. *Opt. Commun.* **2016**, *363*, 80–84. [CrossRef]
32. Xu, Z.; Li, X.; Ling, W.; Zhang, Z. Design of short polarization splitter based on dual-core photonic crystal fiber with ultra-high extinction ratio. *Opt. Commun.* **2015**, *354*, 314–320. [CrossRef]
33. Jiang, H.; Wang, E.; Zhang, J.; Hu, L.; Mao, Q.; Li, Q.; Xie, K. Polarization splitter based on dual-core photonic crystal fiber. *Opt. Express* **2014**, *22*, 30461–30466. [CrossRef] [PubMed]
34. Jiang, L.; Zheng, Y.; Hou, L.; Zheng, K.; Peng, J.; Zhao, X. An ultrabroadband polarization splitter based on square-lattice dual-core photonic crystal fiber with a gold wire. *Opt. Commun.* **2015**, *351*, 50–56. [CrossRef]
35. Sheng, Z.; Wang, J.; Feng, R. Design of a compact polarization splitter based on the dual-elliptical-core photonic crystal fiber. *Infrared Phys. Technol.* **2014**, *67*, 560–565. [CrossRef]
36. Vorobyev, A.Y.; Guo, C. Shot-to-shot correlation of residual energy and optical absorptance in femtosecond laser ablation. *Appl. Phys.* **2007**, *86*, 235–241. [CrossRef]
37. Zinger, O.; Zhao, G.; Schwartz, Z.; Simpson, J.; Wieland, M.; Landolt, D.; Boyan, B. Differential regulation of osteoblasts by substrate microstructural features. *Biomaterials* **2005**, *26*, 1837–1847. [CrossRef]
38. Ali, N.; Bashir, S.; Begum, N.; Rafique, M.S.; Husinsky, W. Effect of liquid environment on the titanium surface modification by laser ablation. *Appl. Surf. Sci.* **2017**, *405*, 298–307. [CrossRef]
39. Feit, M.D.; Fleck, J.A., Jr. Light propagation in graded-index optical fiber. *Appl. Opt.* **1978**, *24*, 3990–3998. [CrossRef]
40. Xiao, J.; Sun, X. A Modified full-vectorial finite-difference beam propagation method based on H-fields for optical waveguides with step-index profiles. *Opt. Commun.* **2006**, *266*, 505–511. [CrossRef]
41. Xie, K.; Boardman, A.D.; Xie, M.; Yang, Y.J.; Jiang, H.M.; Yang, H.J.; Wen, G.J.; Li, J.; Chen, K.; Chen, F.S. A Simulation of longitudinally magnetized three-dimensional magneto-optical devices by a full-vectorial beam propagation method. *Opt. Commun.* **2008**, *281*, 3275–3285. [CrossRef]
42. Sani, E.; Dell'Oro, A. Spectral optical constants of ethanol and isopropanol from ultraviolet to far infrared. *Opt. Mater.* **2016**, *60*, 137–141. [CrossRef]
43. Rakić, A.D.; Djurišic, A.B.; Elazar, J.M.; Majewski, M.L. Optical properties of metallic films for vertical-cavity optoelectronic devices. *Appl. Opt.* **1998**, *37*, 5271–5283. [CrossRef] [PubMed]
44. Saitoh, K.; Sato, Y.; Koshiba, M. Coupling characteristics of dual-core photonic crystal fiber couplers. *Opt. Express* **2003**, *11*, 3188–3195. [CrossRef]
45. Eisenmann, M.; Weidel, E. Single-mode fused biconical coupler optimized for polarization beam splitting. *J. Lightw. Technol.* **1991**, *9*, 853–858. [CrossRef]
46. Saitoh, K.; Sato, Y.; Koshiba, M. Polarization splitter in three-core photonic crystal fibers. *Opt. Express* **2004**, *12*, 3940–3946. [CrossRef]

47. Zhang, Y.; Li, K.; Wang, L.; Ren, L.; Zhao, W.; Miao, R.; Large, M.C.J.; Eijkelenborg, M.A.V. Casting preforms for microstructured polymer optical fibre fabrication. *Opt. Express* **2006**, *14*, 5541–5547. [CrossRef]
48. Chen, X.; Li, M.; Koh, J.; Nolan, D.A. Wide band single polarization and polarization maintaining fibers using stress rods and air holes. *Opt. Express* **2008**, *16*, 12060–12068. [CrossRef]
49. Zhang, S.; Yu, X.; Zhang, Y.; Shum, P.; Zhang, Y. Theoretical study of dual-core photonic crystal fibers with metal wire. *IEEE Photonics J.* **2012**, *4*, 1178–1187. [CrossRef]

Article

Elastic Wave Propagation of Two-Dimensional Metamaterials Composed of Auxetic Star-Shaped Honeycomb Structures

Shu-Yeh Chang [1], Chung-De Chen [1], Jia-Yi Yeh [2] and Lien-Wen Chen [1,*]

[1] Department of Mechanical Engineering, National Cheng Kung University, University Road, Tainan City 701, Taiwan; ccpp993113@gmail.com (S.-Y.C.); cdchen@mail.ncku.edu.tw (C.-D.C.)
[2] Department of Digital Design and Information Management Chung Hwa University of Medical Technology, Tainan City 717, Taiwan; yeh@mail.hwai.edu.tw
[*] Correspondence: chenlw@mail.ncku.edu.tw

Received: 29 January 2019; Accepted: 25 February 2019; Published: 26 February 2019

Abstract: In this paper, the wave propagation in phononic crystal composed of auxetic star-shaped honeycomb matrix with negative Poisson's ratio is presented. Two types of inclusions with circular and rectangular cross sections are considered and the band structures of the phononic crystals are also obtained by the finite element method. The band structure of the phononic crystal is affected significantly by the auxeticity of the star-shaped honeycomb. Some other interesting findings are also presented, such as the negative refraction and the self-collimation. The present study demonstrates the potential applications of the star-shaped honeycomb in phononic crystals, such as vibration isolation and the elastic waveguide.

Keywords: phononic crystal; auxetic structure; star-shaped honeycomb structure; wave propagation

---

1. Introduction

Phononic crystals are artificial crystals composed of a periodic alternation of at last two different materials. These materials were first studied by Sigalas and Economou et al. [1] and Kushwaha et al. [2,3], in which one of the most important properties in the crystals, the band gap, was investigated. They found that for some periodic arrangement, the waves could be stopped by phononic crystals. This unique property was found to have potential to be applied in isolating incident waves, acoustics and vibrations in many studies [4–6]. Sigalas [7] studied the elastic wave propagation in phononic crystals by using the plane wave expansion method. They investigated the defect effects inside the crystal system. The analysis of the band gaps suggested that the wave propagation in phononic crystals can be controlled by the defect states. Other research regarding the defect effects on the wave propagation in phononic crystals was also presented in the literature [8,9]. These studies suggested that their design of defect modes can be used as high-Q narrow band pass acoustical filters.

The concept of locally resonant phononic crystal was first proposed by Liu et al. [10]. The materials, based on the simple realization and composited with locally resonant structural units, can exhibit effective negative elastic constants at certain frequency ranges in their study. The model and experiments showed the spectral gaps with lattice constants two orders of magnitude smaller than the relevant sonic wavelength. Their findings suggested that the locally resonant crystal is the key factor to control the wave propagation with large wave length by small-size phononic crystals. Liu [11] and Wang [12] performed systematic studies to analyze the factors that affect the locally resonant phononic crystals. They also found that the band gap can be manipulated by tuning the elastic modulus. The suitable range of the elastic modulus can be obtained through the design of the local structures in the periodic crystal [13–15]. For the locally resonant phononic crystals, the wave length is much larger than the lattice constant. Therefore, the crystals can be modelled as a homogeneous medium with effective elastic modulus and effective Poisson's ratio [16,17].

Some of the above mentioned phononic crystals are the cellular structures, which have properties such as high stiffness and low density. Some cellular structures have unique properties, such as negative Poisson ratio. The material with the negative Poisson ratio is called the auxetics. Gibson et al. [18] investigated that some cellular structures, such as re-entrance honeycomb, have a negative effective Poisson ratio. Then, Almgren [19] proposed a 3D structure composed of honeycomb structure with a Poisson ratio of -1. The negative poisson's ratios in composites with star-shaped inclusions by numerical homogenization approach was presented by Panagiotopoulus et al. [20]. The wave propagations in the auxetic materials were also investigated. Gonella and Ruzzene [21] considered the plane wave propagation in re-entrant and hexagonal honeycomb structures. Two-dimensional dispersion relations were analyzed to reveal peculiar properties of the re-entrant honeycomb structures. Liebold-Ribeiro and Körner [22] performed the eigenmode analysis to various periodic cellular materials. Then, the star-shaped honeycomb with varied Poisson ratio as presented by Meng et al. [23] and the wave propagation behaviors and the band gaps were also analyzed in their studies. Wojciechowski et al. [24] presented some investigations about the auxetics and other systems of anomalous characteristics. The abovementioned papers showed that the auxetic material with negative values of Poisson ratio have significant effects on the wave motion in elastic solid.

Self-collimation is another important phenomenon of wave propagation in phononic crystals. Qiu et al. [25] proposed a design of a highly-directional acoustic source, formed by placing a line acoustic source inside a planar cavity of two-dimensional phononic crystals. The propagation characteristics of elastic transverse waves emitted by line sources embedded inside two-dimensional phononic crystals were studied by Liu and Su [26]. Cicek and his co-workers [27,28] studied numerically the wave propagation characteristics and self-collimation phenomenon of phononic crystals. Besides this, some experimental results have been performed to validate this phenomenon [29–31].

Recently, we published some research regarding wave propagation in phononic crystals [32,33], in which we performed the wave propagation analysis and had some interesting findings, including self-collimation and negative reflection. In this paper, we propose a novel auxetic material composed of star-shaped honeycomb structure. The Poisson's ratio with various negative value of the structure can be obtained by adjusting the star angle θ. The dispersion of the phononic crystal composed of auxetic star-shaped honeycomb will be analyzed by eigenmode analysis. The self-collimation characteristic of the star-shaped honeycomb is also investigated.

2. Materials and Methods

The dynamic equilibrium equations for an isotropic elastic solid are expressed as [34]:

$$(\lambda + \mu)u_{j,ji} + \mu u_{i,jj} = \rho \ddot{u}_i \quad (1)$$

where u_i is the displacement vector component, λ and μ are the Lame constant, and ρ is the mass density, respectively. The Lame constants can be written in terms of Young's modulus E and Poisson's ratio ν, i.e., $\lambda = E\nu/(1+\nu)(1-2\nu)$ and $\mu = E/2(1-\nu)$. The constitutive equation for a linear isotropic elastic material is denoted as:

$$\tau_{ij} = \lambda \varepsilon_{kk}\delta_{ij} + 2\mu\varepsilon_{ij} \quad (2)$$

in which, τ_{ij}, ε_{ij}, and δ_{ij} are stress, strain tensor, and the unitary tensor, respectively. Then, the strain-displacement relation is as follows:

$$\varepsilon_{ij} = \frac{1}{2}(u_{i,j} + u_{j,i}) \quad (3)$$

The phononic crystal composed of square lattice of steel cylinder in an auxetic matrix is a periodic medium. The displacement field of the periodic structure can be expressed as follows by applying the Bloch theorem.

$$u(r) = u_k(r)e^{i(\omega t - k \cdot r)} \tag{4}$$

where $u(r)$ is the displacement vector, ω is the natural frequency, r is the position vector, k is the wave vector in the first Brillouin zone, and $u_k(r)$ is the periodic vector function of r, respectively. The phononic crystal lattice and u_k have the same periodicity. The wave propagation in the periodic structure can be solved by using Equations (1) and (2). In this paper, the solution procedures are done by the finite element method. The boundary conditions of the unit cell is:

$$u(r + a) = e^{i(k \cdot a)}u(r) \tag{5}$$

Thus, we can obtain the discrete from of finite element eigenvalue system in the unit cell as following form by applying the boundary conditions:

$$\left(K - \omega^2 M\right)u = 0 \tag{6}$$

where K and M are stiffness and mass matrix of the unit cell and a is the lattice constant, respectively.

3. Results

In this section, numerical results will be presented to demonstrate the band structure of the phononic crystal composed of auxetic materials. The two types of phononic crystal systems in this paper are shown in Figure 1a,b, respectively. The lattice constant of phononic crystal is $a = 0.8$ m. The material used for the inclusion is steel with material properties $E_s = 200$ GPa, $\rho_s = 7850$ kg/m^3 and $\nu_s = 0.33$. Two shapes of the cross sections of the inclusion, the circular shape with radius $r = 0.3a$ and rectangular shape with width $w = 14a/15$ and height $h = a/5$, are considered, as shown in Figure 1a. For rectangular shaped cross section as shown in Figure 1b, the incline angle ϕ of the inclusion with rectangular cross section is also considered as a parameter. Therefore, we can realize an auxetic material by using the star-shaped honeycomb structure. The effective Young's modulus and effective Poisson's ratio of the star-shaped honeycomb structure are considered as the material properties of the auxetic matrix.

Figure 2 presents band structures of the elastic wave in the phononic crystals with cylinder inclusions and auxetic matrix, of which three different Poisson's ratios, $\nu = -0.5, -0.7$, and -0.9, are considered. The numerical results show that as the absolute value of the Poisson's ratio, $|\nu|$, increases, the center frequencies of the band gaps increase. The materials with negative Poisson's ratio are rare in the natural materials, especially for high strength materials. Thus, the re-entrant and star-shaped honeycomb structures as shown in Figure 3 are proved to have negative values of Poisson's ratio [23]. In this study, the shar-shaped honeycomb structure is adopted as the matrix in the phononic crystals. The scheme of the phononic crystal, composed of periodic steel inclusion with circular cross section and the star-shaped honeycomb matrix, is shown in Figure 3a. Figure 3b shows the unit cell of the star-shaped honeycomb, the material properties of which is steel, same as the material used for inclusions. The length of the ligament of the star is 0.05 m, the width of the beam is one-tenth of the length, and lattice constant of the star shaped unit cell is $L = 0.16$ m. Then, the effective material properties of the matrix are listed in Table 1 [35] and it can be observed that the Poisson's ratio varies with star angle θ. For star angles smaller than $65°$, the Poisson's ratio is negative, and the matrix can be considered as an auxetic material.

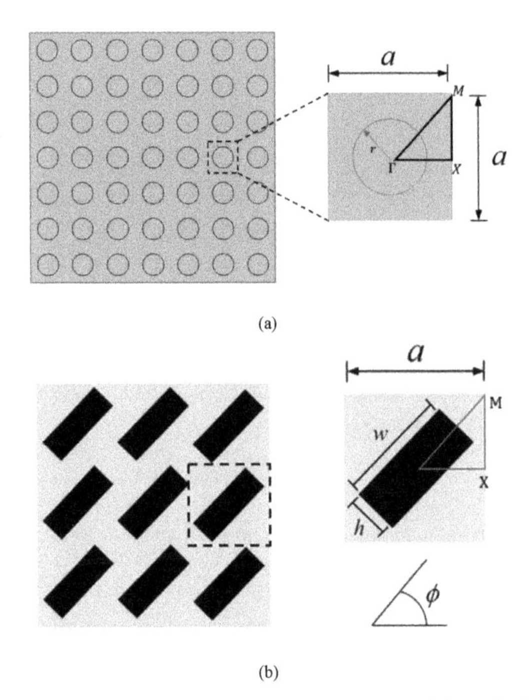

Figure 1. The illustration of phononic crystal. Two cross sections of the inclusions are considered. They are (**a**) circular cross section and (**b**) rectangular cross section.

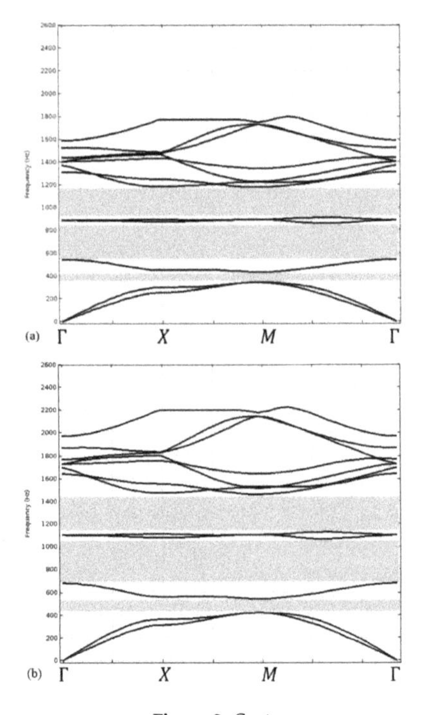

Figure 2. *Cont.*

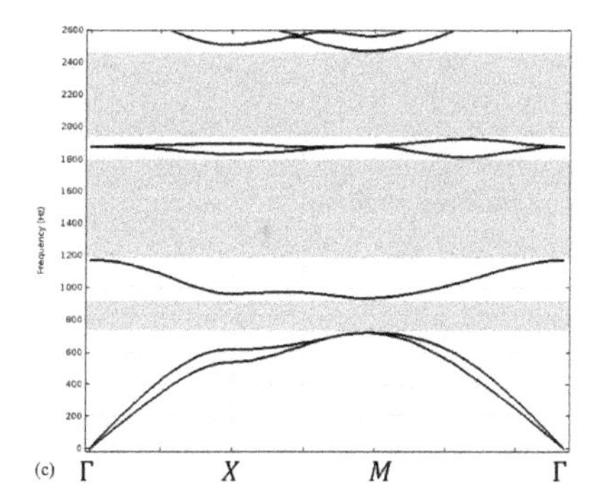

Figure 2. The band structure of the elastic wave in the phononic crystal with auxetic matrix (**a**) $\nu = -0.5$ (**b**) $\nu = -0.7$ (**c**) $\nu = -0.9$.

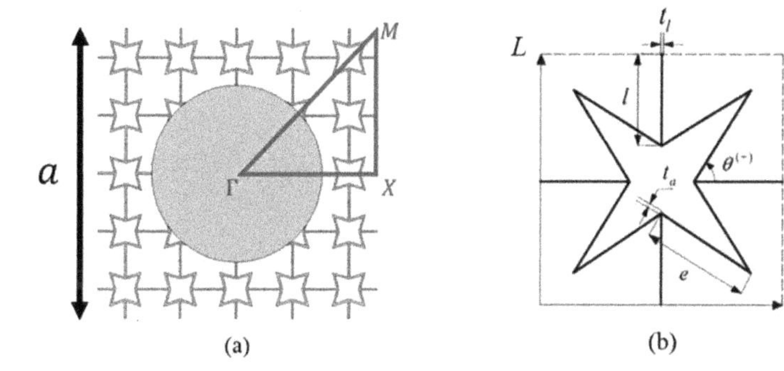

Figure 3. The scheme of the auxetic phononic crystals. (**a**) The unit cell of the phononic crystals composed of star-shaped honeycomb structure and circular shaped inclusion; (**b**) the unit cell structure of the star-shaped honeycomb. The parameter of star-shaped honeycomb is $l = 0.05$ m, $e = 0.05$ m, $t_l = 0.005$ m and $t_a = 0.005$ m. The filling ratio of cylinder are $r = 0.3a$.

Table 1. The effective material properties of steel star-shaped honeycomb, calculated by COMSOL FEM software [35].

STAR ANGLE θ	55°	60°	65°	70°
Young's module E (Gpa)	0.5124	0.3123	0.2305	0.1806
Poisson's ratio ν	−0.53	−0.35	−0.15	0.05
Mass density ρ ($\mathrm{kg/m^3}$)	921	921	921	921

Figure 4 shows the effects of the star angles on the band structures of the phononic crystal composed of steel cylinders in the star-shaped honeycomb matrix. Three different star angles, $\boldsymbol{\theta = 55^\circ}$, $\mathbf{60^\circ}$, and $\mathbf{70^\circ}$ are calculated and the corresponding effective Poisson's ratios are $\mathbf{-0.53}$, $\mathbf{-0.35}$ and $\mathbf{-0.15}$, respectively. The band structures of the phononic crystal system are computed and obtained by the finite element software ***COMSOL***$^{(r)}$. Three band gaps are found in the frequency range from **0 Hz** to **1.3 kHz**. The frequency of the band gap increases if the absolute value of Poisson's ratio decreases. Additionally, it is also observed that the widths of the band gap increase with the decrease of Poisson's ratio. According to the figure, it is evident that the band structures strongly depend on the auxeticity of the matrix.

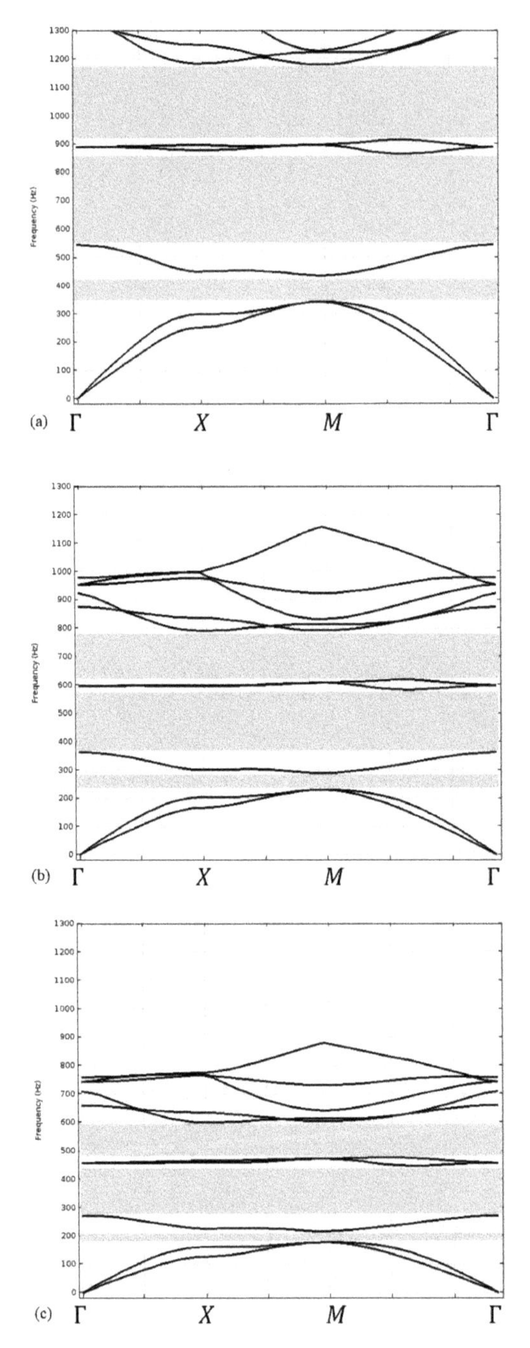

Figure 4. The band structure of phononic crystals composed of steel cylinders and star-shaped honeycombs. **(a)** $\theta = 55^\circ, \nu = -0.53$ **(b)** $\theta = 60^\circ, \nu = -0.35$ **(c)** $\theta = 65^\circ, \nu = -0.15$.

Then, the equi-frequency contour (EFC) of the third band of the star-shaped unit cell for the case of $\boldsymbol{\theta} = \mathbf{55}^\circ$ and $\boldsymbol{\nu} = \mathbf{-0.53}$ is plotted in Figure 5. The dashed black line denotes the construction line and the red arrow indicates the gradient of the contour, which is the direction of the elastic wave propagation in the phononic crystal. It indicates that the negative refraction occurs at this band. The full wave finite element simulation result is shown in Figure 6. A plane elastic wave of **500 Hz** is incident from the left side of the phononic crystal composed of star-shaped honeycomb matrix with star angle $\boldsymbol{\theta} = \mathbf{55}^{\circ}$. The width of the plane wave and the incident angle are $\mathbf{4a}$ and $\mathbf{30}^{\circ}$, respectively. The simulation of the phononic crystal system is also computed and obtained by the finite element software ***COMSOL***$^{(r)}$.The negative refraction is shown clearly in Figure 6. The refraction angle is about $-\mathbf{45}^{\circ}$ and the refraction ratio is about $-\mathbf{0.707}$.

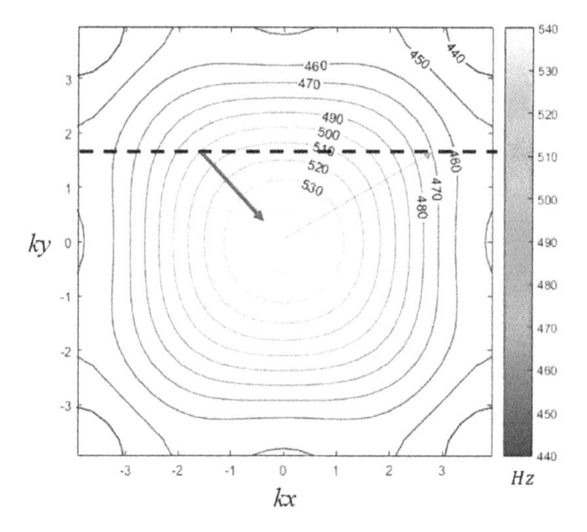

Figure 5. The equi-frequency contour of the third band in Figure 4a. The star angle and the effective Poisson's ratio are $\theta = 55°$ and $\nu = -0.53$, respectively.

Figure 6. The simulation of wave propagation for negative refraction. A plane elastic wave of 500 Hz is incident from the left side of phononic crystal. The width of the wave and the incident angle are $4a$ and $30°$, respectively. The refraction angle is about $-45°$ and the refraction ratio is about -0.707. The material on both sides of the structure is steel.

Figure 7 presents the numerical results of wave propagation of the phononic crystal, composed of inclusion with rectangular shape and the star-shaped honeycomb as the matrix. The star angle is $\boldsymbol{\theta} = \mathbf{55}^{\circ}$, and Poisson's ratio of the matrix is $\boldsymbol{\nu} = -\mathbf{0.53}$, respectively. The geometric parameters of the inclusions are $\boldsymbol{w} = \mathbf{14a/15}$, $\boldsymbol{h} = \boldsymbol{a}/\mathbf{5}$, and $\boldsymbol{\phi} = \mathbf{0}^{\circ}$. Figure 7a is the band structure, in which the $\mathbf{X} - \mathbf{M}$

region exhibit approximately horizontal lines. The equi-frequency contours are shown in Figure 7b by taking the 8th band as an example. The frequency range of the 8th band is from **820 Hz** to **900 Hz**. In the case that the eigen-frequency is greater than **870 Hz**, the equi-frequency contours are more like the vertical straight lines. It suggests that the self-collimation phenomenon occurs with the wide range of the wave incident into the phononic crystal. After that, the simulation of the wave propagation for three different orientation angles of the rectangular inclusions is presented in Figure 8 and the finite element model is a $\mathbf{50 \times 40}$ array of lattices. The incident wave with **885 Hz** is from the left boundary and the incident angle is fixed horizontally for all cases. It is obvious that the wave propagation direction aligns the orientation angle $\boldsymbol{\phi}$. In addition, Figure 9 shows that the simulation of the wave propagation for the model consists of gradually varied orientation angles. The orientation angle is $\boldsymbol{\phi} = 0^{\circ}$ at the left boundary, $\boldsymbol{x} = \mathbf{0}$, and linearly increases to $\boldsymbol{\phi} = \mathbf{50}^{\circ}$ at the right boundary, $\boldsymbol{x} = \mathbf{100a}$. The simulation results reveal that when the wave is horizontally incident from the left boundary to the phononic crystal, the wave propagation direction is automatically collimated along the orientation angles of the rectangular inclusions, resulting in a curved path of the wave propagation. The numerical results shown in Figures 7 and 8 validate the self-collimation phenomenon of the phononic crystal with star-shaped honeycomb matrix and rectangular inclusions.

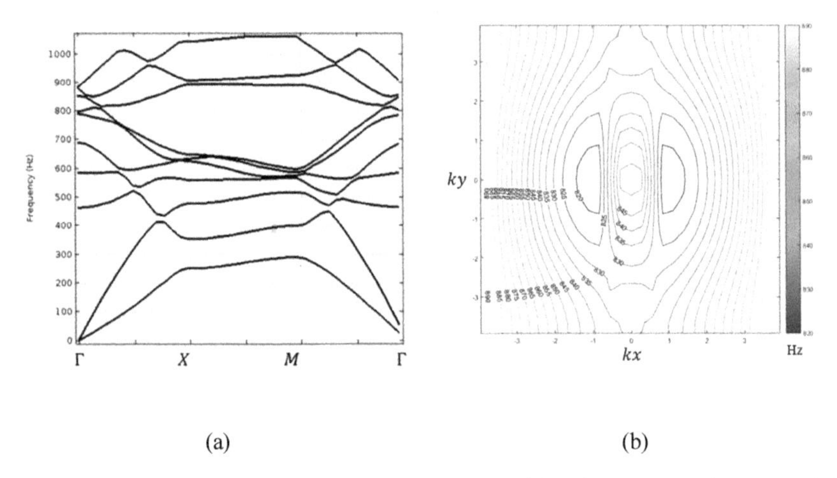

(a) (b)

Figure 7. The numerical results of the wave propagation of the phononic crystals composed of inclusions with rectangular cross section and star-shaped honeycomb matrix. (**a**) the band structures; (**b**) the equi-frequency contour of the 8th band. The star angle is $\theta = 55^{\circ}$, Poisson's ratio of the matrix is $\nu = -0.53$. The geometric parameters of the inclusions are $w = 14a/15$, $h = a/5$ and $\phi = 0^{\circ}$.

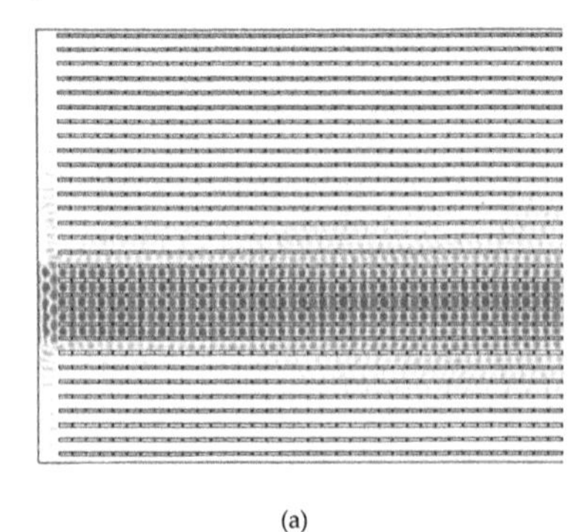

(a)

Figure 8. *Cont.*

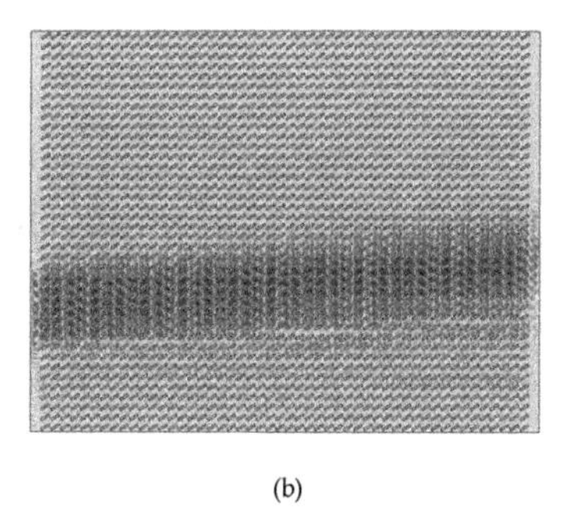

(b)

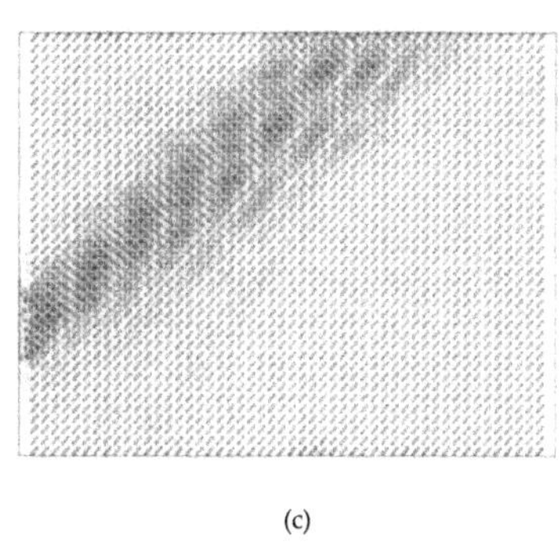

(c)

Figure 8. The displacement field simulations of the wave propagation in phononic crystals composed of star-shaped honeycomb and rectangular inclusions for (**a**) $\phi = 0°$, (**b**) $\phi = 25°$ and (**c**) $\phi = 60°$. The star angle of the matrix is $\theta = 55°$ and the dimensions of the rectangular inclusions are $w = 14/15a$, $h = a/5$.

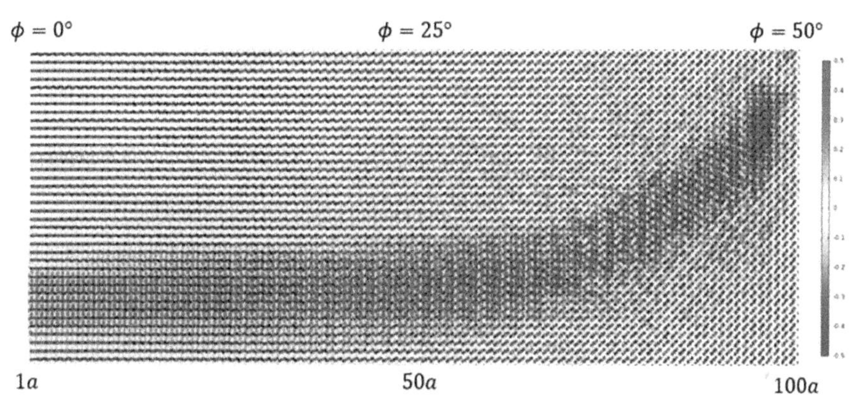

Figure 9. The displacement field simulation of the wave propagation in the phononic crystal composed of star-shaped honeycomb matrix and rectangular inclusions with gradually varied orientations in the x-direction. The orientation angle is $\phi = 0°$ at the left boundary, $x = 0$, and linearly increases to $\phi = 50°$ at the right boundary, $x = 100a$. The orientation angles ϕ are linearly varied by $\phi(x) = (x/100a)50°$.

4. Discussion and Conclusions

In this paper, the elastic wave propagations in phononic crystals composed of steel inclusions and auxetic material matrix are investigated. The star-shaped honeycomb structures are utilized and have been proven to have negative Poisson's ratio. Two types of inclusions with circular and rectangular cross sections are considered and discussed. The band structures of the phononic crystals are calculated and obtained by the finite element method.

The auxeticity of the star-shaped honeycomb is found to have significant effect on the band structure of the phononic crystal. The bandwidths and the mid-frequencies of the band gap increase as the Poisson's ratio decreases. In addition to this, the negative refraction is also found in the phononic crystals for the star-shaped honeycomb as the matrix. The self-collimation phenomenon is also studied. The wide angle and wide band collimation are observed in the phononic crystal with rectangular inclusions. The present study demonstrates the potential applications of the star-shaped honeycomb in phononic crystals, such as vibration isolation and the elastic waveguide.

Author Contributions: Formal analysis, S.-Y.C.; Methodology, S.-Y.C.; Resources, J.-Y.Y.; Supervision, L.-W.C.; Validation, J.-Y.Y.; Writing – review & editing, C.-D.C.

Acknowledgments: The financial support of Ministry of Science and Technology under the contract MOST 104-2221-E-006-113-MY3 is appreciated.

Conflicts of Interest: The authors declare no conflict of interest.

References

1. Sigalas, M.M.; Economou, E.N. Elastic and Acoustic Wave Band Structure. *J. Sound Vib.* **1992**, *158*, 377–382. [CrossRef]
2. Kushwaha, M.S.; Halevi, P.; Dobrzynski, L.; Djafari-Rouhani, B. Acoustic Band Structure of Periodic Elastic Composites. *Phys. Rev. Lett.* **1993**, *71*, 2022. [CrossRef] [PubMed]
3. Kushwaha, M.S.; Halevi, P.; Martinez, G.; Dobrzynski, L.; Djafari-Rouhani, B. Theory of Acoustic Band Structure of Periodic Elastic Composites. *Phys. Rev. B* **1994**, *49*, 2313. [CrossRef]
4. Martinezsala, R.; Sancho, J.; Sanchez, J.V.; Gomez, V.; Llinares, J.; Meseguer, F. Sound Attenuation by Sculpture. *Nature* **1995**, *378*, 241. [CrossRef]
5. Thomas, E.L.; Gorishnyy, T.; Maldovan, M. Colloidal Crystals Go Hypersonic. *Nat. Mater.* **2006**, *5*, 773–774. [CrossRef] [PubMed]
6. Wen, J.; Wang, G.; Yu, D.; Zhao, H.; Liu, Y. Theoretical and Experimental Investigation of Flexural Wave Propagation in Straight Beams with Periodic Structures: Application to a vibration isolation structure. *J. Appl. Phys.* **2005**, *97*, 114907. [CrossRef]
7. Sigalas, M. Elastic Wave Band Gaps and Defect States in Two-dimensional Composites. *J. Acous. Soc. Am.* **1997**, *101*, 1256. [CrossRef]
8. Wu, F.; Hou, Z.; Liu, Z.; Liu, Y. Point Defect States in Two-dimensional Phononic Crystals. *Phys. Lett. A* **2001**, *292*, 198–202. [CrossRef]
9. Zhang, X.; Liu, Z.; Liu, Y.; Wu, F. Defect States in 2D Acoustic Band-gap Materials with Bend-shaped Linear Defects. *Solid State Commun.* **2004**, *130*, 67–71. [CrossRef]
10. Liu, Z.; Zhang, X.; Mao, Y.; Zhu, Y.; Yang, Z.; Chan, C.T.; Sheng, P. Locally Resonant Sonic Materials. *Science* **2000**, *289*, 1734–1736. [CrossRef] [PubMed]
11. Liu, Z.; Chan, C.T.; Sheng, P. Three-component Elastic Wave Band-gap Material. *Phys. Rev. B* **2002**, *65*, 165116. [CrossRef]
12. Gang, W.; Li-Hui, S.; Yao-Zong, L.; Ji-Hong, W. Accurate Evaluation of Lowest Band Gaps in Ternary Locally Resonant Phononic Crystals. *Chin. Phys.* **2006**, *15*, 1843. [CrossRef]
13. Larabi, H.; Pennec, Y.; Djafari-Rouhani, B.; Vasseur, J. Multicoaxial Cylindrical Inclusions in Locally Resonant Phononic Crystals. *Phys. Rev. E* **2007**, *75*, 066601. [CrossRef] [PubMed]
14. Zhang, S.; Cheng, J. Existence of Broad Acoustic Bandgaps in Three-Component Composite. *Phys. Rev. B* **2003**, *68*, 245101. [CrossRef]
15. Zhang, X.; Liu, Y.; Wu, F.; Liu, Z. Large Two-dimensional Band Gaps in Three-component Phononic Crystals. *Phys. Lett. A* **2003**, *317*, 144–149. [CrossRef]
16. Fok, L.; Zhang, X. Negative Acoustic Index Metamaterial. *Phys. Rev. B* **2011**, *83*, 214304. [CrossRef]
17. Wu, Y.; Lai, Y.; Zhang, Z.-Q. Elastic Metamaterials with Simultaneously Negative Effective Shear Modulus and Mass Density. *Phys. Rev. Lett.* **2011**, *107*, 105506. [CrossRef] [PubMed]

18. Gibson, L.J.; Ashby, M.F.; Schajer, G.; Robertson, C.I. The Mechanics of Two-Dimensional Cellular Materials. *Proc. R. Soc. Lond. A* **1982**, *382*, 25–42. [CrossRef]
19. Robert, F.A. An Isotropic Three-dimensional Structure with Poisson's Ratio = −1. *J. Elast.* **1985**, *15*, 427.
20. Theocaris, P.S.; Stavroulakis, G.E.; Panagiotopoulus, P.D. Negative Poisson's ratios in composites with star-shaped inclusions: A numerical homogenization approach. *Arch. Appl. Mech.* **1997**, *67*, 274–286. [CrossRef]
21. Gonella, S.; Ruzzene, M. Analysis of In-plane Wave Propagation in Hexagonal and Re-entrant Lattices. *J. Sound Vib.* **2008**, *312*, 125–139. [CrossRef]
22. Liebold-Ribeiro, Y.; Körner, C. Phononic Band Gaps in Periodic Cellular Materials. *Adv. Eng. Mater.* **2014**, *16*, 328–334. [CrossRef]
23. Meng, J.; Deng, Z.; Zhang, K.; Xu, X.; Wen, F. Band Gap Analysis of Star-shaped Honeycombs with Varied Poisson's Ratio. *Smart Mater. Struct.* **2015**, *24*, 095011. [CrossRef]
24. Wojciechowski, K.W.; Scarpa, F.; Grima, J.N. Auxetics and Other Systems of Anomalous Characteristics. *Phys. Status Solidi B-Basic Solid Status Phys.* **2017**, *254*, 1770266. [CrossRef]
25. Qiu, C.; Liu, Z.; Shi, J.; Chan, C. Directional Acoustic Source Based on the Resonant Cavity of Two-dimensional Phononic Crystals. *Appl. Phys. Lett.* **2005**, *86*, 224105. [CrossRef]
26. Liu, W.; Su, X. Collimation and Enhancement of Elastic Transverse Waves in Two-dimensional Solid Phononic Crystals. *Phys. Lett. A* **2010**, *374*, 29682971. [CrossRef]
27. Cicek, A.; Kaya, O.A.; Ulug, B. Wide-band All-angle Acoustic Self-collimation by Rectangular Sonic Crystals with Elliptical Bases. *J. Phys. D Appl. Phys.* **2011**, *44*, 205104. [CrossRef]
28. Cicek, A.; Kaya, O.A.; Ulug, B. Impacts of Uniaxial Elongation on the Bandstructures of Two-dimensional Sonic Crystals and Associated Applications. *Appl. Acoust.* **2012**, *73*, 28–36. [CrossRef]
29. Cebrecos, A.; Romero-Garcia, V.; Pico, R.; Perez-Arjona, I.; Espinosa, V.; Sanchez-Morcillo, V.; Staliunas, K. Formation of Collimated Sound Beams by Three-dimensional Sonic Crystals. *J. Appl. Phys.* **2012**, *111*, 104910. [CrossRef]
30. Morvan, B.; Tinel, A.; Vasseur, J.; Sainidou, R.; Rembert, P.; Hladky-Hennion, A.-C.; Swinteck, N.; Deymier, P. Ultra-directional Source of Longitudinal Acoustic Waves Based on a Two-dimensional Solid/solid Phononic Crystal. *J. Appl. Phys.* **2014**, *116*, 214901. [CrossRef]
31. Soliveres, E.; Espinosa, V.; Pérez-Arjona, I.; Sánchez-Morcillo, V.J.; Staliunas, K. Self Collimation of Ultrasound in a Three-dimensional Sonic Crystal. *Appl. Phys. Lett.* **2009**, *94*, 164101. [CrossRef]
32. Liu, G.-T.; Tsai, C.-N.; Chang, I.-L.; Chen, L.-W. Elastic Dispersion Analysis of Two Dimensional Phononic Crystals Composed of Auxetic Materials. In Proceedings of the ASME 2015 International Mechanical Engineering Congress and Exposition, Houston, TX, USA, 13–19 November 2015; American Society of Mechanical Engineers: New York, NY, USA; p. V013T16A021.
33. Tsai, C.-N.; Chen, L.-W. The manipulation of self-collimated beam in phononic crystals composed of orientated rectangular inclusions. *Appl. Phys. A* **2016**, *122*, 659. [CrossRef]
34. Graff, K.F. *Wave Motion in Elastic Solids*; Courier Corporation: Chelmsford, MA, USA, 2012.
35. Tang, H.-W.; Chou, W.-D.; Chen, L.-W. Wave propagation in the polymer-filled star-shaped honeycomb periodic structure. *Appl. Phys. A* **2017**, *123*, 523. [CrossRef]

Article

Analysis of the Transmission Characteristic and Stress-Induced Birefringence of Hollow-Core Circular Photonic Crystal Fiber

Jingxuan Yang [1,2,*], Hu Zhang [1,3,*], Xiaoguang Zhang [1], Hui Li [1] and Lixia Xi [1]

1 State Key Laboratory of Information Photonics and Optical Communications, Beijing University of Posts and Telecommunications, Beijing 100876, China; xgzhang@bupt.edu.cn (X.Z.); lihui1206@bupt.edu.cn (H.L.); xilixia@gmail.com (L.X.)

2 College of Electrical Information, Langfang Normal University, Langfang 065000, China

3 School of Ethnic Minority Education, Beijing University of Posts and Telecommunications, Beijing 100876, China

* Correspondence: zhh309@sina.com (H.Z.); yangjingxuan@bupt.edu.cn (J.Y.)

Received: 21 January 2019; Accepted: 26 February 2019; Published: 2 March 2019

Abstract: Orbital angular momentum modes in optical fibers have polarization mode dispersion. The relationship between polarization mode dispersion and the birefringence vector can be deduced using an optical fiber dynamic equation. First, a mathematical model was established to formulate mode dispersion caused by stress-induced birefringence. Second, in the stress-induced birefringence simulation model, the finite element method was used to analyze the transmission characteristics of the hollow-core circular photonic crystal fiber. Finally, mode dispersion caused by stress-induced birefringence was obtained using theoretical derivation and simulation analyses. The results showed that the new fiber type has good transmission characteristics and strong stress sensitivity, which provide key theoretical support for optimizing the structural parameters of optical fiber and designing stress sensors.

Keywords: orbital angular momentum; modal dispersion; stress-induced birefringence; finite element method

---

1. Introduction

With the development of 5G—Internet of things, big data, and real-time data—optical fiber communication systems, it is imperative to expand their capacity, improve transmission stability, and network intelligence [1–3]. Current communication systems' capacity expansion technologies, such as time division multiplexing, frequency division multiplexing, polarization division multiplexing, and high-order modulated, cannot meet future information communication demands [4,5]. Therefore, space division multiplexing, a new capacity expansion technology, is emerging as a solution for this problem both in free space communication and optical fiber communication. Moreover, mode division multiplexing (MDM), a space division multiplexing technology, has been a hot area of research for several years.

Orbital angular momentum (OAM) multiplexing is a MDM technology. When transmitting information independently, it utilizes different modes as transmission channels based on the orthogonality of different OAM modes, and it can therefore spread in the optical fiber at the same time. In 1984, to obtain analytical solutions of some uniform geometrical shape, fiber birefringence—a thermal elastic model of gradient function—was developed by Chu et al. [6]. For fiber composed of uniformly-shaped stress zones, Tsai et al. [7] proposed the idea of superposition in 1991, whereby total stress field distribution is obtained by stress zone superposition. In [8], Yu et al. studied stress field distribution and the size of birefringence from stress microelements, and concluded that stress size and birefringence at the core's center are completely unrelated to the shape and placement direction of stress microelements. Since the first photonic crystal fiber (PCF) was successfully sketched in 1996, nonlinear effects were found to effectively improve with the variety of the cladding structures [9]. In [10], Yang Yue et al. first proposed that PCF can be used for supercontinuum generation of optical vortex modes, but PCF cladding of hexagonal symmetrical structures is only to support two groups of OAM modes, which would cause the problem of large loss and dispersion. Subsequently, Wong et al. twisted photonic crystal fiber to adjust dispersion and nonlinearity of OAM transmission [11]. Then a new type, C-PCF, designed by Zhang et al. was able to transmit OAM mode with low transmission loss and flat dispersion [12]. PCF is of great potential in the optical fiber communication field because it has the characteristics of low loss, small dispersion, and low nonlinear effect, especially for long-distance communication systems. Compared with circular fiber, a series of C-PCFs proposed by our research [13], can achieve a high refractive index difference of fiber cross-section without a complicated doping process. Continuous progress in PCF design methods and manufacturing processes are slowly perfecting it. Therefore, PCF is of great practical value for the future.

The fiber cross-section should be ideally a circle. The propagation constants of two degenerate modes in the orthogonal direction are equal, and the two polarization directions are perpendicular to each other, so that the OAM mode can stably transfer without a birefringence problem. However, residual birefringence exists in optical fiber, which can be divided into two categories: intrinsic and extrinsic birefringence [14]. Extrinsic birefringence is caused by fiber bending, external stress, torsion, environmental temperature change, external electromagnetic field, and vibration, which may occur during fiber cable manufacturing and laying. In the same OAM mode, propagation speeds of odd and even modes are different, leading to the birefringence modal dispersion (MD) effect [15]. Recently, in [16,17], Wang et al. analyzed the phenomenon of birefringence MD based on the polarization mode dispersion of single-mode fibers in OAM fiber. Stress-induction is a key influential factor in birefringence MD; however, to the best of our knowledge, it is rarely considered. Therefore, it is necessary to further research the problem of stress-induced birefringence.

In view of most approaches addressing SMFs birefringence, it can still be used for birefringence MD. The main contribution in this paper is that the transmission characteristics and stress-induced birefringence of the new designed C-PCF. For multiple modes in the OAM fibers, an improved method of dynamic equation is used to solve the problem for birefringence MD. In this article, the solid mechanics module in the COMSOL Multiphysics software, which is based on the finite element method [18], was used to analyze OAM modes the propagation properties supported by our hollow-core C-PCF. The results indicate that the new type hollow-core C-PCF fiber can realize effective segregation and stable transmission of OAM modes. The theoretical derivation of stress-induced birefringence in this paper is based on the cross-section of the fiber with an irregular shape in the stress region. Therefore, the designed hollow-core C-PCF in the application of MDM fiber communication systems might greatly improve channel capacity and spectral efficiency.

2. Theoretical Model

With the rapid development of optical fiber communication networks, the communication capacity of the existing SMFs is close to its Shannon limit, which gradually fails to meet the growing demand for information. The OAM multiplexing technology can solve increasing channel congestion problems of fiber communication. Due to the orthogonality of OAM modes with different topological charges in space—it is theoretically considered that the number of topological charges is infinite—research using OAM modes as information carriers in communication systems has aroused wide public concern [2,3,19]. In this part, we first provide the electric field distribution of the OAM mode in optical fiber. Then, the mode dispersion of birefringence is given. Finally, the mathematical formula on the stress-induced birefringence model is obtained.

2.1. OAM Modes in PCF

According to [19,20], the OAM mode electric field distribution in optical fiber can be expressed as follows:

$$\vec{E}(r,\phi,z,t) = \vec{E}(r)e^{jl\phi}e^{j(\omega t-\beta z)} \tag{1}$$

where $e^{jl\phi}$ is the phase distribution of mode field in the radial direction, $e^{j(\omega t-\beta z)}$ is the mode solutions, and $\vec{E}(r)$ is radial mode profile.

OAM modes in optical fiber are composed of the linear combination of vector eigenmodes [20]. The OAM mode in optical fiber is specifically expressed in the following configuration:

$$\begin{cases} OAM^{\pm}_{\pm l,m} = HE^{even}_{l+1,m} \pm jHE^{odd}_{l+1,m} & l > 1 \\ OAM^{\mp}_{\pm l,m} = EH^{even}_{l-1,m} \pm jEH^{odd}_{l-1,m} & \\ OAM^{\pm}_{\pm l,m} = HE^{even}_{l+1,m} \pm jHE^{odd}_{l+1,m} & l = 1 \\ OAM^{\mp}_{\pm l,m} = TM_{0,m} \pm jTE_{0,m} & \end{cases} \tag{2}$$

where "±" represents the polarization direction of the OAM mode. In OAM fiber, OAM mode can be regarded as the superposition of even and odd modes of HE or EH, and there is $\pi/2$ phase difference between the even and odd modes. When the value of l is larger than 1, there are four OAM modes to compose a OAM mode group for a given l and m, but when l equals 1, only two OAM modes form a OAM mode group. Because $TE_{0,m}$ and $TM_{0,m}$ modes have different propagation constants, which cannot form OAM modes, only even and odd $HE_{2,m}$ modes could be used to form an OAM mode. For the even and odd modes in the same OAM mode, the absolutes of topological charge number are both degenerate. In other words, their effective refractive indices are almost the same, and the electric field distribution of odd and even modes in Equation (2) can be expanded as

$$\begin{cases} HE^{even}_{l,m} = \vec{E}(r)e^{j(\omega t-\beta_e z)}\cos l\phi \\ HE^{odd}_{l,m} = \vec{E}(r)e^{j(\omega t-\beta_o z)}\sin l\phi \end{cases} \tag{3}$$

2.2. Birefringence Mode Dispersion

A dynamic equation is an effective method for analyzing the polarization mode dispersion (PMD) vector with the variation of transmission distance [21,22]. For multiple modes of transmission in OAM fiber, there exists not only inter-mode dispersion but also intra-mode dispersion. In order to facilitate the analysis, we split the fiber into numerous segments, and each segment of birefringence can be considered uniform, specifically

$$\begin{cases} \frac{d\hat{s}^{\pm|l|}}{dz} = \hat{\beta}^{\pm|l|} \times \hat{s}^{\pm|l|} \\ \frac{d\hat{s}^{\pm|l|}}{d\omega} = \hat{\tau}^{\pm|l|} \times \hat{s}^{\pm|l|} \end{cases} \tag{4}$$

where $\hat{\beta}^{\pm|l|}$ is the birefringence vector, $\hat{\tau}^{\pm|l|}$ is the OAM-PMD vector, and $\hat{s}^{\pm|l|}$ is the polarization state. Assuming that the frequency ω and distance z are invariable, we computed the input polarization state $\hat{s}$ partial derivatives about ω and z, respectively. The relationship between the PMD and birefringence vectors can be simplified as:

$$\frac{d\hat{\tau}^{\pm|l|}}{dz} = \frac{d\hat{\beta}^{\pm|l|}}{d\omega} + \hat{\beta}^{\pm|l|} \times \hat{\tau}^{\pm|l|} \tag{5}$$

when fiber is disturbed by an external force or electromagnetic effect, its cross-section becomes slightly deformed, and the propagation constant of two linear polarization modes changes. Meanwhile, the state of the two polarization modes is not degenerate as before. The study on OAM-PMD vector τ can be started with the birefringence vector β.

As birefringence exists in the OAM fiber, and the refractive index of odd and even modes is different, and, therefore, the propagation constant is different. Generally speaking, stress-induced birefringence is much larger than that caused by fiber core manufacturing errors of noncircular deformation. Since fiber is a type of good elastic medium, and the photoelastic effect occurs under the action of external force, which will cause refractive index changes. In order to study fiber stress distribution under external forces, fiber stress distribution is obtained as follows:

As shown in Figure 1, stresses in each direction can be decomposed into the sum of the stresses in the three axes. Hence, the stress distribution matrix under external force is as follows:

$$\begin{bmatrix} \sigma_x & \tau_{xy} & \tau_{xz} \\ \tau_{yx} & \sigma_y & \tau_{yz} \\ \tau_{zx} & \tau_{zy} & \sigma_z \end{bmatrix} \tag{6}$$

According to the equivalent theory of shearing stress [23], the above matrix is symmetric. Optical fiber changes are described by positive and shear strains. A positive strain is indicated by 'ε', shear strain is indicated by 'γ', and the strain matrix is expressed as follows:

$$\begin{bmatrix} \varepsilon_x & \gamma_{xy} & \gamma_{xz} \\ \gamma_{yx} & \varepsilon_y & \gamma_{yz} \\ \gamma_{zx} & \gamma_{zy} & \varepsilon_z \end{bmatrix} \tag{7}$$

where the x component of strain ε_x is related to the x component of displacement u, and the y component strain ε_y is related to the y component displacement v.

$$\begin{cases} \varepsilon_x = \frac{\partial u}{\partial x} \\ \varepsilon_y = \frac{\partial v}{\partial y} \\ \gamma_{xy} = \frac{\partial u}{\partial y} + \frac{\partial v}{\partial x} \end{cases} \tag{8}$$

The size of z direction is considered much larger than that of the other two directions, and the vertical external force on the cross-section of optical fiber. We only compute displacement vectors at all points parallel to the x and I planes. Since any optical fiber cross-section that can be treated is symmetric, Equation (7) can be transformed into a plane strain problem.

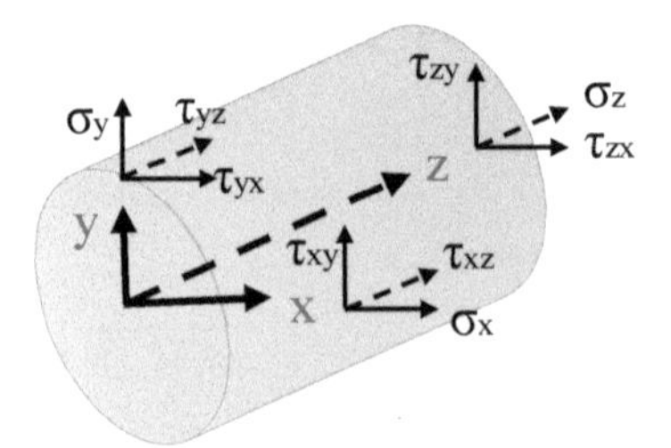

Figure 1. Diagram of stress decomposition.

$$\begin{cases} \varepsilon_x = \varepsilon_x(x,y) \\ \varepsilon_y = \varepsilon_y(x,y) \\ \gamma_{xy} = \gamma_{yx} = \gamma_{xy}(x,y) \end{cases} \tag{9}$$

In the case of complete elasticity and isotropy, the above equation of physical properties is conformed to Hooke's law within the range of elastic response.

$$\begin{cases} \varepsilon_x = \frac{1+\mu}{E}\left[(1-\mu)\sigma_x - \mu\sigma_y\right] \\ \varepsilon_y = \frac{1+\mu}{E}\left[(1-\mu)\sigma_y - \mu\sigma_x\right] \\ \gamma_{xy} = \frac{2(1+\mu)}{E}\tau_{xy} \end{cases} \tag{10}$$

where E is the modulus of elasticity in stress direction, and μ is the ratio of Poisson, which indicates the coefficient of shrinkage. We can introduce Airy's stress function F to solve this equation as follows:

$$\begin{cases} \sigma_x = \frac{\partial^2 F}{\partial y^2} \\ \sigma_y = \frac{\partial^2 F}{\partial x^2} \\ \tau_{xy} = -\frac{\partial^2 F}{\partial x \partial y} \end{cases} \quad (11)$$

where σ_x and σ_y are the x and y components of stress distribution in the homogeneous fiber, respectively. The Airy stress function is the solution for the biharmonic Equation [23]:

$$\nabla^4 F = 0 \quad (12)$$

To achieve the calculation of the Airy stress function, we used Fourier series expansion of polar coordinates to denote F, and the expansion coefficients a_n and b_n were determined from boundary conditions.

$$F = b_0 r^2 + a_0 + \sum \left(a_n r^n + b_n r^{n+2} \right) \cos n\theta \quad (13)$$

The stress components of Equation (11) were transformed into polar coordinates as follows:

$$\begin{cases} \sigma_r = \frac{-E}{1+\mu} \frac{1}{r} \left(\frac{\partial F}{\partial r} + \frac{\partial^2 F}{r \partial \theta^2} \right) \\ \sigma_\theta = \frac{-E}{1+\mu} \frac{\partial^2 F}{\partial r^2} \\ \tau_{r\theta} = \frac{E}{1+\mu} \frac{\partial}{\partial r} \left(\frac{1}{r} \frac{\partial F}{\partial \theta} \right) \end{cases} \quad (14)$$

The corresponding stress-induced birefringence B is given using the stress distribution formula:

$$B = C(\sigma_x - \sigma_y) \quad (15)$$

where C is the stress-optic coefficient of the fiber, which depends on the wavelength and material. The stress-optic coefficient is proportional to the difference in the material's refractive index. The electric field distribution depends on the normalized frequency, which can be defined as:

$$v = \frac{2\pi}{\lambda} na\sqrt{2\Delta} \quad (16)$$

where λ is the wavelength, n is the refractive index, a is the fiber core radius, and Δ is the relative index difference between core and cladding, respectively. Equation (15) can be rewritten as:

$$B = C \frac{\int_0^{2\pi} \int_0^{\infty} [\sigma_x(r,\theta) - \sigma_y(r,\theta)] |E(r,\theta,v)|^2 r dr d\theta}{\int_0^{2\pi} \int_0^{\infty} |E(r,\theta,v)|^2 r dr d\theta} \quad (17)$$

where $E(r,\theta,v)$ is the electric field's intensity distribution, while $\sigma_x(r,\theta)$ and $\sigma_y(r,\theta)$ represent anisotropic stress distributions in the fiber. The stress-induced birefringence formula can be used to calculate the cross-section of any photonic crystal fiber with an irregular shape.

3. Simulation and Analysis

Hollow-core circular photonic crystal fiber (C-PCF) can support 26 OAM modes, and the numerical analysis shows that the proposed fiber possesses very good fiber parameter values. In Reference [24], a novel ring-core photonic crystal fiber was fabricated, which supports four groups of OAM modes with 2.13×10^{-3} minimum difference of effective indices at 1550 nm. The realized fiber is expected to be a good platform for OAM mode transmission. In this section, we analyzed the stress distribution of hollow-core C-PCF. Then, the transmission characteristics and the resulting birefringence of OAM modes were analyzed using finite element methods. Finally, the simulation results were obtained using the solid mechanics module in the COMSOL Multiphysics software.

3.1. Model of Stress-Induced Field

We used [25,26] hollow-core C-PCF as the stress field model. The simulation parameters are given in Table 1. This structure's adjustable parameters in Figure 2(a) include: center air hole radius r, the cladding inner radius R, outer air hole diameter d_n (n = 2, 3, 4, 5. . .), the distance between two consecutive circles Λ, and the number of outer air hole rings N. In this paper, we set r = 4.4 μm, R = 6 μm, Λ = 2.2 μm, N = 4, d_n/Λ = 0.8, and d = 1.76 μm. Figure 2b,c indicate that hollow-core C-PCF deformed under external force, the deformation not only included air hole spacing but also change in air hole shape. Therefore, photonic crystal fiber symmetry changed, Figure 2b,c are oriented toward qualitative analysis and have little quantitative analysis on the deformation of the stress field [27,28]. Due to fiber symmetry, only the propagation characteristics simulation results of stress in the vertical direction are given.

Table 1. Main parameter settings in the simulation.

parameter	Value	Units
Refractive index	1.444	—
Young's modulus	7.8×10^{10}	Pa
Poisson's ratio	0.17	—
wavelength	1.55	μm

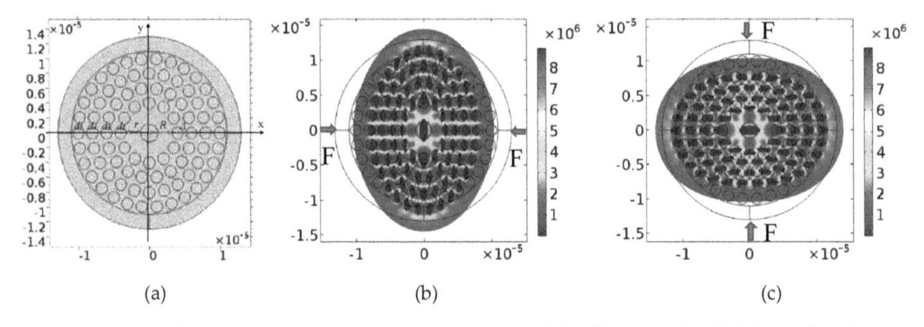

Figure 2. (**a**) Hollow-core C-PCF cross section without stress; (**b**) hollow-core C-PCF deformed under x direction external force; (**c**) hollow-core C-PCF deformed under y direction external force

Effective Refractive Index

Considering the actual situation of pressure in the process of optical fiber production and laying, we chose a 1–200 MPa pressure range in the simulation calculation. Figure 3a shows an effective refractive index as a function of stress for different modes, and Figure 3b shows the effective index difference between even and odd modes as a function of stress. Compared to the effective refractive index difference of fiber under no stress, we obtained the effective refractive index difference of odd and even modes due to stress. The effective refractive index increased with stress, as did the effective index difference between even and odd modes. The difference of effective refractive index between different modes was much larger than that between odd and even modes within the same mode with different stress. When external stress was applied to fiber, the fiber's cross-section was not symmetrical. The index difference between even and odd modes in the lower order modes was larger than that in the higher order modes. Electric field distribution of fiber depends on the normalized frequency defined by effective index difference between even and odd modes.

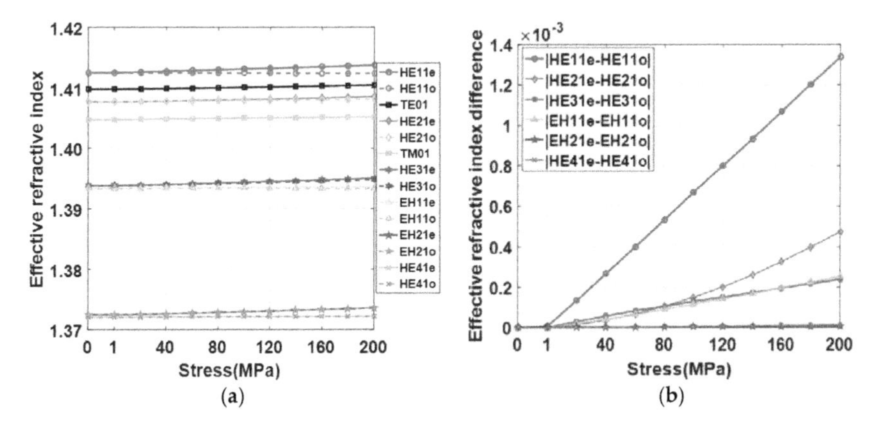

Figure 3. (**a**) Effective refractive index as a function of stress for different modes; (**b**) effective index difference between even and odd modes as a function of stress.

Since the stress-optic coefficient of fiber was determined by the wavelength of the transmitted beam, analyzed the change of effective refractive index with wavelength. Figure 4a shows that the effective refractive index changed with wavelength variation for different modes under 100 MPa stress in the y direction. Figure 4b shows that the effective refractive index difference between even and odd modes changed with wavelength variation. The difference of effective refractive index between different modes was much larger than that between odd and even modes within the same mode with different wavelengths. The effective refractive index decreased with wavelength variation, and the difference between even and odd modes in the lower order modes was larger than those in the higher order modes. Although these changes were small, it may be very important that they impact the value of modal birefringence.

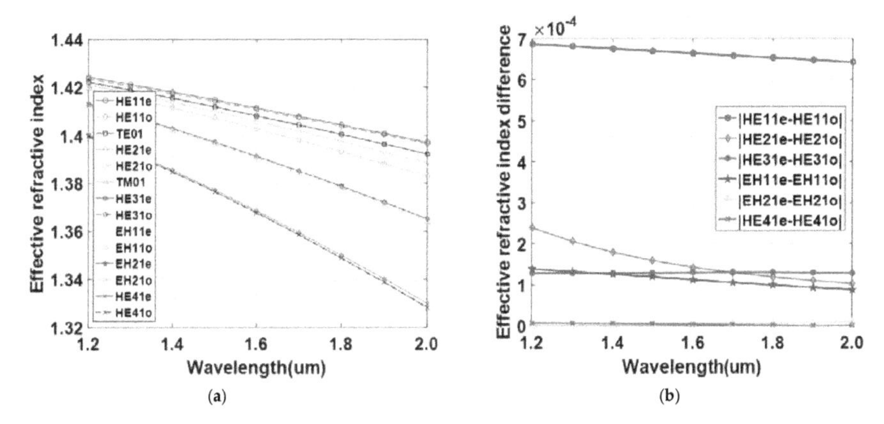

Figure 4. (**a**) Effective refractive index as a function of wavelength for different modes; (**b**) effective index difference between even and odd modes as a function of wavelength under 100 MPa stress in the y direction.

3.2. Transmission Characteristics of OAM Modes

This section simulates the transmission characteristics of hollow-core C-PCF under longitudinal pressure, including intensity, phase, polarization, and loss. There are two ways to compose an OAM mode in Equation (2), which can be obtained by the linear superposition of the odd and even modes of HE mode or EH mode. In Figure 3, the effective refractive index difference between EH mode and

HE mode that make up the same mode group is much larger than that between the odd and even mode within the same mode, so we only consider the dispersion effect within the mode. Here we use the transmission characteristics of HE mode to represent the transmission characteristics of an OAM mode. When the pressure was less than 1 MPa, transmission characteristics changes are unobvious, and the maximum pressure in the fiber's practical application was less than 200 MPa. Therefore, we exerted 1–200 MPa stress. The intensity distribution of each OAM mode transmitted in optical fiber was annular, and the phase distribution was related to the topological charge. Here we only selected the HE31 mode as an example to reflect the transmission characteristics of each mode.

3.2.1. Intensity and Phase

Figure 5 shows the normalized intensity distribution of HE31 under 1–200 MPa stress in the y direction. The xy plane represents the fiber core's cross-section, and the z-axis represents the normalized intensity of the HE31 mode. Figure 6 illustrates the intensity distribution of the axis cross-section in HE31 mode under the same situation. The intensity distribution of HE31 under the different stresses were almost the same, but the peak value of intensity distribution in the middle cross-section changed with increasing stress. Since the ringlike energy distribution changed with increasing stress, the ringlike intensity distribution of any single-mode vortex beam was no longer regular, and it generated distortion.

Figures 7 and 8 illustrate the normalized intensity distribution for even and odd modes of HE31 under 1–200 MPa stress in the y direction. The intensity distribution for even and odd modes have the same shape for the one OAM mode, but the peak value in the middle of the even and modes' cross-section changed in opposite tendency with increasing stress. The strength lobes of odd and even modes were complementary, thus the strength distribution of the OAM mode was comprised by superposition, which presented a ring distribution. Figure 9 shows the intensity distribution in the axis cross-section of HE31 under the same situation, HE31e and HE31o are even and odd modes, respectively. The abscissa represents the fiber's core scale, and the ordinate represents mode intensity. As stress is impacted in the vertical direction, HE31e surface strength was no longer flat. The quartered petal-shaped distribution gradually separated with increasing stress, and the top of the quartered distribution became irregular.

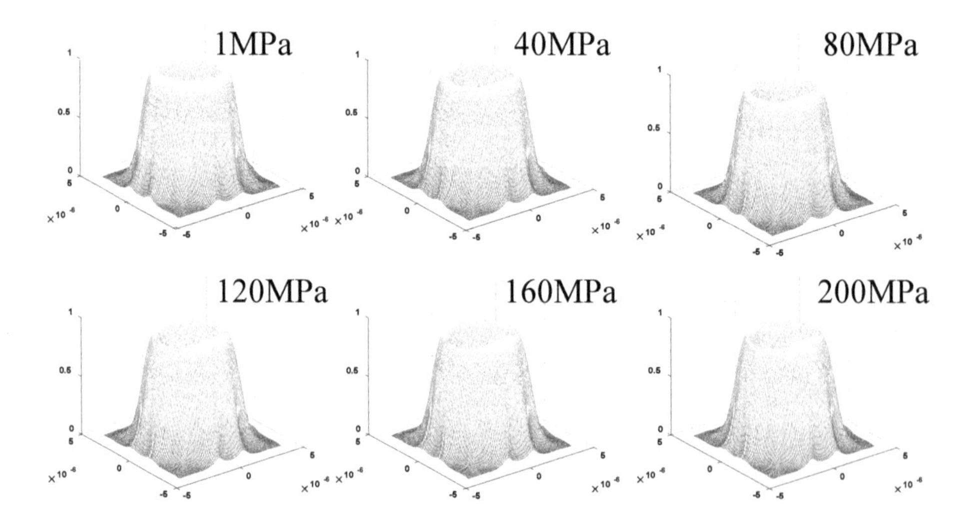

Figure 5. The normalized intensity distribution of HE31 under 1–200 MPa stress in the y direction.

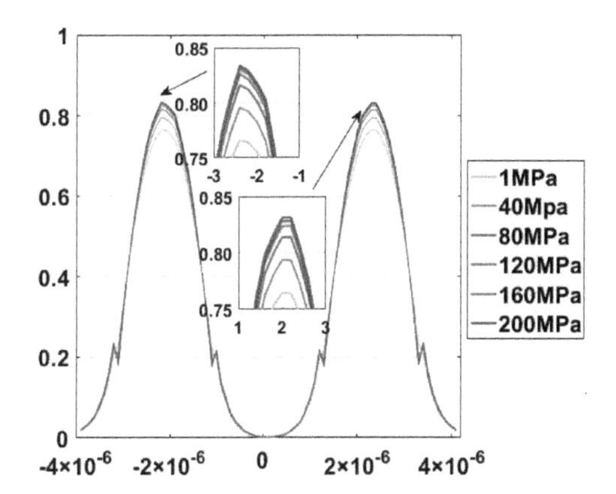

Figure 6. Intensity distribution of the axis cross-section in HE31 mode under the 1–200 MPa stress in the y direction.

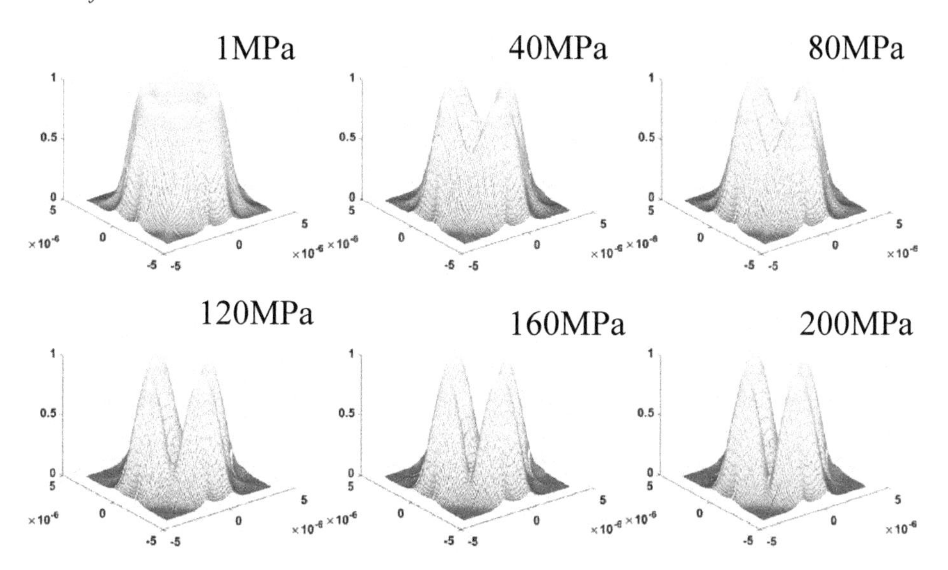

Figure 7. The normalized intensity distribution for the even mode of HE31 under 1–200 MPa stress in the y direction.

Figure 10a–f shows the phase distribution of HE31 mode under 1–200 MPa stress in the y direction. As the pressure increases, the isophase line becomes gradually irregular. However, phase distribution of HE31 satisfied phase characteristics of the OAM mode, and we concluded that the phase is insensitive to stress disturbances. Figure 11a–f shows the phase distribution of HE31 even and odd modes under the same situation. The first row represents even mode phase results, and the last row represents odd mode phase results. Figure 11 shows that phase distortion of HE31 even and odd modes worsened with increasing stress, and the phase difference between odd and even modes was $\pi/4$. The isophase lines in phase distribution were no longer smooth. With stress increase, the phase ambiguity of optical fiber became gradually serious, indicating that phase distortion was more serious. The phase distribution diagram shows that optical fiber can maintain stable OAM mode transmission within the selected stress range.

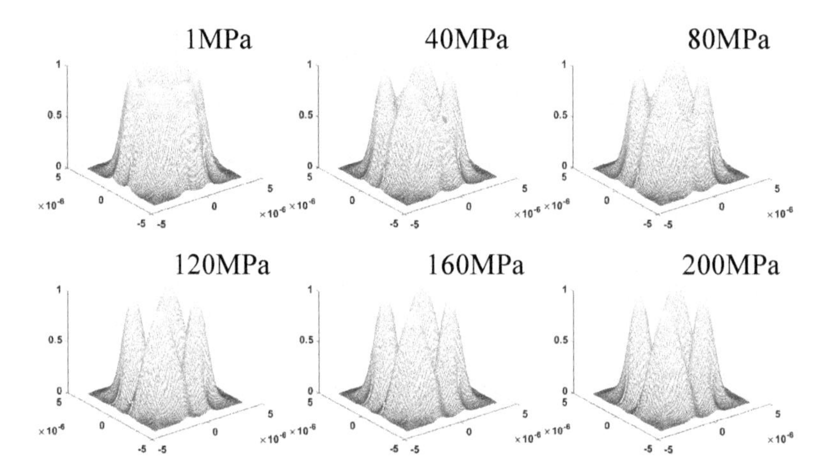

Figure 8. The normalized intensity distribution for the odd mode of HE31 under 1–200 MPa stress in the y direction.

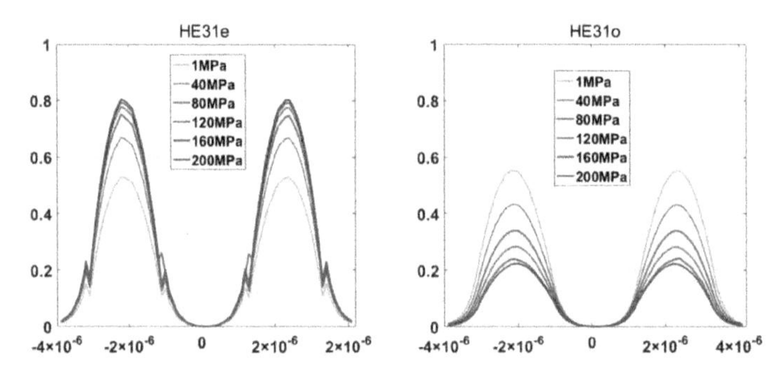

Figure 9. The intensity distribution of the axis cross-section in even and odd modes of HE31 under 1–200 MPa stress in the y direction.

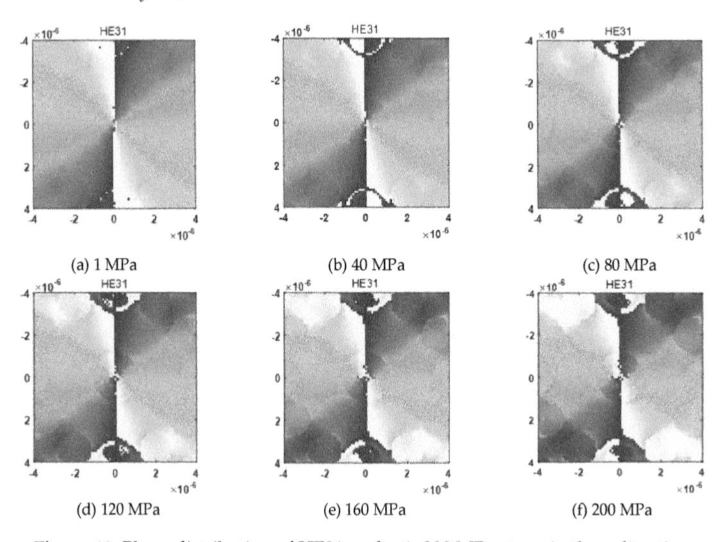

Figure 10. Phase distribution of HE31 under 1–200 MPa stress in the y direction.

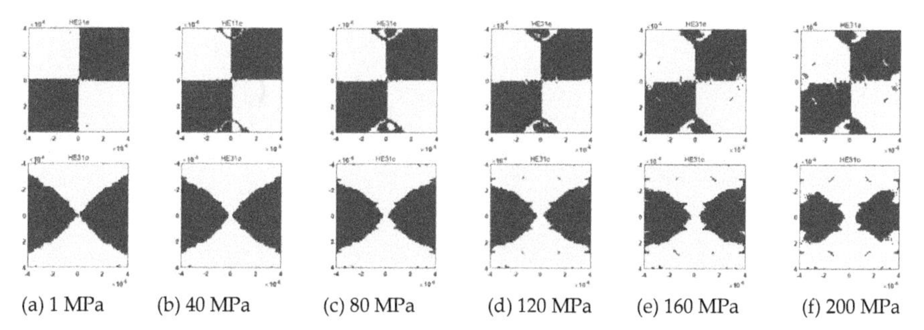

(a) 1 MPa (b) 40 MPa (c) 80 MPa (d) 120 MPa (e) 160 MPa (f) 200 MPa

Figure 11. Phase distribution of HE31 even and odd modes under the stress from 1 MPa to 200 MPa in the *y* direction.

3.2.2. Polarization

Figure 12 shows the polarization distribution of HE31 mode under stress from 1MPa to 200 MPa in the *y* direction. In Figure 12, ellipticity represents polarization intensity [29], and change of polarization intensity in the vertical direction was faster than that in the horizontal direction. In Figure 12a, the ellipse shapes were almost perfect circles, indicating that the OAM mode synthesized by HEe ± iHEo in the fiber was almost only LCP light (or the RCP light). However, the ellipticities of the ellipses in Figure 12b were slightly larger, meaning that the light synthesized by HE31e ± iHE31o in the fiber was composed of both LCP light and RCP light. Thus, this synthesized light contains SAM and OAM. Ellipticities in Figure 4c–f are obvious, and were even larger, meaning that polarization of the synthesized OAM mode in fiber gradually increased with stress.

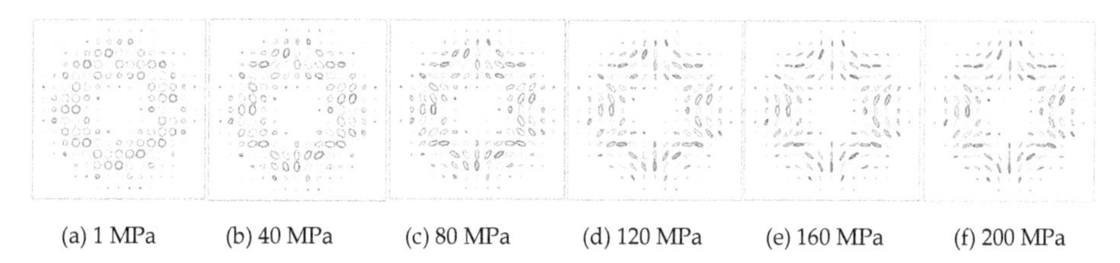

(a) 1 MPa (b) 40 MPa (c) 80 MPa (d) 120 MPa (e) 160 MPa (f) 200 MPa

Figure 12. Polarization distribution of HE31 mode under stress from 1MPa to 200 MPa in the *y* direction.

3.2.3. Loss Characteristics

Dispersion is an important factor causing optical fiber transmission signal distortion, which leads to communication quality decline and limited communication capacity. In optical fiber design, dispersion characteristics of optical fiber should be as flat as possible. According to [30], dispersion can be calculated using the following formula:

$$D = -\frac{\lambda}{c}\frac{d^2\mathrm{Re}\left[n_{eff}\right]}{d\lambda^2} \tag{18}$$

Figure 13 shows dispersion as a function of wavelength for different modes under different stress values. Dispersion of the new type C-PCF increased with stress variation, and the higher order mode was larger than the lower order mode. Dispersion is insensitive to pressure changes (the same color represent the same mode under different stress), so the designed optical fiber could keep good dispersion properties.

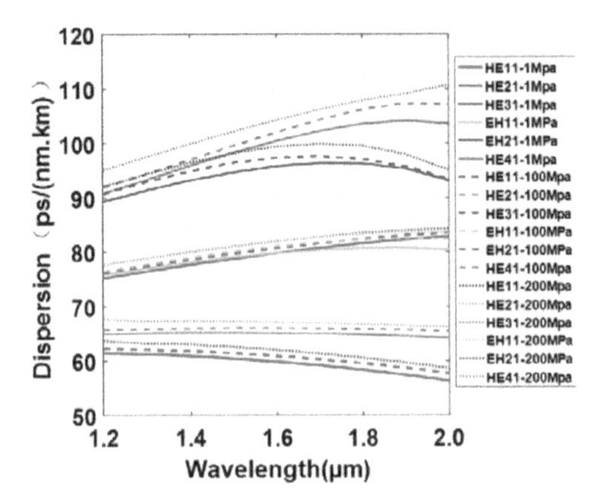

Figure 13. Dispersion as a function of wavelength for different modes.

In the PCF structure, due to transverse distribution limitation, there is confinement loss in fiber structure, which is mainly determined by the imaginary part of the effective refractive index [31,32]. The calculation formula is as follows:

$$L = \frac{2\pi}{\lambda} \frac{20}{\ln(10)} 10^6 \mathrm{Im}(n_{eff})(dB/m) \tag{19}$$

Figure 14a shows confinement loss as a function of stress for different modes, and Figure 14b shows confinement loss as a function of length for different modes under 100 MPa stress in the y direction. Confinement loss increases with stress variation, and increases with wavelength variation. The loss in the lower order mode was smaller than that in higher order mode, indicating that the lower order mode was more stable. In Figure 14, the solid line shows fiber confinement loss without stress, while the dotted line shows confinement loss under 100 MPa stress. Fiber has a low confinement loss and can provide good quality OAM modes.

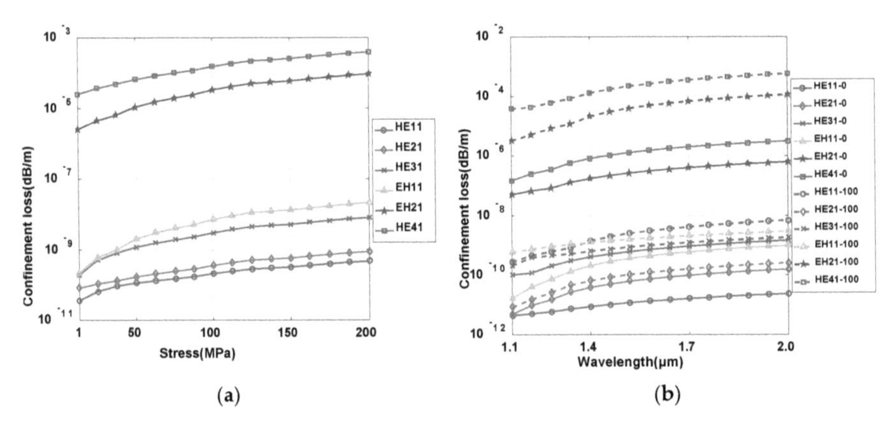

Figure 14. (**a**) Confinement loss as a function of stress for different modes, and (**b**) confinement loss as a function of length for different modes under 100 MPa stress in the y direction.

The nonlinear effect in optical fiber is also one of the main factors affecting the quality of mode transmission. A large effective mode field area is conducive to restraining the nonlinear effect in optical fiber. Therefore, it is necessary to generate fiber structure with a large effective mode field area as far as possible, so as to improve the quality of mode transmission. The area of the effective mode field can be obtained by the following formula [33]:

$$A_{eff} = \frac{\left(\int\int \left| \vec{E}(x,y) \right|^2 dxdy \right)^2}{\int\int \left| \vec{E}(x,y) \right|^4 dxdy} \tag{20}$$

The formula for calculating the nonlinear coefficient is:

$$\gamma = \frac{2\pi n_2}{\lambda A_{eff}} \tag{21}$$

Figure 15a shows effective mode area as a function of stress for different modes, and Figure 15b shows an effective mode area as a function of wavelength for different modes under 100 MPa stress in the y direction. Effective mode area decreased with increasing stress, and increased with increasing wavelength. Compared to the higher order mode, the effective mode area of the lower order mode was not sensitive stress. Figure 16a shows the nonlinear parameter as a function of stress for different modes, and Figure 16b shows the nonlinear parameter as a function of wavelength for different modes under 100 MPa stress in the y direction. Nonlinear parameter increased with increasing stress, while it decreased with increasing wavelength. The variation trend of the nonlinear parameter was the opposite to that of the effective mode area. Higher order modes were more sensitive to wavelength.

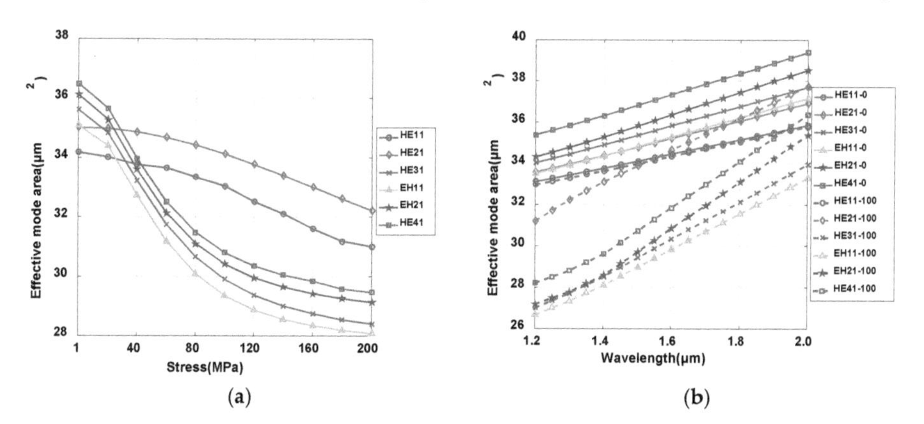

Figure 15. (**a**) Effective mode area as a function of stress for different modes, and (**b**) effective mode area as a function of wavelength for different modes under 100 MPa stress in the y direction.

3.3. Stress-Induced Birefringence

The modal birefringence is an important parameter for reflecting OAM fiber performance. Figure 17 shows the stress-induced birefringence of the vector modes as a function of stress. It shows that stress-induced birefringence increased with increasing stress. Birefringence in the lower order mode was larger than that in the higher order mode. The difference of effective refractive indices between even and odd modes decides the stress-induced birefringence. The higher order model had smaller birefringence compared to the lower order model, so it had better stress resistance. Birefringence curves of the same mode group were close to each other, while birefringence curves of different mode groups were separated. Therefore, the effect of stress-induced birefringence should be considered in MDM systems.

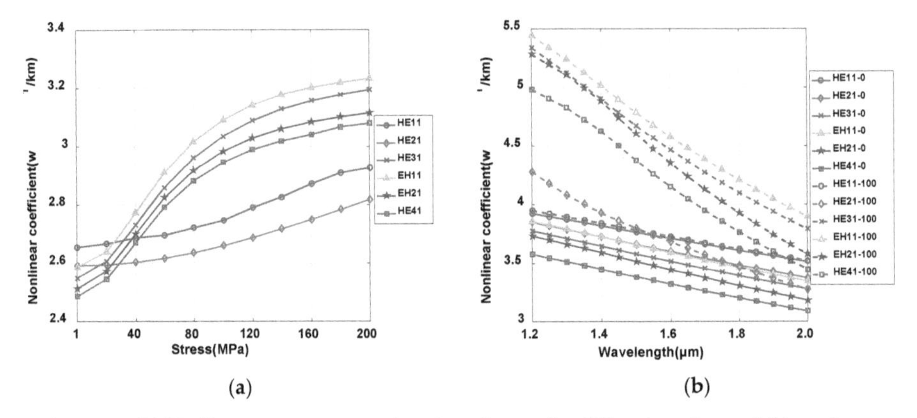

Figure 16. (**a**) Nonlinear parameter as a function of stress for different modes, and (**b**) nonlinear parameter as a function of wavelength for different modes under 100 MPa stress in the *y* direction.

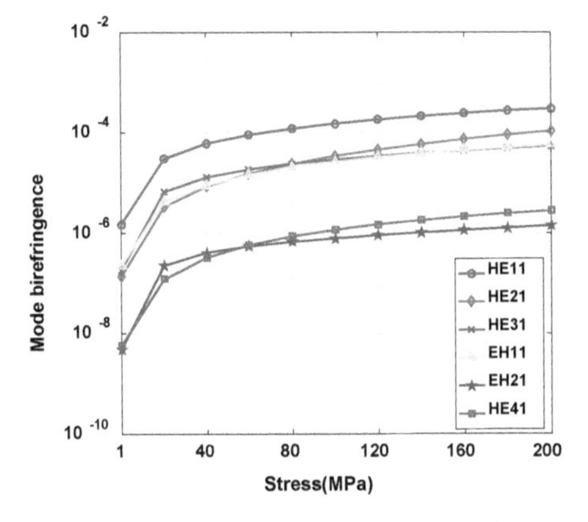

Figure 17. Modal birefringence of vector modes as a function of stress in the *y* direction.

4. Conclusions

In this article, a complete numerical dispersion theory framework, in relation to processing PCF fiber birefringence patterns, is established. We deducted the mathematical model for the OAM mechanism, which can now be used to calculate stress birefringence of any irregularly shaped fiber core. Furthermore, the model was simulated by structural mechanics and wave optics modules using the COMSOL software. Considering that fiber core symmetry changed due to fiber force, conclusions can be drawn through the above theoretical calculation and simulation analysis, and verified by the transmission characteristics of hollow-core C-PCF. At first, the intensity and phase of the OAM modes remain relatively stable within the range of actual stress. For another, this kind of new designed hollow-core C-PCF could keep a low loss under stress. Modal birefringence for the hollow-core C-PCF is sensitive to stress. The study of transmission characteristics provides basic theoretical research for mode division multiplexing, and improves the quality and performance of optical fiber communication systems.

Author Contributions: Methodology, X.Y.; software, H.Z.; validation, G.Z.; formal analysis, X.Y.; resources, G.Z. and H.Z.; writing—original draft preparation, X.Y.; funding acquisition, G.Z. X.X. and H.Z.; writing-review and editing, X.Y. and H.L.

Funding: This work was supported by the National Natural Science Foundation of China (NSFC) [61571057, 61527820, 61575082].

Conflicts of Interest: The authors declare no conflict of interest.

References

1. Essiambre, R.J.; Kramer, G.; Winzer, P.J.; Foschini, G.J.; Goebel, B. Capacity limits of optical fiber networks. *J. Light. Technol.* **2010**, *28*, 662–701. [CrossRef]
2. Chen, H.; Zhuang, Y. Research progress on key technologies in mode division multiplexing system. *J. Nanjing Univ. Posts Telecommun.* **2018**, *38*, 37–44.
3. Huang, H.; Milione, G.; Lavery, M.P.; Xie, G.; Ren, Y.; Cao, Y.; Ahmed, N.; Nguyen, T.A.; Nolan, D.A.; Li, M.-J.; et al. Mode division multiplexing using an orbital angular momentum mode sorter and MIMO-DSP over a graded-index few-mode optical fiber. *Sci. Rep.* **2015**, *5*, 14931. [CrossRef] [PubMed]
4. Wang, L.; Vaity, P.; Messaddeq, Y.; Rusch, L.; LaRochelle, S. Orbital-angular-momentum polarization mode dispersion in optical fibers and its measurement technique. In Proceedings of the 2015 European Conference on Optical Communication (ECOC), Valencia, Spain, 27 September–1 October 2015.
5. Puttnam, B.J.; Luís, R.S.; Klaus, W.; Sakaguchi, J.; Mendinueta, J.-M.D.; Awaji, Y.; Wada, N.; Tamura, Y.; Hayashi, T.; Hirano, M.; et al. 2.15 Pb/s transmission using a 22 core homogeneous single-mode multi-core fiber and wideband optical comb. In Proceedings of the 2015 European Conference on Optical Communication (ECOC), Valencia, Spain, 27 September–1 October 2015.
6. Chu, P.; Sammut, R. Analytical method for calculation of stresses and material birefringence in polarization-maintaining optical fiber. *J. Light. Technol.* **1984**, *2*, 650–662.
7. Tsai, K.H.; Kim, K.S.; Morse, T.F. General solutions for stress-induced polarization in optical fibers. *J. Light. Technol.* **1991**, *9*, 7–17. [CrossRef]
8. Yu, P.; Ji, M. Stress element analysis method of the birefringence in stress- induced fiber. *Laser Optoelectron. Prog.* **2014**, *52*, 020604.
9. Philip, R. Photonic crystal fibers. *Science* **2003**, *299*, 358–362.
10. Yue, Y.; Zhang, L.; Yan, Y.; Ahmed, N.; Yang, J.; Huang, H.; Ren, Y.; Dolinar, S.; Tur, M.; Willner, A.E. Octave-spanning supercontinuum generation of vortices in an As2S3 ring photonic crystal fiber. *Opt. Lett.* **2012**, *37*, 1889–1891. [CrossRef] [PubMed]
11. St Russell, P.; Beravat, R.; Wong, G.K.L. Helically twisted photonic crystal fibres. Philosophical Transactions of the Royal Society A: Mathematical. *Phys. Eng. Sci.* **2017**, *375*, 20150440. [CrossRef] [PubMed]
12. Zhang, L.; Wei, W.; Zhang, Z.; Liao, W.; Yang, Z.; Fan, W.; Li, Y. Propagation properties of vortex beams in a ring photonic crystal fiber. *Acta Phys. Sin.* **2017**, *66*, 014205.
13. Tian, W.; Zhang, H.; Zhang, X.; Xi, L.; Zhang, W.; Tang, X. A circular photonic crystal fiber supporting 26 OAM modes. *Opt. Fiber Technol.* **2016**, *30*, 184–189. [CrossRef]
14. Park, Y.; Paek, U.; Kim, D.Y. Determination of stress-induced intrinsic birefringence in a single-mode fiber by measurement of the two-dimensional stress profile. *Opt. Lett.* **2002**, *27*, 1291–1293. [CrossRef] [PubMed]
15. Galtarossa, A.; Palmieri, L.; Schiano, M.; Tambosso, T. Statistical characterization of fiber random birefringence. *Opt. Lett.* **2000**, *25*, 1322–1324. [CrossRef] [PubMed]
16. Wang, L.; Vaity, P.; Chatigny, S.; Messaddeq, Y.; Rusch, L.A.; LaRochelle, S. Orbital-Angular-Momentum Polarization Mode Dispersion in Optical Fibers. *J. Light. Technol.* **2014**, *34*, 1661–1671. [CrossRef]
17. Brunet, C.; Ung, B.; Wang, L.; Messaddeq, Y.; LaRochelle, S.; Rusch, L.A. Design of a family of ring-core fibers for OAM transmission studies. *Opt. Express* **2015**, *23*, 10553–10563. [CrossRef] [PubMed]
18. Bréchet, F.; Marcou, J.; Pagnoux, D.; Roy, P. Complete analysis of the characteristics of propagation into photonic crystal fibers, by the finite element method. *Opt. Fiber Technol.* **2000**, *6*, 181–191. [CrossRef]
19. Willner, A.E.; Huang, H.; Yan, Y.; Ren, Y.; Ahmed, N.; Xie, G.; Bao, C.; Li, L.; Cao, Y.; Zhao, Z.; Wang, J.; et al. Optical communications using orbital angular momentum beams. *Adv. Opt. Photonics* **2015**, *7*, 66–106. [CrossRef]
20. Brunet, C.; Vaity, P.; Messaddeq, Y.; LaRochelle, S.; Rusch, L.A. Design, fabrication and validation of an OAM fiber supporting 36 states. *Opt. Express* **2014**, *22*, 26117–26127. [CrossRef] [PubMed]
21. Menyuk, C.R.; Wai, P.K.A. Polarization evolution and dispersion in fibers with spatially varying birefringence. *J. Opt. Soc. Am. B* **1994**, *11*, 1288–1296. [CrossRef]

22. Bakhshali, A.; Chan, W.Y.; Cartledge, J.C.; O'Sullivan, M.; Laperle, C.; Borowi, A. Frequency-Domain Volterra-Based Equalization Structures for Efficient Mitigation of Intrachannel Kerr Nonlinearities. *J. Light. Technol.* **2016**, *34*, 1770–1777. [CrossRef]
23. Shibata, N.; Okamoto, K.; Tateda, M.; Seikai, S.; Sasaki, Y. Modal birefringence and polarization mode dispersion in single-mode fibers with stress-induced anisotropy. *IEEE J. Quant. Electron.* **1989**, *19*, 1110–1115. [CrossRef]
24. Zhang, H.; Zhang, W.; Xi, L.; Tang, X.; Zhang, X.; Zhang, X. A New Type Circular Photonic Crystal Fiber for Orbital Angular Momentum Mode Transmission. *IEEE Photonics Technol. Lett.* **2016**, *28*, 1426–1429. [CrossRef]
25. Zhang, H.; Zhang, X.; Li, H.; Deng, Y.; Zhang, X.; Xi, L.; Tang, X.; Zhang, W. A design strategy of the circular photonic crystal fiber supporting good quality orbital angular momentum mode transmission. *Opt. Commun.* **2017**, *397*, 59–66. [CrossRef]
26. Li, H.; Zhang, H.; Zhang, X.; Deng, Y.; Xi, L.; Tang, X.; Zhang, W. Design tool for circular photonic crystal fibers supporting orbital angular momentum modes. *Appl. Opt.* **2018**, *57*, 2474–2481.
27. Zhu, Z.; Brown, T.G. Stress-induced birefringence in micro structured optical fibers. *Opt. Lett.* **2003**, *28*, 2306–2308. [CrossRef] [PubMed]
28. Schreiber, T.; Schultz, H.; Schmidt, O.; Röser, F.; Limpert, J.; Tünnermann, A. Stress-induced birefringence in large-mode-area micro-structured optical fibers. *Opt. Express* **2005**, *13*, 3637–3646. [CrossRef] [PubMed]
29. Zhang, Z.; Gan, J.; Heng, X.; Wu, Y.; Li, Q.; Qian, Q.; Chen, D.; Yang, Z. Optical fiber design with orbital angular momentum light purity higher than 99.9%. *Opt. Express* **2015**, *23*, 29331–29341. [CrossRef] [PubMed]
30. Xu, H.; Wu, J.; Xu, K.; Dai, Y.; Xu, C.; Lin, J. Ultra-flattened chromatic dispersion control for circular photonic crystal fibers. *J. Opt.* **2011**, *13*, 994–1001. [CrossRef]
31. Varnham, M.P.; Payne, D.N.; Barlow, A.J.; Birch, R.D. Analytic solution for the birefringence produced the thermal stress in polarization-maintaining optical fibres. *J. Light. Technol.* **1983**, *LT-1*, 332–339. [CrossRef]
32. Maji, P.S.; Chaudhuri, P.R. Circular photonic crystal fibers: Numerical analysis of chromatic dispersion and losses. *ISRN Opt.* **2013**, *2013*, 986924. [CrossRef]
33. Tandjè, A.; Yammine, J.; Bouwmans, G.; Dossou, M.; Vianou, A.; Andresen, E.R.; Bigot, L. Design and Fabrication of a Ring-Core Photonic Crystal Fiber for Low-Crosstalk Propagation of OAM Modes. In Proceedings of the 2018 European Conference on Optical Communication (ECOC), Rome, Italy, 23–27 September 2018.

Article

Two-Layer Erbium-Doped Air-Core Circular Photonic Crystal Fiber Amplifier for Orbital Angular Momentum Mode Division Multiplexing System

Hu Zhang [1,2,*], Di Han [1], Lixia Xi [1,*], Zhuo Zhang [3], Xiaoguang Zhang [1], Hui Li [1] and Wenbo Zhang [1,4]

1 State Key Laboratory of Information Photonics and Optical Communications, Beijing University of Posts and Telecommunications, Beijing 100876, China; hd2016@bupt.edu.cn (D.H.); xgzhang@bupt.edu.cn (X.Z.); lihui1206@bupt.edu.cn (H.L.); zhangwb@bupt.edu.cn (W.Z.)
2 School of Ethnic Minority Education, Beijing University of Posts and Telecommunications, Beijing 100876, China
3 Institute of Electronics, Chinese Academy of Sciences, Beijing 100190, China; zhangzhuo@mail.ie.ac.cn
4 School of Sciences, Beijing University of Posts and Telecommunications, Beijing 100876, China
* Correspondence: zhh309@bupt.edu.cn (H.Z.); xilixia@bupt.edu.cn (L.X.); Tel.: +86-010-6119-8077 (H.Z.); +86-010-6119-8077 (L.X.)

Received: 25 February 2019; Accepted: 12 March 2019; Published: 15 March 2019

Abstract: Orbital angular momentum (OAM) mode-division multiplexing (MDM) has recently been under intense investigations as a new way to increase the capacity of fiber communication. In this paper, a two-layer Erbium-doped fiber amplifier (EDFA) for an OAM multiplexing system is proposed. The amplifier is based on the circular photonic crystal fiber (C-PCF), which can maintain a stable transmission for 14 OAM modes by a large index difference between the fiber core and the cladding. Further, the two-layer doped region can balance the amplification performance of different modes. The relationship between the performance and the parameters of the amplifier is analyzed numerically to optimize the amplifier design. The optimized amplifier can amplify 18 modes (14 OAM modes) simultaneously over the C-band with a differential mode gain (DMG) lower than 0.1 dB while keeping the modal gain over 23 dB and noise figure below 4 dB. Finally, the fabrication tolerance and feasibility are discussed. The result shows a relatively large fabrication tolerance in the OAM EDFA parameters.

Keywords: mode-division multiplexing; Erbium-doped fiber amplifier; photonic crystal fibers; orbital angular momentum

1. Introduction

Optical communication technology has developed rapidly in the past decades. All kinds of multiplexing and high order modulation technological have greatly increased the capacity of single fibers. The transmission capacity is gradually reaching the limit in that a standard single-mode fiber (SMF) cannot carry more than about 100 $\text{Tbit}\cdot\text{s}^{-1}$ of data in the C + L band [1]. To meet the stupendously increasing demands for transmission capacity, mode-division multiplexing (MDM), which is one of space-division multiplexing (SDM), has been proposed. MDM utilizing the orthogonality among different orbital angular momentum (OAM) states as the multiplex method has exhibited promising prospects in recent years. An OAM beam is characterized by a helical phase front exp (ilφ) (in polar coordinates, l is the topological charge, and φ is the azimuthal angle), which is an optical vortex beam [2–4]. Theoretically, l can be any integer value (that is, OAM has an infinite number of orthogonal eigenstates), which means that OAM has great potential to increase the transmission capacity in a MDM system [5]. For communications using OAM beams, there are two key problems to be solved. One is the optical vortex generation, which is the most fundamental technique to implement OAM multiplexing. The optical vortex can be generated by a variety of schemes, such as spatial light modulator [6] and modified interference of different modes [7]. A liquid-crystal spatial light modulator is another promising scheme for generating OAM modes, which exhibits some good properties including simplicity, high efficiency, and reconfiguration [8–10]. Another is the optical vortex beam transmission, which needs the fiber supporting OAM modes. In current research, the ring fiber was designed mostly to transmit OAM modes [11–15]. The circular photonic crystal fiber (C-PCF) was also proposed as a potential OAM fiber structure [16–19]. One fiber can transmit several dozens of mutually orthogonal modes. To date, terabit data transmission based on an OAM fiber has been demonstrated [20]. The transmission distance of the OAM mode in fibers has already reached 50 km [21]. However, many techniques are still needed for handling the implementation of long-haul MDM systems based on OAM modes. OAM carrier signal amplification is one of the critical techniques.

The Erbium-doped fiber amplifier (EDFA) not only allows the light signal to be amplified online directly but also possesses the same range between amplification wavelength and the low loss wavelength of optical fibers, which is suitable for MDM systems based on OAM modes. However, unlike the single-mode fiber amplifier, there are tens (even dozens) of modes in the OAM fiber amplifier. The transverse distribution of each mode is different. The biggest difference of these mode gains was defined as differential mode gain (DMG) [22]. DMG should be low enough to ensure approximately equal amplification of the multimode in the fiber.

In recent studies, most designs of OAM-EDFA are based on the ordinary circular air-core fiber with a one-layer doped region [23–26]. These OAM-EDFAs perform well in the modal gains but the DMG still needs to be improved. A new design of a two-layer Erbium-doped fiber amplifier transmitting OAM modes was introduced in reference [27] and provides a feasible way to deal with the DMG. They reduce DMG by adjusting the concentrations in the two doped regions. In addition, a one-layer doped region OAM fiber amplifier based on the C-PCF has also been presented by our research team. The amplifier can provide DMG lower than 0.2 dB, which is lower than that of the ordinary circular air-core fiber amplifier [28]. Naturally, the C-PCF structure combined with a two-layer doped region would be a more optimal design for OAM-EDFA. Compared with previous work, this OAM-EDFA can provide better performance in both modal gain and DMG. In addition, our design exhibits a higher tolerance for application.

In this paper, we propose and design a new OAM-EDFA with two-layer Erbium doped based on the C-PCF. The intensity distribution of OAM modes is studied to adjust the doped regions reasonably. The widths and the doped concentrations of the two layers are considered. The performance of the OAM-EDFA with different parameters is analyzed to get an optimal design. All 14 OAM modes are amplified as equally as possible to achieve long-haul transmission.

2. Theory

Erbium ion has a unique three-level system, as shown in Figure 1. It includes a ground state, metastable state, and excited state, which is suitable to realize the population inversion between the metastable state and ground state to amplify a signal light whose wavelength is around the 1550 nm window (C-band). We used the Giles and Desurvire model to carry out the analyses upon the population conversion, which is widely used in simulations for fiber amplifiers [29–33]. Owing to the lifetime of the Erbium ion in the metastable state, which is much larger than that in the excited state, the particles of the excited state can be ignored; therefore, the model is considered as a two-level system. Then, we can describe the changing rate of the Erbium ion concentration in the metastable state by the following equations:

$$\frac{dn_2(r,\varphi,z)}{dt} = \sum_k \frac{P_k i_k \sigma_{ak}}{h\nu_k} n_1(r,\varphi,z) - \sum_k \frac{P_k i_k \sigma_{ek}}{h\nu_k} n_2(r,\varphi,z) - \frac{n_2(r,\varphi,z)}{\tau} \quad (1)$$

$$n_t(r,\varphi,z) = n_1(r,\varphi,z) + n_2(r,\varphi,z) \quad (2)$$

where n_1 and n_2 are the Erbium ion concentrations of the ground state and the metastable state, respectively; P_k is the power of light; $k = s,p,a$ corresponds to the light power of the signal and pump as well as the amplifier spontaneous emission (ASE) noise, respectively; i_k is the normalized light intensity; σ_{ak} and σ_{ek} are the absorption and emission cross-sections (which are the attributes of Er^{3+}) respectively; h is Planck constant; and τ is the lifetime of the metastable state. Equation (2) denotes the particle conservation in the two-level system, where n_t is the total Erbium ion concentration. When the EDFA is in a stable state, the number of particles in the metastable state remains unchanged, and Equation (1) is equal to zero. Then, the particle inversion can be given by:

$$n_2(r,\varphi,z) = n_t \frac{\sum_k \frac{\tau P_k i_k \sigma_{ak}}{h\nu_k}}{1+\sum_k \frac{\tau P_k i_k (\sigma_{ak}+\sigma_{ek})}{h\nu_k}} \tag{3}$$

The transmission equation in the optical fiber can be expressed by the following propagation equation:

$$\begin{aligned} \frac{dP_k}{dz} &= \sigma_{ek}(P_k(z) + mh\nu_k \Delta\nu_k) \int_0^{2\pi} \int_0^{\infty} i_k(r,\Phi)\, n_2(r,\Phi,z) r dr d\Phi \\ &-\sigma_{ak}\, P_k(z) \int_0^{2\pi} \int_0^{\infty} i_k(r,\Phi)\, n_1(r,\Phi,z) r dr d\Phi \end{aligned} \tag{4}$$

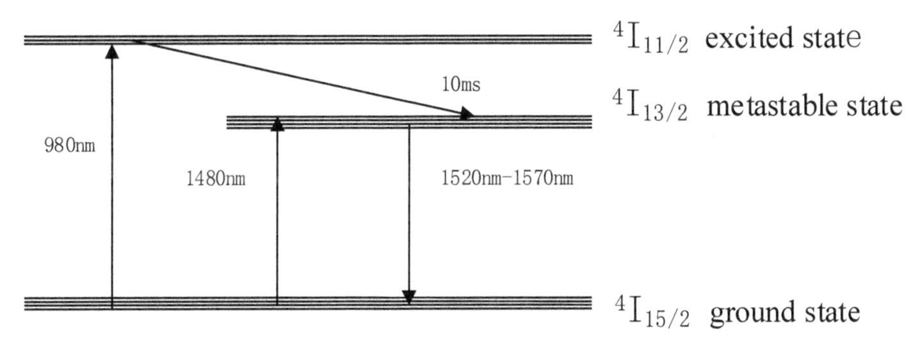

Figure 1. Three-level diagram of Erbium ion.

In the EDFA, the amplification occurs in the process of energy exchange between the pump light and Erbium ions. Therefore, the overlap of the pump light and the doped region determines mainly the performance of EDFA. Therefore, we define the overlap factor as the dimensionless integral overlap between the normalized optical intensity distribution and the normalized Erbium ion distribution:

$$\Gamma_k = \frac{\int_0^{2\pi} \int_a^b i_k(r,\varphi) n_2(r,\varphi,z) r dr d\varphi}{\overline{n_2}} \tag{5}$$

where a and b are the inner and outer radius of the fiber doped region. To obtain efficient amplification, it is necessary to make the signal light and the doped area perfectly overlapped, and the signal light and pump light matched greatly. Thus, a correction factor η is presented to evaluate the overlap between the signal light and the pump light [28]:

$$\eta = \frac{\left| \iint E_s^* \cdot E_p dxdy \right|^2}{\left| \iint E_s \cdot E_s^* dxdy \right| \left| \iint E_p \cdot E_p^* dxdy \right|} \tag{6}$$

where E_s and E_p is the signal and pump electric field, respectively. For an accurate description, the overlap factor is modified as $\Gamma_k{}' = \Gamma_k \times \eta$. The noise figure is another property of EDFA and can be calculated by:

$$F_n = 2n_{sp}(G-1)/G \tag{7}$$

where n_{sp} and G represent the spontaneous emission factor and the mode gain, respectively. The Equations (1)–(7) are combined to analyze the performance of the amplifier based on the OAM fiber.

3. Modeling of the OAM-EDFA

The light intensity distributions of OAM modes transmitted in fiber are ring-shaped. Therefore, the circular fiber structure, such as C-PCF, is a good choice for an OAM fiber. The OAM modes are formed by linearly combining orthogonal even and odd vector modes with a $\pi/2$ phase shift by the following equations:

$$\begin{cases} \left.\begin{array}{l} OAM^{\pm}_{\pm l,m} = HE^{even}_{l+1,m} \pm jHE^{odd}_{i+1,m} \\ OAM^{\mp}_{\pm l,m} = EH^{even}_{l-1,m} \pm jEH^{odd}_{i-1,m} \end{array}\right\} (l>1) \\ \left.\begin{array}{l} OAM^{\pm}_{\pm 1,m} = HE^{even}_{2,m} \pm jHE^{odd}_{2,m} \\ OAM^{\mp}_{\pm 1,m} = TM_{0m} \pm jTE_{0m} \end{array}\right\} (l=1) \end{cases} \tag{8}$$

where l is called the topological charge representing the number of the azimuthal period, and m is the radial order giving the number of concentric rings. The superscript "$\pm$" denotes the polarization state of the OAM mode. OAM modes combined of eigenmode HE possess a circular polarization in the same direction as the OAM rotation, while OAM modes formed of eigenmode EH exhibit a circular polarization in the opposite direction as the OAM rotation [12]. For a given topological charge ($l > 1$) and radial order, four OAM modes form an OAM family. The $OAM_{1,1}$ family composed of even and odd mode of $HE_{2,1}$ have two OAM modes because the $OAM_{1,1}$ composed by $TE_{0,m}$ and $TM_{0,m}$ mode is unstable [34].

The C-PCF fiber structure without the two doped regions in Figure 2a was proposed by our group [17]. It supports the 14 OAM mode transmission and exhibits some good features, such as wide bandwidth, flat dispersion, and low confinement loss. The C-PCF is formed by a large air-hole located at the fiber center, solid circular ring region, and four rings of air-hole arrays as the photonic crystal cladding. The substrate material is pure silica with a refractive index of 1.444 (at 1.55 μm), and the air region contributes to the refractive index of 1. Thus, the refractive index of the cladding constituted by the period photonic crystal structure is determined by the weight of the two materials, which can be changed by tailoring the air-filling fraction of the photonic crystal cladding [35]. Between the central air-hole and the cladding air-hole arrays is a ring-shaped area, which acts as the high index fiber ring-core to confine the OAM modes well within it. If we arrange the two-layer Erbium-doped regions as shown in Figure 2a, it would be a good OAM-EDFA structure for both transmission and amplification. The spatial lattice positions on the x–y plane are given by:

$$x = \Lambda N \cos\left(\frac{2n\pi}{6N}\right),\ y = \Lambda N \sin\left(\frac{2n\pi}{6N}\right),\ n = 1-6N \tag{9}$$

where Λ and N are the lattice constant and number of concentric lattice periods, respectively; r denotes the inner radius of the ring-shaped area with a high index (that is the radius of a large air-hole in the fiber center), and d_2 to d_5 are the diameters of the cladding air holes, φ is the azimuthal angle. We set the lattice constant of Λ = 2 μm, the diameter of a large air hole of d_0 = 2.4 μm, and parameters d_n/Λ = 0.8, $d_2 = d_3 = d_4 = d_5$ = 1.6 μm.

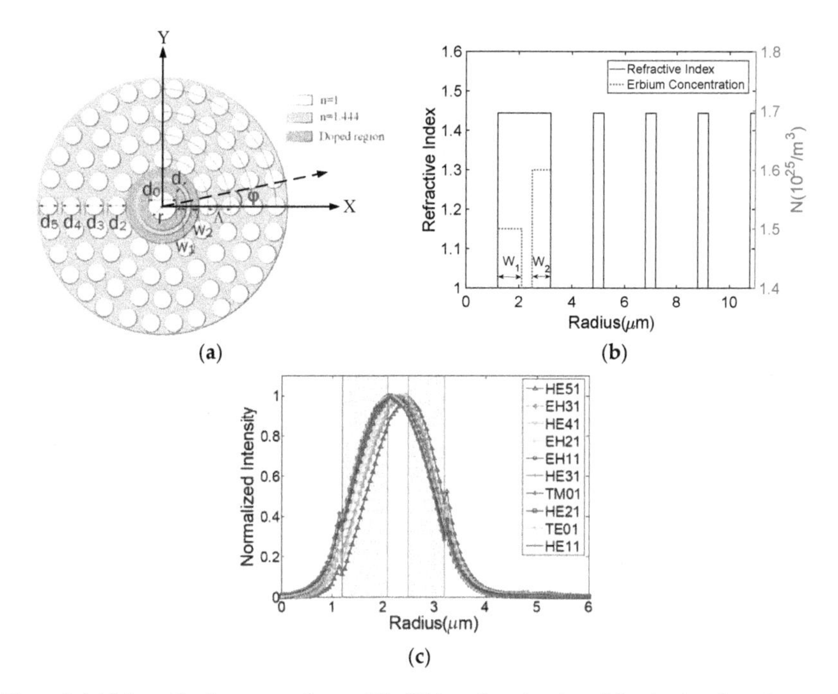

Figure 2. (**a**) Schematic of a cross-section and the Erbium-doped region of the circular photonic crystal fiber (C-PCF); (**b**) the refractive index and Erbium concentration profile of the fiber; (**c**) normalized intensity profile of the modes in the C-PCF.

The schematic of the two doping regions of EDFA is shown in Figure 2a covered with a red grid. The refractive index and Erbium concentration profile of the fiber is shown in Figure 2b and the two doping regions correspond to the two grey areas in Figure 2c. The normalized intensity distribution of the 14 OAM modes is shown in Figure 2c. We can see that the low order OAM modes tend to distribute in the inner area of the high index region, while the high order OAM modes fall in the outer area. According to Equation (5), the amplification of different OAM modes is determined by the overlapping region between the doped region and the mode field distributions, which will cause different DMGs.

To reduce this difference, we arranged two-layer doped Erbium regions at the inner and outer sides of the high index region. The inner doped region overlaps more with the low order OAM modes, while the outer doped region overlaps more with the high order OAM modes. Therefore, this two-layer Erbium-doped arrangement balances the amplification difference between the lower and higher order OAM modes, and hence minimizes the DMG. Then, the doped widths and Erbium ion concentrations in the two regions should be optimized to obtain the optimal performance.

In the one-layer structure, the design can only be adjusted at the boundary of the doped region where the difference among the intensity of modes is the biggest, which means that the tolerance for application is low. Little doped boundary unconformity will cause a big difference for the performance. However, in the two-layer structure, the design is adjusted where intensities of different modes are very close. The tolerance of our design will be much higher.

4. Analysis and Optimization

To design an OAM-EDFA with good features, we need to study the influence of the parameters on the performance of the amplifier. The equilibrium amplification of different OAM modes and the conversion efficiency should be considered to realize low DMG and high gain. Figure 2a shows the structural parameters of the proposed OAM-EDFA, where w_1 and w_2 are the widths of the inner and outer ring doped region, respectively, which defines the performance of the amplifier. d denotes the spacing of the two-layer doped region, which provides the total effective amplifier area of the EDFA, and, furthermore, determines the conversion efficiency. Next, we investigate the selection of the parameters. First, w_1 and w_2 are set as equal, and d is swept from 0.8 μm to 0.2 μm with a step of 0.2 μm. Figure 3 shows the gains of different order modes versus d at 1550 nm wavelength. When d is equal to 0.4 μm, the gains of different order modes are almost identical, so the DMG is the lowest.

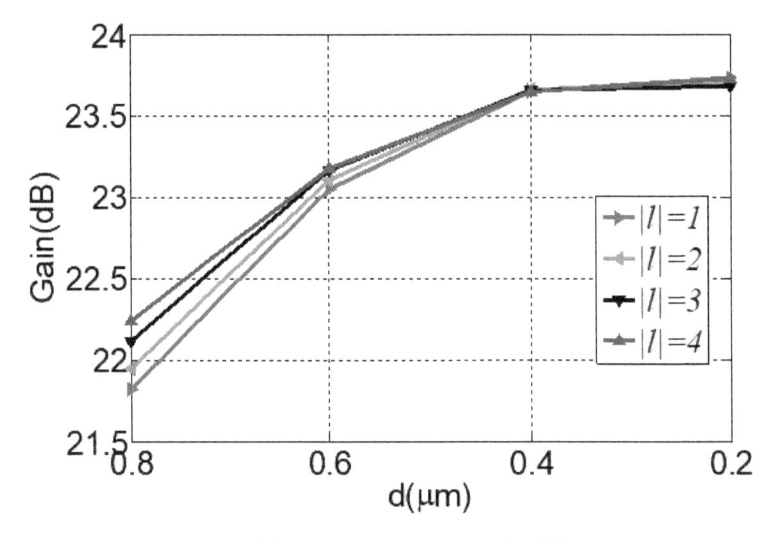

Figure 3. Gain as a function of spacing d at 1550 nm.

Figure 2c shows that the overlap between the field distribution of the lower order modes and the doped region is larger than that of higher order modes when w_1 is increased. Thus, the gain of lower order modes will increase faster than that of higher order modes. On the contrary, when w_2 is increased, the increases in the gain of lower order modes will be slower than that of higher order modes. We can balance the amplification of different modes by increasing w_1 and decreasing w_2; accordingly, the DMG can be reduced. The final parameters are tailored to w_1 = 0.9 μm and w_2 = 0.7 μm while d = 0.4 μm through further optimization, where the w_1 is nearly at the point of the intersection of different modes in Figure 2c.

Besides the widths of the doped regions, the doping concentrations and the length of the EDFA also affect the performance. The relationship between the two concentrations of the doped regions and DMG (Erbium-doped profile is assumed uniform in each area) is shown in Figure 4a. The concentration in the circle can be set with an acceptable DMG. However, when the doped concentration is higher than the value of 1.8×10^{25} m^{-3}, the noise figure is higher than 4dB, as shown in Figure 4b. Considering to balance the DMG and noise figure, N_1 is set at 1.5×10^{25} m^{-3} and N_2 is set at 1.6×10^{25} m^{-3} as shown in Figure 2b, providing a great result for both low DMG and noise. The length of EDFA is swept from 0 m to 15m as shown in Figure 4c. The mode gains increase rapidly and then gradually saturates with the increase in fiber length, while the noise figure (NF) increases slowly and then rises rapidly. To compromise the mode gains and the noise figure, the length is set to 7 m.

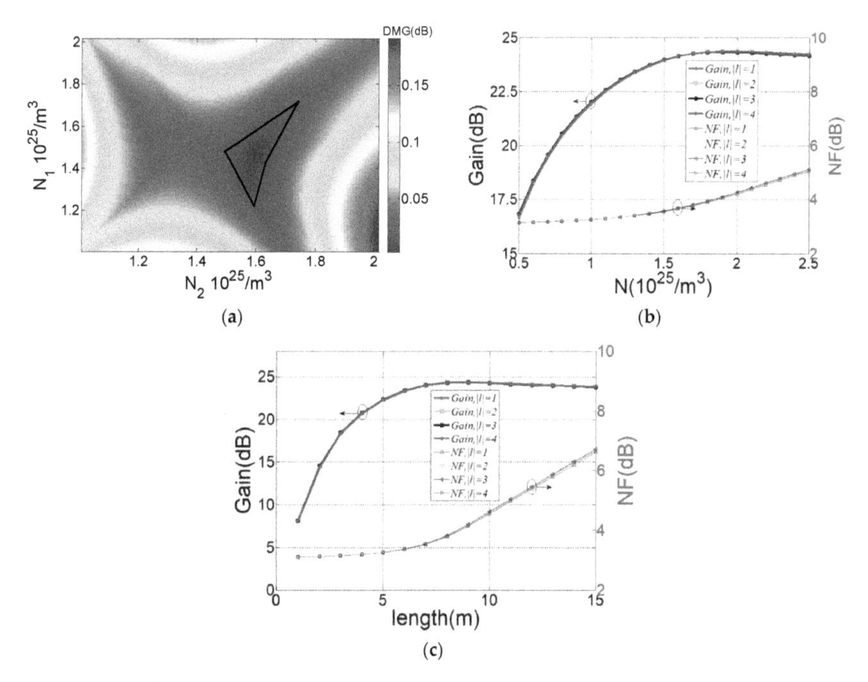

Figure 4. (**a**) Differential mode gain (DMG) as a function of the doping concentration N_1 and N_2, (**b**) Gain and noise figure (NF) as a function of the doping concentration N; (**c**) Gain and NF versus fiber length.

When the parameters of the structure are settled, the mode gains and DMG as functions of the pump power and signal power are shown in Figure 5. The signal wavelength and the pump wavelength are set to 1550 nm and 980 nm, respectively. The pump power is swept from 50 mW to 300 mW. The gain first increases rapidly and then gradually becomes saturated, and the DMG changes similarly, as shown in Figure 5a. To get a relatively high gain and low DMG, the pump power is chosen as 150 mW by a trade-off method where the gain is close to saturation and the DMG is low enough. Figure 5b shows the gain and DMG as the function of input signal power. Considering the trade-off between gain and DMG, we can take the value of the input signal power to be −15 dBm, which provides the gain is larger than 20 dB, and DMG is less than 0.05 dB.

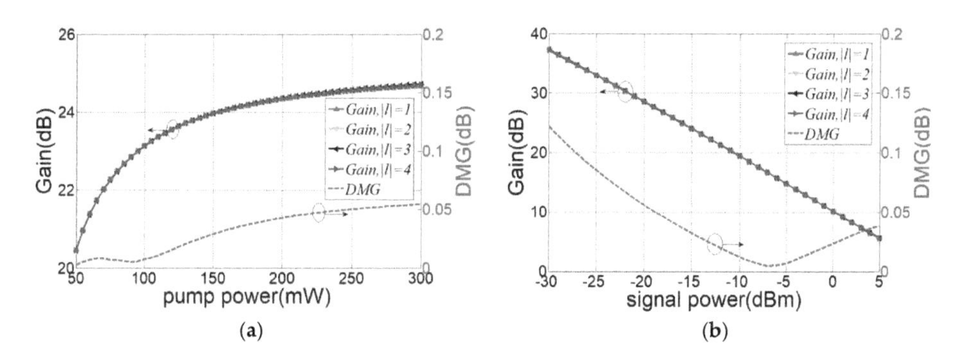

Figure 5. (**a**) Gain and DMG versus pump power; (**b**) Gain and DMG versus signal power.

Figure 6 exhibits the performance of the OAM-EDFA with optimal parameters. The parameters within the doped region are set to w_1 = 0.9 μm, w_2 = 0.7 μm, d = 0.4 μm, $N_1 = 1.5 \times 10^{25}$ m^{-3}, and $N_2 = 1.6 \times 10^{25}$ m^{-3}. The length of OAM-EDFA is selected as 7 m. The pump power and signal power are set to 150 mW and −15 dBm, respectively. The DMG of 14 OAM modes is lower than 0.1 dB, and the noise figures are below 4 dB over the whole C-band, as shown in Figure 6a. As well, over 23 dB gain across the C-band is obtained for the OAM-EDFA, as shown in Figure 6b, which can meet the need of MDM systems well.

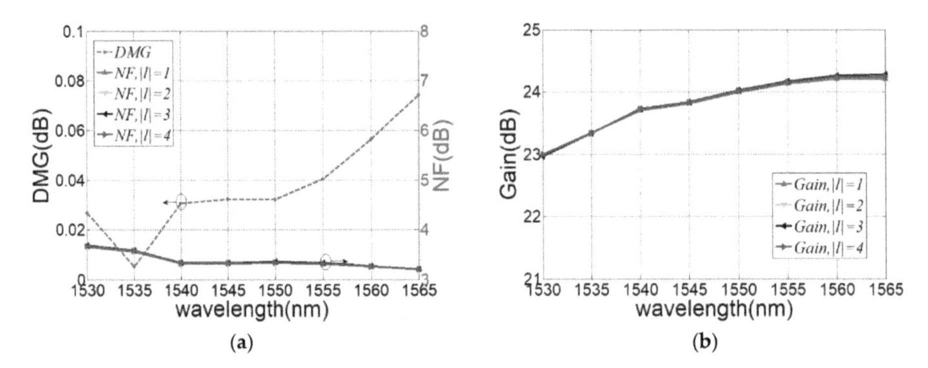

Figure 6. (**a**) DMG and NF versus wavelength; (**b**) Gain versus wavelength.

In Figure 4a, we can see that the DMG would be acceptable when the doping concentrations are in the marked area, which means a large range for concentration tolerance. To obtain a balance between avoiding the radially higher order modes and achieving a good quality of OAM modes, a value of d0 can be selected from 2.2 μm to 3.6 μm, whose large range indicates a large fabrication tolerance [17]. Besides the concentration and the diameter of a large air-hole, the doped width tolerance should also be considered. Owing to the two-layer structure, OAM-EDFA can equalize the gain of the high order OAM modes and the lower order OAM modes. Therefore, the width between w_1 and w_2 can be tolerated within a larger range. At present, the fabrication of PCF and the doping process of the ion have been matured. A fiber similar to the C-PCF structure has been drawn and experimented on [36–38]. Therefore, it is feasible to manufacture the OAM-EDFA based on C-PCF.

5. Conclusions

We have presented a new design of OAM-EDFA with a two-layer doped profile to stably maintain 18 eigenmodes (14 OAM modes) for the MDM system. The amplifier is based on the C-PCF supporting OAM modes and adopts the core-pumping scheme. Parameters that affect the amplifier performance, such as the width of the two doped regions, the doping concentrations, the length of the amplifier fiber, and the pump power and signal power, are optimized by a trade-off scheme. The fabrication tolerance of the EDFA is also discussed, and the results show a relatively large tolerance. The two-layer doped region can balance the amplification performance of different modes to minimize the DMG of 14 OAM modes below 0.08 dB. The optimal OAM-EDFA designed will theoretically achieve a gain over 23 dB and a noise figure less than 4 dB for all 14 OAM modes across the full C-band.

Author Contributions: H.Z. and L.X. conceived and designed the manuscript, H.Z. wrote the manuscript, and L.X. modified the manuscript and confirmed the final version to be submitted; D.H. and H.L. calculated and analyzed the data; X.Z., Z.Z., and W.Z. investigated the literature and presented formal analyses.

Funding: This research was funded partly by the National Natural Science Foundation of China (NSFC), grant number 61571057, 61527820, 61575082.

Conflicts of Interest: The authors declare no conflict of interest.

References

1. Richardson, D.J.; Fini, J.M.; Nelson, L.E. Space-division multiplexing in optical fibres. *Nat. Photonics* **2013**, *7*, 354–362. [CrossRef]
2. Allen, L.; Beijersbergen, M.W.; Spreeuw, R.; Woerdman, J. Orbital angular momentum of light and the transformation of Laguerre-Gaussian laser modes. *Phys. Rev. A* **1992**, *45*, 8185–8189. [CrossRef] [PubMed]
3. Ramachandran, S.; Kristensen, P. Optical vortices in fiber. *Nanophotonics* **2013**, *2*, 455–474. [CrossRef]
4. Brunet, C.; Rusch, L.A. Optical fibers for the transmission of orbital angular momentum modes. *Opt. Fiber Technol.* **2016**, *31*, 172–177. [CrossRef]
5. Yao, A.M.; Padgett, M.J. Orbital angular momentum: Origins, behavior and applications. *Adv. Opt. Photonics* **2011**, *3*, 161–204. [CrossRef]
6. Chen, H.; Hao, J.; Zhang, B.; Xu, J.; Ding, J.; Wang, H. Generation of vector beam with space-variant distribution of both polarization and phase. *Opt. Lett.* **2011**, *36*, 3179–3181. [CrossRef] [PubMed]
7. Robert, D.N.; Mark, E.S.; Juliet, T.G. Continuously tunable orbital angular momentum generation using a polarization-maintaining fiber. *Opt. Lett.* **2016**, *41*, 3213–3216.
8. José, F.A.; Urruchi, V.; Braulio, G.; José, M.S. Generation of optical vortices by an ideal liquid crystal spiral phase plate. *IEEE Electron Device Lett.* **2014**, *35*, 856–858. [CrossRef]
9. Andrey, S.O.; Rickenstorff-Parrao, C.; Arrizón, V. Generation of the "perfect" optical vortex using a liquid-crystal spatial light modulator. *Opt. Lett.* **2013**, *38*, 534–536.
10. Kotova, S.P.; Mayorova, A.M.; Samagin, S.A. Formation of ring-shaped light fields with orbital angular momentum using a modal type liquid crystal spatial modulator. *J. Opt.* **2018**, *20*, 055604. [CrossRef]
11. Yan, Y.; Yue, Y.; Huang, H.; Yang, J.; Chitgarha, M.R.; Ahmed, N.; Tur, M.; Dolinar, S.J.; Willner, A.E. Efficient generation and multiplexing of optical orbital angular momentum modes in a ring fiber by using multiple coherent inputs. *Opt. Lett.* **2012**, *37*, 3645–3647. [CrossRef] [PubMed]
12. Li, S.; Wang, J. Multi-orbital-angular-momentum multi-ring fiber for high-density space-division multiplexing. *IEEE Photonics J.* **2013**, *5*. [CrossRef]
13. Yue, Y.; Yan, Y.; Ahmed, N.; Yang, J.; Zhang, L.; Ren, Y.; Huang, H.; Birnbaum, K.M.; Erkmen, B.I.; Dolinar, S.; et al. Mode properties and propagation effects of optical orbital angular momentum (OAM) modes in a ring fiber. *IEEE Photonics J.* **2012**, *4*, 535–543.
14. Brunet, C.; Vaity, P.; Messaddeq, Y.; LaRochelle, S.; Rusch, L.A. Design, fabrication and validation of an OAM fiber supporting 36 states. *Opt. Express* **2014**, *22*, 26117–26127. [CrossRef] [PubMed]
15. Wang, L.; Nejad, R.M.; Corsi, A.; Lin, J.; Messaddeq, Y.; LaRochelle, S.; Rusch, L.A. Linearly polarized vector modes: Enabling MIMO-free mode-division multiplexing. *Opt. Express* **2017**, *25*, 11736–11749. [CrossRef] [PubMed]
16. Zhang, H.; Zhang, X.; Li, H.; Deng, Y.; Xi, L.; Tang, X.; Zhang, W. The orbital angular momentum modes supporting fibers based on the photonic crystal fiber structure. *Crystals* **2017**, *7*, 286. [CrossRef]
17. Zhang, H.; Zhang, W.; Xi, L.; Tang, X.; Zhang, X.; Zhang, X. A new type circular photonic crystal fiber for orbital angular momentum mode transmission. *IEEE Photonics Technol. Lett.* **2016**, *28*, 1426–1429. [CrossRef]
18. Zhang, H.; Zhang, W.; Xi, L.; Tang, X.; Zhang, X. A New Design of a Circular Photonic Crystal Fiber Supporting 42 OAM Modes. In Proceedings of the Australian Conference on Optical Fibre Technology, Sydney, Australia, 5–8 September 2016; pp. 1–2.
19. Zhang, H.; Zhang, X.; Li, H.; Deng, Y.; Zhang, X.; Xi, L.; Tang, X.; Zhang, W. A design strategy of the circular photonic crystal fiber supporting good quality orbital angular momentum mode transmission. *Opt. Commun.* **2017**, *397*, 59–66. [CrossRef]
20. Bozinovic, N.; Yue, Y.; Ren, Y.; Tur, M.; Kristensen, P.; Huang, H.; Willner, A.E.; Ramachandran, S. Terabit scale orbital angular momentum mode division multiplexing in fibers. *Science* **2013**, *340*, 1545–1548. [CrossRef] [PubMed]
21. Wang, A.; Zhu, L.; Chen, S.; Du, C.; Mo, Q.; Wang, J. Characterization of LDPC-coded orbital angular momentum modes transmission and multiplexing over a 50 km fiber. *Opt. Express* **2016**, *24*, 11716–11726. [CrossRef] [PubMed]
22. Qian, X.; Boucouvalas, A.C. Propagation Characteristics of single-mode optical fibers with arbitrary complex index profiles. *IEEE J. Quantum Elect.* **2006**, *40*, 771–777. [CrossRef]

23. Kang, Q.; Gregg, P.; Jung, Y.; Lim, E.L.; Alam, S.; Ramachandran, S.; Richardson, D.J. Amplification of 12 OAM Modes in an air-core erbium doped fiber. *Opt. Express* **2015**, *23*, 28341–28348. [CrossRef] [PubMed]
24. Kang, Q.; Lim, E.L.; Jung, Y.; Poletti, F.; Alam, S.; Richardson, D.J. Design of Four-Mode Erbium Doped Fiber Amplifier with Low Differential Modal Gain for Modal Division Multiplexed Transmissions. In Proceedings of the Optical Fiber Communication Conference and Exposition and the National Fiber Optic Engineers Conference IEEE, Anaheim, CA, USA, 17–21 March 2013; pp. 1–3.
25. Kang, Q.; Lim, E.L.; Poletti, F.; Jung, Y.; Baskiotis, C.; Alam, S.; Richardson, D.J. Minimizing differential modal gain in cladding-pumped EDFAs supporting four and six mode groups. *Opt. Express* **2014**, *22*, 21499–21507. [CrossRef] [PubMed]
26. Jung, Y.; Kang, Q.; Sidharthan, R.; Ho, D.; Yoo, S.; Gregg, P.; Ramachandran, S.; Alam, S.; Richardson, D.J. Optical orbital angular momentum amplifier based on an air-hole erbium-doped fiber. *J. Lightwave Technol.* **2017**, *35*, 430–436. [CrossRef]
27. Ma, J.; Xia, F.; Li, S.; Wang, J. Design of Orbital Angular Momentum (OAM) Erbium Doped Fiber Amplifier with Low Differential Modal Gain. In Proceedings of the Optical Fiber Communication Conference and Exposition/National Fiber Optic Engineers Conference IEEE, Los Angeles, CA, USA, 22–26 March 2015; pp. 1–3.
28. Deng, Y.; Zhang, H.; Li, H.; Tang, X.; Zhang, X.; Xi, L.; Zhang, W.; Zhang, X. Erbium-doped amplification in circular photonic crystal fiber supporting orbital angular momentum modes. *Appl. Opt.* **2017**, *56*, 1748–1752. [CrossRef] [PubMed]
29. Giles, C.R.; Desurvire, E. Modeling erbium-doped fiber amplifiers. *J. Lightwave Technol.* **1991**, *9*, 271–283. [CrossRef]
30. Bigot, L.; Le Cocq, G.; Quiquempois, Y. Few-mode erbium-doped fiber amplifiers: A review. *J. Lightwave Technol.* **2015**, *33*, 588–596. [CrossRef]
31. Varshney, S.K.; Saitoh, K.; Koshiba, M.; Pal, B.P.; Sinha, R.K. Design of S-band erbium-doped concentric dual-core photonic crystal fiber amplifiers with ASE suppression. *J. Lightwave Technol.* **2009**, *27*, 1725–1733. [CrossRef]
32. Le Cocq, G.; Quiquempois, Y.; Le Rouge, A.; Bouwmans, G.; El Hamzaoui, H.; Delplace, K.; Bouazaoui, M.; Bigot, L. Few mode Er^{3+}-doped fiber with micro-structured core for mode division multiplexing in the C-band. *Opt. Express* **2013**, *21*, 31646–31659. [CrossRef] [PubMed]
33. Saitoh, S.; Saitoh, K.; Kashiwagi, M.; Matsuo, S.; Dong, L. Design optimization of large-mode-area all-solid photonic bandgap fibers for high-power laser applications. *J. Lightwave Technol.* **2014**, *32*, 440–449. [CrossRef]
34. Zhou, G.; Zhou, G.; Chen, C.; Xu, M.; Xia, C.; Hou, Z. Design and analysis of a microstructure ring fiber for orbital angular momentum transmission. *IEEE Photonics J.* **2017**, *8*, 1–12. [CrossRef]
35. Li, H.; Zhang, H.; Zhang, X.; Zhang, Z.; Xi, L.; Tang, X.; Zhang, W.; Zhang, X. Design tool for circular photonic crystal fibers supporting orbital angular momentum modes. *Appl. Opt.* **2018**, *57*, 2474–2481. [CrossRef] [PubMed]
36. Tandje, A.; Yammine, J.; Bouwmans, G.; Dossou, M.; Vianou, A.; Andresen, E.R.; Bigot, L. Design and Fabrication of a Ring-Core Photonic Crystal Fiber for Low-Crosstalk Propagation of OAM Modes. In Proceedings of the European Conference on Optical Communication IEEE, Rome, Italy, 23–25 September 2018; pp. 1–3.
37. Chen, Y.; Liu, Z.; Sandoghchi, S.R.; Jasion, G.T.; Bradley, T.D.; Fokoua, E.N.; Hayes, J.R.; Wheeler, N.V.; Gray, D.R.; Mangan, B.J.; et al. Multi-kilometer long, longitudinally uniform hollow core photonic bandgap fibers for broadband low latency data transmission. *J. Lightwave Technol.* **2016**, *34*, 104–113. [CrossRef]
38. Foroni, M.; Passaro, D.; Poli, F.; Cucinotta, A.; Selleri, S.; Lægsgaard, J.; Bjarklev, A.O. Guiding properties of silica/air hollow-core bragg fibers. *J. Lightwave Technol.* **2008**, *26*, 1877–1884. [CrossRef]

Article

Flexible Photonic Nanojet Formed by Cylindrical Graded-Index Lens

Cheng-Yang Liu

Department of Biomedical Engineering, National Yang-Ming University, Taipei City 11221, Taiwan; cyliu66@ym.edu.tw; Tel.: +886-2-28267020

Received: 18 March 2019; Accepted: 5 April 2019; Published: 7 April 2019

Abstract: Photonic nanojets formed in the vicinity of the cylindrical graded-index lens with different types of index grading are numerically investigated based on the finite-difference time-domain method. The cylindrical lens with 1600 nm diameter is assembled by eighty-seven hexagonally arranged close-contact nanofibers with 160 nm diameter. Simulation and analysis results show that it is possible to engineer and elongate the photonic nanojet. Using differently graded-index nanofibers as building elements to compose this lens, the latitudinal and longitudinal sizes of the produced photonic nanojet can be flexibly adjusted. At an incident wavelength of 532 nm, the cylindrical lens with index grading = 2 can generate a photonic nanojet with a waist about 173 nm (0.32 wavelength). This lens could potentially contribute to the development of a novel device for breaking the diffraction limit in the field of optical nano-scope and bio-photonics.

Keywords: cylindrical lens; photonic nanojet; graded-index

---

1. Introduction

Optical super-resolution has become significant for many applications including optical imaging [1,2], optical trapping and manipulation [3,4], nano-patterning and lithography [5,6], spectroscopy [7], and data storage [8]. Because the traditional objective lens has a diffraction-limited light spot, many investigations have been devoted to finding a practical way to obtain a small focusing spot beyond the diffraction limit [9]. One of the practical ways is the photonic nanojet (PNJ). The PNJ generated by an illuminated dielectric microcylinder is introduced and numerically demonstrated by Chen et al. in 2004 [10]. The mechanism of super-resolution imaging by dielectric microcylinders and microspheres have been increasingly attractive to researchers [11–15]. The PNJ is a high-intensity narrow focusing spot in the near-field of transparent microcylinder. When the diameter of transparent microcylinder is larger than the incident wavelength, the PNJ is generated due to the interferences between the scattering and illuminating fields. The main property of the PNJ is that it is a non-resonant phenomenon with low divergence and a small waist on the sub-wavelength scale. To generate a PNJ, it has been investigated that the refractive index contrast between the single microcylinder and its surrounding medium performs a critical role in the characters of PNJ [16]. This feature of microcylinder-based PNJ restricts the selection of transparent materials.

In order to optimize key parameters (focal length, waist, and intensity) of the PNJ, several studies indicate that the PNJ distributions depend on the geometric shape and refractive index of the microcylinder [17–24]. Moreover, the microcylinder or microsphere consisted of a concentric core-shell structure with different refractive indices for adjusting the propagation length and width of the PNJ [25–33]. The PNJ phenomenon can be changed significantly by applying shell materials with refractive index > 2. The PNJ length formed by the core-shell microcylinder is increased to approximately 20 wavelengths. However, the price for this PNJ elongation is the waist widening and the intensity attenuation. The fabrication process of layered inhomogeneous core-shell microcylinder is very difficult and rather costly. Therefore, the new and simple procedure is an interesting research issue for PNJ shaping. The transparent medium in other geometries, such as micro-cuboids, micro-axicons, nanofibers, and optical fiber tips, are presented for the formation of PNJ [34–39]. These novel structures of transparent medium cause special features of PNJ-like intensity distributions and are highly probable to develop new applications.

In this paper, the combination of the metamaterial concept with the PNJ by plane wave illumination is proposed and numerically investigated. The graded-index nanofibers are used as building blocks to assemble the artificial cylindrical lens. By varying the graded-index type of the compositional nanofibers, the focusing properties of the lens are able to modulate according to our requirements. Using the finite-difference time-domain (FDTD) method, we simulated the optical field propagation of a plane wave passing through the cylindrical lens assembled by hexagonally arranged nanofibers in the air medium. The physical modeling is given in Section 2 for cylindrical graded-index lens. The effects of the graded-index types on the shape, focal length, full-width at half-maximum (FWHM), and intensity of PNJs are presented and discussed in Section 3. Finally, the conclusions of this investigation are summarized in Section 4.

2. Physical Modeling

Several numerical methods have theoretically studied optical intensity distribution in the vicinity of a transparent core-shell microcylinder or microsphere illuminated by a light source [26–29]. These studies suggest that the refractive index contrast between different shells plays a critical role in the formation of a PNJ. In order to verify the influence of the graded-index nanofibers, we performed FDTD calculations for modeling computational electromagnetics [40]. The schematic diagram of a PNJ generated by the cylindrical graded-index lens is shown in Figure 1. Geometrically, this cylindrical lens is constructed by multiple hexagonally arranged close-contact nanofibers which fully fill a cylindrical area with a particular diameter of 1600 nm. The number and diameter of these nanofibers are 87 and 160 nm. The proposed cylindrical lens is normally illuminated by a transverse electric plane wave propagating along x direction with the electric field polarized along the z direction. The length of this lens along the z direction is defined as infinitely long for guaranteeing the accuracy and speed of FDTD calculation. The grid size of the FDTD mesh is chosen to be 10 nm after the convergence verification. The boundary conditions at the x and y directions added enough space to deliver the power flow distributions of optical beam in the background medium. The background medium is air with a refractive index of 1.

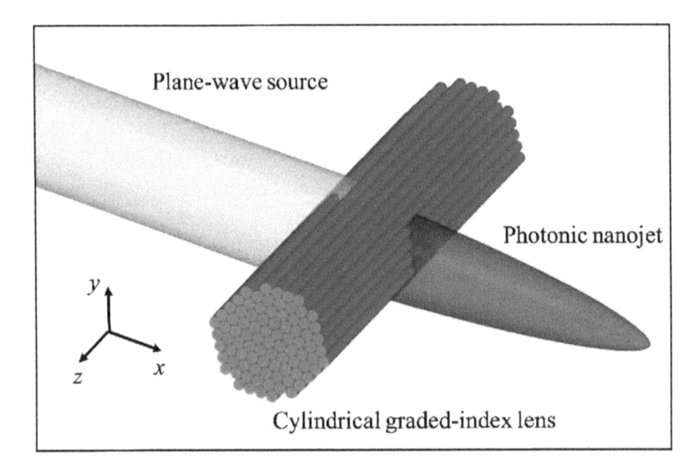

Figure 1. Schematic diagram of the cylindrical graded-index lens for photonic nanojet.

Since the location and the intensity of PNJ depend on the refractive index contrast between each nanofiber layer, we consider a micrometer size cylindrical lens consisting of several concentric nanofibers with equal diameter. Figure 2a shows the graded-index model of the hexagonally arranged nanofibers. Every nanofiber layer with a number s is a homogeneous material and is defined by the refractive index n_s (s = 0 to N). In order to specify the refractive index variation from layer to layer, the refractive index contrast of the cylindrical lens is expressed as $n_s/n_0 = (n_N/n_0)^{(s/N)t}$ [26]. The index grading type parameter is t and the dielectric central core is $s = 0$. The $t > 0$ indicates that the refractive index grading starts from the central nanofiber and terminates in the outermost nanofiber, which has the lowest value of the refractive index. Figure 2b shows the different index grading types of refractive index n_s in the graded-index lens. The t value defines the variety of refractive index grading including linear ($t = 1$), concave ($t = 0.2$ and $t = 0.5$), and convex ($t = 1.5$ and $t = 2$) types. In the present lens, the refractive index grading is realized with N = 5 distinct concentric nanofibers. When the t value is 1, the refractive index grading is the linear layer-by-layer variation with the constant contrast. The maximum value of refractive index at the central nanofiber is 1.5 and it decreases in the radial direction to a minimum value of 1.05 at the outer nanofiber. This choice of index values is based on the practicability of modern micro-scale coating technology of objects with a thin film that has adjustable refractive indices [41,42]. The refractive indices in the range of 1.05 to 2.0 for material synthesis have been realized by controlling the porosity of silica glass. This manufacturing process is also possible to use for the graded-index photonic crystal structure and fiber.

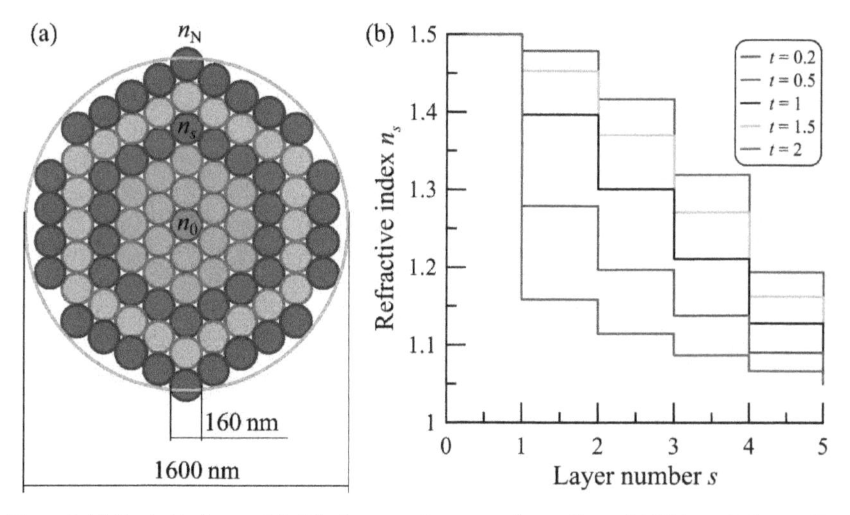

Figure 2. (**a**) Graded-index model of the hexagonally arranged nanofibers; (**b**) Different index grading types of refractive index n_s in the graded-index lens.

3. Results and Discussion

The PNJ produced by a microcylinder has been found to present several important properties. First, the PNJ intensity is several hundred times higher than the incident light power. Second, the PNJ has a smaller waist than the classical diffraction limit. Using high-resolution FDTD calculation, we have simulated the intensity distributions of the cylindrical lens at different index grading types. The incident beam is linearly polarized with a wavelength of 532 nm. Figure 3 shows the spatial intensity distributions of PNJs formed in the vicinity of cylindrical graded-index lens with homogeneous material ($n = 1.5$), $t = 0.2$, $t = 0.5$, $t = 1$, $t = 1.5$, and $t = 2$. Figure 3a represents the reference model at the same modeling conditions, which are the cylindrical lens with all homogeneous nanofibers. It demonstrated that an optical beam propagates from the top of the lens and the significant near-field focusing effect is observed at the bottom of the lens. Accordingly, an intensity peak of the electric field is known as the PNJ. The intensity peak in Figure 3a is 3.6 compared to the incident intensity of 1, and the FWHM of PNJ is 165 nm smaller than the incident wavelength of 532 nm. The similar intensity distributions representing a cylindrical lens with nanofibers of five different index grading types are shown in Figure 3b–f. We could see that the PNJ gradually shifts from the outside to the inside of the lens when the index grading type parameter t increases from 0.2 to 2. The focusing effect plays a significant role in the propagation of the light wave in the lens and the PNJ is located at the interior of the nanofibers. Compared to the homogeneous model, PNJs created by different graded-index nanofibers assembled lens have stronger modulation of intensity peak and FWHM.

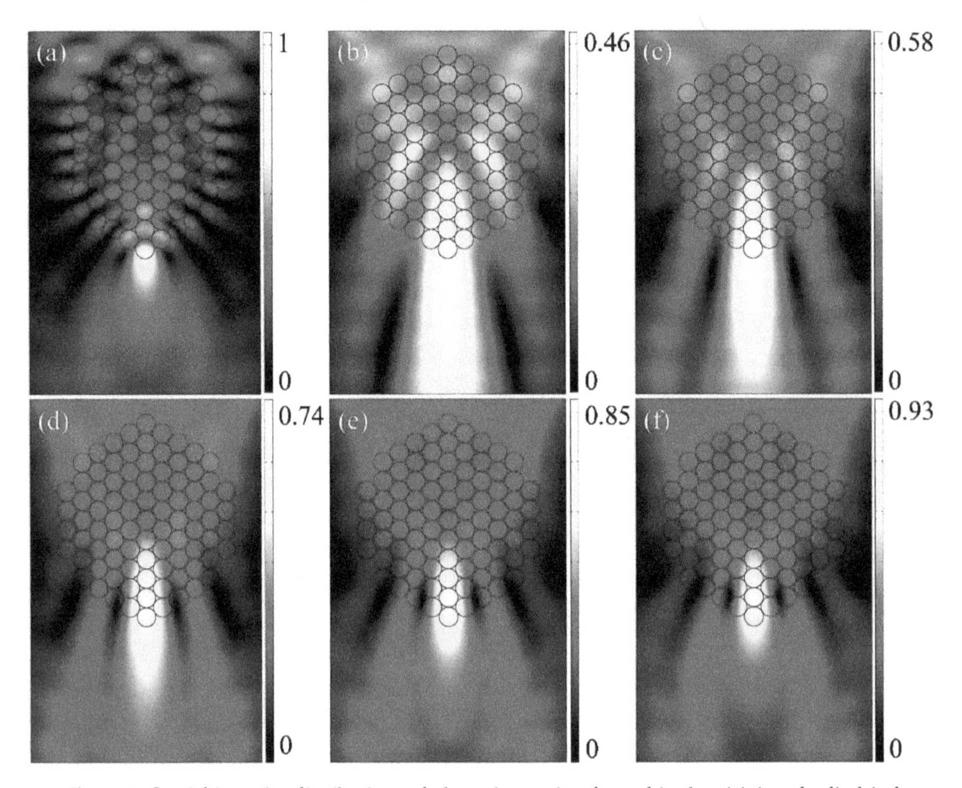

Figure 3. Spatial intensity distributions of photonic nanojets formed in the vicinity of cylindrical graded-index lens with (**a**) homogeneous material ($n = 1.5$), (**b**) $t = 0.2$, (**c**) $t = 0.5$, (**d**) $t = 1$, (**e**) $t = 1.5$, and (**f**) $t = 2$.

Figure 4a shows the normalized intensity distributions of PNJ for cylindrical graded-index lens along the propagation axis (x axis). The longitudinal profile in Figure 4a is acquired as a two-dimensional cross-section of the intensity distribution by the straight line located at the center of the lens. The dashed line is the edge of the lens. According to Figure 4a, the position of intensity peak for PNJ decreases from 1035 nm to 603 nm as the index grading type parameter increases. The transversal profiles at the highest intensity peak are plotted along the y axis in Figure 4b. The FWHMs are 326 nm, 261 nm, 213 nm, 177 nm, and 173 nm corresponding to $t = 0.2$, 0.5, 1, 1.5, and 2, respectively. The FWHM of the PNJ monotonically increases with the growth of the index grading type parameter as well. These indicate that the graded-index nanofibers are able to focus a light spot smaller than the Abbe diffraction limit. The smallest FWHM (173 nm) of the PNJ achieved by the model at $t = 2$ is 35% smaller than the half of incident wavelength. Meanwhile, the highest intensity peak of the PNJ is delivered by the same graded-index nanofibers at $t = 2$. These graded-index lens can be used in combination with a traditional objective lens for super-resolution imaging applications.

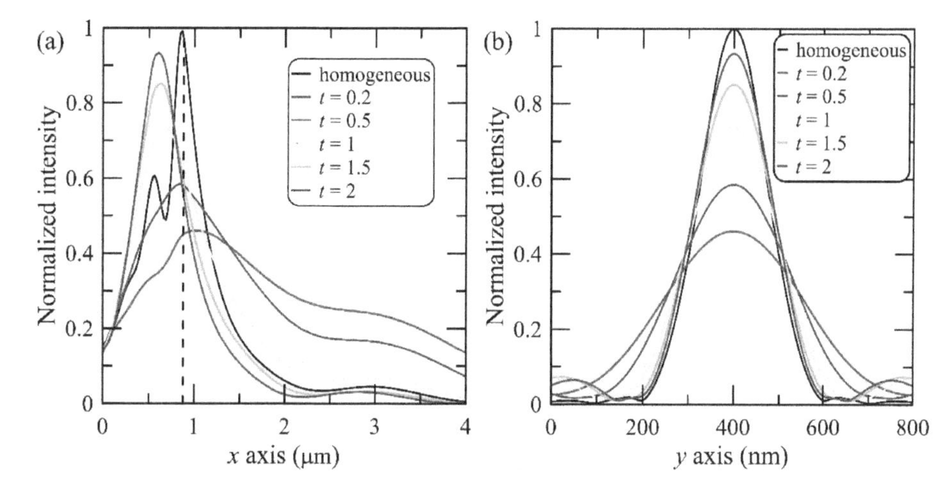

Figure 4. Normalized intensity distributions of photonic nanojet for cylindrical graded-index lens along (**a**) the propagation axis (x axis) and (**b**) the transversal axis (y axis). The dashed line is the edge of the lens.

If the PNJ is focused inside the lens, a magnified real image is formed as in the cases of Figure 3d–f. It can be noted that the cylindrical graded-index lens produces a one-dimensional super-resolution image along the nanofiber axis. Therefore, we may obtain a complete two-dimensional super-resolution image in a large area by rotating the cylindrical lens in a circular mode. It is clear that the improved PNJ properties in the proposed graded-index lens originate from the introduced inhomogeneity of the refractive index. This graded-index lens is essentially a compact compound scattering media with altering refractive index along the propagation direction. Figure 5 shows the focal length and FWHM as a function of the index grading type parameter for cylindrical lens. Apparently, decreasing grading parameter t results in elongated PNJ, accompanied by an expanded FWHM and decreased peak intensity. The evolution of the FWHM with the intensity peak with grading parameter t increasing from 326 nm to 173 nm. Therefore, the key way to manipulate PNJ is to find an optimum graded-index configuration. If the grading parameter is $t > 1$, the optical contrast of the serial nanofiber layers increases with their layer number. The central nanofiber plays the major function in the transformation of the PNJ inside the lens. When the grading parameter is $t >> 1$, the graded-index lens according to its optical properties becomes similar to a homogeneous lens with high refractive index.

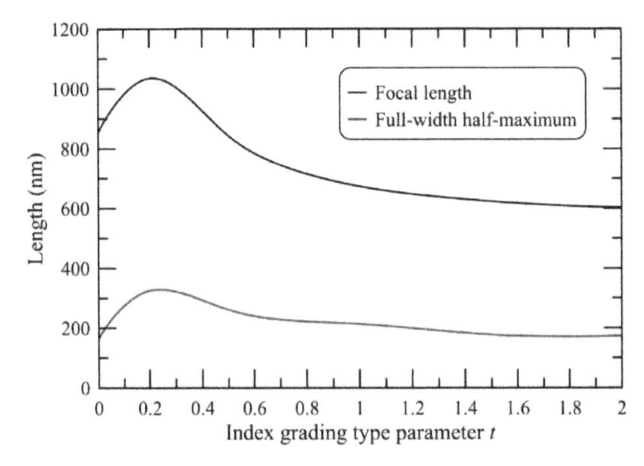

Figure 5. Focal length and full-width half-maximum as a function of the index grading type parameter for a cylindrical graded-index lens.

It can be seen from Figure 5 that the PNJ length decreases as t value increases. The intensity peak is placed inside the nanofibers at the value of $t = 2$. Combining basic properties of the PNJ, a modified quality criterion Q is expressed as $Q = (L \times I) / FWHM$ [25]. The effective length, maximum peak intensity, and FWHM of the PNJ are L, I, and FWHM, respectively. The usability of a PNJ can be estimated by using this quality criterion in the solution of practical problems. When the Q value is high, the peak intensity of nanojet is high, its FWHM is small, and the effective length is long. Figure 6 shows the quality criterion as a function of the index grading type parameter for cylindrical graded-index lens. At the value of $t = 0.2$, the cylindrical graded-index lens optimally combines the high spatial localization with high intensity. The super-resolution and the relationship between the nanofibers and the light beam in the graded-index lens may have a physical connection with photonic crystal [43]. An individual nanofiber operates like a single nanolens. Due to the hexagonal arrangement, the grading refractive index is capable of guiding the intensity flow to the bottom nanofiber and generating a strong focus with a high-intensity peak. The nanofiber-assembled graded-index lens has some singular points which could be used to focus more power on the same phase. Therefore, this graded-index lens with the selected refractive index is suitable for nano-scale imaging.

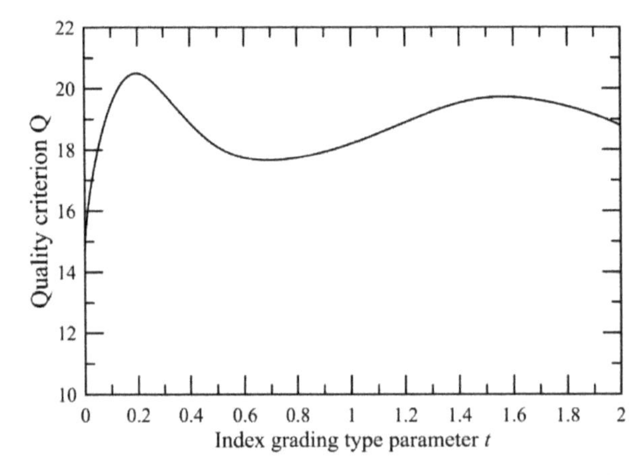

Figure 6. Quality criterion as a function of the index grading type parameter for the cylindrical graded-index lens.

In order to compare the PNJ properties of our lens assembly to those of a single uniform-index microcylinder, we also performed FDTD calculation for a single microcylinder. The refractive index of a single microcylinder is 1.5. Spatial intensity distribution, normalized intensity distributions along the propagation axis and the transversal axis for PNJ formed by a single microcylinder with 1600 nm diameter are shown in Figure 7. In comparison with PNJ formed by cylindrical graded-index lens, it is noted that maximal intensity for the graded-index lens along the propagation axis is larger than peak intensity for uniform single microcylinder. Moreover, the FWHM for uniform single microcylinder is 277 nm, but the FWHM for the graded-index lens at $t = 0.2$ is 173 nm. The focal length for uniform single microcylinder is 65 nm and the normalized peak intensity is 0.76. The location of PNJ is close to the surface of the microcylinder.

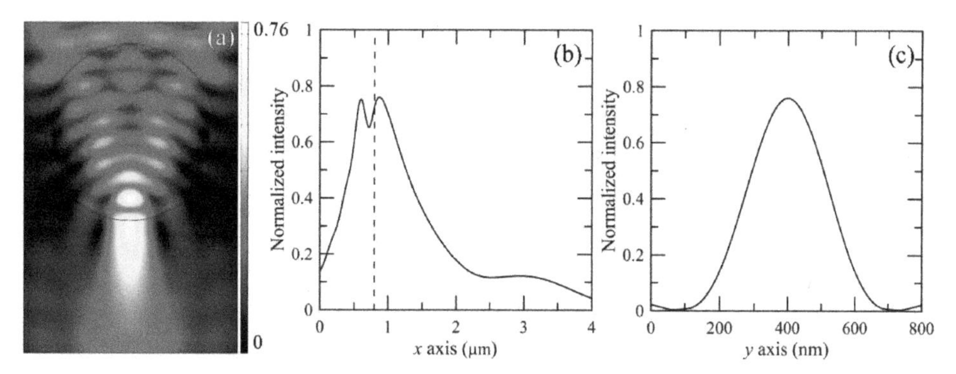

Figure 7. PNJ formed by a single microcylinder with 1600 nm diameter: (**a**) spatial intensity distribution, (**b**) normalized intensity distribution along the propagation axis, and (**c**) normalized intensity distribution along the transversal axis. The dashed line is the edge of the microcylinder.

4. Conclusions

In conclusion, the cylindrical graded-index lens assembled by hexagonally arranged transparent nanofibers is reported. The effective refractive index of the nanofibers can be changed by tuning the index grading type parameter. We are able to modulate the PNJ by varying the graded-index type of the compositional nanofibers. Using high-resolution FDTD calculation, we indicate that the PNJ is dynamically switched by the graded-index lens. Moreover, we present an optimization demonstration which pursues better focusing characters of the PNJ. The cylindrical graded-index lens can successfully achieve lateral resolution beyond the diffraction limit under the plane wave illumination of 532 nm wavelength. The hexagonal arrangement of the graded-index nanofibers leads to an alternating change of refractive index that effectively collects evanescent waves accompanied by near-field coupling of scattering light. Such a mechanism for PNJ manipulation may bring about new applications for optical imaging with super-resolution.

Funding: This research was funded by Ministry of Science and Technology of Taiwan, grant number MOST 107-2221-E-032-033.

Acknowledgments: The author would like to thank Igor V. Minin for his invaluable discussion.

Conflicts of Interest: The authors declare no conflict of interest.

References

1. Wang, Z.; Guo, W.; Li, L.; Lukyanchuk, B.; Khan, A.; Liu, Z.; Chen, Z.; Hong, M. Optical virtual imaging at 50 nm lateral resolution with a white-light nanoscope. *Nat. Commun.* **2011**, *2*, 218. [CrossRef] [PubMed]
2. Upputuri, P.; Pramanik, M. Microsphere-aided optical microscopy and its applications for super-resolution imaging. *Opt. Commun.* **2017**, *404*, 32–41. [CrossRef]

3. Li, Y.; Xin, H.; Liu, X.; Zhang, Y.; Lei, H.; Li, B. Trapping and detection of nanoparticles and cells using a parallel photonic nanojet array. *ACS Nano* **2016**, *10*, 5800–5808. [CrossRef] [PubMed]
4. Shakhov, A.; Astaflev, A.; Nadtochenko, V. Microparticle manipulation using femtosecond photonic nanojet-assisted laser cavitation. *Opt. Lett.* **2018**, *43*, 1858–1861. [CrossRef]
5. Wu, W.; Katsnelson, A.; Memis, O.G.; Mohseni, H. A deep sub-wavelength process for the formation of highly uniform arrays of nanoholes and nanopillars. *Nanotechnology* **2007**, *18*, 485302. [CrossRef]
6. Jacassi, A.; Tantussi, F.; Dipalo, M.; Biagini, C.; Maccaferri, N.; Bozzola, A.; Angelis, F. Scanning probe photonic nanojet lithography. *ACS Appl. Mater. Interfaces* **2017**, *9*, 32386–32393. [CrossRef] [PubMed]
7. Li, X.; Chen, Z.; Taflove, A.; Backman, V. Optical analysis of nanoparticles via enhanced backscattering facilitated by 3-D photonic nanojets. *Opt. Express.* **2005**, *13*, 526–533. [CrossRef] [PubMed]
8. Kong, S.; Sahakian, A.V.; Heifetz, A.; Taflove, A.; Backman, A. Robust detection of deeply subwavelength pits in simulated optical data-storage disks using photonic jets. *Appl. Phys. Lett.* **2008**, *92*, 211102. [CrossRef]
9. Betzig, E.; Trautman, J.K. Near-field optics: Microscopy, spectroscopy, and surface modification beyond the diffraction limit. *Science* **1992**, *257*, 189–195. [CrossRef] [PubMed]
10. Chen, Z.; Taflove, A.; Backman, V. Photonic nanojet enhancement of backscattering of light by nanoparticles: a potential novel visible-light ultramicroscopy technique. *Opt. Express* **2004**, *12*, 1214–1220. [CrossRef]
11. Ferrand, P.; Wenger, J.; Devilez, A.; Pianta, M.; Stout, B.; Bonod, N.; Popov, E.; Rigneault, H. Direct imaging of photonic nanojets. *Opt. Express* **2008**, *16*, 6930–6940. [CrossRef]
12. Kim, M.; Scharf, T.; Mühlig, S.; Rockstuhl, C.; Herzig, H. Engineering photonic nanojets. *Opt. Express* **2011**, *19*, 10206–10220. [CrossRef] [PubMed]
13. Darafsheh, A.; Walsh, G.; Dal Negro, L.; Astratov, V. Optical super-resolution by high-index liquid-immersed microspheres. *Appl. Phys. Lett.* **2012**, *101*, 141128. [CrossRef]
14. Liu, C.; Wang, Y. Real-space observation of photonic nanojet in microspheres. *Phys. E* **2014**, *61*, 141–147. [CrossRef]
15. Darafsheh, A.; Bollinger, D. Systematic study of the characteristics of the photonic nanojets formed by dielectric microcylinders. *Opt. Commun.* **2017**, *402*, 270–275. [CrossRef]
16. Darafsheh, A. Influence of the background medium on imaging performance of microsphere-assisted super-resolution microscopy. *Opt. Lett.* **2017**, *42*, 735–738. [CrossRef] [PubMed]
17. Jalalia, T.; Ernib, D. Highly confined photonic nanojet from elliptical particles. *J. Mod. Opt.* **2014**, *61*, 1069–1076. [CrossRef]
18. Wu, M.; Huang, B.; Chen, R.; Yang, Y.; Wu, J.; Ji, R.; Chen, X.; Hong, M. Modulation of photonic nanojets generated by microspheres decorated with concentric rings. *Opt. Express* **2015**, *23*, 20096–20103. [CrossRef] [PubMed]
19. Wu, M.; Chen, R.; Soh, J.; Shen, Y.; Jiao, L.; Wu, J.; Chen, X.; Ji, R.; Hong, M. Super-focusing of center-covered engineered microsphere. *Sci. Rep.* **2016**, *6*, 31637. [CrossRef]
20. Liu, C.; Lin, F. Geometric effect on photonic nanojet generated by dielectric microcylinders with non-cylindrical cross-sections. *Opt. Commun.* **2016**, *380*, 287–296. [CrossRef]
21. Wu, M.; Chen, R.; Ling, J.; Chen, Z.; Chen, X.; Ji, R.; Hong, M. Creation of a longitudinally polarized photonic nanojet via an engineered microsphere. *Opt. Lett.* **2017**, *42*, 1444–1447. [CrossRef] [PubMed]
22. Yang, J.; Twardowski, P.; Gerard, P.; Duo, Y.; Fontaine, J.; Lecler, S. Ultra-narrow photonic nanojets through a glass cuboid embedded in a dielectric cylinder. *Opt. Express* **2018**, *26*, 3723–3731. [CrossRef]
23. Zhou, Y.; Gao, H.; Teng, J.; Luo, X.; Hong, M. Orbital angular momentum generation via a spiral phase microsphere. *Opt. Lett.* **2018**, *43*, 34–37. [CrossRef]
24. Chen, L.; Zhou, Y.; Wu, M.; Hong, M. Remote-mode microsphere nano-imaging: new boundaries for optical microscopes. *Opto-Electron. Adv.* **2018**, *1*, 170001. [CrossRef]
25. Kong, S.; Taflove, A.; Backman, V. Quasi one-dimensional light beam generated by a graded-index microsphere. *Opt. Express* **2009**, *17*, 3722–3731. [CrossRef]
26. Geints, Y.; Zemlyanov, A.; Panina, E. Photonic nanojet calculations in layered radially inhomogeneous micrometer-sized spherical particles. *J. Opt. Soc. Am. B* **2011**, *28*, 1825–1830. [CrossRef]
27. Liu, C. Superenhanced photonic nanojet by core-shell microcylinders. *Phys. Lett. A* **2012**, *376*, 1856–1860. [CrossRef]
28. Shen, Y.; Wang, L.; Shen, J. Ultralong photonic nanojet formed by a two-layer dielectric microsphere. *Opt. Lett.* **2014**, *39*, 4120–4123. [CrossRef]

29. Wu, P.; Li, J.; Wei, K.; Yue, W. Tunable and ultra-elongated photonic nanojet generated by a liquid-immersed core–shell dielectric microsphere. *Appl. Phys. Express* **2015**, *8*, 112001. [CrossRef]
30. Minin, I.; Minin, O. Terahertz artificial dielectric cuboid lens on substrate for super-resolution images. *Opt. Quant. Electron.* **2017**, *49*, 326. [CrossRef]
31. Xing, H.; Zhou, W.; Wu, Y. Side-lobes-controlled photonic nanojet with a horizontal graded-index microcylinder. *Opt. Lett.* **2018**, *43*, 4292–4295. [CrossRef] [PubMed]
32. Liu, C.; Yen, T.; Minin, O.; Minin, I. Engineering photonic nanojet by a graded-index micro-cuboid. *Phys. E* **2018**, *98*, 105–110. [CrossRef]
33. Geints, Y.; Zemlyanov, A.; Minin, O.; Minin, I. Systematic study and comparison of photonic nanojets produced by dielectric microparticles in 2D- and 3D-spatial configurations. *J. Opt.* **2018**, *20*, 065606. [CrossRef]
34. Kotlyar, V.; Stafeev, S. Modeling the sharp focus of a radially polarized laser mode using a conical and a binary microaxicon. *J. Opt. Soc. Am. B* **2010**, *27*, 1991–1997. [CrossRef]
35. Geints, Y.; Zemlyanov, A.; Panina, E. Microaxicon generated photonic nanojets. *J. Opt. Soc. Am. B* **2015**, *32*, 1570–1574. [CrossRef]
36. Liu, C. Photonic jets produced by dielectric micro cuboids. *Appl. Opt.* **2015**, *54*, 8694–8699. [CrossRef]
37. Yue, L.; Yan, B.; Wang, Z. Photonic nanojet of cylindrical metalens assembled by hexagonally arranged nanofibers for breaking the diffraction limit. *Opt. Lett.* **2016**, *41*, 1336–1339. [CrossRef]
38. Yue, L.; Yan, B.; Monks, J.; Wang, Z.; Tung, N.; Lam, V.; Minin, O.; Minin, I. Production of photonic nanojets by using pupil-masked 3D dielectric cuboid. *J. Phys. D: Appl. Phys.* **2017**, *50*, 175102. [CrossRef]
39. Pierron, R.; Zelgowski, J.; Pfeiffer, P.; Fontaine, J.; Lecler, S. Photonic jet: key role of injection for etchings with a shaped optical fiber tip. *Opt. Lett.* **2017**, *42*, 2707–2709. [CrossRef] [PubMed]
40. Taflove, A.; Hagness, S. *Computational Electrodynamics: The Finite Difference Time Domain Method*; Artech House: Boston, MA, USA, 1998.
41. Poco, J.; Hrubesh, L. Method of producing optical quality glass having a selected refractive index. U.S. Patent 6,158,244, 2008.
42. Zhu, Q.; Fu, Y. Graded index photonic crystals: A review. *Ann. Phys.-Berlin* **2015**, *527*, 205–218. [CrossRef]
43. Ruskuc, A.; Koehler, P.; Weber, M.; Andres-Arroyo, A.; Frosz, M.; Russell, P.; Euser, T. Excitation of higher-order modes in optofluidic photonic crystal fiber. *Opt. Express* **2018**, *26*, 30245–30254. [CrossRef] [PubMed]

Article

A Magnetic-Dependent Vibration Energy Harvester Based on the Tunable Point Defect in 2D Magneto-Elastic Phononic Crystals

Tian Deng [1,2], Shunzu Zhang [1,2] and Yuanwen Gao [1,2,*]

[1] Key Laboratory of Mechanics on Disaster and Environment in Western China attached to the Ministry of Education of China, Lanzhou University, Lanzhou 730000, China; dengt17@lzu.edu.cn (T.D.); zhangshz16@lzu.edu.cn (S.Z.)
[2] Department of Mechanics and Engineering Sciences, College of Civil Engineering and Mechanics, Lanzhou University, Lanzhou 730000, China
* Correspondence: ywgao@lzu.edu.cn; Tel.: +86-931-8914560; Fax: +86-931-8914560

Received: 22 April 2019; Accepted: 16 May 2019; Published: 19 May 2019

Abstract: In this work, an innovative vibration energy harvester is designed by using the point defect effect of two-dimensional (2D) magneto-elastic phononic crystals (PCs) and the piezoelectric effect of piezoelectric material. A point defect is formed by removing the central Tenfenol-D rod to confine and enhance vibration energy into a spot, after which the vibration energy is electromechanically converted into electrical energy by attaching a piezoelectric patch into the area of the point defect. Numerical analysis of the point defect can be carried out by the finite element method in combination with the supercell technique. A 3D Zheng-Liu (Z-L) model which accurately describes the magneto-mechanical coupling constitutive behavior of magnetostrictive material is adopted to obtain variable band structures by applied magnetic field and pre-stress along the z direction. The piezoelectric material is utilized to predict the output voltage and power based on the capacity to convert vibration energy into electrical energy. For the proposed tunable vibration energy harvesting system, numerical results illuminate that band gaps (BGs) and defect bands of the in-plane mixed wave modes (*XY* modes) can be adjusted to a great extent by applied magnetic field and pre-stress, and thus a much larger range of vibration frequency and more broad-distributed energy can be obtained. The defect bands in the anti-plane wave mode (Z mode), however, have a slight change with applied magnetic field, which leads to a certain frequency range of energy harvesting. These results can provide guidance for the intelligent control of vibration insulation and the active design of continuous power supply for low power devices in engineering.

Keywords: vibration energy harvester; phononic crystal; defect bands; piezoelectric material; magnetostrictive material; output voltage and power

---

1. Introduction

With the ever-increasing development of self-powered wireless transmitters and embedded systems, the demands of independent power supply and extended lifespans become more and more intense [1,2]. In fact, energy harvesters of renewable and clear resources have attracted increasing attention of worldwide research communities with the increase of environmental issues caused by traditional resources. The various forms of those resources include sunlight, waste heat, flowing water, wind, and mechanical vibration, etc. [3–5]. Vibration energy especially, as a broad-distributed source, is the most prevalent within energy harvesting research [4], with numerous vibration energy generators of piezoelectric [6], electromagnetic [7], and electrostatic [8] conversion having been investigated. The piezoelectric generator, as one of the most effective collection devices, can harvest higher output power owing to a better capability of electrical-mechanical coupling and higher strain for a given size. Considering the intrinsic advantage of high energy density in point defected phononic crystals (PCs), vibration energy can be accurately localized and enhanced at the point defect area, which can provide an excellent capability to achieve energy conversion through the direct piezoelectric effect.

As a typical composite material, PCs have different periodic structures, which can generate elastic/acoustic wave band gaps (BGs) where the wave propagation is forbidden in some ranges of wave frequency. With the introduction of some defects into perfect PCs, much attention has been concentrated on defect modes of PCs. Based on the characteristics of defect bands trapped in the BGs, where the elastic/acoustic wave can be localized and enhanced in the defect or propagate along the defected direction, defected PCs have extensive applications in engineering, such as being vibration isolators and noise suppressors [9], acoustic filters [10], and waveguides [11]. In recent years, a great deal of effort has been paid to obtain the formation mechanisms and influence factors of the point defected PCs. Khelif et al. [12,13] have theoretically and experimentally studied the characteristics of band structures and the localization effect of a 2D point-defected elastic PC. An accurate interferometric setup has been used by Romero-García et al. [14] for observing the symmetric and antisymmetric vibrational patterns in sonic crystals with double-point defects. Additionally, the bending wave propagation characteristics of a 2D point-defected PC thin plate have been discussed by Yao et al. [15]. Wu et al. [16] have reported the effects of superlattice configuration and defect shapes (square, circular, and rectangular) on defect bands of PCs. In fact, the control of defect bands for PCs with elastic materials has a significant dependence on not only the shapes of inclusions, lattice arrangement, and filling fraction, but also the different physical properties of components.

With the swift development of technology, considering the rich physics characteristics and extraordinary capabilities of smart materials and that their geometric and physical parameters can be changed greatly by external stimuli (magnetic field, pre-stress, temperature, and electric field, etc.), it is desirable and inevitable to construct dynamic tunable PCs with smart materials (magnetostrictive materials and piezoelectric materials, etc.) [17–21]. Regarding PCs with magnetostrictive material, researchers have experienced compelling interest in the modulation of elastic wave propagation and band structures in perfect magneto-elastic PCs by extrinsic motivation [22–25]. For example, Bou Matar et al. [22] have studied the wave propagation characteristics of 2D Terfenol-D/epoxy PCs, in which the BGs are tuned by adjusting the orientation and magnitude of the magnetic field. Considering the mechanical-magnetic coupling effect, Ding et al. [23] have presented longitudinal wave propagation characteristics in 1D Ni6/Terfenol-D PC rods. Zhang et al. [24] have found that the maximum width of BGs can reach an optimal orientation angle of 45° in the magnetic field for a magneto-elastic PC thin plate. However, for magneto-elastic PCs with defect modes, only a few studies have been reported in the existing literature. Gu and Jin [25] have explored the band structure of a 2D point-defected PC composed of Terfenol-D rods embedded in a polymethyl methacrylate (PMMA) matrix tuned by an applied magnetic field and pre-stress along the z-axis. The numerical results indicated that an applied suitable magnetic field can enlarge the first band gap (FBG) and capture a new band gap (NBG) in the in-plane modes (*XY* modes).

Owing to the wave localization and enhancement phenomena on account of point defect modes inside the BGs, vibration energy can be easily converted into electric energy by placing a ceramic piezoelectric patch within the defect state. There is considerable interest in the research of vibration energy harvesting by point-defected PCs [26–31]. Wu et al. [26] primarily used a polyvinylidene fluoride (PVDF) beam to harvest acoustic energy through a point-defected PC consisting of PMMA cylinders within the air background, and found that the maximum energy is collected at the resonance frequency of the point-defected PC and piezoelectric beam. In addition, when considering a 2D point-defected PC with a solid-solid system, Lv et al. [27] demonstrated experimentally that the output voltage and power harvesting efficiency are 421 and 177241 times larger than that in a rubber block, respectively. Based on the characteristics of wave focusing and energy localization, Carrara et al. [28] have demonstrated that acoustic wave energy harvesting can be produced by stub-plate acoustic metamaterials. Yang et al. [29] studied a coupled resonance structure between two PC resonators in order to expand the acoustic wave localization and enhancement effects, with the experimental results revealing that the maximum pressure magnification of a coupled structure is three times larger than that of an individual PC. For the vibration energy harvester with point-defected PCs consisting of with elastic materials, the frequency of the defect band is a certain value. In fact, it is necessary to explore ways to enlarge the BGs and adjust the position of the defect bands, allowing the broad-distributed vibration energy to be converted into electrical energy. Hence, considering the magneto-electro-elastic coupling interaction and introducing smart materials (i.e., magnetostrictive material and piezoelectric material) into point-defected PCs is a good way to realize broad-frequency energy harvesting.

The aim of this paper is to present a way to modulate the range of vibration energy harvesting using magneto-elastic point-defected PCs with piezoelectric material, considering the tunability of the BGs and defect bands via an applied magnetic field and pre-stress. Two analytical approaches (band structure and vibration energy conversion) are proposed to quantitatively discuss magneto-electric conversion efficiency and further obtain the optimized output voltage and power of a vibration energy harvesting system. More specifically, a 3D nonlinear magneto-elastic coupling constitutive relationship in combination with the supercell technique is adopted to calculate band structure, and the direct piezoelectric effect is used to predict output voltage and power for energy harvesting by the finite element method (FEM) implemented by COMSOL Multiphysics 5.3a. [32]. These results provide a feasible way to broaden BGs and expand the frequency range of vibration energy harvesting simultaneously. This paper is organized as follows: In Section 2, the setup of the magneto-electro-elastic coupling theoretical model and calculation method are briefly presented. In Section 3, two schemes, i.e., the in-plane modes and the out-plane mode, are discussed. Finally, in Section 4 we give a conclusion.

2. Theoretical Model and Calculation Method

We consider a vibration energy harvester by inserting piezoelectric material into a 2D magneto-elastic PC with point defect from a 5×5 supercell. As shown in Figure 1a, the square lattice assumes that the lattice constant is a and the radius of rod is r. The corresponding regions of rod, matrix and piezoelectric patch, indicated by A, B, and C, represent the magnetostrictive phase, elastic phase, and the piezoelectric phase. Figure 1b shows the first irreducible Brillouin zone of the supercell (the triangular area $\Gamma - X - M - \Gamma$) in the square lattice. A diagram of the magnetostrictive rod with applied pre-stress and magnetic field is shown in Figure 1c. The center of the rod is at the origin of the Cartesian coordinate system (o-xyz), with the x-y plane being located on the cross section of the rod and the z-axis being parallel to the length direction. Note that the pre-stress and magnetic field are applied along the z-axis of the Terfenol-D rod to tune the range of energy harvesting. The piezoelectric phase connecting to an electric circuit is shown in Figure 1d, where the piezoelectric patch is deformed by pressure difference p, so that the mechanical strain energy can be converted into electrical energy.

For the magnetostrictive phase, the Terfenol-D rod was chosen as the inclusion because it is widely used in high-precision actuators and smart devices due to their complex nonlinear magneto-mechanical coupling effect and sensitivity to external stimuli (magnetic field and pre-stress). The 3D Z-L nonlinear magneto-mechanical coupling constitutive equation is written as [33]:

$$\varepsilon_{ij} = \frac{1}{E}\Big[(1+\upsilon)\sigma_{ij} - \upsilon\sigma_{kk}\delta_{ij}\Big] + \frac{\lambda_s}{M_s^2}\Big[\frac{3}{2}M_iM_j - M_kM_k\Big(\frac{1}{2}\delta_{ij} + \widetilde{\sigma}_{ij}/\sigma_s\Big)\Big], \tag{1a}$$

$$H_k = \Big\{\frac{1}{kM}f^{-1}\Big(\frac{M}{M_s}\Big)\delta_{kl} - \frac{\lambda_s}{\mu_0 M_s^2}\Big[2\widetilde{\sigma}_{kl} - \big(I_\sigma^2 - 3II_\sigma\big)\delta_{kl}/\sigma_s\Big]\Big\}M_l. \tag{1b}$$

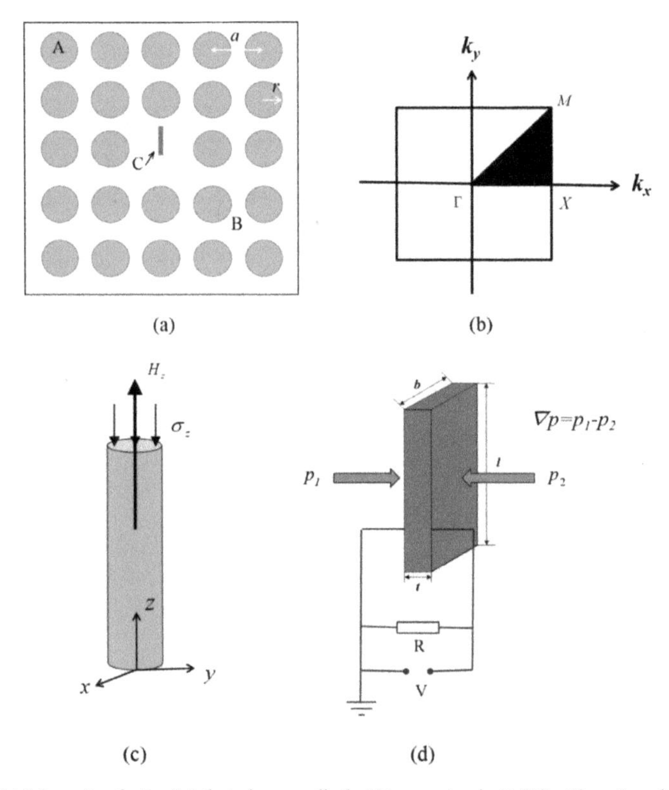

Figure 1. (**a**) Schematic of a 5 × 5 defected supercell of a 2D magneto-elastic PC with a piezoelectric patch. (**b**) The first irreducible Brillouin zone (triangular area $\Gamma - X - M - \Gamma$). (**c**) The Terfenol-D rod is affected by applied pre-stress and the magnetic field. (**d**) The structure and circuit connection of a piezoelectric patch.

In our study, by assuming that the Terfenol-D rod is infinite in magneto-elastic PCs, the demagnetization effect of the magnetostrictive material was able to be neglected [23]. Hence, the pre-stress and magnetic field were applied easily along the z-axis, that is, $\sigma_x = 0, \sigma_y = 0, H_x = 0, H_y = 0$. The corresponding constitutive relationship from Equation (1) is

$$\begin{pmatrix} \varepsilon_x \\ \varepsilon_y \\ \varepsilon_z \\ \gamma_{yz} \\ \gamma_{zx} \\ \gamma_{xy} \end{pmatrix} = \begin{pmatrix} 1/E & -v/E & -v/E & 0 & 0 & 0 \\ -v/E & 1/E & -v/E & 0 & 0 & 0 \\ -v/E & -v/E & 1/E & 0 & 0 & 0 \\ 0 & 0 & 0 & 1/G & 0 & 0 \\ 0 & 0 & 0 & 0 & 1/G & 0 \\ 0 & 0 & 0 & 0 & 0 & 1/G \end{pmatrix} \begin{pmatrix} \sigma_x \\ \sigma_y \\ \sigma_z \\ \tau_{yz} \\ \tau_{zx} \\ \tau_{xy} \end{pmatrix} + \frac{\lambda_s}{M_s^2} \begin{pmatrix} -1/2 - \widetilde{\sigma}_x/\sigma_s \\ -1/2 - \widetilde{\sigma}_y/\sigma_s \\ 1 - \widetilde{\sigma}_z/\sigma_s \\ -2\widetilde{\tau}_{yz}/\sigma_s \\ -2\widetilde{\tau}_{zx}/\sigma_s \\ -2\widetilde{\tau}_{xy}/\sigma_s \end{pmatrix} M_z^2 \quad (2a)$$

$$H_z = \frac{1}{k} f^{-1}\left(\frac{M_z}{M_s}\right) - \frac{\lambda_s}{\mu_0 M_s^2}\left[2\widetilde{\sigma}_z - \left(I_\sigma^2 - 3II_\sigma\right)/\sigma_s\right] \cdot M_z \quad (2b)$$

where σ_{ij} and ε_{ij} represent the elastic stress and strain tensors, respectively, and $G = E/2(1+v)$ represents the shear modulus with E and v being Young's modulus and Poisson's ratio. $M = \sqrt{M_k M_k}$ is the magnetization intensity, and λ_s, M_s, and σ_s represent, respectively, the saturation magnetostrictive coefficient, magnetization, and magnetostrictive stress. H_z and M_z represent the magnetic field and magnetization intensity along the z-axis, respectively. $k = 3\chi_m/M_s$ is the relaxation factor, where χ_m is the susceptibility in the initial linear region. For $\mathrm{I}_\sigma^2 - 3\mathrm{II}_\sigma = 2\widetilde{\sigma}_{\mathrm{ij}}\widetilde{\sigma}_{\mathrm{ij}}/3, \widetilde{\sigma}_{ij} = 3\sigma_{ij}/2 - \sigma_{kk}\delta_{ij}/2$ represents 3/2 times the deviatoric stress σ_{ij}, and δ_{ij} is the Kronecker delta. $\mu_0 = 4\pi \times 10^{-7}(\mathrm{H/m})$ represents the vacuum permeability. The nonlinear scalar function $f(x) = \coth(x) - 1/x$ represents the Langevin function, which is based on Boltzmann statistics and can give a better calculation for the magnetization curve [33].

It is difficult to directly incorporate the nonlinear constitutive equations into the mechanical equations because the equations contain inverse functions and implicit expressions. For simplicity, we can rewrite the above nonlinear constitutive equations of magnetostrictive material as the general form of the linear-like constitutive equations with variable equivalent coefficients, which are widely used in much magneto-elastic PCs research and reflect the nonlinear constitutive relations effectively in macroscopic scenarios [17–19].

$$\varepsilon_{kl}^{m} = s_{ijkl}^{m}(\boldsymbol{\sigma},\boldsymbol{H})\sigma_{ij}^{m} - d_{mkl}^{m}(\boldsymbol{\sigma},\boldsymbol{H})^{T}H_{m}^{m}, \tag{3a}$$

$$B_{n}^{m} = d_{nij}^{m}(\boldsymbol{\sigma},\boldsymbol{H})\sigma_{ij}^{m} + \mu_{nm}^{m}(\boldsymbol{\sigma},\boldsymbol{H})H_{m}^{m}, \tag{3b}$$

where the right superscript m denotes the magnetostrictive phase. $B_{n}^{m} = \mu_{0}^{m}(H_{m}^{m} + H_{n}^{m})$ represents the magnetic induction vector, and $s_{ijkl}^{m}(\boldsymbol{\sigma},\boldsymbol{H})$, $d_{mkl}^{m}(\boldsymbol{\sigma},\boldsymbol{H})$, and $\mu_{nm}^{m}(\sigma,H)$ represent, respectively, the flexibility matrix, the piezomagnetic coupling matrix and the magnetic permeability matrix, which are the functions of the magnetic field and pre-stress. $(.)^T$ denotes the transposition of the matrix. $\boldsymbol{\sigma}$ and $\boldsymbol{H}$ represent, respectively, the stress and magnetic field intensity vectors, which are the independent variables. The effective material coefficients and exact expressions can be found in Ref. [25].

For the piezoelectric phase, piezoelectric material as the vibration energy generator has received wide attention in research because of its high efficiency of electrical-mechanical conversion and high output voltage. The 31 piezoelectric mode is adopted for electricity generation of piezoelectric patches, and the pressure difference p applied across the two sides of the patch acts as the external force to drive the vibration of the piezoelectric patch, which converts the vibration energy into electrical energy. Because the load resistance of the external loading circuit has a significant effect on the piezoelectric energy harvesting, the electrical impedance is in series at the two sides of the piezoelectric patch to find the optimal output voltage and power [31]. In the development of the generator structure for the piezoelectric patch, the established 3D linear piezoelectric constitutive equation in reduced-matrix form is [34]

$$\begin{bmatrix} D^{p} \\ \sigma^{p} \end{bmatrix} = \begin{bmatrix} \varepsilon^{s} & e^{p} \\ -(e^{p})^{T} & c^{p} \end{bmatrix} \begin{bmatrix} E^{p} \\ \varepsilon^{p} \end{bmatrix} \tag{4}$$

where the right superscript p denotes the piezoelectric phase. σ^p and D^p are the stress and electric displacement vectors, respectively. ε^p and E^p are the strain and electric field vectors, respectively. c^p, e^p, and ε^s are the elastic coefficient at constant electric field, piezoelectric stress coefficient, and dielectric constant at constant stress field, respectively. $(.)^T$ denotes the transposition of the matrix.

According to the [31], both output voltage and power are functions of the external load resistance. It can be found that the maximum value of output power can be reached when the optimal loading resistance is yielded as

$$R_{opt} = \frac{1}{\omega C_p} \frac{2\zeta}{\sqrt{4\zeta^2 + k^4}} \tag{5}$$

where ω, ζ, k, and C_p are the forcing frequency of the piezoelectric patch, damping radio, piezoelectric coupling coefficient, and capacitance of the piezoelectric patch, respectively.

In order to obtain the band structure of point defected PCs, a periodic boundary condition based on the Bloch-Floquet theorem was applied on the interfaces of both the x and y directions between the adjacent supercell systems.

$$u_i = (x + a_1, y + a_1) = u_i(x,y)e^{i(k_x a_1 + k_y a_1)}, (i = x,y,z) \tag{6}$$

where u_i represents the elastic displacement vector, $a_1 = 5 \times a$ denotes the lattice constant of the supercell, and k_x and k_y are the Bloch vectors. The supercell can be meshed by the quadratic Lagrange triangular element, and a group of eigenvalues and eigenmodes are obtained by scanning the Bloch wave vector k from the first irreducible Brillouin zone along the path of $\Gamma - X - M - \Gamma$. Hence, the defect bands of magnetic-elastic PCs can be obtained by applying different magnetic fields and pre-stress.

In this study, the FEM is utilized to analyze and calculate the system of vibration energy harvesting based on magneto-elastic PCs with point defects. The structural mechanics and AC/DC modules are used to model the coupling problems of the mechanical, magnetic, and electrostatic fields and to estimate the output voltage and power from the piezoelectric generators. The relationship between the magnetic-electro-mechanical coupling is displayed in Figure 2. The structural mechanics module is used to calculate the band structure of the 5×5 supercell with a point defect under a different magnetic field and pre-stress. The Partial Differential Equation (PDE) module is adopted to solve the implicit function problem in the effective material parameters of magnetostrictive material. Using the AC/DC module, the piezoelectric patch connecting to an electric circuit can be utilized to convert the mechanical vibration into electrical energy, and then the series resistance is added in the circuit module to obtain the maximum output power. Thus, the output voltage and power can be harvested from point defected magneto-elastic PCs with piezoelectric material.

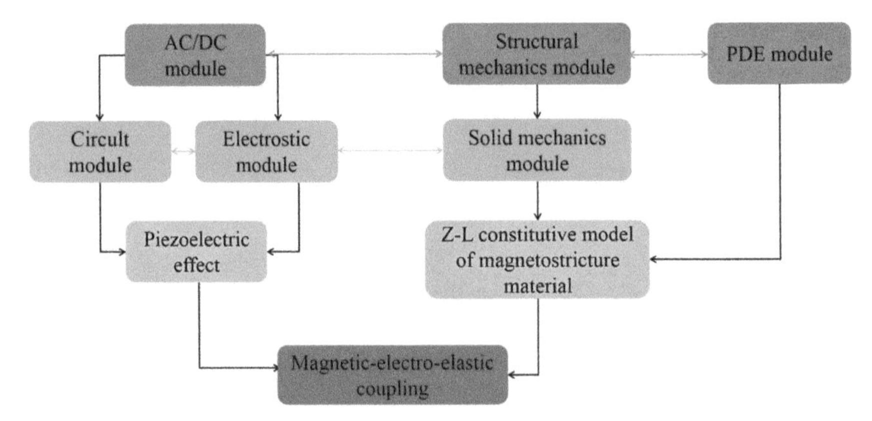

Figure 2. The relationships between the magnetic-electro-mechanical coupling of each module.

3. Numerical Results and Discussion

3.1. Verification of Point-Defect Bands

In this section, we adequately investigate the tunable vibration energy harvesting system of a 2D magnetic-elastic (M-E) PC including point defect modes (in-plane modes and out-plane modes) under different magnetic and stress fields. It is noteworthy that the magneto-elastic system in this paper can be degenerated directly to the study of a conventional point-defected PC with elastic material by excluding the magnetostrictive material. The band structure of the point-defected PC consisting of a steel rod and rubber matrix is calculated in Figure 3 and compared with the results of Ref. [35] in order to validate the accuracy of the proposed model. The results show the BG occurred in the frequency range of 564–1038 Hz and two defect bands appeared at 648.2 and 648.5 Hz in the frequency range of 0–1100 Hz. It can be seen that the results of the PC with elastic material (black lines) are consistent with the previous results (red dots) qualitatively and quantitatively, which verifies the accuracy of the current numerical model. As a specific case, the M-E PC material parameters of Terfenol-D [36] and PMMA [37] are listed in Table 1. The physical parameters of the piezoelectric patch (PZT-5A) [38] are shown in Table 2. The length (l), width (b), and thickness (t) of the piezoelectric patch are set as 20 mm, 7 mm, and 0.7 mm, respectively. The damping ratio ξ is chosen to be 0.025. The lattice constant a and the radius of scatter cylinder r are 15.5 mm and 5.5 mm, meaning the corresponding filling fraction is $f = \pi r^2/a^2 = 39.6\%$.

Table 1. Material parameters of Terfenol-D [36] and polymethyl methacrylate (PMMA) [37].

Materials	$\rho\left(\mathrm{kg/m^3}\right)$	$E(\mathrm{GPa})$	υ	C_{11} (GPa)	C_{44} (GPa)	λ_s (ppm)	χ_m	σ_s (GPa)	$\mu_0 M_s$ (T)
Terfenol-D	9200	60	0.3	—	—	1950	20.4	200	0.96
PMMA	1200	—	—	7.11	2.03	—	—	—	—

Table 2. Physical parameters of the piezoelectric patch (PZT-5A) [38].

Physical Parameters	Elastic Coefficient (GPa)					Piezoelectric Coefficient (10^{-12} C/m^2)			Dielectric Constant (10^{-9} F/m)	
	c^p_{11}	c^p_{12}	c^p_{13}	c^p_{33}	c^p_{44}	e^p_{31}	e^p_{33}	e^p_{15}	ε^s_{11}	ε^s_{33}
PZT-5A	121	75.40	75.20	111	21.1	-5.4	15.8	12.3	8.107	7.346

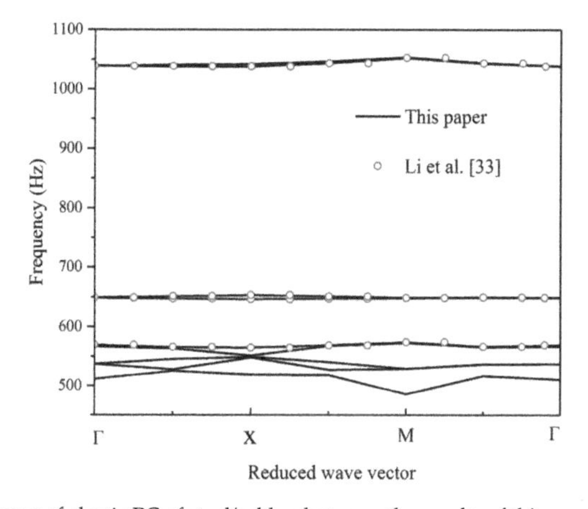

Figure 3. Comparisons of elastic PC of steel/rubber between the results of this paper (black lines) and those of Ref. [35] (red dots).

3.2. Output Voltage and Power for in-Plane Modes (XY Modes)

The effects of the magnetic field, pre-stress, and the piezoelectric patch on the band structure (*XY* modes) of the M-E PC with point defect are depicted in Figure 4. The transverse coordinate represents the reduced wave vector and the vertical coordinate represents the normalized frequency $(\omega a/2\pi c_t)$. ω represents the frequency and $c_t = 1300\ \mathrm{m/s}$ represents the transverse wave velocity of PMMA. It can be observed in Figure 4a that one BG appears in the band structure and no defect band exists in the PC without the point defect mode being $\sigma_z = 0$ MPa, $H_z = 0$ kOe. It can be seen from Figure 4b that the range of the first band gap (FBG) is 0.545006–1.050319, where three defect bands (original defect bands) are observed when $\sigma_z = 0$ MPa, $H_z = 0$ kOe. Figure 4c shows the band structure of the M-E PC with a point defect computed by considering the structural effect of the piezoelectric patch. By comparing Figure 4b,c, we can confirm that the piezoelectric patch has no obvious effect on the BGs and defect bands. When the stress is set to 0 MPa, and the magnetic field changes from 0 to 1 kOe, as shown Figure 4d, the lower edge of the FBG is slightly changed, while the upper edge FBG increases to 1.177729, and three new defect bands appear above the original defect bands, meaning that the width of the FBG increases by 0.127410. Note that a new band gap (NBG) appears in range 1.337568–1.450522, in which the four defect bands are trapped. When $H_z = 1$ kOe, the compressive pre-stress changes from 0 to 20 MPa, as shown Figure 4e, and the range of the FBG is 0.545954–1.061739. In comparison with Figure 4d, the upper edge of the FBG decreases, and new defect bands of FBG and NBG are closed. Some displacement fields of the eigenmodes (NBG1st, NBG2nd, NBG3rd, and NBG4th) display the vibration confinement and the double degenerate modes on the point defect states of the

M point, as shown in Figure 5, when the $\sigma_z = 0$ MPa and $H_z = 1$ kOe. Hence, it can be concluded that under the action of the magnetic field and pre-stress, the material parameters of Terfenol-D are changed, which leads to change in the BGs and the appearance of new defect bands; similar results can be found in Ref. [25]. In the following calculation, we will pay more attention to the investigation of the tunable vibration energy harvester used by the M-E PC with point defect.

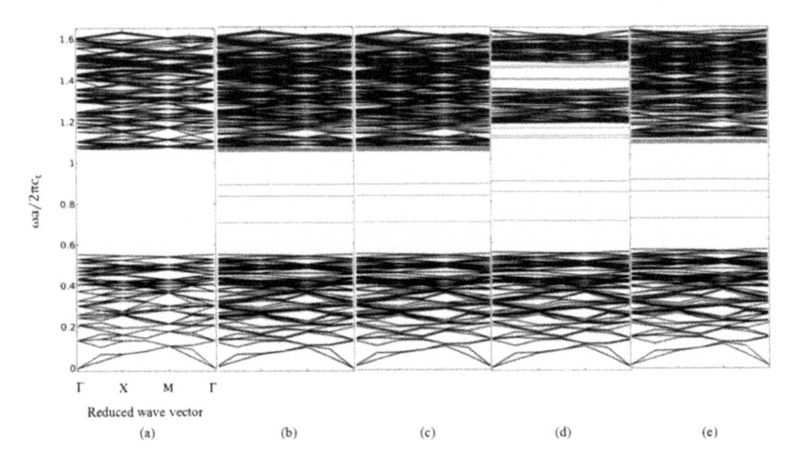

Figure 4. The band structure (*XY* modes) of a magnetic-elastic (M-E) PC with point defect under different magnetic fields and pre-stress values. (**a**) perfected M-E PC at $\sigma_z = 0$ MPa, $H_z = 0$ kOe. (**b**) $\sigma_z = 0$ MPa, $H_z = 0$ kOe without PZT-5A. (**c**) $\sigma_z = 0$ MPa, $H_z = 0$ kOe with PZT-5A. (**d**) $\sigma_z = 0$ MPa, $H_z = 1$ kOe with PZT-5A. (**e**) $\sigma_z = -20$ MPa, $H_z = 1$ kOe with PZT-5A.

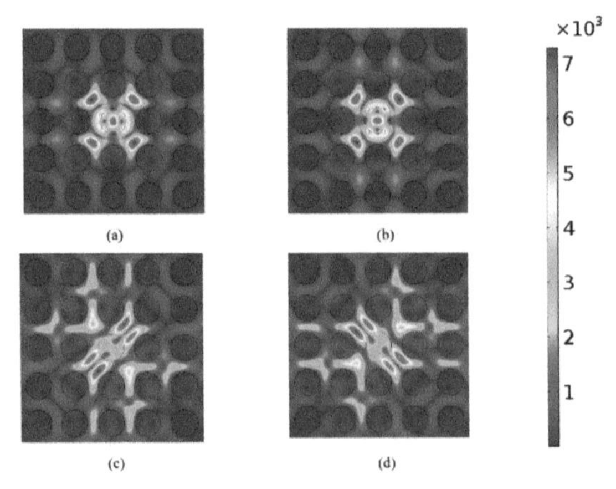

Figure 5. Displacement distributions of the point defect at the M point for $\sigma_z = 0$ MPa, $H_z = 1$ kOe. (**a**) NBG1st is at $\omega a/2\pi c_t = 1.385799$, (**b**) NBG2nd is at $\omega a/2\pi c_t = 1.388351$, (**c**) NBG3rd is at $\omega a/2\pi c_t = 1.439993$, and (**d**) NBG4th is at $\omega a/2\pi c_t = 1.447252$.

Figure 6 displays the variation of BGs and defect bands with the magnetic field in an M-E point-defected PC as $\sigma_z = 0$ MPa and $\sigma_z = -20$ MPa, respectively. The lower and upper edge of the FBG represent the start and cut off frequency, respectively, of the first band gap. The six defect bands trapped in the FBG are expressed as 1st, 2nd, 3rd, 4th, 5th, and 6th, respectively. The NBG and those corresponding to defect bands have the same representation. When the magnetic field increases from 0 to 4 kOe and $\sigma_z = 0$ MPa, Figure 6a shows that the position of the lower edge of the FBG and the original defect bands of the FBG (FBG1st, FBG2nd, and FBG3rd) stay nearly unchanged, and the upper edge of the FBG increases and tends to a stable state as H_z increases. It is noteworthy that the new defect bands of the FBG are opened at H_z = 0.3 kOe, and then gradually increase and tend to a fixed value with the increase of H_z. Moreover, the NBG and corresponding defect bands are opened up when H_z = 0.5 kOe, and the width and more defect bands of NBG increase with the rise of H_z. It is interesting that the defect bands in the frequency ranges 1.07–1.19 and 1.31–1.47 gradually separate and become flatter. When increasing the compressive pre-stress to 20 MPa, by comparing Figure 6a,b, it can be seen that the new defect bands of the FBG and NBG open in the higher magnetic field (1.1 kOe), and the edges and width of all these BGs and defect bands increase gradually and finally reach a certain constant at the saturated magnetic field. These results show that not only does the magnetic field have a great effect on the BGs and defect bands, but that pre-stress also has a significant effect on them. From the viewpoint of the magnetic domain, when H_z is in the low and intermediate field, the physics mechanism for these changes is that the magnetic domain rotation of the Terfenol-D rod is along the direction of easy magnetization with the rise of the magnetic field, resulting in the expansion range of the FBG and the generation of the NBG. Then, the magnetization of the magnetostrictive material is up to saturation, resulting in the BGs and corresponding defect bands remaining constant when H_z reaches the high field. In addition, when H_z is in the low and intermediate field, the applied pre-stress can make the magnetic domain rotate toward the direction of hard magnetization. In order to achieve the same magnetization intensity under larger pre-stress, a larger magnetic field must be applied. It is difficult for pre-stress to affect the saturation magnetization when H_z reaches the high field.

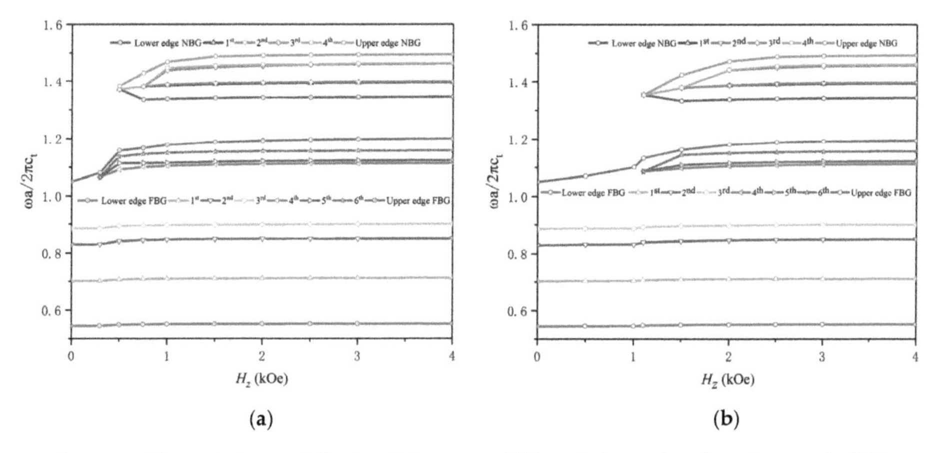

Figure 6. The variation in defect band frequency (*XY* modes) as a function of magnetic field. (**a**) σ_z = 0 MPa and (**b**) σ_z = −20 MPa. Legend: NBG, new band gap; FBG, first band gap.

Figure 7 presents the output voltage of defect bands in the FBG and NBG as a function of magnetic field in which σ_z = 0 MPa and R = 20 kΩ. Here, the output voltages at the defect bands of the FBG are expressed as FBG1st, FBG2nd, FBG3rd, FBG4th, FBG5th, and FBG6th, respectively. Output voltages at the defect bands of NBG are expressed as NBG1st, NBG2nd, NBG3rd, and NBG4th, respectively. The vibration energy can be converted into electrical energy by applied acceleration of 1 m/s^2 on the left-hand side of the supercell, incident from the x-axis. It can be seen from Figure 7a that the output voltage of the FBG1st is 11 mV and that the output voltages of the FBG4th, FBG5th, and FBG6th begin to appear simultaneously in H_z = 0.3 kOe. The output voltages of FBG2nd and FBG3rd become constant at 0.2 mV with the increases of the magnetic field. The output voltages of FBG4th are approximately in the range of 1.1–1.5 mV when H_z is in the low and intermediate field and tend to 1.1 mV in the high field. The output voltages of the FBG5th and FBG6th tend to 0.2 mV with the rise of the magnetic field. It is shown in Figure 7b that the output voltage of the defect bands in the NBG are opened up with the magnetic field at 0.5 kOe, and the output voltages of NBG1st, NBG2nd, and NBG3rd rapidly decrease with the magnetic field in the low and intermediate field. The output voltages of NBG1st and NBG2nd reach a steady state of about 5 mV, and those of NBG3rd reach a plateau of about 10 mV in the high field. The output voltage of NBG4th gradually increases to a maximum value of 52 mV as the magnetic field increases. Hence, we can conclude that the magnetic field changes the displacement field distributions of the defect bands, which leads to variation of the collected voltage in the magnetic field in the low and intermediate field. However, the displacement field of the defect bands is unchanged at the high field, which results in the stability of the output voltage. It can be concluded that the higher output voltage of 52 mV can be collected in the NBG compared to the FBG. In order to find the larger output voltage and power, in the following work we focus on the vibration energy harvesting range of the NBG.

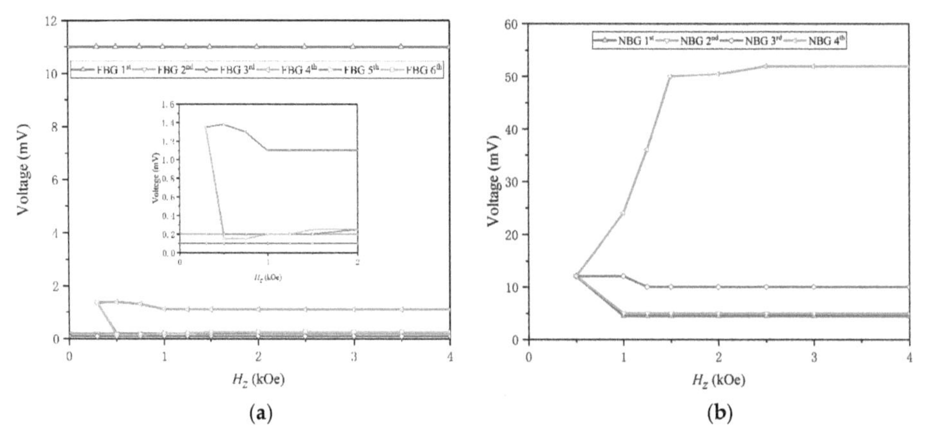

Figure 7. The output voltage of defect bands (*XY* modes) versus the magnetic field when $\sigma_z = 0$ MPa, R = 20 kΩ. (**a**) The output voltage of the FBG. (**b**) The output voltage of the NBG.

Figure 8 illustrates the dependence of the output voltage of the NBG on the frequency under different magnetic fields and pre-stress values. The effect of the magnetic field on the output voltage for the defect bands is illustrated in Figure 8a. The voltage can hardly be collected at $H_z = 0$ kOe. The output voltage has three peak values when $H_z = 1.1$ kOe, with the corresponding frequencies being 1.388752, 1.445999, and 1.456722, respectively. As a result of the frequencies of NBG1st and NBG2nd being very close, there is an output voltage peak in the voltage-frequency curve, and other frequencies corresponding to the peak voltage are consistent with the frequencies of the defect bands. The frequencies of the three peaks output voltage for $H_z = 2$ kOe can be found at 1.392721, 1.455566, and 1.462921, respectively, which also agree well with the defect band frequencies in the NBG. The output voltage versus the frequency of the defect bands in NBG at $\sigma_z = -20$ MPa is shown in Figure 8b. The voltage can hardly be collected at $H_z = 0$ kOe, the output voltage only has one peak with the frequency of 1.412125 at $H_z = 1.1$ kOe, and the output voltage has three peak values at $H_z = 2$ kOe, with the corresponding frequencies being 1.387582, 1.444213, and 1.448955, respectively, in agreement with the frequencies of the defect bands. In order to express more clearly and intuitively the effect of pre-stress on the output voltage, a curve of the output voltage with frequency under different pre-stress values at $H_z = 2.5$ kOe is given in Figure 9. The results show that the frequency corresponding to the peak voltage moves to the lower frequency region and that the highest peak voltage of the NBG gradually decreases with the incensement of compressive pre-stress. On the other hand, these phenomena show that the localized and enhanced characteristics of point-defected PC can be used to harvest the highest voltage of the piezoelectric patch.

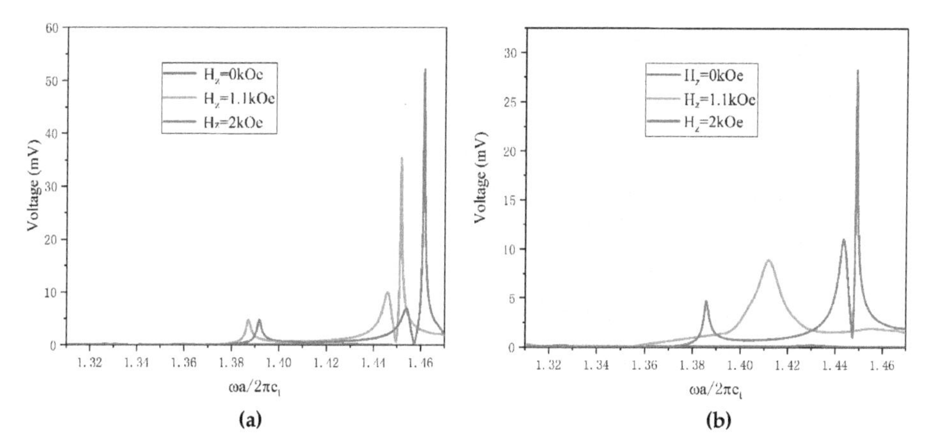

Figure 8. Output voltage versus the normalized frequency in NBG with R = 20 kΩ. (**a**) At σ_Z = 0 MPa. (**b**) At σ_Z = −20 MPa.

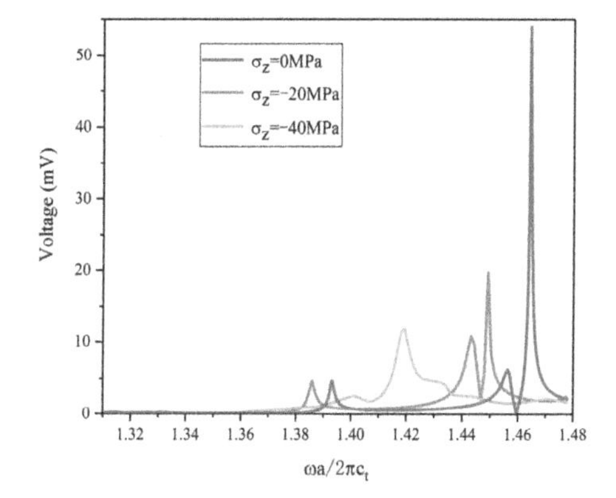

Figure 9. The influence of pre-stress on the output voltage in the NBG when the magnetic field is 2.5 kOe.

Figures 9 and 10 show the output voltage spectrum as a function of frequency under different magnetic fields and pre-stress values, respectively. Hence, quality factor can be introduced in this study, where the quality factor is defined by the ratio of the frequency in the defect band over the bandwidth between the half-voltage point ($\Delta\omega$) [39]. Note that the frequency of defect band corresponds to the highest peak voltage and the $\Delta\omega$ can be easily obtained from the output voltage spectrum to calculate the quality factor. Hence, the quality factor of the M-E PC with point defect under different magnetic fields and pre-stress values is presented in Table 3. We can see that when the pre-stresses are 0 and −20 MPa, the quality factor decreases gradually with a rise in magnetic field, which is opposite to the highest peak voltage. In particular, the quality factor reaches the lowest value because of the lower frequency corresponding to a higher bandwidth at H_z = 1.1 kOe, σ_Z= −20 MPa. Moreover, under the high magnetic field, the quality factor increases gradually with the increases of pre-stress.

Figure 10 shows the output voltage and power as a function of load resistance at defect band frequencies of the NBG (NBG1st, NBG2nd, NBG3rd, and NBG4th) when σ_Z = 0 MPa, H_z = 1 kOe. It is shown in Figure 10a that the output voltage at defect band frequencies increases rapidly to reach a fixed value with an increasing of load resistance initially. It can be easily seen in Figure 10b that the output power at defect band frequencies gradually increases with a rise of load resistance until an extreme value, followed by a decrease and finally a stabilization. It is demonstrated that there is an optimal load resistance of R = 11 kΩ corresponding to a maximum output power of P = 17 nW when the frequency is in NBG4th. Note that the optimal load resistance agrees with the result of Equation (5).

Table 3. The achieved $\Delta\omega$ and quality factor results from different magnetic fields and pre-stress values.

Magnetic Field H_z (kOe)	Pre-Stress σ_z (MPa)	Frequency of Defect Band (Hz)	Bandwidth ($\Delta\omega$)	Quality Factor
1.1	0	121,410	809	150.074
2	0	122,384	960	127.483
2.5	0	122,579	1215	100.888
1.1	−20	118,401	1838	64.418
2	−20	120,948	739	163.664
2.5	−20	122,078	920	132.693

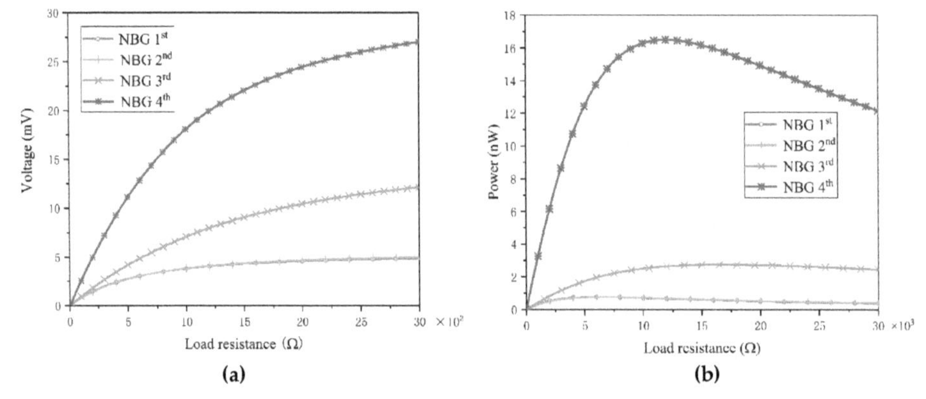

Figure 10. The output voltage (**a**) and power (**b**) versus the load resistance at the frequencies of the defect bands in the NBG at σ_z = 0 MPa, H_z = 1 kOe.

Figure 11 illustrates that the output voltage is a sinusoidal function of time at defect bands of the NBG when σ_z = 0 MPa, H_z = 1 kOe, R = 20 kΩ. The period (T) of the defect band frequency can be represented by $T = 1/\omega$. It can be verified that the values of T are 8.61 μs, 8.60 μs, 8.27 μs, and 8.24 μs, which correspond to the frequencies of the four defect bands in the NBG, the 1st, 2nd, 3rd and 4th, respectively. It is shown that the corresponding amplitudes of the output voltage at the four defect band frequencies are 4.5 mV, 4.6 mV, 10.5 mV, and 22.5 mV, respectively, which is in agreement with the result of Figure 10a at R = 20 kΩ. Furthermore, Figure 12 presents the output voltage and power as a function of the load resistances at the fourth defect band (NBG4th) for the piezoelectric patch placed in three different positions. The positions 1, 2, and 3 are shown in Figure 12a, with the position in the y direction of the gravity center for the piezoelectric patch being located at 6.25 mm, 5.25 mm, and 4.25 mm above that of the supercell, respectively, and the position along the x direction of the gravity center adhering consistently with that of the supercell. It can be found from Figure 12b,c that the different positions of the piezoelectric patch lead to different output voltage and power. Hence, the harvesting of vibrational energy using point defect modes is highly dependent on the position of the piezoelectric patch because of the distribution of the maxima and minima of the localized displacement field, where the displacement field structure of point defects is extremely complex. The different position of the localized field is distributed with different pressure, so the different positions produce a different pressure difference p between the two sides of the patch with different degrees of deformation when the piezoelectric patch is placed in different positions of the localized displacement field. Finally, the greater the pressure difference applied, the larger the output voltage harvested. However, the position of the piezoelectric patch does not influence the optimal load resistance corresponding to the maximum power. Because we can see from Equation (5) that the optimal load resistance is determined by vibration frequency, capacitance of the piezoelectric patch, the piezoelectric coupling coefficient, and the damping ratio, the external stimuli do not affect the value of the optimum resistance, although only if the material parameters are not varied.

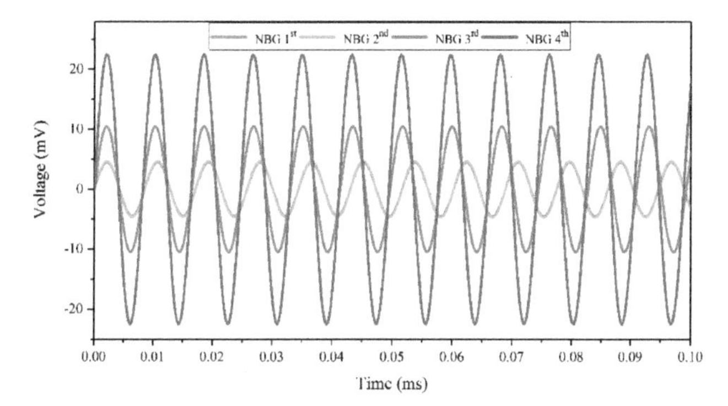

Figure 11. The output voltage from the piezoelectric material versus the time at defect bands of the NBG at $\sigma_z = 0$ MPa, $H_z = 1$ kOe, R = 20 kΩ.

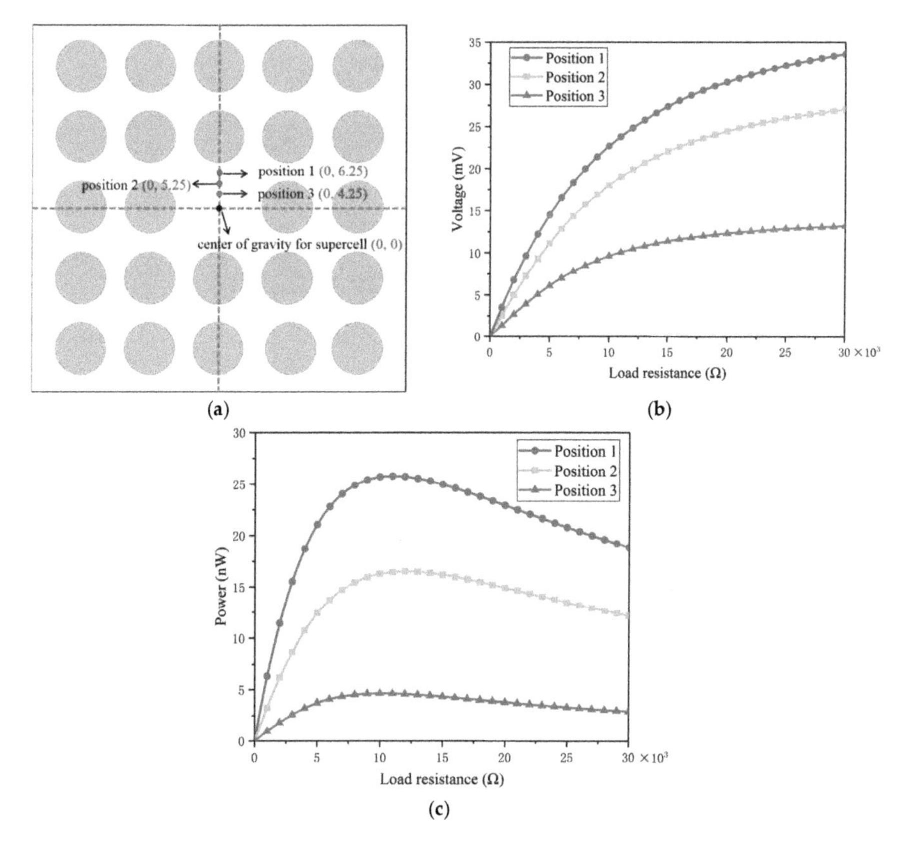

Figure 12. (**a**) Diagram of the piezoelectric patch placed in three different positions of the M-E PC with point defect. (**b**,**c**) show the output voltage and power of NBG4th versus the load resistance for different positions of the piezoelectric patch at $\sigma_z = 0$ MPa, $H_z = 1$ kOe.

3.3. Output Voltage for Anti-Plane Mode (Z Mode)

Figure 13 depicts the effects of the magnetic field and piezoelectric patch on the band structure (Z mode) of the M-E PC with point defect. It can be observed in Figure 13a that two BGs appear in the band structure and no defect band exists in the BGs of the PC without a point defect, as $\sigma_Z = 0$ MPa, $H_z = 0$ kOe. It can be seen in Figure 13b that the ranges of the FBG and second band gap (SBG) are 0.333796–0.790031 and 1.128115–1.365773, respectively. When a perfect PC is transformed into a point defect PC, three defect bands are trapped in the FBG and five defect bands are trapped in the SBG as $\sigma_Z = 0$ MPa, $H_z = 0$ kOe. By comparing Figure 13b,c, it can be found that the piezoelectric patch has no obvious effect on the BGs and defect bands. As magnetic field H_z increases from 0 to 1 kOe, we can see from Figure 13d that the lower edge of the FBG has slightly changed, while the upper edge of the FBG decreases to 0.736598, but that the corresponding defect bands of the FBG are invariant. Note that the positions of the lower and upper edge in the SBG decline obviously but that the width of the SBG is almost unchanged and the corresponding defect bands of the SBG decline significantly.

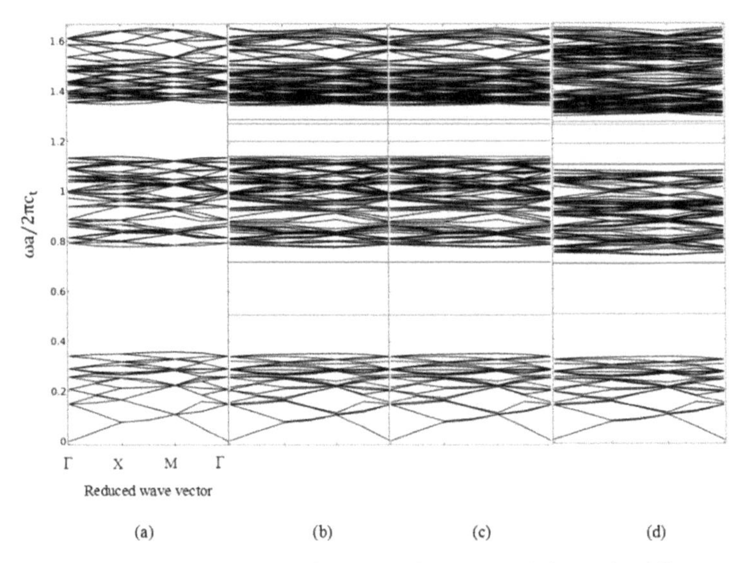

Figure 13. The band structure (Z mode) of the M-E PC with point defect under different magnetic fields. (**a**) perfected M-E PC at $\sigma_Z = 0$ MPa, $H_z = 0$ kOe. (**b**) $\sigma_Z = 0$ MPa, $H_z = 0$ kOe without PZT-5A. (**c**) $\sigma_Z = 0$ MPa, $H_z = 0$ kOe with PZT-5A. (**d**) $\sigma_Z = 0$ MPa, $H_z = 1$ kOe with PZT-5A.

Figure 14 presents the relation between the frequency of the defect bands and magnetic field in the M-E PC (Z mode) with point defect for the magnetic field changing from 0 to 4 kOe. The lower and upper edge FBG represent the start and cut off frequency of first band gap, respectively. Three defect bands trapped in the FBG are expressed as the 1st, 2nd, and 3rd respectively. The SBG and those corresponding to the defect bands have the same representation. We can easily find that the upper edge of the FBG declines clearly and then tends to a stable value at the saturated magnetic field, but that the lower edge FBG is unchanged with a rise in H_z, which leads to the width of the FBG decreasing in the low and intermediate magnetic field. Differently from the case of the *XY* modes, the defect bands of the FBG for the Z mode are nearly unchanged, and new BG cannot be opened up as H_z increases. It is noteworthy that the positions of the lower and upper edges of the SBG and the corresponding defect bands decrease dramatically in the low and intermediate magnetic field and then tend to remain constant in the high field. The physical mechanism of these phenomena is determined by the certain effective material parameters of magnetostrictive material given in Ref. [25], where those parameters show a complicated variation when the magnetic field is in the low and intermediate field, and then tend to a fixed value in the high field.

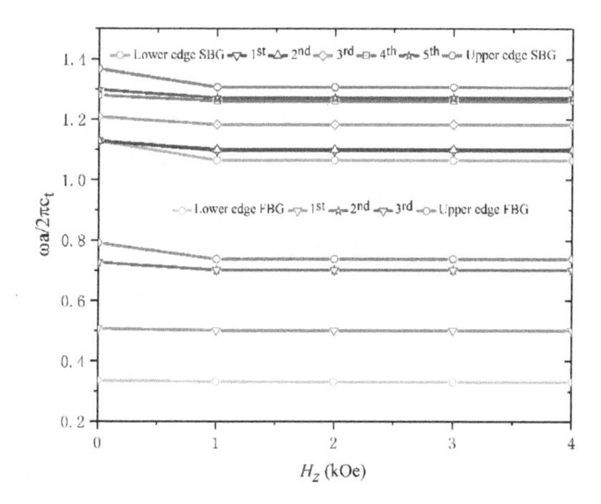

Figure 14. Variation in defect band frequency (Z mode) as a function of the magnetic field with given $\sigma_Z = 0$ MPa.

Figure 15 presents the output voltage of the defect bands (Z mode) as a function of the magnetic field when $\sigma_Z = 0$ MPa, R = 20 kΩ. The output voltages at the defect bands of the FBG are expressed as FBG1st, FBG2nd, and FBG3rd, respectively. Output voltages at the defect bands of the SBG are expressed as SBG1^{st,} SBG2nd, SBG3rd, SBG4th, and SBG5th, respectively. The acceleration excitation agrees with the *XY* modes. One can find that the output voltage does not change significantly with the increasing magnetic field due to a slight effect of the magnetic field on the defect bands. It can be seen that when the Terfenol-D is not applied to the magnetic field, the piezoelectric patch can collect the highest voltage of 19.2 mV in SBG5th, meaning the output voltages of FBG3rd and SBG4th are 3.4 mV and 7.2 mV, respectively. The output voltages of other defect bands are less than 1mV. The output voltage corresponding to the defect bands rapidly decreases in the low and intermediate magnetic fields and finally remains at a stable value in the high field. Hence, we can conclude that the SBG5th for the Z mode harvests the highest voltage of 19.2 mV at $\sigma_Z = 0$ MPa, $H_z = 0$ kOe. Because the magnetic field has very little effect on the output voltage of the Z mode and the maximum energy is harvested when the magnetic field is not applied on the Terfenol-D rod, increasing efforts have been expended on the vibration energy harvesting of the point-defected M-E PC in *XY* modes by applying different magnetic fields and pre-stress values.

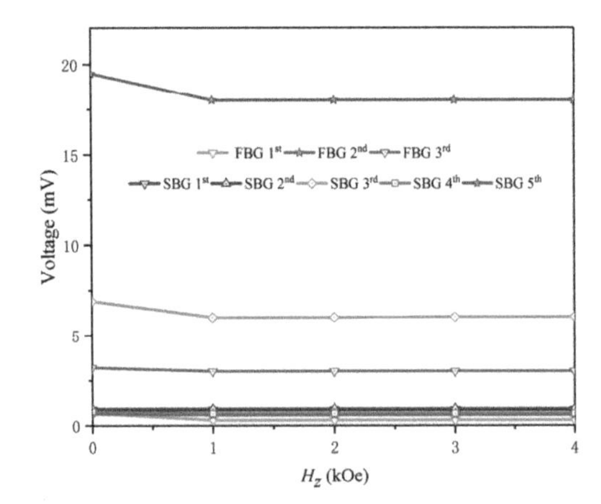

Figure 15. The output voltage of defect bands (Z mode) versus the magnetic field when $\sigma_Z = 0$ MPa, R = 20 kΩ.

4. Conclusions

In summary, a tunable vibration energy harvesting generator based on a 2D point-defected PC with magnetostrictive material and piezoelectric material under different magnetic fields and pre-stress values has been investigated. Owing to the fact that the elastic wave can be localized and enhanced in the point defect of PC, higher efficiency vibration energy conversion can be realized by attaching a piezoelectric patch to the point defect area. It has been found that the piezoelectric patch has no obvious influence on the band structure of point defect modes when BGs are in the higher region. However, the locations of the piezoelectric patch at the point defect position have a significant effect on energy harvesting. For the in-plane modes (*XY* modes), the intelligent tunability of the output voltage of the defect band is realized by applied magnetic field and pre-stress, which not only achieves vibration isolation and noise control at different frequencies, but also increases the frequency range of vibration energy harvesting simultaneously. Moreover, it is easy to expand the bandwidth of energy conversion and seek the optimal output voltage and power, which makes available higher energy harvesting efficiency. The tunability effect of the magnetic field on the output voltage of defect bands is unremarkable for the anti-plane mode (*Z* mode). The introduction of magnetostrictive material into a point-defected PC expands the intelligent regulation of elastic wave propagation behavior in complex multi-field environments. This tunable energy harvester used by a M-E PC provides an efficient method for active collection of broad-distributed vibration energy, which will present an important approach to apply to self-powered and low-powered devices in engineering, such as wireless sensors, medical implants, and the Internet of Things.

Author Contributions: Y.G. conceived of the main ideas and supervised the research. T.D. performed numerical simulations, discussed the results, and wrote the manuscript. S.Z. gave constructive advice on the calculations and revised the manuscript. Subsequently, all authors made improvements to the manuscript.

Funding: This work was supported by the National Natural Science Foundation of China (grant nos. 11872194 and 11572143). The costs of publishing this paper were borne by the National Natural Science Foundation of China (grant nos. 11872194 and 11572143).

Conflicts of Interest: The authors declare no conflict of interest.

References

1. Mitcheson, P.D.; Green, T.C.; Yeatman, E.M.; Holmes, A.S. Architectures for vibration-driven micropower generators. *J. Microelectromech. Syst.* **2004**, *13*, 429–440. [CrossRef]
2. Liu, H.C.; Zhong, J.W.; Lee, C.K.; Lee, S.W.; Lin, L.W. A comprehensive review on piezoelectric energy harvesting technology: Materials, mechanisms, and applications. *Appl. Phys. Rev.* **2018**, *5*, 041306. [CrossRef]

3. Kaldellis, J.K.; Zafirakis, D. The wind energy (r)evolution: A short review of a long history. *Renew. Energ.* **2011**, *36*, 1887–1901. [CrossRef]
4. Beeby, S.P.; Tudor, M.J.; White, N.M. Energy harvesting vibration sources for microsystems applications. *Meas. Sci. Technol.* **2006**, *17*, R175. [CrossRef]
5. Parameshwarana, R.; Kalaiselvam, S.; Harikrishnan, S.; Elayaperumal, A. Sustainable thermal energy storage technologies for buildings: A review. Renew. *Sust. Energ. Rev.* **2012**, *16*, 2394–2433.
6. Ottman, G.K.; Hofmann, H.F.; Lesieutre, G.A. Optimized piezoelectric energy harvesting circuit using step-down converter in discontinuous conduction mode. *IEEE Trans. Power Electron.* **2003**, *18*, 1988–1994. [CrossRef]
7. Ching, N.N.H.; Wong, H.Y.; Li, W.J.; Leong, P.H.W.; Wen, Z. A laser-micromachined multi-modal resonating power transducer for wireless sensing systems. *Sens. Actuator A-Phys.* **2002**, *97*, 685–690. [CrossRef]
8. Meninger, S.; Mur-Miranda, J.O.; Amirtharajah, R.; Chandrakasan, A.P.; Lang, J.H. Vibration-to-electricenergy conversion. IEEE Trans. *Very Large Scale Integr.* **2001**, *9*, 64–76. [CrossRef]
9. Cui, Z.Y.; Chen, T.N.; Wu, J.H.; Chen, H.L.; Zhang, B. Measurements and calculations of two-dimensional band gap structure composed of narrowly slit tubes. *Appl. Phys. Lett.* **2008**, *93*, 377. [CrossRef]
10. Pennec, Y.; Djafari-Rouhani, B.; Vasseur, J.O.; Khelif, A.; Deymier, P.A. Tunable filtering and demultiplexing in phononic crystals with hollow cylinders. *Phys. Rev. E* **2004**, *69*, 046608. [CrossRef] [PubMed]
11. Khelif, A.; Wilm, M.; Laude, V.; Ballandras, S.; Djafari-Rouhani, B. Guided elastic waves along a rod defect of a two-dimensional phononic crystal. *Phys. Rev. E* **2004**, *69*, 279–307. [CrossRef]
12. Khelif, A.; Djafarirouhani, B.; Vasseur, J.O.; Deymier, P.; Lambin, P.; Dobrzynski, L. Transmittivity through straight and stublike waveguides in a two-dimensional phononic crystal. *Phys. Rev. B* **2002**, *65*, 174308. [CrossRef]
13. Khelif, A.; Choujaa, A.; Djafari-Rouhani, B.; Wilm, M.; Ballandras, S.; Laude, V. Trapping and guiding of acoustic waves by defect modes in a full-band-gap ultrasonic crystal. *Phys. Rev. B* **2006**, *68*, 214301. [CrossRef]
14. Romero-García, V.; Sánchez Pérez, J.V.; García Raffi, L.M. Propagating and evanescent properties of double-point defect in sonic crystals. *New J. Phys.* **2010**, *12*, 083024. [CrossRef]
15. Yao, Z.J.; Yu, G.L.; Wang, Y.S.; Wang, Y.S.; Shi, Z.F. Propagation of bending waves in phononic crystal thin plates with a point defect. *Int. J. Solids Struct.* **2009**, *46*, 2571–2576. [CrossRef]
16. Wu, F.G.; Liu, Z.Y.; Liu, Y.Y. Splitting and tuning characteristics of the point defect modes in two-dimensional phononic crystals. *Phys. Rev. E* **2004**, *69*, 066609. [CrossRef] [PubMed]
17. Su, X.L.; Gao, Y.W.; Zhou, Y.H. The influence of material properties on the elastic band structures of one-dimensional functionally graded phononic crystals. *J. Appl. Phys.* **2012**, *112*, 73–83. [CrossRef]
18. Bayat, A.; Gordaninejad, F. Dynamic response of a tunable phononic crystal under applied mechanical and magnetic loadings. *Smart Mater. Struct.* **2015**, *24*, 065027. [CrossRef]
19. Zhang, S.Z.; Yao, H.; Gao, Y.W. A 2D mechanical-magneto-thermal model for direction-dependent magnetoelectric effect in laminates. *J. Magn. Magn. Mater.* **2017**, *428*, 437–447. [CrossRef]
20. Zou, X.Y.; Chen, Q.; Liang, B.; Cheng, J.C. Control of the elastic wave bandgaps in two-dimensional piezoelectric periodic structures. *Smart Mater. Struct.* **2007**, *17*, 015008. [CrossRef]
21. Wang, G.; Wang, J.W.; Chen, S.B.; Wen, J.H. Vibration attenuations induced by periodic arrays of piezoelectric patches connected by enhanced resonant shunting circuits. *Smart Mater. Struct.* **2011**, *20*, 125019. [CrossRef]
22. Bou Matar, O.; Robillard, J.F.; Vasseur, J.O.; Hladky-Hennion, A.C.; Deymier, P.A.; Pernod, P.; Preobrazhensky, V. Band gap tunability of magneto-elastic phononic crystal. *J. Appl. Phys.* **2012**, *111*, 141. [CrossRef]
23. Ding, R.; Su, X.L.; Zhang, J.J.; Gao, Y.W. Tunability of longitudinal wave band gaps in one dimensional phononic crystal with magnetostrictive material. *J. Appl. Phys.* **2014**, *115*, 074104. [CrossRef]
24. Zhang, S.Z.; Shi, Y.; Gao, Y.W. Tunability of band structures in a two-dimensional magnetostrictive phononic crystal plate with stress and magnetic loadings. *Phys. Lett. A.* **2017**, *381*, 1055–1066. [CrossRef]
25. Gu, C.L.; Jin, F. Research on the tunability of point defect modes in a two-dimensional magneto-elastic phononic crystal. *J. Phys. D-Appl. Phys.* **2016**, *49*, 175103. [CrossRef]
26. Wu, L.Y.; Chen, L.W.; Liu, C.M. Acoustic energy harvesting using resonant cavity of a sonic crystal. *Appl. Phys. Lett.* **2009**, *95*, 57. [CrossRef]

27. Lv, H.Y.; Tian, X.Y.; Wang, M.Y.; Li, D.C. Vibration energy harvesting using a phononic crystal with point defect states. *Appl. Phys. Lett.* **2013**, *102*, 034103. [CrossRef]
28. Carrara, M.; Cacan, M.R.; Toussaint, J.; Leamy, M.J.; Ruzzene, M.; Erturk, A. Metamaterial-inspired structures and concepts for elastoacoustic wave energy harvesting. *Smart Mater. Struct.* **2013**, *22*, 065004. [CrossRef]
29. Yang, A.C.; Li, P.; Wen, Y.M.; Lu, C.J.; Peng, X.; Zhang, J.T.; He, W. Enhanced acoustic wave localization effect using coupled sonic crystal resonators. *Appl. Phys. Lett.* **2014**, *104*, 151904. [CrossRef]
30. Wang, W.C.; Wu, L.Y.; Chen, L.W.; Liu, C.M. Acoustic energy harvesting by piezoelectric curved beams in the cavity of a sonic crystal. *Smart Mater. Struct.* **2010**, *19*, 126–134. [CrossRef]
31. Qi, S.B.; Oudich, M.; Li, Y.; Assouar, B. Acoustic energy harvesting based on a planar acoustic metamaterial. *Appl. Phys. Lett.* **2016**, *108*, 263501. [CrossRef]
32. *COMSOL Multiphysics 5.3a, Manual, comsol AB*; COMSOL Inc.: Stockholm, Sweden, 2017.
33. Liu, X.E.; Zheng, X.J. A nonlinear constitutive model for magnetostrictive materials. *Acta Mech. Sin.* **2005**, *21*, 278–285. [CrossRef]
34. Roundy, S.; Wright, P.K. A piezoelectric vibration based generator for wireless electronics. *Smart Mater. Struct.* **2004**, *13*, 1131. [CrossRef]
35. Li, Y.G.; Chen, T.N.; Wang, X.P.; Ma, T.; Jiang, P. Acoustic confinement and waveguiding in two-dimensional phononic crystals with material defect states. *J. Appl. Phys.* **2014**, *116*, 024904. [CrossRef]
36. Gao, Y.W.; Zhang, J.J. Nonlinear magnetoelectric transient responses of a circular-shaped magnetoelectric layered structure. *Smart Mater. Struct.* **2012**, *22*, 015015. [CrossRef]
37. Wu, T.T.; Huang, Z.G. Level repulsions of bulk acoustic waves in composite materials. *Phys. Rev. B* **2004**, *70*, 155–163. [CrossRef]
38. Butt, Z.; Pasha, R.A.; Qayyum, F.; Anjum, Z.; Ahmad, N.; Elahi, H. Generation of electrical energy using lead zirconate titanate (PZT-5A) piezoelectric material: Analytical, numerical and experimental verifications. *J. Mech. Sci. Technol.* **2016**, *30*, 3553–3558. [CrossRef]
39. Kinsler, L.E.; Frey, A.R.; Coppens, A.B.; Sanders, J.V. *Fundamentals of Acoustics*, 4th ed.; John Wiley and Sons: New York, NY, USA, 1999.

Article

A Numerical Method for Flexural Vibration Band Gaps in A Phononic Crystal Beam with Locally Resonant Oscillators

Xu Liang [1], Titao Wang [1,*], Xue Jiang [2,*], Zhen Liu [1], Yongdu Ruan [1] and Yu Deng [1]

1 Ocean College, Zhejiang University, Zhoushan 316021, China; liangxu@zju.edu.cn (X.L.); 17858802443@163.com (Z.L.); RUANYongdu@zju.edu.cn (Y.R.); dengyu_oe@zju.edu.cn (Y.D.)
2 Department of Naval Architecture, Ocean and Marine Engineering, University of Strathclyde, Glasgow G1 5AE, UK
* Correspondence: titao_wang@zju.edu.cn (T.W.); xue.jiang@strath.ac.uk (X.J.); Tel.: +86-1385-804-8340 (X.J.)

Received: 27 April 2019; Accepted: 3 June 2019; Published: 5 June 2019

Abstract: The differential quadrature method has been developed to calculate the elastic band gaps from the Bragg reflection mechanism in periodic structures efficiently and accurately. However, there have been no reports that this method has been successfully used to calculate the band gaps of locally resonant structures. This is because, in the process of using this method to calculate the band gaps of locally resonant structures, the non-linear term of frequency exists in the matrix equation, which makes it impossible to solve the dispersion relationship by using the conventional matrix-partitioning method. Hence, an accurate and efficient numerical method is proposed to calculate the flexural band gap of a locally resonant beam, with the aim of improving the calculation accuracy and computational efficiency. The proposed method is based on the differential quadrature method, an unconventional matrix-partitioning method, and a variable substitution method. A convergence study and validation indicate that the method has a fast convergence rate and good accuracy. In addition, compared with the plane wave expansion method and the finite element method, the present method demonstrates high accuracy and computational efficiency. Moreover, the parametric analysis shows that the width of the 1st band gap can be widened by increasing the mass ratio or the stiffness ratio or decreasing the lattice constant. One can decrease the lower edge of the 1st band gap by increasing the mass ratio or decreasing the stiffness ratio. The band gap frequency range calculated by the Timoshenko beam theory is lower than that calculated by the Euler-Bernoulli beam theory. The research results in this paper may provide a reference for the vibration reduction of beams in mechanical or civil engineering fields.

Keywords: phononic crystal; locally resonant; band gap; differential quadrature method

1. Introduction

Phononic crystals (PCs) are periodic composites or structures which modify the band structure in some way. The band gap is a frequency range in which the propagation of elastic waves in phononic crystal structures is suppressed. By adjusting the parameters of the artificial periodic structure, the position and width of the band gap and its ability to suppress wave propagation can be artificially regulated. In engineering practice, structures can be designed as phononic crystals, and then the band gap characteristics can be used in vibration and noise reduction.

There are two formation mechanisms of the elastic band gap in PCs: one is the Bragg scattering mechanism [1–18], and the other is the locally resonant mechanism. The wave length corresponding to the elastic band gap formed by Bragg scattering is generally equal to the lattice size or lattice constant, which restricts its application in engineering practice. In 1999, Liu et al. [19] proposed the locally resonant mechanism of the elastic band gap. It was found that the corresponding wave length of this locally resonant phononic crystal is far larger than the lattice size, which breaks through the limitation of the Bragg scattering mechanism and opens up a wide potential for application in low-frequency wave band. Since then, investigations on the locally resonant mechanism of phononic crystals have been carried out continuously [20–23].

The study of the band gap mechanism of PCs depends on the solving methods of elastic wave band gaps. At present, computational methods of vibration band gaps mainly include the transfer-matrix method (TM), multiple-scattering theory (MST), finite difference time-domain method (FDTD), lumped-mass method (LM), plane wave expansion method (PWE), differential quadrature method (DQM), finite element method (FEM), extended plane wave expansion method (EPWE) [24], wave finite element method [25], and boundary element method (BE) [26]. The transfer-matrix method [27,28] is widely applied to calculate the band gap characteristics of 1-D PCs. Although the analytical solution can be obtained quickly, it is not suitable for the study of the vibration dispersion relations of 2-D and 3-D periodic structures, since the transfer matrix can usually only be transmitted in one direction. The PWE method [29,30] is the most basic and common method to calculate the band gap characteristics of phononic crystals. It can be used to solve the elastic band gap of all-dimensional PCs. However, when the material parameters vary greatly, or the filling rate is too high or too low, it is difficult to achieve convergence [31]. It is worth mentioning that in order to overcome the limitation of the convergence of PWE, the improved plane wave expansion (IPWE) method [26] is proposed. The FDTD method [32,33] can calculate the transmission, reflection, and energy band characteristics of infinite structures. However, the computational amount is large, and the large elastic constant difference may lead to numerical instability and divergence. The multiple scattering theory [34,35] can not only calculate the dispersion curves of periodic structures, but also disordered structures. It has fast convergence and high accuracy and is easy to deal with the problem of elastic mismatch, but only some kinds of scatterers with regular shape can be dealt with. At present, only phononic crystals composed of cylindrical (two-dimensional) and spherical (three-dimensional) scatterers can be calculated. The lumped mass method [36,37] converges fast, but there are some difficulties in dealing with multi-field coupling problems. The FEM method [38–40] is a commonly used method in engineering with good applicability, wide application range, and good convergence. There are various kinds of mature commercial software, such as Comsol, ANSYS, etc., which facilitate the modeling and analysis of complex periodic structures. The finite element method can be used to accurately calculate the band gap characteristics of PCs of various dimensions and various shapes of scatterers.

The differential quadrature method can approximate the value of each derivative of the function at the node with the weighted sum of the function values at all nodes in the domain and transforms the problem of the continuous system into a discrete problem. It has strong convergence and high precision. To the authors' knowledge, the DQM was only applied to solve the elastic wave band gap of a beam or plate structure with the Bragg scattering mechanism [8,9]. However, the Bloch boundary conditions for locally resonant structures lead to nonlinear fundamental equations when using DQM, which causes difficulties in solving. Hence, numerical solutions of band gaps in LR structures based on DQM have not been reported up to now.

In this work, through applying an unconventional matrix-partitioning method and a variable substitution method, we transform the problem of solving the dispersion relation into a quadratic eigenvalue problem and propose a numerical method based on the differential quadrature method to calculate the bending vibration band gap of locally resonant beams. The purpose is to improve calculation accuracy and computational efficiency. Moreover, a parametric study is undertaken to investigate the effects of shear deformation and rotary inertia, the lumped mass, the spring stiffness coefficient, and the lattice size on the 1st band gap. Some major novelties of the contribution are pointed out as follows:

(1) Based on the differential quadrature method, we propose a numerical method for calculating the flexural vibration band gaps of a locally resonant beam. In this paper, the differential quadrature method is applied to calculate the band gaps of locally resonant structures for the first time.
(2) By using an unconventional matrix-partitioning method and a variable substitution method, we can transform the problem of solving the dispersion relation into a standard quadratic eigenvalue problem, and the problem of the non-linear term after using the DQM method can be solved easily.
(3) Compared with the plane wave expansion method and the finite element method, the high accuracy and computational efficiency of the present method are demonstrated.
(4) The proposed method has high precision and a rapid speed of convergence.

2. Method

2.1. Differential Quadrature Method

The basic idea of the differential quadrature method is to use the sum of the weighted values at all discrete points in the computational domain to approximate the unknown function values and their derivatives at any discrete points, so that the continuous system can be transformed into a discrete system for solving as follows.

$$f(\varepsilon) = \sum_{j=1}^{N_x} p_j(\varepsilon) f_j, \tag{1}$$

$$\frac{\partial^r f(\varepsilon_i)}{\partial x^r} = \sum_{i=1}^{N_x} A_{ij}^{(r)} \cdot f(\varepsilon_i), \tag{2}$$

where N_x is the total number of discrete points in the calculation domain; $i, j = 1, 2, \ldots, N_x$; $f(\varepsilon_j) = f_j$; $p_j(\varepsilon)$ are the Lagrange interpolation polynomials, and A_{ij}^r is the weight coefficient of the rth derivative defined by [41]:

$$A_{ij}^{(1)} = \frac{\prod\limits_{r=1, r\neq i}^{N_x} (\varepsilon_i - \varepsilon_r)}{(\varepsilon_i - \varepsilon_j) \prod\limits_{r=1, r\neq j}^{N_x} (\varepsilon_j - \varepsilon_r)}, \tag{3}$$

$$A_{ij}^{(r)} = r\left(A_{ij}^{(r-1)} A_{ij}^{1} - \frac{A_{ij}^{(r-1)}}{x_i - x_j}\right), \tag{4}$$

where $r = 1, 2, \ldots, N_{x-1}$; $i,j = 1, 2, \ldots, N_x$, $(i \neq j)$; and $A_{ii}^{(r)}$ are defined as:

$$A_{ii}^{(r)} = -\sum_{j=1, j\neq i}^{N_x} A_{ij}^{(r)}, \tag{5}$$

In order to obtain higher accuracy and faster convergence, the non-uniform mesh partition (Gauss-Lobatto-Chebyshev pattern) [42] is adopted in this paper. The coordinates of discrete points are as follows:

$$\varepsilon_i = -\cos\left(\frac{\pi(i-1)}{N_x - 1}\right), \; i = 1, 2, \ldots, N_x, \tag{6}$$

2.2. Unconventional Matrix-Partitioning Method & Variable Substitution Method

In the process of using the DQM method to calculate the band gaps of a locally resonant structure, a non-linear term of frequency exists in the matrix equation, which makes it impossible to solve the dispersion relationship by using the conventional matrix-partitioning method. Hence, we propose an unconventional matrix-partitioning method and a variable substitution method to calculate the flexural band gap of a locally resonant beam. The idea is to separate the nonlinear term from the matrix by unconventional matrix partitioning and variable substitution, and then transform the matrix equation into a quadratic eigenvalue problem. The proposed method is applicable to both the Euler–Bernoulli beam model and the Timoshenko beam model. The detailed application process is shown in Section 3.2. The following is a concise and general formulation of the proposed method.

$$\begin{bmatrix} K_{qq}(\omega) & \mathbf{K}_{qb} & \mathbf{K}_{qd} \\ \mathbf{K}_{bq} & \mathbf{K}_{bb} & \mathbf{K}_{bd} \\ \mathbf{K}_{dq} & \mathbf{K}_{db} & \mathbf{K}_{dd} \end{bmatrix} \cdot \begin{pmatrix} U_q \\ \mathbf{U}_b \\ \mathbf{U}_d \end{pmatrix} - \omega^2 \begin{bmatrix} 0 & 0 & 0 \\ 0 & 0 & 0 \\ 0 & 0 & \mathbf{M}_{dd} \end{bmatrix} \cdot \begin{pmatrix} U_q \\ \mathbf{U}_b \\ \mathbf{U}_d \end{pmatrix} = \mathbf{0}, \tag{7}$$

where $\mathbf{U}$ is a vector of independent variables. Subscripts "q", "b", and "d" refer to the shear boundary condition, other boundary conditions, and the domain, respectively. $K_{qq}(\omega)$ is a scalar with the non-linear term. By variable substitutions and matrix operations, Equation (7) can be transformed into a quadratic eigenvalue problem.

$$r_z^2 \mathbf{H}_2 \cdot \mathbf{U}_d + r_z \mathbf{H}_1 \cdot \mathbf{U}_d + \mathbf{H}_0 \cdot \mathbf{U}_d = 0, \tag{8}$$

$$\mathbf{H}_f = -\begin{bmatrix} \mathbf{0} & \mathbf{H}_2 \\ -\mathbf{I} & \mathbf{0} \end{bmatrix}^{-1} \cdot \begin{bmatrix} \mathbf{H}_0 & \mathbf{H}_1 \\ \mathbf{0} & \mathbf{I} \end{bmatrix}, \tag{9}$$

$$\omega^2 = (-1/m_z c_4) r_z + (k_z/m_z + c_1/m_z c_4), \tag{10}$$

For any given wave vector k in the first Brillouin zone, r_z can be obtained by calculating the eigenvalue of matrix H_f. One can get the circular frequency ω due to Equation (10), thus the dispersion relation and bending vibration band gaps can be plotted.

3. Locally Resonant (LR) Beam Models and Solutions

Figure 1 shows the configuration of a straight elastic metamaterial beam with periodical locally resonant (LR) oscillator structures. Harmonic locally resonant oscillators are periodically connected along the x-axis to the infinite Euler beam. Each oscillator is formed by a lumped mass m_z and a spring with stiffness k_z, taking the distance between two adjacent LR oscillators as lattice size a. By extending the Euler-Bernoulli beam theory and the Timoshenko beam theory, one can obtain the theoretical model of the LR beam. Since the structure has infinite periodicity, we can apply the Floquet-Bloch theorem to simplify the whole model into a unit cell.

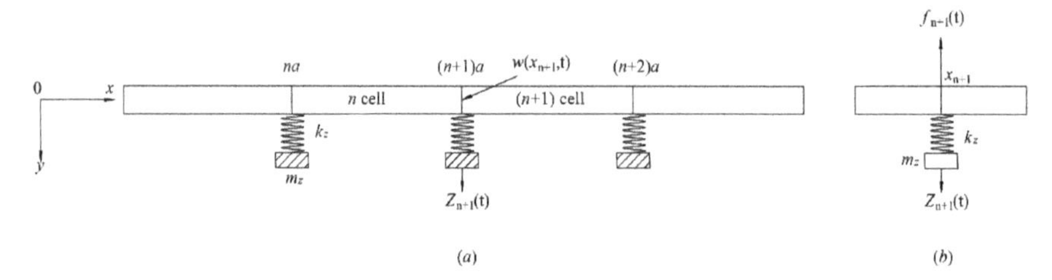

Figure 1. (**a**) Configuration of a straight elastic metamaterial beam with locally resonant (LR) oscillators. (**b**) Diagram of the force equilibrium of the $(n+1)^{\text{th}}$ LR oscillator.

3.1. Euler–Bernoulli Model & Solution Procedures

When the length of each beam unit cell is much larger than its height and width, the Euler-Bernoulli approximation is satisfied, thus the influences of shear force and rotary inertia can be ignored. The governing equation for the bending vibration of the nth cell is shown below [43]:

$$EI\frac{\partial^4 w_n(x,t)}{\partial x^4} + \rho A\frac{\partial^2 w_n(x,t)}{\partial t^2} = 0, \tag{11}$$

where ρ is the density; E is Young's modulus; A is the cross-section area; I is the area moment of inertia with respect to the axis perpendicular to the beam axis, and $w(x,t)$ is the lateral displacement at x. By assuming $w(x,t) = W(x)\exp(-i\omega t)$, where $W(x)$ is the vibration amplitude of the beam at x, and ω is the circular frequency, Equation (11) can be rewritten as follows:

$$EI\frac{\partial^4 W_n(x)}{\partial x^4} - \omega^2 \rho A W_n(x) = 0, \tag{12}$$

For the $(n+1)^{\text{th}}$ locally resonant oscillator, consider the balance of forces in the y-axis direction, we can get:

$$f_{n+1}(t) - m_z \ddot{Z}_{n+1}(t) = 0, \tag{13}$$

where $f_{n+1}(t)$ is the interaction force between the beam and the oscillator at node x_{n+1}, $Z_{n+1}(t) = V_{n+1}\exp(-i\omega t)$ is the displacement of the lumped mass of the $(n+1)^{\text{th}}$ oscillator, and the vibration amplitude of the $(n+1)^{\text{th}}$ oscillator is denoted by the absolute value of V_{n+1}.

According to Hooke's law, $f_{n+1}(t)$ can be expressed as follows:

$$f_{n+1}(t) = k_z[w(x_{n+1},t) - Z_{n+1}(t)] = k_z[W_{n+1}(0) - V_{n+1}]\exp(-i\omega t) \triangleq F_{n+1}\exp(-i\omega t), \tag{14}$$

Substituting Equation (14) into Equation (13), one can get:

$$V_{n+1} = \frac{k_z}{k_z - m_z\omega^2} W_{n+1}(0), \tag{15}$$

Ignoring the stress concentration between two adjacent units, the following boundary conditions can be listed according to the continuity of displacement, angle of rotation, bending moment, and shear force at the node x_{n+1}.

$$W_{n+1}(0) = W_n(a), \tag{16}$$

$$W'_{n+1}(0) = W'_n(a), \tag{17}$$

$$EIW''_{n+1}(0) = EIW''_n(a), \tag{18}$$

$$EIW'''_{n+1}(0) - F_{n+1} = EIW'''_n(a), \tag{19}$$

According to the Floquet-Bloch theorem [44], Equations (16)–(19) can be rewritten as follows:

$$e^{ika} W_n(0) = W_n(a), \tag{20}$$

$$e^{ika} W'_n(0) = W'_n(a), \tag{21}$$

$$e^{ika} EIW''_n(0) = EIW''_n(a), \tag{22}$$

$$e^{ika} EIW'''_n(0) - e^{ika} F_n = EIW'''_n(a), \tag{23}$$

where k is the Bloch wave vector, also known as the wave number.

To facilitate the application of DQM conveniently, the computational domain of each cell needs to be converted to a standardized computational domain [[−1,1] by the reversible transformation below:

$$\varepsilon = \frac{x - x_n^l}{L_x} - 1, \tag{24}$$

where n is the number of the cell; x_n^l represents the coordinates of the left side of the nth beam unit; L_x is equal to $0.5a$, and ε is the local coordinate in the normalized computational domain.

Substituting Equation (24) into Equation (12) and Equations (20)–(23), the governing equation and boundary conditions in the normalized computational domain can be obtained.

$$\frac{EI}{L_x^4}\frac{\partial^4 W_n(\varepsilon)}{\partial \varepsilon^4} - \omega^2 \rho A W_n(\varepsilon) = 0, \tag{25}$$

$$e^{ika} W_n(-1) - W_n(1) = 0, \tag{26}$$

$$e^{ika} W_n'(-1) - W_n'(1) = 0, \tag{27}$$

$$e^{ika} W_n''(-1) - W_n''(1) = 0, \tag{28}$$

$$e^{ika} W_n'''(-1) - W_n'''(1) - e^{ika}\frac{L_x^3}{EI}\cdot\frac{m_z k_z \omega^2}{k_z - m_z\omega^2} W_n(-1) = 0, \tag{29}$$

Substituting Equations (1)–(6) into Equations (25)–(29), the boundary conditions and governing equations discretized by DQM can be obtained as below:

$$\frac{EI}{L_x^4}\sum_{j=1}^{N_x} A_{ij}^{(4)} W_j - \omega^2 \rho A W_i = 0,\ i = 3, 4, \ldots, N_x - 2, \tag{30}$$

$$e^{ika} W_1 - W_{Nx} = 0, \tag{31}$$

$$e^{ika}\sum_{j=1}^{N_x} A_{1j}^{(1)} W_j - \sum_{j=1}^{N_x} A_{Nxj}^{(1)} W_j = 0, \tag{32}$$

$$e^{ika}\sum_{j=1}^{N_x} A_{1j}^{(2)} W_j - \sum_{j=1}^{N_x} A_{Nxj}^{(2)} W_j = 0, \tag{33}$$

$$e^{ika}\sum_{j=1}^{N_x} A_{1j}^{(3)} W_j - \sum_{j=1}^{N_x} A_{Nxj}^{(3)} W_j - e^{ika}\frac{L_x^3}{EI}\cdot\frac{m_z k_z \omega^2}{k_z - m_z\omega^2} W_1 = 0, \tag{34}$$

Equations (30)–(34) can be expressed as a matrix equation form as shown below. Because the third term on the left side of Equation (34) is the non-linear term of ω^2, it is impossible to solve the relationship between wave vector k and ω by using the conventional matrix-partitioning method. Next, we use the proposed unconventional matrix-partitioning method and the variable substitution method to solve this problem.

$$\begin{bmatrix} K_{qq}(\omega) & \mathbf{K}_{qb} & \mathbf{K}_{qd} \\ \mathbf{K}_{bq} & \mathbf{K}_{bb} & \mathbf{K}_{bd} \\ \mathbf{K}_{dq} & \mathbf{K}_{db} & \mathbf{K}_{dd} \end{bmatrix} \cdot \begin{pmatrix} U_q \\ \mathbf{U}_b \\ \mathbf{U}_d \end{pmatrix} - \omega^2 \begin{bmatrix} 0 & 0 & 0 \\ 0 & 0 & 0 \\ 0 & 0 & \mathbf{M}_{dd} \end{bmatrix} \cdot \begin{pmatrix} U_q \\ \mathbf{U}_b \\ \mathbf{U}_d \end{pmatrix} = 0, \tag{35}$$

$$\mathbf{M}_{dd} = \rho A \mathbf{I}, \tag{36}$$

where $U_q = W_1$; $\mathbf{U}_b = (W_2, W_{N-1}, W_N)^T$; $U_d = (W_3, W_4, \ldots, W_{N-2})^T$; $\mathbf{K}_{qb}$ is a 1×3 vector; $\mathbf{K}_{qd}$ is a $1 \times (N-4)$ vector; $\mathbf{K}_{bq}$ is a 3×1 vector; $\mathbf{K}_{bb}$ is a 3×3 matrix, which is always invertible; $\mathbf{K}_{bd}$ is a $3 \times (N-4)$ matrix; $\mathbf{K}_{dq}$ is a $(N-4) \times 1$ vector; $\mathbf{K}_{db}$ is a $(N-4) \times 3$ matrix; $\mathbf{K}_{dd}$ is a $(N-4) \times (N-4)$ matrix; $\mathbf{M}_{dd}$ is a $(N-4) \times (N-4)$ matrix; K_{qq} is a scalar as shown below:

$$K_{qq} = \frac{c_1}{p_z} + c_2, \tag{37}$$

where $p_z = k_z - m_z\omega^2$; $c_1 = -e^{ika} k_z^2 L_x^3 / EI$; $c_2 = e^{ika} A_{1,1}^{(3)} - A_{N,1}^{(3)} + e^{ika} k_z L_x^3 / EI$; Performing a partitioned matrix operation on Equation (35), the following three matrix equations can be obtained:

$$\left(\frac{c_1}{p_z}+c_2\right)U_q+\mathbf{K}_{qb}\cdot\mathbf{U}_b+\mathbf{K}_{qd}\cdot\mathbf{U}_d=0, \tag{38}$$

$$\mathbf{K}_{bq}U_q+\mathbf{K}_{bb}\cdot\mathbf{U}_b+\mathbf{K}_{bd}\cdot\mathbf{U}_d=0, \tag{39}$$

$$\mathbf{K}_{dq}U_q+\mathbf{K}_{db}\cdot\mathbf{U}_b+\mathbf{K}_{dd}\cdot\mathbf{U}_d-\omega^2\rho A\mathbf{I}\cdot\mathbf{U}_d=0, \tag{40}$$

Performing matrix operation and simplification on Equation (39), and using $\mathbf{U}_d$ and U_q to represent $\mathbf{U}_b$.

$$\mathbf{U}_b=\mathbf{S}_1\cdot U_q+\mathbf{S}_2\cdot\mathbf{U}_d, \tag{41}$$

where $\mathbf{S}_1=-\mathbf{K}_{bb}^{-1}\cdot\mathbf{K}_{bq}$; $\mathbf{S}_2=-\mathbf{K}_{bb}^{-1}\cdot\mathbf{K}_{bd}$, substituting Equation (41) into Equation (38) for calculation and simplification, the following equation can be obtained:

$$U_q=\left(\frac{1}{r_z}\mathbf{S}_5+\mathbf{S}_6\right)\cdot\mathbf{U}_d, \tag{42}$$

where $\mathbf{S}_5=(c_1/c_4)\mathbf{S}_4$; $\mathbf{S}_6=(-1/c_4)\mathbf{S}_4$; $r_z=c_4p_z+c_1$; $c_4=c_2+\mathbf{K}_{qb}\cdot\mathbf{S}_1$; and $\mathbf{S}_4=\mathbf{K}_{qd}+\mathbf{K}_{qb}\cdot\mathbf{S}_2$. Substituting Equation (42) into Equation (41), one can get:

$$\mathbf{U}_b=\left(\frac{1}{r_z}\mathbf{S}_7+\mathbf{S}_8\right)\cdot\mathbf{U}_d, \tag{43}$$

where $\mathbf{S}_7=\mathbf{S}_1\cdot\mathbf{S}_5$; and $\mathbf{S}_8=\mathbf{S}_1\cdot\mathbf{S}_6+\mathbf{S}_2$. Substituting Equation (42) and (43) into Equation (40), one can obtain a standard quadratic eigenvalue equation after simplification.

$$r_z^2\mathbf{H}_2\cdot\mathbf{U}_d+r_z\mathbf{H}_1\cdot\mathbf{U}_d+\mathbf{H}_0\cdot\mathbf{U}_d=0, \tag{44}$$

$$\mathbf{H}_f=-\begin{bmatrix}\mathbf{0} & \mathbf{H}_2\\ -\mathbf{I} & \mathbf{0}\end{bmatrix}^{-1}\cdot\begin{bmatrix}\mathbf{H}_0 & \mathbf{H}_1\\ \mathbf{0} & \mathbf{I}\end{bmatrix}, \tag{45}$$

$$\omega^2=(-1/m_zc_4)r_z+(k_z/m_z+c_1/m_zc_4), \tag{46}$$

where $\mathbf{H}_2=(\rho A/m_zc_4)\mathbf{I}$; $\mathbf{H}_1=\mathbf{K}_{dq}\cdot\mathbf{S}_6+\mathbf{K}_{db}\cdot\mathbf{S}_8+\mathbf{K}_{dd}-\rho A(k_z/m_z+c_1/m_zc_4)\mathbf{I}$; and $\mathbf{H}_0=\mathbf{K}_{dq}\cdot\mathbf{S}_5+\mathbf{K}_{db}\cdot\mathbf{S}_7$. For any given wave vector k in the first Brillouin zone, r_z can be got by calculating the eigenvalue of matrix $\mathbf{H}_f$. One can get the circular frequency ω due to Equation (46), thus the dispersion relation and bending vibration band gaps can be plotted.

3.2. Timoshenko Model & Solution Procedures

For deep beams, the effects of transverse shear deformation and rotary inertia must be considered. Based on the Timoshenko beam theory, the governing equation for the bending vibration of the nth cell is shown below [8]:

$$k_sGA\left(\frac{\partial\varphi_n(x)}{\partial x}-\frac{\partial^2W_n(x)}{\partial x^2}\right)-\omega^2\rho AW_n(x)=0, \tag{47}$$

$$k_sGA\left(\varphi_n(x)-\frac{\partial W_n(x)}{\partial x}\right)-EI\frac{\partial^2\varphi_n(x)}{\partial x^2}-\omega^2\rho I\varphi_n(x)=0, \tag{48}$$

where k_s is the shear coefficient; G is the shear modulus, and φ is the rotation of the cross-section. Ignoring the stress concentration between two adjacent units, the following boundary conditions can be listed according to the continuity of displacement, angle of rotation, bending moment, and shear force at the node x_{n+1}.

$$W_{n+1}(0)=W_n(a), \tag{49}$$

$$\varphi_{n+1}(0)=\varphi_n(a), \tag{50}$$

$$EI\varphi'_{n+1}(0) = EI\varphi'_n(a), \tag{51}$$

$$k_sGA\left[\varphi_{n+1}(0) - W'_{n+1}(0)\right] - F_{n+1} = k_sGA[\varphi_n(a) - W'_n(a)], \tag{52}$$

According to the Floquet–Bloch theorem [44], Equations (49)–(52) can be rewritten as follows:

$$e^{ika}W_n(0) = W_n(a), \tag{53}$$

$$e^{ika}\varphi_n(0) = \varphi_n(a), \tag{54}$$

$$e^{ika}EI\varphi'_n(0) = EI\varphi'_n(a), \tag{55}$$

$$e^{ika}k_sGA[\varphi_n(0) - W'_n(0)] - e^{ika}F_n = k_sGA[\varphi_n(a) - W'_n(a)], \tag{56}$$

Substituting Equation (24) into Equation (47), (48) and Equations (53)–(56), the governing equation and boundary conditions in the normalized computational domain can be obtained.

$$k_sGA\left(\frac{1}{L_x}\frac{\partial\varphi_n(\varepsilon)}{\partial\varepsilon} - \frac{1}{L_x^2}\frac{\partial^2W_n(\varepsilon)}{\partial\varepsilon^2}\right) - \omega^2\rho AW_n(\varepsilon) = 0, \tag{57}$$

$$k_sGA\left(\varphi_n(\varepsilon) - \frac{1}{L_x}\frac{\partial W_n(\varepsilon)}{\partial\varepsilon}\right) - \frac{EI}{L_x^2}\frac{\partial^2\varphi_n(\varepsilon)}{\partial\varepsilon^2} - \omega^2\rho I\varphi_n(\varepsilon) = 0, \tag{58}$$

$$e^{ika}W_n(-1) - W_n(1) = 0, \tag{59}$$

$$e^{ika}\varphi_n(-1) - \varphi_n(1) = 0, \tag{60}$$

$$e^{ika}\varphi'_n(-1) - \varphi'_n(1) = 0, \tag{61}$$

$$e^{ika}\left[\varphi_n(-1) - \frac{1}{L_x}W'_n(-1)\right] - \left[\varphi_n(1) - \frac{1}{L_x}W'_n(1)\right] - e^{ika}\frac{1}{k_sGA}\frac{m_zk_z\omega^2}{k_z - m_z\omega^2}W_n(-1) = 0, \tag{62}$$

Substituting Equations (1)–(6) into Equations (57)–(62), the boundary conditions and governing equations discretized by DQM can be obtained as below:

$$\frac{k_sGA}{L_x}\sum_{j=1}^{N_x}A_{ij}^{(1)}\varphi_j - \frac{k_sGA}{L_x^2}\sum_{j=1}^{N_x}A_{ij}^{(2)}W_j - \omega^2\rho AW_i = 0,\ i = 2,3,\ldots,N_x - 1, \tag{63}$$

$$k_sGA\varphi_i - \frac{k_sGA}{L_x}\sum_{j=1}^{N_x}A_{ij}^{(1)}W_j - \frac{EI}{L_x^2}\sum_{j=1}^{N_x}A_{ij}^{(2)}\varphi_j - \omega^2\rho I\varphi_i = 0,\ i = 2,3,\ldots,N_x - 1, \tag{64}$$

$$e^{ika}W_1 - W_{Nx} = 0, \tag{65}$$

$$e^{ika}\varphi_1 - \varphi_{Nx} = 0, \tag{66}$$

$$e^{ika}\sum_{j=1}^{N_x}A_{1j}^{(1)}\varphi_j - \sum_{j=1}^{N_x}A_{Nxj}^{(1)}\varphi_j = 0, \tag{67}$$

$$e^{ika}\left[\varphi_1 - \frac{1}{L_x}\sum_{j=1}^{N_x}A_{1j}^{(1)}W_j\right] - \left[\varphi_{N_x} - \frac{1}{L_x}\sum_{j=1}^{N_x}A_{N_xj}^{(1)}W_j\right] - e^{ika}\frac{1}{k_sGA}\frac{m_zk_z\omega^2}{k_z - m_z\omega^2}W_1 = 0, \tag{68}$$

Equations (63)–(68) can be expressed as a matrix equation form as shown below. Because the third term on the left side of Equation (68) is the non-linear term of ω^2, it is impossible to solve the relationship between wave vector k and ω by using the conventional matrix-partitioning method. Next, we use the proposed unconventional matrix-partitioning method and the variable substitution method to solve this problem.

$$\begin{bmatrix} K_{qq}(\omega) & \mathbf{K}_{qb} & \mathbf{K}_{qd} \\ \mathbf{K}_{bq} & \mathbf{K}_{bb} & \mathbf{K}_{bd} \\ \mathbf{K}_{dq} & \mathbf{K}_{db} & \mathbf{K}_{dd} \end{bmatrix} \cdot \begin{pmatrix} U_q \\ \mathbf{U}_b \\ \mathbf{U}_d \end{pmatrix} - \omega^2 \begin{bmatrix} 0 & 0 & 0 \\ 0 & 0 & 0 \\ 0 & 0 & \mathbf{M}_{dd} \end{bmatrix} \cdot \begin{pmatrix} U_q \\ \mathbf{U}_b \\ \mathbf{U}_d \end{pmatrix} = \mathbf{0}, \tag{69}$$

$$\mathbf{M}_{dd} = \begin{bmatrix} \rho A\mathbf{I} & \mathbf{0} \\ \mathbf{0} & \rho I\mathbf{I} \end{bmatrix}, \tag{70}$$

where $U_q = W_1$; $\mathbf{U}_b = (W_N, \varphi_1, \varphi_N)^T$; $U_d = (W_2, W_3, \ldots, W_{N-1}, \varphi_2, \varphi_3, \ldots, \varphi_{N-1})^T$; $\mathbf{K}_{qb}$ is a 1×3 vector; $\mathbf{K}_{qd}$ is a $1 \times (2N-4)$ vector; $\mathbf{K}_{bq}$ is a 3×1 vector; $\mathbf{K}_{bb}$ is a 3×3 matrix; $\mathbf{K}_{bd}$ is a $3 \times (2N-4)$ matrix; $\mathbf{K}_{dq}$ is a $(2N-4) \times 1$ vector; $\mathbf{K}_{db}$ is a $(2N-4) \times 3$ matrix; $\mathbf{K}_{dd}$ is a $(2N-4) \times (2N-4)$ matrix; $\mathbf{M}_{dd}$ is a $(2N-4) \times (2N-4)$ matrix; K_{qq} is a scalar as shown below:

$$K_{qq} = \frac{c_1}{p_z} + c_2, \tag{71}$$

where $p_z = k_z - m_z\omega^2$; $c_1 = -e^{ika}k_z^2/k_sGA$; $c_2 = -e^{ika}A_{1,1}^{(1)}/L_x - A_{N_x,1}^{(1)}/L_x + e^{ika}k_z/k_sGA$; Performing a partitioned matrix operation on Equation (69), the following three matrix equations can be obtained:

$$\left(\frac{c_1}{p_z} + c_2\right)U_q + \mathbf{K}_{qb} \cdot \mathbf{U}_b + \mathbf{K}_{qd} \cdot \mathbf{U}_d = 0, \tag{72}$$

$$\mathbf{K}_{bq}U_q + \mathbf{K}_{bb} \cdot \mathbf{U}_b + \mathbf{K}_{bd} \cdot \mathbf{U}_d = 0, \tag{73}$$

$$\mathbf{K}_{dq}U_q + \mathbf{K}_{db} \cdot \mathbf{U}_b + \mathbf{K}_{dd} \cdot \mathbf{U}_d - \omega^2\mathbf{M}_{dd} \cdot \mathbf{U}_d = 0, \tag{74}$$

The following procedure is similar to the Euler model. Substituting Equations (41)–(43) into Equations (72)–(74), one can obtain a standard quadratic eigenvalue equation after simplification.

$$r_z^2\mathbf{H}_2 \cdot \mathbf{U}_d + r_z\mathbf{H}_1 \cdot \mathbf{U}_d + \mathbf{H}_0 \cdot \mathbf{U}_d = 0, \tag{75}$$

For any given wave vector k in the first Brillouin zone, r_z can be got by calculating the eigenvalue of matrix $\mathbf{H}_f$. One can get the circular frequency ω from the Equation (46), thus the dispersion relation and bending vibration band gaps can be plotted.

4. Numerical Results and Discussions

In this section, we first gave a study on the convergence of the proposed method. Next, the correctness and accuracy of the present method were verified by comparing the results with the existing literature. It is worth mentioning that, compared with the plane wave expansion method and the finite element method, the present method demonstrated high accuracy and computational efficiency. Finally, the parameter analysis was carried out to discuss the effects of shear deformation and rotary inertia, the lumped mass m_z, the spring stiffness coefficient k_z, and the lattice size a on the band gap.

Yu et al. [45] used the transfer-matrix method to calculate the bending vibration band gap of a LR beam. For the convenience of comparison and analysis, Yu et al.'s [45] geometry and material parameters of the LR beam are adopted. Figure 2 illustrates a straight beam with locally resonant oscillators. The material of the LR beam is aluminum, and the shape of the cross-section is a ring with outer radius and inner radius $r_1 = 1 \times 10^{-2}$ m and $r_0 = 7 \times 10^{-3}$ m, respectively. Each locally resonant oscillator consists of a rubber ring and a copper ring coated on the outside, and their outer radii are $r_2 = 1.5 \times 10^{-2}$ m and $r_3 = 1.95 \times 10^{-2}$ m, respectively. Both rings have the same width $l_z = 1 \times 10^{-2}$ m, and the lattice constant of the LR beam is taken as $a = 7.5 \times 10^{-2}$ m. Material parameters of the LR beam are listed in Table 1. Unless otherwise specified, values of the parameters in the following studies are consistent with those here.

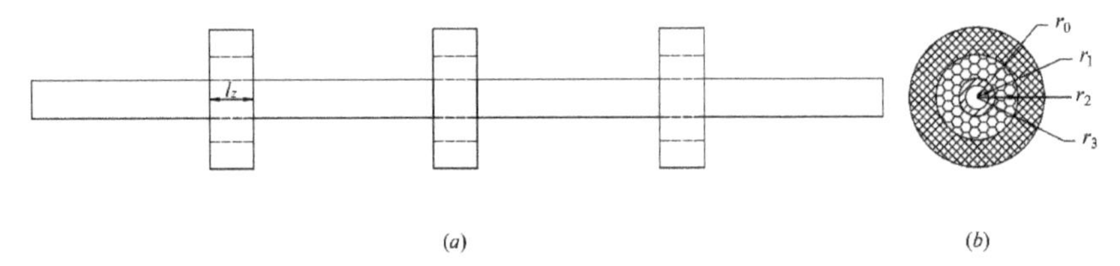

Figure 2. (**a**) Illustration of a straight beam with locally resonant oscillators. (**b**) The sketch of the locally resonant oscillator.

Table 1. Material parameters.

Parameters	Items	Values
ρ_{Al}	Aluminum density (kg/m^3)	2600
E_{Al}	Elastic modulus of aluminum (Pa)	7×10^{10}
ρ_{rubber}	Rubber density (kg/m^3)	1300
E_{rubber}	Elastic modulus of rubber (Pa)	7.7×10^5
G_{rubber}	Shear modulus of rubber (Pa)	2.6×10^5
ρ_{Cu}	Copper density (kg/m^3)	8950

For a ring rubber, the radial equivalent stiffness can be expressed as follows:

$$k_z = \frac{\pi(3.29H_z^2 + 5)G_{rubber}l_z}{\ln(r_2/r_1)}, \tag{76}$$

where $H_z = 1/(r_1 + r_2)ln(r_2/r_1)$ is the shape coefficient.

4.1. Convergence Study

To study the convergence of the present method proposed in this work, Table 2 shows the fundamental frequencies of the locally resonant beam. We give a series of results corresponding to various numbers of discrete points N_x. It is found that when the number of sampling points $N_x \geq 8$, the frequency converges rapidly. In the rest part of this paper, 15 discrete points are selected to calculate and analyze the bending vibration band gap characteristics of LR beams in order to ensure good accuracy.

Table 2. Fundamental frequency of a locally resonant (LR) beam.

Wave Vector k	Present Frequency (Hz)				
	$N_x = 6$	$N_x = 7$	$N_x = 8$	$N_x = 9$	$N_x = 15$
0.0	0.000	0.000	0.000	0.000	0.000
0.3	277.547	277.635	277.638	277.638	277.638
0.5	304.968	304.972	304.973	304.973	304.973
0.7	308.056	308.057	308.057	308.057	308.057
1.0	308.722	308.722	308.722	308.722	308.722

4.2. Validation

As Figure 3 shows, the curves and scatters represent the dispersion relationship of a straight beam with LR oscillator structures. The dispersion relationship curve does not cover the full frequency range and the frequency range through which no dispersion curve covers are represented by a shadow zone, which is the band gap. The bending wave in the band gap frequency range cannot propagate through the beam. The 1st band gap locates between 308.722–478.943 H_z, and the range obtained by Yu et al. [45] is 309.1–479.4 H_z. From a qualitative point of view, the scatter points are all on the curve.

From a quantitative point of view, the relative error between the present results and Yu et al.'s is within 0.13%. The above proves that our numerical solutions match well with those in the existing literature.

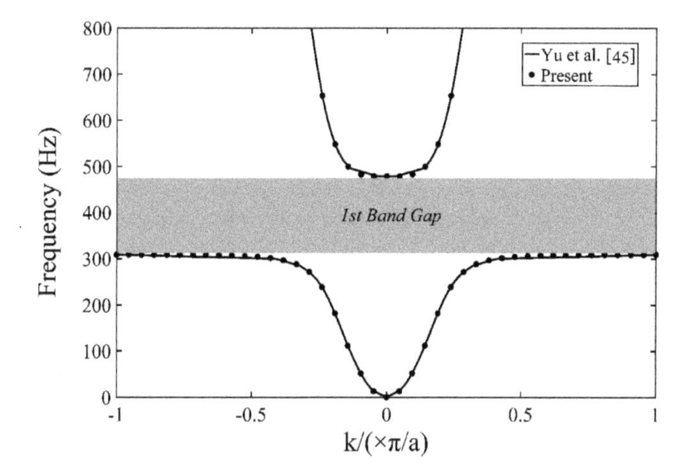

Figure 3. Dispersion relationship and band gaps for a locally resonant (LR) beam.

Consider a finite locally resonant beam consisting of eight preceding periodic cells, in which a harmonic displacement excitation in the y-direction between 0 and 800 Hz is applied at one end and the frequency response function (FRF) at the other end is shown in Figure 4a. The solid line represents the frequency response function and the dashed line represents the input spectrum (0 dB). Within the frequency range marked by a double arrow, the FRF has a maximum attenuation of more than 60 dB. For infinite structures, the imaginary part of k causes attenuating vibration. The larger the absolute value of the imaginary part, the stronger the spatial attenuation of the evanescent wave. As shown in Figure 4b, the band gap frequency range of the infinite structure is in good agreement with that of the finite structure, which also proves the correctness of the present method from the perspective of the evanescent modes.

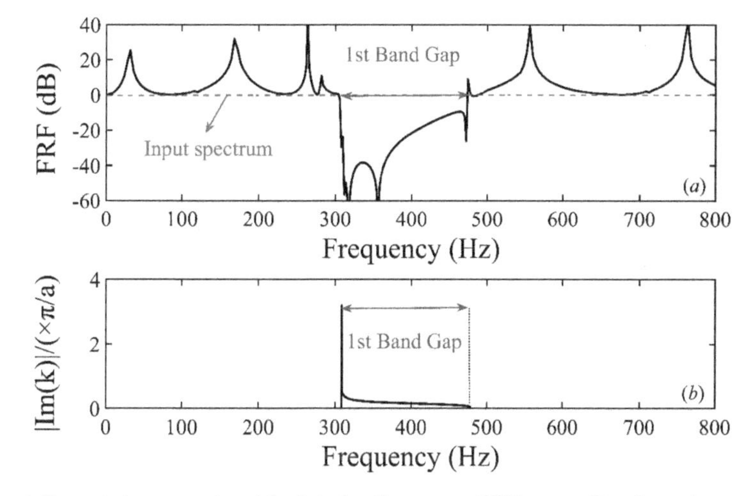

Figure 4. Transmission properties of the finite locally resonant (LR) beam and band gap characteristics of the corresponding infinite beam in the range of 0–800 Hz: (**a**) Frequency response function of the finite beam. (**b**) The imaginary part of wave vectors of the infinite beam.

4.3. Method Advantages

To demonstrate the advantages of the present approach, the following case is compared with the PWE method and the FEM method. Han et al. [11] used the modified transfer matrix method (MTM) and the PWE method to calculate the band gaps of a PC beam. The frequency ranges of the first three band gaps obtained by MTM, PWE, FEM, and the present method are listed in Table 3. As with the TM method, the MTM obtains an analytical solution; therefore, the closer the results of the other three numerical methods to the results obtained by the MTM, the higher the accuracy. The relative errors between FEM and DQM are within 0.014%, and both methods show higher accuracy than the PWE method. The following compares the methods from the perspective of computational efficiency. A total of 40 wave vectors are selected at equal intervals in the first Brillouin zone, and the same computer and software are used for the calculation. Both the number of the DQM discrete points and the number of the FEM nodes are taken as 30. The relevant simulation times and computational resources (memory) are listed in Table 4. It can be seen that the present method has a shorter simulation time and requires less memory than the FEM method. Therefore, by comparing with two classical, widely-used methods, the high accuracy and computational efficiency of the present method are demonstrated.

Table 3. The first three band gaps in a phononic crystals (PC) beam.

Method	Band Gaps (Hz)					
	First		Second		Third	
	Lower	Upper	Lower	Upper	Lower	Upper
Present	616.378	1097.33	3038.92	6334.27	9601.44	11929.7
FEM	616.379	1097.33	3038.94	6334.57	9602.36	11931.3
MTM	616	1098	3038	6335	9601	11930
PWE	617	1103	3053	6358	9662	11931

Table 4. Simulation times and memory required of the present method and finite element method (FEM) (Intel (R) Xeon(R) E5620 CPU, Mathematica 11.1, simulation time is in second, memory is in KB).

Method	Present	FEM
Simulation time	8.8438	17.0781
Memory	154584	160380

4.4. Parameter Studies

In this part, we conducted a parametric analysis to investigate the effect of shear deformation and rotary inertia, the lumped mass m_z, the spring stiffness coefficient k_z, and the lattice size a on the 1st band gap of the locally resonant beam. The emphasis is on changes in the lower edge as well as the width of the 1st band gap. Since most of the vibrations presented in the project are low-frequency vibrations, for engineering purposes, the following research focuses on ways to decrease the corresponding frequency and widen the band gap.

4.4.1. Effects of Shear Deformation and Rotary Inertia

In this case, the material and geometric parameters of a LR beam are as follows: $E = 4.35 \times 10^9$ Pa, $\rho = 1180$ kg/m^3, $G = 5 \times 10^8$ Pa, $A = 1 \times 10^{-4}$ m^2, $I = 8.333 \times 10^{-10}$ m^4, $k_s = 0.8333$, $a = 0.075$ m. Both the Euler-Bernoulli beam model (EB) and the Timoshenko beam model (TB) are used to calculate the band structure of the LR beam. The results of the two models are plotted in Figure 5. The 1st band gap locates between 260.301–743.608 H_z by EB, and 253.44–734.597 H_z by TB, respectively. The shear deformation will reduce the stiffness of the beam, and the rotary inertia will increase the inertia of the beam, both of which will reduce the natural frequency of the beam. Therefore, the band gap frequency range calculated by TB is lower than that calculated by EB.

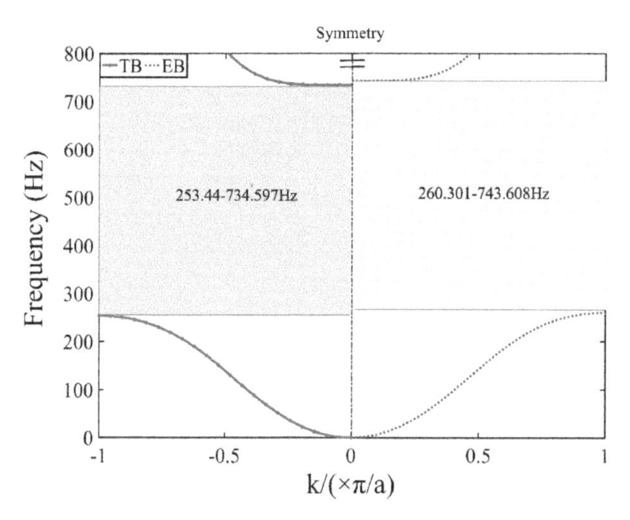

Figure 5. Dispersion relationship and band gaps for a locally resonant (LR) beam calculated by Timoshenko beam model (TB) and Euler-Bernoulli beam model (EB). Half of the dispersion relationship is plotted because of its symmetry.

4.4.2. Effects of Lumped Mass m_z

The mass ratio m_z/m is introduced here to characterize the relative magnitude of the lumped mass, where $m = \rho Aa$ is the mass of the LR beam per period. Figure 6 manifests the changes in the lower edge and width of the 1st band gap over the mass ratio m_z/m from 0 to 500. As is demonstrated in Figure 6a that the lower edge falls considerably when the mass ratio (m_z/m) is small. Since then, it remains more or less stable and finally tends to 0 Hz. According to Figure 6b, the width of the 1st band gap grows rapidly when the mass ratio (m_z/m) is small. After that, it reaches a plateau and tends to be stable. According to the above, we have found that it may be possible to simultaneously decrease the corresponding frequency range and widen the band gap by increasing the lumped mass m_z within a certain range.

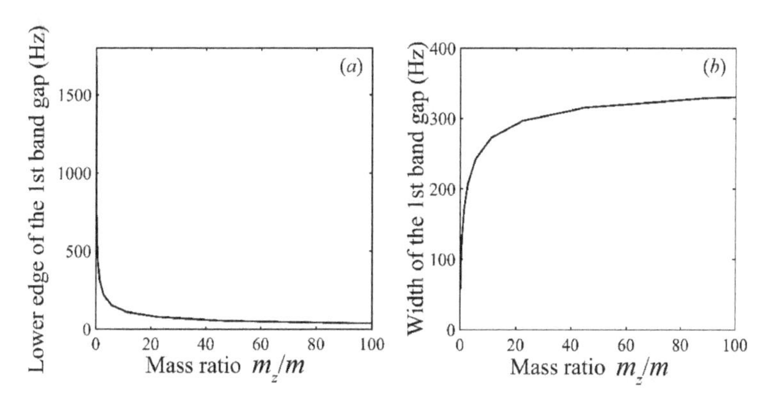

Figure 6. Effects of lumped mass m_z on the 1st band gap: (**a**) Lower edge of the 1st band gap. (**b**) Width of the 1st band gap.

4.4.3. Effects of the Spring Stiffness Coefficient k_z

In order to facilitate the description of the relative magnitude of the spring stiffness coefficient k_z, a stiffness ratio k_z/k is introduced here, where $k = EI/a$ is the equivalent stiffness of the LR beam in a period. Figure 7 illustrates the changes in the lower edge and width of the 1st band gap over the stiffness ratio k_z/k from 0 to 250. It is apparent from Figure 7 that both the width and the lower edge experience a steady growth with the stiffness ratio k_z/k increasing. Therefore, we can conclude that it is hard to simultaneously decrease the corresponding frequency range and widen the band gap by merely changing the stiffness coefficient of the LR oscillator.

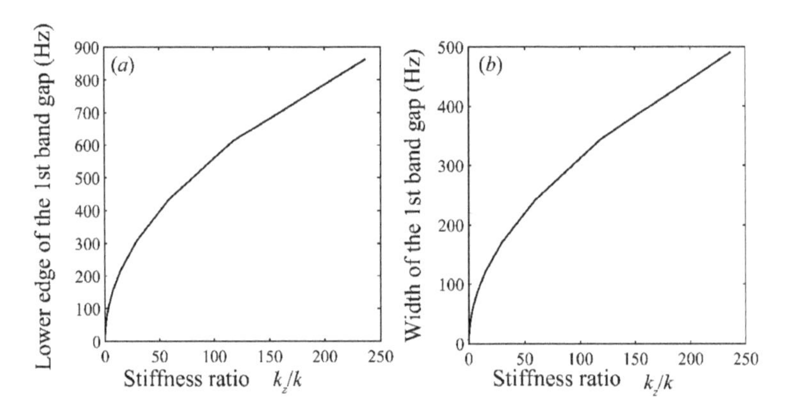

Figure 7. Effects of the spring stiffness coefficient k_z on the 1st band gap: (**a**) Lower edge of the 1st band gap. (**b**) Width of the 1st band gap.

4.4.4. Effects of the Lattice Constant *a*

When propagating in the locally resonant beam, the elastic wave is reflected at the nodes, which satisfies the Bragg band gap mechanism. Thus, there are also Bragg band gaps in the LR beam. In the case of small lattice constant, the LR band gap is separated from the Bragg band gap frequency range. Here, we mainly focus on the effects on the 1st LR band gap, so the lattice size *a* selected here is between (0, 0.075], which can separate the LR band gap from the Bragg band gap. As is shown in Figure 8, with the lattice constant *a* increasing, the band gap width decreases gradually, showing an inverse correlation (Figure 8b), but the lower edge always levels off at about 309 Hz (Figure 8a). It shows that the starting frequency of the locally resonant band gap is independent of the lattice constant because its frequency is determined by $f_0 = \sqrt{k_z/m_z}/(2\pi)$ [46], which is the resonant frequency of the locally resonant oscillator. Altering the lattice constant can only affect the band gap width.

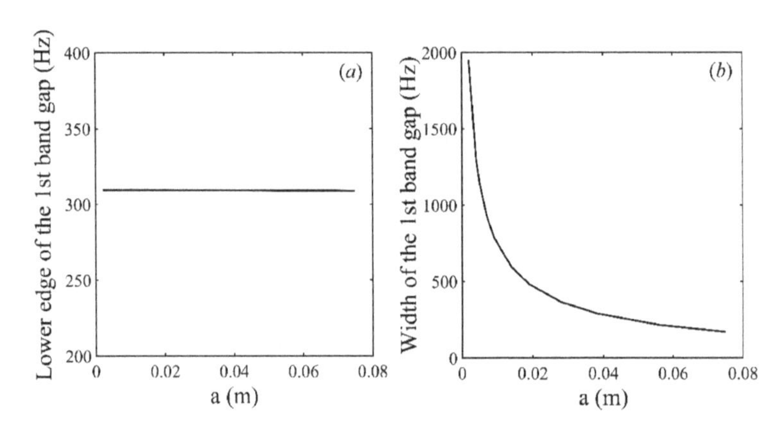

Figure 8. Effects of the lattice constant *a* on the 1st band gap: (**a**) Lower edge of the 1st band gap. (**b**) Width of the 1st band gap.

5. Conclusions

The differential quadrature method has been developed to calculate the elastic band gaps from the Bragg reflection mechanism in periodic structures efficiently and accurately. However, there have been no reports on the successful use of this method to calculate the band gaps of locally resonant structures. Hence, this paper proposes a numerical method to calculate and study the flexural vibration band gap of a locally resonant beam. The proposed method is based on the DQM method, an unconventional matrix-partitioning method, and a variable substitution method. According to the analysis and research in this work, the following conclusions can be obtained.

(1) The governing equation and periodic boundary conditions are discretized by the DQM method. The matrix equation is transformed into a standard quadratic eigenvalue problem by the partitioned matrix operation and variable substitution. Thus, the dispersion relationship and the band gap characteristics of a locally resonant beam can be solved. By extending the proposed method further, it can also be suitable for 2-dimensional or 3-dimensional LR structures.
(2) By comparing with the results of the existing literature, the validity of the proposed method is developed from both propagation modes and evanescent modes. Convergence studies indicate that accurate enough results could be got when the number of discrete points $N_x \geq 8$.
(3) By comparing with the plane wave expansion method and the finite element method, the high accuracy and computational efficiency of the present method are demonstrated.
(4) The parametric analysis shows that the width of the 1st band gap can be widened by increasing the mass ratio or the stiffness ratio or decreasing the lattice constant. In addition, one can decrease the lower edge of the 1st band gap by increasing the mass ratio or decreasing the stiffness ratio. The starting frequency of the LR band gap has nothing to do with the lattice constant because it is defined by the resonant frequency of the LR oscillator. The band gap frequency range calculated by the Timoshenko beam theory is lower than that calculated by the Euler–Bernoulli beam theory.

Referring to the method and solving process in this paper, future research can apply the proposed method to study the band gap characteristics of 2-D or 3-D locally resonant structures.

Author Contributions: Conceptualization, X.L. and T.W.; methodology, X.L. and Y.D.; software, T.W. and Z.L.; validation, Y.R.; formal analysis, T.W.; writing—original draft preparation, T.W.; writing—review and editing, T.W.; visualization, X.J.; supervision, X.L. and X.J.; project administration, X.L.; funding acquisition, X.L.

Acknowledgments: This work was supported by the National Natural Science Foundation of China (Grant No. 51879231, 51679214, 51338009, 51409228) and the Zhejiang Provincial Key Research and Development Program (2018C03031).

Conflicts of Interest: The authors declare no conflict of interest.

References

1. Sigalas, M.M.; Economou, E.N. Elastic and acoustic wave band structure. *J. Sound Vibr* **1992**, *158*, 377–382. [CrossRef]
2. Kushwaha, M.S.; Halevi, P.; Dobrzynski, L.; Djafari-Rouhani, B. Acoustic band structure of periodic elastic composites. *Phys. Rev. Lett.* **1993**, *71*, 2022–2025. [CrossRef] [PubMed]
3. Martínez-Sala, R.; Sancho, J.; Sánchez, J.V.; Gómez, V.; Llinares, J.; Meseguer, F. Sound attenuation by sculpture. *Nature* **1995**, *378*, 241. [CrossRef]
4. Kushwaha, M.S.; Halevi, P.; Martínez, G.; Dobrzynski, L.; Djafari-Rouhani, B. Theory of acoustic band structure of periodic elastic composites. *Phys. Rev. B* **1994**, *49*, 2313–2322. [CrossRef] [PubMed]
5. Vasseur, J.O.; Deymier, P.A.; Chenni, B.; Djafari-Rouhani, B.; Dobrzynski, L.; Prevost, D. Experimental and theoretical evidence for the existence of absolute acoustic band gaps in two-dimensional solid phononic crystals. *Phys. Rev. Lett.* **2001**, *86*, 3012. [CrossRef] [PubMed]
6. Tanaka, Y.; Tomoyasu, Y.; Tamura, S.I. Band structure of acoustic waves in phononic lattices: Two-dimensional composites with large acoustic mismatch. *Phys. Rev. B* **2000**, *62*, 7387–7392. [CrossRef]

7. Wu, F.G.; Hou, Z.L.; Liu, Z.Y.; Liu, Y.Y. Acoustic band gaps in two-dimensional rectangular arrays of liquid cylinders. *Solid State Commun.* **2002**, *123*, 239–242. [CrossRef]
8. Xiang, H.J.; Shi, Z.F. Analysis of flexural vibration band gaps in periodic beams using differential quadrature method. *Comput. Struct.* **2009**, *87*, 1559–1566. [CrossRef]
9. Cheng, Z.B.; Xu, Y.G.; Zhang, L.L. Analysis of flexural wave bandgaps in periodic plate structures using differential quadrature element method. *Int. J. Mech. Sci.* **2015**, *100*, 112–125. [CrossRef]
10. Miranda Jr, E.J.P.; Dos Santos, J.M.C. Flexural wave band gaps in phononic crystal euler-bernoulli beams using wave finite element and plane wave expansion methods. *Mater. Res.-Ibero-Am. J. Mater.* **2017**, *20*, 729–742. [CrossRef]
11. Han, L.; Zhang, Y.; Ni, Z.Q.; Zhang, Z.M.; Jiang, L.H. A modified transfer matrix method for the study of the bending vibration band structure in phononic crystal Euler beams. *Phys. B Condens. Matter* **2012**, *407*, 4579–4583. [CrossRef]
12. Hajhosseini, M.; Rafeeyan, M.; Ebrahimi, S. Vibration band gap analysis of a new periodic beam model using GDQR method. *Mech. Res. Commun.* **2017**, *79*, 43–50. [CrossRef]
13. Zhang, Y.; Ni, Z.Q.; Han, L.; Zhang, Z.M.; Jiang, L.H. Flexural vibrations band gaps in phononic crystal timoshenko beam by plane wave expansion method. *Optoelectron. Adv. Mater.-Rapid Commun.* **2012**, *6*, 1049–1053.
14. De Miranda Júnior, E.J.P.; Dos Santos, J.M.C. Band structure in carbon nanostructure phononic crystals. *Mater. Res.* **2017**, *20*, 572–579. [CrossRef]
15. Chen, H.; Fung, K.H.; Ma, H.; Chan, C.T. Polarization gaps and negative group velocity in chiral phononic crystals: Layer multiple scattering method. *Phys. Rev. B* **2008**, *77*, 224304. [CrossRef]
16. Lu, Y.; Srivastava, A. Combining plane wave expansion and variational techniques for fast phononic computations. *J Eng. Mech.* **2017**, *143*. [CrossRef]
17. Yao, L.; Huang, G.; Chen, H.; Barnhart, M.V. A modified smoothed finite element method (M-SFEM) for analyzing the band gap in phononic crystals. *Acta Mech.* **2019**, 1–15. [CrossRef]
18. Wormser, M.; Wein, F.; Stingl, M.; Körner, C. Design and additive manufacturing of 3D phononic band gap structures based on gradient based optimization. *Materials* **2017**, *10*, 1125. [CrossRef]
19. Liu, Z.Y.; Zhang, X.; Mao, Y.; Zhu, Y.Y.; Yang, Z.; Chan, C.T.; Sheng, P. Locally resonant sonic materials. *Science* **2000**, *289*, 1734–1736. [CrossRef]
20. Wang, X.; Wang, M.Y. An analysis of flexural wave band gaps of locally resonant beams with continuum beam resonators. *Meccanica* **2016**, *51*, 171–178. [CrossRef]
21. Yu, D.L.; Liu, Y.Z.; Zhao, H.G.; Wang, G.; Qiu, J. Flexural vibration band gaps in Euler-Bernoulli beams with locally resonant structures with two degrees of freedom. *Phys. Rev. B* **2006**, *73*, 064301. [CrossRef]
22. Wang, G.; Wen, X.S.; Wen, J.H.; Liu, Y.Z. Quasi-one-dimensional periodic structure with locally resonant band gap. *J. Appl. Mech.* **2006**, *73*, 167–170. [CrossRef]
23. Wang, G.; Yu, D.L.; Wen, J.H.; Liu, Y.Z.; Wen, X.S. One-dimensional phononic crystals with locally resonant structures. *Phys. Lett. A* **2004**, *327*, 512–521. [CrossRef]
24. Miranda, E.J.P.; Dos Santos, J.M.C. Evanescent Bloch waves and complex band structure in magnetoelectroelastic phononic crystals. *Mech. Syst. Signal Process.* **2018**, *112*, 280–304. [CrossRef]
25. Nobrega, E.D.; Gautier, F.; Pelat, A.; Dos Santos, J.M.C. Vibration band gaps for elastic metamaterial rods using wave finite element method. *Mech. Syst. Signal Process.* **2016**, *79*, 192–202. [CrossRef]
26. Li, F.L.; Wang, Y.S.; Zhang, C.; Yu, G.L. Boundary element method for band gap calculations of two-dimensional solid phononic crystals. *Eng. Anal. Bound. Elem.* **2013**, *37*, 225–235. [CrossRef]
27. Camley, R.E.; Djafari Rouhani, B.; Dobrzynski, L.; Maradudin, A.A. Transverse elastic waves in periodically layered infinite, semi-infinite, and slab media. *J. Vac. Sci. Technol. B* **1983**, *1*, 371–375. [CrossRef]
28. Sigalas, M.M.; Soukoulis, C.M. Elastic-wave propagation through disordered and/or absorptive layered systems. *Phys. Rev. B* **1995**, *51*, 2780–2789. [CrossRef] [PubMed]
29. Kafesaki, M.; Penciu, R.S.; Economou, E.N. Air bubbles in water: A strongly multiple scattering medium for acoustic waves. *Phys. Rev. Lett.* **2000**, *84*, 6050–6053. [CrossRef]
30. Economou, E.N.; Zdetsis, A. Classical wave propagation in periodic structures. *Phys. Rev. B* **1989**, *40*, 1334–1337. [CrossRef]
31. Cao, Y.; Hou, Z.; Liu, Y. Convergence problem of plane-wave expansion method for phononic crystals. *Phys. Lett. A* **2004**, *327*, 247–253. [CrossRef]

32. Sigalas, M.M.; Garcı, A.N. Theoretical study of three dimensional elastic band gaps with the finite-difference time-domain method. *J. Appl. Phys.* **2000**, *87*, 3122–3125. [CrossRef]
33. Kafesaki, M.; Sigalas, M.M.; García, N. Frequency modulation in the transmittivity of wave guides in elastic-wave band-gap materials. *Phys. Rev. Lett.* **2000**, *85*, 4044–4047. [CrossRef] [PubMed]
34. Liu, Z.Y.; Chan, C.T.; Sheng, P.; Goertzen, A.L.; Page, J.H. Elastic wave scattering by periodic structures of spherical objects: Theory and experiment. *Phys. Rev. B* **2000**, *62*, 2446–2457. [CrossRef]
35. Psarobas, I.E.; Stefanou, N.; Modinos, A. Scattering of elastic waves by periodic arrays of spherical bodies. *Phys. Rev. B* **2000**, *62*, 278–291. [CrossRef]
36. Wang, G.; Wen, X.S.; Wen, J.H.; Shao, L.H.; Liu, Y.Z. Two-dimensional locally resonant phononic crystals with binary structures. *Phys. Rev. Lett.* **2004**, *93*, 154302. [CrossRef]
37. Wang, G.; Wen, J.H.; Liu, Y.Z.; Wen, X.S. Lumped-mass method for the study of band structure in two-dimensional phononic crystals. *Phys. Rev. B* **2004**, *69*, 184302. [CrossRef]
38. Huang, Y.; Li, J.; Chen, W.; Bao, R. Tunable bandgaps in soft phononic plates with spring-mass-like resonators. *Int. J. Mech. Sci.* **2019**, *151*, 300–313. [CrossRef]
39. Zhao, H.J.; Guo, H.W.; Gao, M.X.; Liu, R.Q.; Deng, Z.Q. Vibration band gaps in double-vibrator pillared phononic crystal plate. *J. Appl. Phys.* **2016**, *119*. [CrossRef]
40. Khelif, A.; Aoubiza, B.; Mohammadi, S.; Adibi, A.; Laude, V. Complete band gaps in two-dimensional phononic crystal slabs. *Phys. Review E* **2006**, *74*. [CrossRef]
41. Liang, X.; Kou, H.L.; Wang, L.Z.; Palmer, A.C.; Wang, Z.Y.; Liu, G.H. Three-dimensional transient analysis of functionally graded material annular sector plate under various boundary conditions. *Compos. Struct.* **2015**, *132*, 584–596. [CrossRef]
42. Malekzadeh, P.; Vosoughi, A.R. DQM large amplitude vibration of composite beams on nonlinear elastic foundations with restrained edges. *Commun. Nonlinear Sci. Numer. Simul.* **2009**, *14*, 906–915. [CrossRef]
43. Doyle, J.F. *Wave Propagation in Structures*, 2nd ed.; Springer: New York, NY, USA, 1997; p. 335.
44. Kittel, C. *Introduction to Solid State Physics*, 8th ed.; John Wiley & Son: New York, NY, USA, 2005; p. 406.
45. Yu, D.L.; Liu, Y.Z.; Wang, G.; Zhao, H.G.; Qiu, J. Flexural vibration band gaps in Timoshenko beams with locally resonant structures. *J. Appl. Phys.* **2006**, *100*, 124901. [CrossRef]
46. Yu, D.L.; Wen, J.H.; Zhao, H.G.; Liu, Y.Z.; Wen, X.S. Vibration reduction by using the idea of phononic crystals in a pipe-conveying fluid. *J. Sound Vibr.* **2008**, *318*, 193–205. [CrossRef]

Article

Deterministic Insertion of KTP Nanoparticles into Polymeric Structures for Efficient Second-Harmonic Generation

Dam Thuy Trang Nguyen and Ngoc Diep Lai *

Laboratoire de Photonique Quantique et Moléculaire, UMR 8537, École Normale Supérieure Paris-Saclay, Centrale Supélec, CNRS, Université Paris-Saclay, 61 avenue de Président Wilson, 94235 Cachan, France

* Correspondence: ngoc-diep.lai@ens-paris-saclay.fr; Tel.: +33-(01)-47-40-55-59

Received: 27 May 2019; Accepted: 11 July 2019; Published: 17 July 2019

Abstract: We investigate theoretically and experimentally the creation of virtually any polymer-based photonic structure containing individual nonlinear $KTiOPO_4$ nanoparticles (KTP NPs) using low one-photon absorption (LOPA) direct laser writing (DLW) technique. The size and shape of polymeric microstructures and the position of the nonlinear KTP crystal inside the structures, were perfectly controlled at nanoscale and on demand. Furthermore, we demonstrated an enhancement of the second-harmonic generation (SHG) by a factor of 90 when a KTP NP was inserted in a polymeric pillar. The SHG enhancement is attributed to the resonance of the fundamental light in the cavity. This enhancement varied for different KTP NPs, because of the random orientation of the KTP NPs, which affects the light/matter interaction between the fundamental light and the NP as well as the collection efficiency of the SHG signal. The experimental result are further supported by a simulation model using Finite-Difference Time-Domain (FDTD) method.

Keywords: direct laser writing; KTP; nonlinear optics; photonic coupling

---

1. Introduction

Second-harmonic generation (SHG) allows the generation of new photons with a double frequency (2ω) [1]. This is a second-order nonlinear optical process, thus does not provoke any absorption. Therefore, the SHG can avoid a bleaching effect and allows observations over long durations with no decrease in the signal quality, which is usually observed with fluorescent and luminescent nanoparticles (NPs) [2,3]. Moreover, the SH spectrum (2ω) is separated from the fundamental spectrum (ω), allowing spectrally filtering out the fundamental beam, resulting in an excellent contrast in nonlinear microscopy. In SHG process, a large range of excitation wavelengths can be used, depending on the nonlinear material used. Thus, for many applications, the fundamental wavelength can be selected in ranges where the absorption and scattering effects are low to avoid the losses and the photodegradation. This allows SHG to be observed in thick materials, which are very useful for bio-applications [4–8].

SHG could be performed with nonlinear materials at macro and micro scales. For some applications, such as biomarkers and nanosensors, nonlinear NPs should be used [4,8]. Several works have been devoted to fundamental studies of nonlinear properties of NPs, such as NPs of CMONS (Cyano-MethOxy-Nitro-Stilbene) [9–11] and hybrid organic/inorganic-based MnPS3 NPs [12]. Recently, Le Xuan et al. [2] have investigated $KTiOPO_4$ (KTP) NPs, with the size in the range of 30 and 100 nm. This nonlinear crystal is commonly used due to its important nonlinear response and its transparency in visible and infrared ranges [13]. The nonlinear analyses of these NPs show that the SH signal is very high as compared to a massive KTP crystal and perfectly stable. The SHG of nonlinear NPs in general is efficient because no phase-matching is required, and those NPs could be easily functionalized with other materials, suggesting numerous applications particularly in the biology domain.

Direct laser writing (DLW) is an excellent method allowing fabrication of any one-, two-, and three-dimensional (1D, 2D, and 3D) structures at a sub-micrometer scale [14,15]. In this method, a femtosecond laser beam is usually used. The laser beam strongly focused into a photoresist by using a large numerical aperture (NA) objective lens (OL). Only the photoresist located inside the focusing spot is polymerized/depolymerized thanks to the multi-photon absorption (MPA) effect. An arbitrary structure then can be created by moving the focusing spot inside the photoresist. However, the MPA-based DLW is rather expensive and complicated since it requires a femtosecond laser and a complex optical system. Recently, a very simple and inexpensive technique called low one-photon absorption (LOPA) DLW is demonstrated as a robust technique [16,17], allowing realization of multidimensional structures as what realized by MPA-based DLW. In the LOPA setup, a simple and low-cost, continuous-wave (cw) and low power laser is used as an excitation laser, similar to the case of conventional OPA. However, this laser wavelength is located in an ultralow absorption range of the photoresist, thus can penetrate deeply inside the material. This LOPA effect is then compensated by the high intensity at the focusing spot, resulting in a compressed polymerized spot. By moving the focusing in 3D, any structure can be created as in the case of MPA-based DLW. Moreover, it was demonstrated that LOPA DLW technique enables the creation of desirable polymeric structures containing a single gold NP [18], that cannot be obtained by a standard MPA-based DLW.

In this work, we demonstrate the use of a modified LOPA system to embed a single KTP NP in a polymeric photonic structure in order to enhance the SHG of KTP NP as well as to manipulate the SH propagation.

2. Fabrication Procedure and Characterization of KTP NPs

2.1. Lopa-Based Dlw Setup

We added a pulsed 1064 nm laser (1 ns, 24.5 kHz) into the LOPA DLW setup. This laser allowed us to investigate the SHG signal of KTP NPs, before and after coupling to the photonic structure. This infrared laser is aligned with the 532 nm laser before being focused into the sample (see Figure 1a). The system can work with the 1064 nm laser or with the 532 nm laser by switching the mirror M1. Another half-wave plate for 1064 nm–wavelength was inserted after the PBS to adjust the polarization of the fundamental laser. To measure the SH signal (532 nm), a KG5 filter is used to cut off the fundamental beam (1064 nm). The SH signal was detected by either the APD to produce a SH image, or by a spectrometer to obtain a SH spectrum. This flexible setup allows both fabrication of desired structure containing a single KTP NP at LOPA regime and characterization of SHG enhancement. The whole process is illustrated in Figure 1b.

2.2. Shg Experimental Setup

In order to measure and to characterize the SHG signal of KTP NPs, we added a pulsed 1064 nm laser (pulse duration of 1 ns and repetition rate of 24.5 kHz) into the LOPA DLW setup. This second laser is aligned so that the laser beam coincides with that of the 532 nm laser (see Figure 1a). A flip-flop mirror allows us to switch easily from one laser to the other. A half-wave plate (not shown in the Figure) for 1064 wavelength can be placed after the polarizer to control the polarization of the fundamental laser beam. To collect SHG signal at 532 nm, an infrared filter is used to cut off the fundamental light at 1064 nm. The signal is then detected by either the APD which gives a SHG image, or a spectrometer which gives a SHG spectrum of the KTP NPs. This home-built setup is very flexible for signal detection as well as structure fabrication at LOPA regime, as illustrated in Figure 1b.

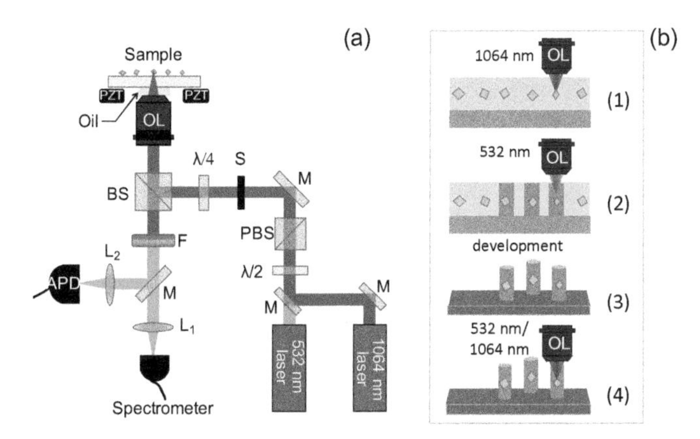

Figure 1. (**a**) Experimental setup of the integrated optical system for fabrication of desired polymeric structures and for characterization of SHG of a single KTP NP. PZT: piezoelectric translator, OL: oil immersion microscope objective, $\lambda/4$: quarter-wave plate, $\lambda/2$: half-wave plate, BS: unpolarizing beam splitter, PBS: polarizing beam splitter, $M_{1,2,3,4}$: mirrors, S: electronic shutter, $L_{1,2}$: lenses, F: infrared filter, APD: avalanche photodiode. (**b**) Fabrication process of polymeric structures containing single KTP NPs. (1) Identification of KTP NPs positions by 1064 nm laser (SHG mapping); (2) Fabrication of photonic structures containing single KTP NP by 532 nm laser; (3) Development of samples; (4) Characterization (SHG by 1064 nm laser or fluorescence by 532 nm laser) of coupled structure.

2.3. *Ktp Nps Preparation*

KTP NPs were obtained from KTP powders, which were extracted from the polishing process of a bulk KTP crystal. The synthesis process of KTP NPs can be described as follows [2]: The KTP powders were first mixed with a polyvinylpyrrolidone polymer (PVP) using a propanol solvent (PrOH) at a controlled concentration. The solution is then placed in an ultrasonic bath for 20 min in order separate KTP NPs assembly and to create a thin layer of PVP polymer around them, which then avoiding the particles aggregation. To obtain particles with a uniform size, the solution was first passed through a filter having a hole diameter of 0.45 μm, in order to eliminate big particles, and then centrifuged with a typical speed of 11,000 rpm for 15 min. This allowed obtaining a quasi-monodisperse solution with particles size between 30 and 100 nm. We dispersed these particles again in the 0.1% PVP/PrOH mixture, which was again placed in an ultrasonic bath to obtain the final colloidal KTP solution.

2.4. *Sample Preparation and Fabrication Process*

To obtain a monolayer of individual KTP NPs for SHG characterization or for fabrication, we mixed the colloidal KTP solution with ethanol with a solution/ethanol ratio of 1:5 using ultrasonication for 30 min. The solution was then spin-coated on a clean glass substrate (for SHG characterization) or on photoresist thin films (for coupling), resulting in a monolayer of well separated KTP NPs.

For fabrication of coupled structures, the photoresist sample was prepared as following. First, a thin layer (500 nm or 5000 nm) of SU8 was spin-coated on a cleaned cover glass. Then, a well dispersed KTP NPs solution was spin-coated on the surface of the first SU8 layer. Finally, a second thin layer (500 nm) of SU8 was deposited on top of the KTP NPs monolayer. To remove the solvents, the sample was placed on a hot plate at 65 °C for 3 min and at 95 °C for 5 min. We note that SU8 2000.5 photoresist was used to obtain a film of 500 nm thickness and SU8 2005 photoresist for a film of 5000 nm thickness. The choice of each thickness depends on the desired structures, 2D or 3D.

The fabrication of photonic structures containing a single KTP NP is shown in Figure 1b and consists of two main steps:

- Determination of the position of a single KTP NP by detecting SHG signal using a pulsed 1064 nm laser.
- Fabrication of the polymeric structure containing a single NP by LOPA-based DLW using a cw 532 nm laser.

We note that the two lasers have been perfectly aligned before the OL so that their focusing spots have the same transverse position, i.e., x and y coordinates. However, these spots are slightly different along the axial axis, due to the difference in wavelength. Practically, this was overcome by adjusting the z position in between the mapping (by 1064 nm laser) and the fabrication (by 532 nm laser) programs. After developing the sample to wash away all un-exposed photoresist, the fabricated structures were then placed again on the LOPA system in order to characterize the SH signal of KTP NPs and to compare it with that obtained before fabrication.

2.5. SHG of a Single KTP NP

We first characterized the SHG of individual KTP NPs spin-coated on glass substrate, by using the 1064 nm laser with a typical average laser power of 0.3 mW. The SH signal is very intense with respect to background noise (signal-to-background ratio of about 100), and a SHG image of an individual KTP NP is shown in Figure 2a. The intensity profile in the x-direction (Figure 2b) shows a full width at half maximum (FWHM) of approximately 500 nm, which agrees with the diffraction limit of the focusing spot at the 1064 nm wavelength (numerical calculation is about 480 nm). This allowed us to identify an individual KTP NP with a great precision. Figure 2c shows the spectral analysis of the SHG of a single KTP NP, obtained by adding an infrared filter to cut off the fundamental wavelength. A clear peak at 532 nm confirms the SHG of the 1064 nm laser. Figure 2d shows the temporal tracking of the SH signal. It shows a perfect stability of the SH signal for a long duration. The small variation of the SH signal is probably due to the mechanical instability of the optical setup. This excellent SHG stability allows us to optically address the same KTP NP at different moments, separated by several months, and to perform different types of measurements.

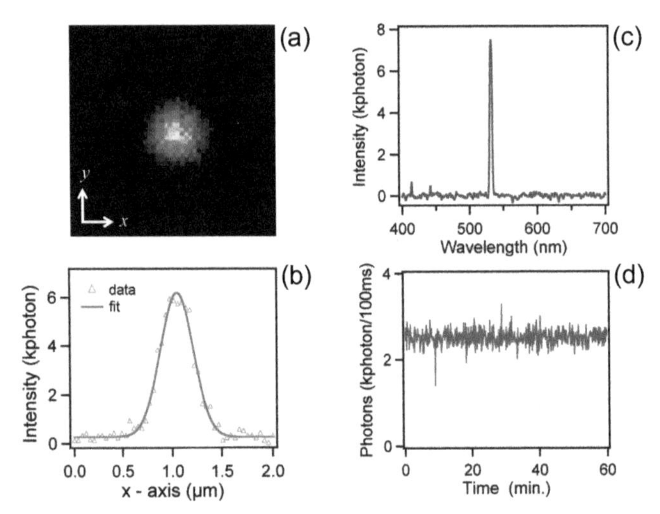

Figure 2. (**a**) Second-harmonic image of a KTP NP with a size of about 100 nm, obtained by scanning with a pulsed 1064 nm laser. (**b**) The SHG intensity profile in the x-direction, extracted from (**a**), showing a FWHM of 500 nm. (**c**) SHG spectrum. (**d**) Time evolution of the SHG signal obtained by an average power of 0.3 mW. The signal is stable for a long duration.

3. Deterministic Coupling of a Single KTP Particle into a Polymeric Structure

In order to fabricate a desired photonic structure containing a single KTP NP, we first determined the position of a KTP NP by scanning the sample with the 1064 nm laser as explained above. Experimentally, the position of a KTP NP was determined with a precision of about 10 nm. Then, the fabrication of desired polymeric structures containing the NP with well-known position was realized. For fabrications, the 1064 nm laser was off and the cw 532 nm laser was turned on. We demonstrated that the modified LOPA-based DLW technique allowed a full control of the coupling of a single KTP NP and an arbitrary polymeric structure.

3.1. Control of Shape of Structures

We first investigated the introduction of KTP NPs to structures having different shapes. Figures 3a,d show scanning electron microscope (SEM) images (top view) of a square and a triangular structures, fabricated by a laser power of 9 mW and a writing velocity of 3 μm/s. These structures were fabricated by scanning the laser spot through the SU8 film, point by point to form square/triangular structures. The distance between two adjacent points was set at 150 nm, which is smaller than the diffraction limit, resulted in continuous photoresist lines/films. Figures 3b,e show the fluorescence images of the square and triangular shapes, respectively, obtained by 532 nm laser scanning (P_{532} = 50 μW). Figures 3c,f represent the SHG images obtained by scanning with the 1064 nm pulsed laser. A bright spot at the center of the structure corresponds to the SHG signal emitted by the KTP NP.

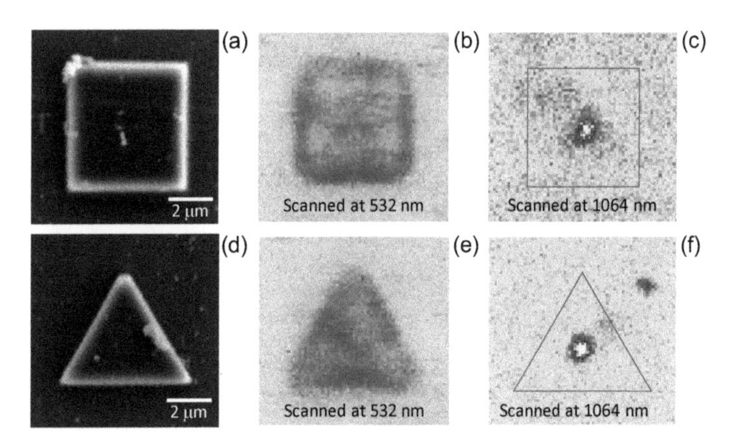

Figure 3. (**a**) SEM image of a square structure containing a KTP NP at the center. Fluorescence image (**b**) and SHG image (**c**) of the corresponding structure, obtained by scanning with 532 nm laser and 1064 nm laser, respectively. The SHG image shows clearly that the polymeric square structure contains a KTP NP at the center. (**d–f**): similar SEM, fluorescence et SHG images of a triangular structure containing a KTP NP.

The first conclusion is that the LOPA-DLW allowed us to realize any structure on demand containing a single nonlinear NP. Secondly, as compared to metallic NPs [18], the nonlinear NPs do not induce a thermal effect, which may influence the structure size and shape. This is an excellent advantage which can be further exploited to embed this KTP NP to different kinds of polymeric structure.

3.2. Control of NP Position in Polymeric Structures

The control of position of the KTP NP in photonic structures is very importance, because the light intensity is usually not uniform in a resonant photonic structure. We therefore also demonstrated that LOPA based DLW allowed precisely controlling and manipulating the NP position in any polymeric structures. Several polymeric structures (e.g., triangle and circular disk) containing KTP NPs at different positions have been fabricated. Figure 4 shows fluorescence and SHG images of a series of polymeric microdisks embedding KTP NPs at different positions, from center to edge. The KTP NP position was experimentally controlled with a precision of 10 nm, that is mainly due to the diffraction limit of the optical system.

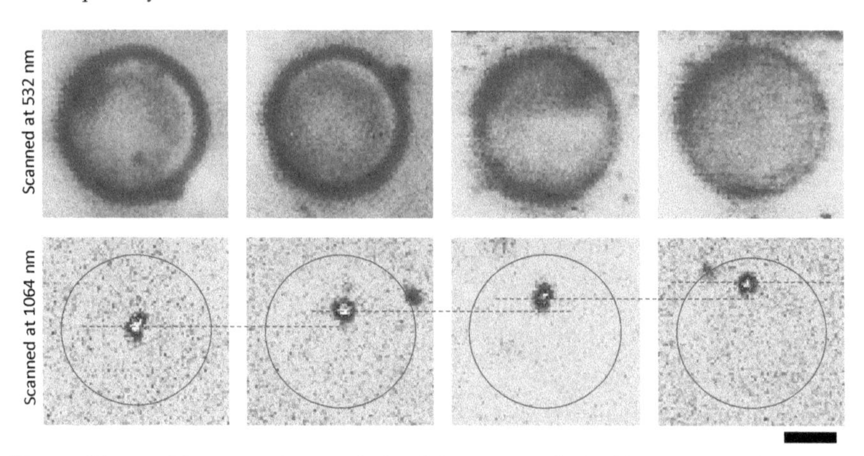

Figure 4. Top row: Fluorescence images of microdisk structures, obtained by scanning with a 532 nm cw laser. **Bottom row**: SH images of corresponding structures obtained by scanning with 1064 nm laser. The bright spots correspond to the SHG of KTP NPs, which are embedded at different positions of the microdisk, from center to edge. Scale bars: 1 μm.

3.3. Control of Structure Size

It was also demonstrated that by controlling the exposure dose, the size of the structure containing single KTP NPs is adjusted on demand. We fabricated various two-dimensional (2D) structures consisting of individual pillars among them the central one contains a single NP. Each pillar was fabricated by scanning the laser spot along the z–axis and through the photoresist film thickness (total scanning distance of 2 μm). Figure 5 shows the SEM images of fabricated structures. All pillars were fabricated with the same exposure dose (P = 6 mW, v = 5 μm/s), except the central one containing the KTP NP. For that, the doses were adjusted as: (top row) P = 6 mW, v = 4 μm/s; 3 μm/s; 2 μm/s; v = 1 μm/s; (bottom row) P = 9 mW, v= 4 μm/s; 3 μm/s; 2 μm/s; and 1 μm/s, respectively. In this example, the diameter of the central pillars increases with a step of about 50 nm. This can be controlled with a smaller step by finely controlling the laser power or the scanning speed, thanks to the non-explosion effect of the KTP NP in LOPA regime.

Thus, the second conclusion is that the LOPA DLW is a powerful technique to embed a single nonlinear NP into any PS with controlled size and shape, without any thermal effect. The position of the NP is also adjusted on demand inside the polymeric structures, which is a great advantage for photonic coupling.

3.4. Coupling of KTP NP to 3D Structures

Furthermore, we demonstrated that LOPA DLW enables the fabrication of three-dimensional (3D) structures containing a single KTP NP. For that, we used SU8 2005 (5 μm–thickness) as the structure material. Figure 6a shows the SEM image of a hexagonal air-hole membrane containing a KTP NP at the center. To support the membrane, six polymeric legs were fabricated as shown in the inset of the Figure 6a. The membrane was fabricated (laser power of 10 mW; scanning speed of 3 μm/s) by scanning the laser spot in a xy-plane with some missing exposures in a hexagonal configuration. Figures 6b,c show the fluorescence and SHG images of the center part of the membrane. The bright

SHG spot confirms the presence of the KTP NP in the 3D membrane structure. It is worth mentionning that it is more challenging for fabrication of 3D structures containing NPs as compared to the case of 2D structures, because the 3D structure is thicker and therefore it is more difficult to find out the position of the NPs. For this moment, we didn't find any SHG enhancement of the coupled KTP NP. This is due to the fact that the refractive index of the SU8 material is low and the periodicity of the 3D photonic membrane is not optimized to open a photonic bandgap, which is important to have a photonic coupling. This requires a long investigation and is out of the scope of this paper. However, the success of embedding a KTP NP into a 3D photonic structure is very encourageable, which suggests further investigations for fascinating applications.

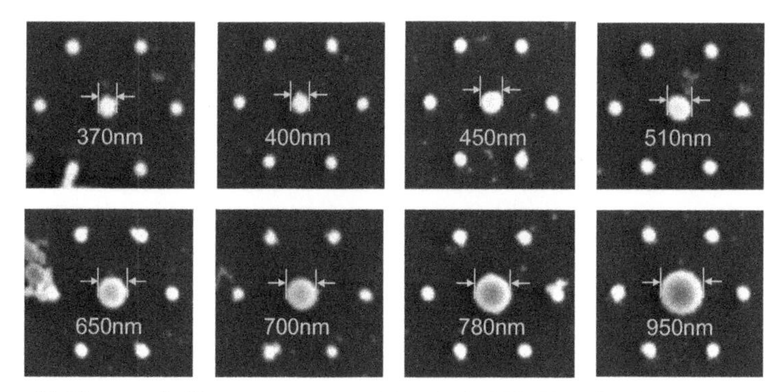

Figure 5. SEM images of pillars structures containing single KTP NPs. All pillars were fabricated by the same exposure dose, except the central one, which was fabricated by the following parameters. **Top row**: from left to right: P = 6 mW, v = 4 μm/s; 3 μm/s; 2 μm/s; and 1 μm/s, respectively. **Bottom row**: from left to right: P = 9 mW, v = 4 μm/s; 3 μm/s; 2 μm/s; and 1 μm/s, respectively.

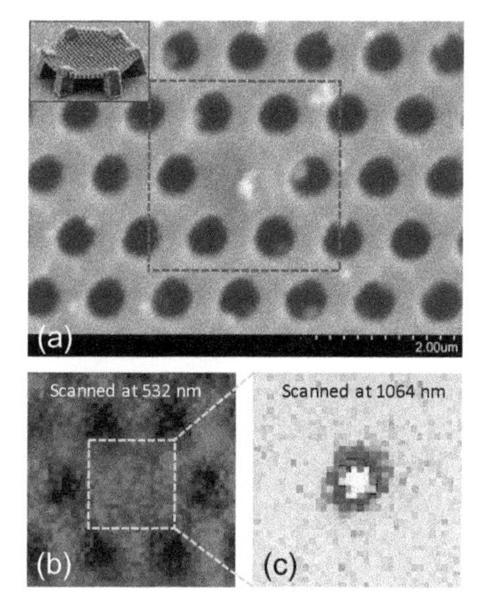

Figure 6. 3D hexagonal membrane structure containing a KTP NP. (**a**) SEM image; Inset: full side view. (**b**) Fluorescence image obtained by scanning with a 532 nm laser with low power. (**c**) SHG image, obtained by scanning with 1064 nm laser, showing the existence of a KTP NP at at center of structure.

4. SHG Signal Enhancement

In this section, as a demonstration of the coupling, we present our experimental investigation on the enhancement of the SH signal of a single KTP NP embedded in a polymeric submicropillar. This is further confirmed by a simulation model based on the FDTD method.

4.1. Experimental Measurement

In order to demonstrate the coupling effect of the fabricated structure, we have compared the SH signal of the same KTP NP obtained before and after being inserted into a polymeric pillar. The comparison was done by using the same experimental conditions, such as excitation power and measurement time.

Figure 7a shows the SH images (by a pulsed 1064 nm laser) of a single KTP NP obtained before (with polymeric film) and after fabrication (with polymeric pillar), and a fluorescence image (by a cw 532 nm laser) of the fabricated structure. It can be seen from the images that the NP exists well inside the micropillar. We note that SU8 does not emit fluorescent or SH signal when excited by 1064 nm laser, therefore only the SH signal of the KTP NP was detected. Comparing the SH signal obtained before and after fabrication, we can clearly see a great enhancement. Figure 7b shows the comparison of the SHG intensities of the same KTP NP located in a polymeric film (red curve) and in a polymeric pillar (green curve). When embedded in a pillar (diameter = 800 nm; height = 1.3 μm), the SHG intensity is enhanced up to 90 times. We have also realized many similar structures with the same fabrication parameters but containing different KTP NPs to verify the orientation dependence of the KTP NP inside the polymeric pillar. Indeed, while a bulk KTP crystal has a unique and well-defined orientation, a KTP NP is randomly oriented in space. This first affects the nonlinear interaction between the fundamental light and the nonlinear crystal, which depends strongly on the relative angle between the electric field of the fundamental beam and the main axis of the KTP crystal. Also, the amount of SH light collected by the microscope objective strongly depends on the orientation of the SH dipole and the surrounding medium. The detected SH signal is therefore the result of the photons number emitted by the nonlinear dipole and the collection efficiency of the optical system, which is defined as the ratio of the photons collected by the microscope to the total photons emitted by the dipole. Clearly, the polymeric pillar strongly enhanced the SHG emission towards the detection system thus enhancing the detected SHG signal. But the enhancement value varied from this NP to the other due to the random orientation of the KTP NP inside the pillar cavity.

4.2. Numerical Simulations

We have then performed different simulations using FDTD method to find out the answers for different questions: How the polymeric pillar cavity enhanced the incident fundamental light (at 1064 nm); How the polymeric pillar guided out the SH signal (at 532 nm) emitted from the embedded NP, and if the coupling out is affected by the NP's orientation. For these simulations, we assumed that the SH signal is emitted from an electric dipole.

We first investigated the enhancement of the fundamental light in the polymeric pillar cavity. Figure 8a shows a model used for simulation. The polymeric pillar cavity was placed on a glass substrate and a plane-wave light source was inserted inside the glass substrate, which emits a 1064 nm–wavelength in upward direction (*z*-axis). This incident light field was monitored in the (*xz*)– or (*yz*)–plane. Similar simulations were also realized for a thin SU8 film deposited on a glass substrate to verify the field enhancement in polymeric cavity. The cavity parameters such as diameter and height are swept in a large range to find out the optimum configuration of the cavity. We found that the pillar possessing parameters similar to experimental structures induces a strong enhancement of the fundamental light. Figure 8b shows the intensity of the fundamental light inside a polymeric pillar with a diameter of 800 nm and a height of 1.3 μm. Figure 8c shows the field intensity when the 1064 nm light passed though a thin film with a thickness of 1.3 μm. It is clear that, the fundamental field is amplified and localized inside the pillar cavity, resulting in an amplification of 7 to 13 times, as compared to that inside the thin film. Therefore, when a KTP NP is embedded inside the polymeric pillar, the SHG will be enhanced by a factor in between 49 and 169. Furthermore, it was observed that a small change of NP position inside the cavity leads to a significant change of the SHG signal due to the modulation of the fundamental light field inside the cavity, as seen in Figure 8b. This suggests that the experimental determination of the NP position in the structures is very importance to obtain an optimum SHG enhancement.

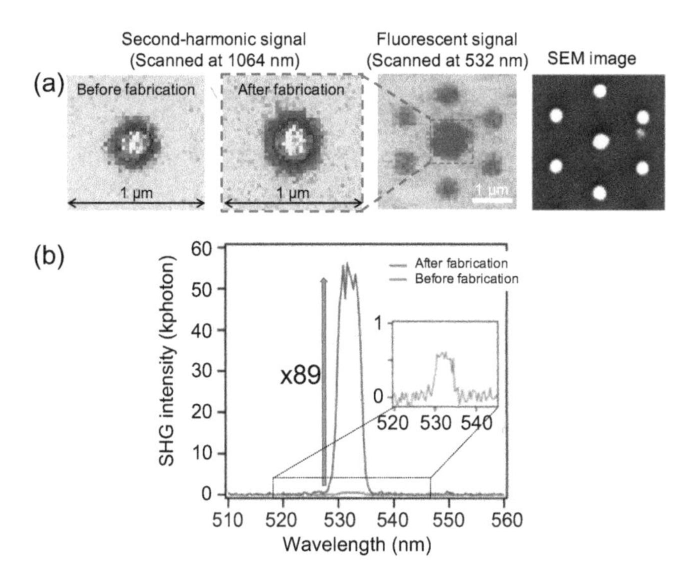

Figure 7. (**a**) SHG images of a KTP NP obtained before and after fabrication, and a fluorescence image of the fabricated structure. In the right: SEM of corresponding fabricated structure. (**b**) Comparison of the SHG spectra intensities of the same KTP NP showing an enhancement factor of about 90; Inset: zoom-in of the SHG spectrum before fabrication.

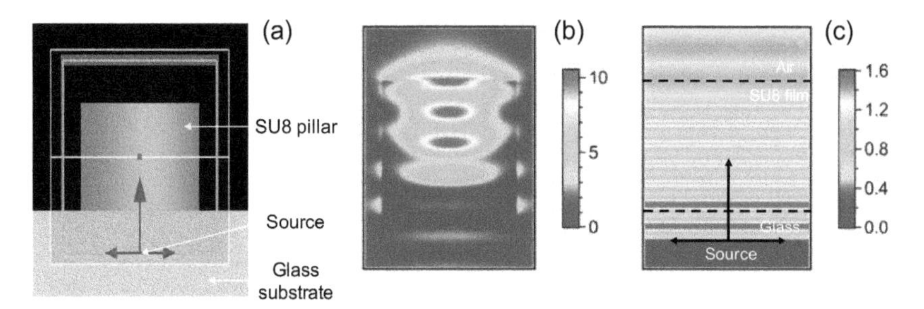

Figure 8. (**a**) Illustration of the model used to simulate the coupling of a light beam and a polymeric pillar. A light beam at 1064 nm–wavelength is emitted from the source and sent towards the cavity ($n_{\text{glass}} = 1.5$, $n_{\text{SU8}} = 1.58$). (**b**) Simulation of light intensity inside a polymeric pillar (diameter = 800 nm, height = 1.3 µm). (**c**) Simulation of light intensity at 1064 nm–wavelength propagating through a polymeric film (thickness = 1.3 µm).

We then investigated how the SH light is coupled out of the cavity. For simplification, we considered the nonlinear NP as a single oscillating electric dipole. Naturally, we found that the best orientation of the emitting dipole should be parallel to the glass substrate surface. We have then compared the two particular configurations: a single KTP NP embedded in a polymeric film (thickness = 1.3 μm) and a polymeric pillar (diameter = 800 nm; height = 1.3 μm). Furthermore, the emitting dipole is assumed to be embedded in polymeric material at center of the thin film or of the pillar, i.e., at 650 nm from the glass interface. The SHG detector is placed inside the glass substrate. Figure 9a shows the SH intensity distribution in the (*xz*) and (*yz*)–planes of a single emitting dipole embedded in a polymeric film (top) and in a polymeric pillar (bottom). It is evident that, thanks to the pillar cavity, the emitted light is compressed in a small cone and mostly directed towards the detector (downward direction). In contrast, in the case of the film, the light is generated in a large angle, which cannot be totally detected by the detector. To be more precise, we have numerically evaluated the collection efficiency of the emitted light as a function of the microscope objective, a main component of the detection system. Figure 9b shows the simulation result of the collection efficiency as a function of the half angle, defined as arcsin(NA/*n*), where NA is the numerical aperture of the objective lens and *n* is the refractive index of the immersion oil. In case of an objective lens with NA = 1.3 (half angle = 58.8°), 55% of the photons emitted by the NP embedded in a polymeric pillar can be collected while this is only 11% in the case of polymeric film. The polymeric pillar cavity thus does not only amplify the fundamental light intensity, resulting in an enhancement of the nonlinear optical process, but also guide the emitted signal into the detection angle, hence enhancing the collection efficiency. Theoretically, when the KTP NP is embedded at the middle of the pillar, the best collection efficiency is obtained with a pillar having a diameter of 800 nm and a height of 1300 nm, which is quite consistent with the experimental result.

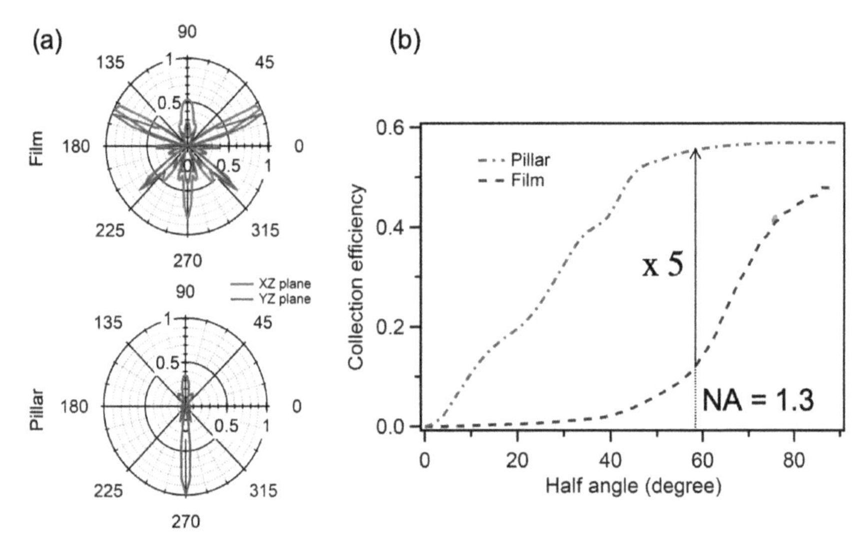

Figure 9. (**a**) Simulation of radiation patterns, in (*xz*) and (*yz*)–plane, of a single dipole embedded in different configurations. Top: emission diagram of a single dipole embedded in a thin film. Bottom: emission diagram of a single dipole embedded in a polymeric pillar. For both cases, the dipole was assumed to be in *x*-axis. (**b**) Calculation of the collection efficiency of the SHG signal as a function of the half angle of the objective lens for two cases. An enhancement factor of 5 is obtained.

5. Conclusions

In this work, we have first investigated the SHG of individual KTP NP with a size of about 100 nm, under a pulsed laser excitation at 1064 nm. The KTP NP shows a high nonlinear coefficient, as compared to a bulk one, and the SHG is perfectly stable at room temperature. Its transparency for both wavelengths of the fundamental and SH wavelengths leads to a high signal to noise ratio. We then demonstrated the LOPA-based DLW as an excellent technique to realize desired polymeric photonic structures containing a single KTP NP. Thanks to its transparency at 532 nm, KTP NPs did not induce explosion or damage during fabrication. Different kinds of 2D and 3D polymeric structures containing individual KTP NPs were successfully demonstrated. The size and shape of the polymeric structures and the position of the KTP NP inside the structure were demonstrated to be controlled at nanoscale. When the KTP NP is embedded in a polymeric pillar, the SHG is enhanced by a factor of 90. The SHG enhancement was varied for different KTP NPs, because of the random orientation of the NP inside the cavity. This affects the nonlinear interaction between the fundamental beam and the KTP crystal as well as the collection efficiency of the optical system. Different simulations using FDTD method have been done and shown very good agreement with the experimental results.

Author Contributions: D.T.T.N. and N.D.L. conceived and designed the experiments; D.T.T.N. performed the experiments, analyzed the data, and did simulations. All authors wrote, reviewed and approved the final version of manuscript.

Funding: This research is funded by a public grant overseen by the French National Research Agency in the frame of GRATEOM project (ANR-17-CE09-0047-01).

Acknowledgments: We thank L. Le Xuan and J.-F. Roch for providing nonlinear nanoparticles, I. Ledoux-Rak for fruitful discussion about nonlinear optics, and Phoebe Marcus-Porter for English revision.

Conflicts of Interest: The authors declare no conflict of interest.

References

1. Boyd, R. *Nonlinear Optics*, 3rd ed.; Academic Press: Cambridge, MA, USA, 2008; pp. 1–640.
2. Le Xuan, L.; Zhou, C.; Slablab, S.; Chauvat, D.; Tard, C.; Perruchas, S.; Gacoin, T.; Villeval, P.; Roch, J.-F. Photostable Second-Harmonic Generation from a Single $KTiOPO_4$ Nanocrystal for Nonlinear Microscopy. *Small* **2008**, *4*, 1332–1336. [CrossRef] [PubMed]
3. Zielinski, M.; Oron, D.; Chauvat, D.; Zyss, J. Second-Harmonic Generation from a Single Core-Shell Quantum Dot. *Small* **2009**, *5*, 2835–2840. [CrossRef] [PubMed]
4. Gao, X.; Cui, Y.; Levenson, R.-M.; Chung, L.-W.; Nie, S. In vivo cancer targeting and imaging with semiconductor quantum dots. *Nat. Biotechnol.* **2004**, *22*, 969–976. [CrossRef] [PubMed]
5. Kachynski, A.V.; Kuzmin, A.N.; Nyk, M.; Roy, I.; Prasad, P.N. Zinc Oxide Nanocrystals for Nonresonant Nonlinear Optical Microscopy in Biology and Medicine. *J. Phys. Chem. C* **2008**, *112*, 10721–10724. [CrossRef] [PubMed]
6. Extermann, J.; Bonacina, L.; Cuna, E.; Kasparian, C.; Mugnier, Y.; Feurer, T.; Wolf, J.-P. Nanodoublers as deep imaging markers for multi-photon microscopy. *Opt. Express* **2009**, *17*, 15342–15349. [CrossRef] [PubMed]
7. Grange, R.; Lanvin, T.; Hsieh, C.-L.; Pu, Y.; Psaltis, D. Imaging with second-harmonic radiation probes in living tissue. *Biomed. Opt. Express* **2011**, *2*, 2532–2539. [CrossRef] [PubMed]
8. Staedler, D.; Magouroux, T.; Hadji, R.; Joulaud, C.; Extermann, J.; Schwung, S.; Passemard, S.; Kasparian, C.; Clarke, G.; Gerrmann, M.; et al. Harmonic Nanocrystals for Biolabeling: A Survey of Optical Properties and Biocompatibility. *ACS Nano* **2012**, *6*, 2542–2549. [CrossRef] [PubMed]
9. Ibanez, A.; Maximov, S.; Guiu, A.; Chaillout, C.; Baldeck, P.L. Controlled Nanocrystallization of Organic Molecules in Sol-Gel Glasses. *Adv. Mater.* **1998**, *10*, 1540–1543. [CrossRef]
10. Sanz, N.; Baldeck, P.L.; Nicoud, J.-F.; Le Fur, Y.; Ibanez, A. Polymorphism and luminescence properties of CMONS organic crystals: Bulk crystals and nanocrystals confined in gel-glasses. *Solid State Sci.* **2001**, *3*, 867–875. [CrossRef]
11. Treussart, F.; Botzung-Appert, E.; Ha-Duong, N.-T.; Ibanez, A.; Roch, J.-F.; Pansu, R. Second Harmonic Generation and Fluorescence of CMONS Dye Nanocrystals Grown in a Sol-Gel Thin Film. *ChemPhysChem* **2003**, *4*, 757–760. [CrossRef] [PubMed]
12. Delahaye, E.; Tancrez, N.; Yi, T.; Ledoux, I.; Zyss, J.; Brasselet, S.; Clement, R. Second harmonic generation from individual hybrid MnPS3-based nanoparticles investigated by nonlinear microscopy. *Chem. Phys. Lett.* **2006**, *429*, 533–537. [CrossRef]
13. Driscoll, T.A.; Hoffman, H.J.; Stone, R.E.; Perkins, P.E. Efficient second-harmonic generation in KTP crystals. *J. Opt. Soc. Am. B* **1986**, *3*, 683–686. [CrossRef]

14. Hohmann, J.K.; Renner, M.; Waller, E.H.; von Freymann, G. Three- Dimensional μ-Printing: An Enabling Technology. *Adv. Opt. Mater.* **2015**, *3*, 1488–1507. [CrossRef]
15. Maruo, S.; Fourkas, J. Recent progress in multiphoton microfabrication. *Laser Photonics Rev.* **2008**, *2*, 100–111. [CrossRef]
16. Li, Q.; Do, M.T.; Ledoux-Rak, I.; Lai, N.D. Concept for three-dimensional optical addressing by ultralow one-photon absorption method. *Opt. Lett.* **2013**, *38*, 4640–4643. [CrossRef] [PubMed]
17. Do, M.T.; Nguyen, T.T.N.; Li, Q.; Benisty, H.; Ledoux-Rak, I.; Lai, N.D. Submicrometer 3d structures fabrication enabled by one-photon absorption direct laser writing. *Opt. Express* **2013**, *21*, 20964–20969. [CrossRef] [PubMed]
18. Do, M.T.; Nguyen, D.T.T.; Ngo, H.M.; Ledoux-Rak, I.; Lai, N.D. Controlled coupling of a single nanoparticle in polymeric microstructure by low one-photon absorption–based direct laser writing technique. *Nanotechnology* **2015**, *26*, 105301. [CrossRef] [PubMed]

Article

Highly Localized and Efficient Energy Harvesting in a Phononic Crystal Beam: Defect Placement and Experimental Validation

Xu-Feng Lv, Xiang Fang, Zhi-Qiang Zhang, Zhi-Long Huang and Kuo-Chih Chuang *

School of Aeronautics and Astronautics, Institute of Applied Mechanics, Zhejiang University, Key Laboratory of Soft Machines and Smart Devices of Zhejiang Province, Hangzhou 310027, China

* Correspondence: chuangkc@zju.edu.cn; Tel.: +86-139-6712-7543

Received: 25 June 2019; Accepted: 27 July 2019; Published: 30 July 2019

Abstract: We study energy harvesting in a binary phononic crystal (PC) beam at the defect mode. Specifically, we consider the placement of a mismatched unit cell related to the excitation point. The mismatched unit cell contains a perfect segment and a geometrically mismatched one with a lower flexural rigidity which serves as a point defect. We show that the strain in the defect PC beam is much larger than those in homogeneous beams with a defect segment. We suggest that the defect segment should be arranged in the first unit cell, but not directly connected to the excitation source, to achieve efficient less-attenuated localized energy harvesting. To harvest the energy, a polyvinylidene fluoride (PVDF) film is attached on top of the mismatched segment. Our numerical and experimental results indicate that the placement of the mismatched segment, which has not been addressed for PC beams under mechanical excitation, plays an important role in efficient energy harvesting based on the defect mode.

Keywords: energy harvesting; defect modes; phononic crystals (PCs)

---

1. Introduction

With the growing need for portable, wireless, or wearable electronic devices, highly efficient alternative power generation has increasingly become a necessity. In the past decade, harvesting ambient energy from environmental sources, such as vibrations, fluid flow, or thermal gradients has gained much attention. Up to now, broadband energy harvesting using piezoelectric materials from ambient vibrations based on cantilever configurations have been the main focus because they can produce high dynamic strain and are compatible to microelectromechanical systems fabrication [1–4].

Energy harvesting using sonic crystals, phononic crystals (PCs), or acoustic/elastic metamaterials based on wave focusing or wave localization have attracted increasing attention [5–9]. PCs are periodic composite structures with frequency band gaps capable of forbidding elastic wave propagations [10,11]. Elastic metamaterials are generally structures with periodic local resonators designed to achieve sub-wavelength band gaps and non-traditional manipulations of wave propagations [12–14]. Energy harvesting using PCs or metamaterials has been achieved at the defect modes through a resonant cavity (or an inclusion) in the airborne sound configuration or under mechanical excitation [15–18]. Here, the defect mode stands for a bandgap resonance mode which comes from the presence of mismatched unit cells. Recently, energy harvesting using a finite phononic crystal beam with a point defect was investigated considering thermal effects, but without experimental validations [9]. In fact, defect modes are one of the important characteristics of imperfect periodic structures that have the potential to be used for energy harvesting due to their capability of localizing sound or elastic waves around the defects [19–22]. However, contrary to the airborne sound configuration, placement of the point defect in finite PCs or metamaterials under mechanical excitation has not been addressed [15]. Placement of the mismatched unit cell should play an important role in energy harvesting for minimizing the influence of wave attenuation inside the frequency band gaps. Since PC beams have received great attention due to their engineering importance, the proposed energy harvesting configuration will provide another practical application for PC beams, and the experimental validations will provide useful information for future related studies of energy harvesting using sonic or phononic crystals [14,23–27].

Although propagation of elastic waves is completely forbidden inside the band gaps in ideally infinite PC beams, it is only attenuated away from the excitation point in practical finite ones. Thus, most of the flexural wave energy is locally confined in the neighborhood of the excitation point. In this work, we simultaneously combine this fact (i.e., wave confinement near the excitation source) with the characteristics of the defect modes (i.e., wave localization around the point defect) in a PC beam to achieve highly efficient bandgap energy harvesting. One of the unit cells in the PC beam has a perfect segment that is connected to the excitation source and a geometrically mismatched flexible one that is connected to the rest of the PC beam. We mainly focus on the placement of the mismatched unit cell related to the mechanical excitation point and the associated energy harvesting at the defect modes. Thus, we note that the energy harvesting is not carried out at ordinary passband resonance frequencies or compared to those of general straight beam-type energy harvesters.

In this work, energy harvesting using a typical but representative binary PC beam inside the band gaps is investigated. A polyvinylidene fluoride (PVDF) film is bonded on the mismatched flexible segment as an energy harvester. A point-wise fiber Bragg grating (FBG) displacement sensing system is set up in advance to detect the displacement transmission and determine the defect-mode frequencies. We will demonstrate that the combination of wave confinement near the excitation point and the localization of the flexural waves at the defect modes enable highly localized, less attenuated, and efficient energy harvesting.

2. Piezoelectric Energy Harvesting

Figure 1 shows a finite binary PC beam containing 10 unit cells made of two different materials (1) and (2). The material of the segment marked as (3) is the same as that of the segment (2), but segment (3) has a lower height to represent a geometrically mismatched segment. We considered three placement conditions of the mismatched unit cell with respect to the excitation point.

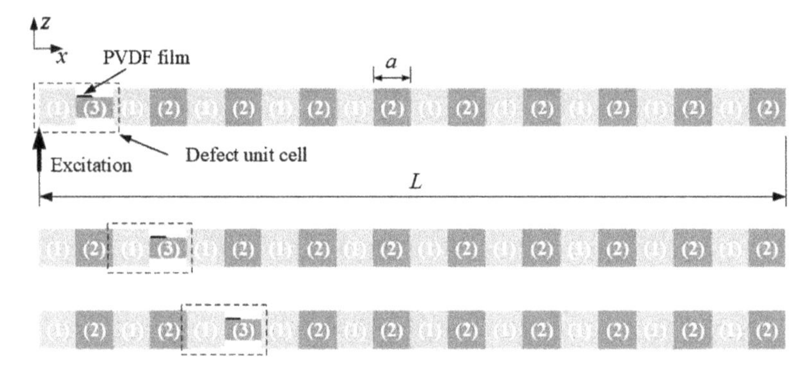

Figure 1. Schematic diagram of the phononic crystal (PC) beam with a mismatched segment (color online).

We attached the easily-deformed PVDF film on the mismatched segment close to the excitation point as an energy harvester, as shown in Figure 1. The electrical displacement (i.e., charge density) is expressed by the piezoelectric constitutive relations as follow:

$$D_3 = d_{31} Y \varepsilon_x + \varepsilon_{33} E_3, \quad (1)$$

where ε_x is the strain on the mismatched segment along the beam, d_{31} is the piezoelectric constant in the 31 coupling direction, Y is Young's modulus of the PVDF film, ε_{33} is the dielectric constant, and E_3 is the transverse electric field in the PVDF. The collected electric charge on the electrode surface can be expressed as the integral of the electrical displacement over the area of the surface as follows:

$$Q = \int_A D_3 \mathrm{d}A = b_t \int_{l_0}^{l_1} (d_{31} Y \varepsilon_x + \varepsilon_{33} E_3) \mathrm{d}x, \tag{2}$$

where b_t is the width of the PVDF. Assuming a uniform electrical field, the charge collected on the electrode surface can be expressed as [28]

$$Q = b_t l_t d_{31} Y \overline{\varepsilon}_x - C_p U, \tag{3}$$

where U is the potential difference, l_t is the length of the PVDF film, $\overline{\varepsilon}_x$ is the average strain over the surface of the piezoelectric layer, and C_p is the capacitance expressed by

$$C_p = \frac{b_t l_t \varepsilon_{33}}{\Delta}, \tag{4}$$

where Δ is the thickness of the piezoelectric layer. The root mean square amplitude of the output harmonic steady-state voltage can be determined as

$$U = \frac{1}{\sqrt{2}} \frac{\omega b_t l_t d_{31} Y \overline{\varepsilon}_x}{1 + \omega b_t l_t \varepsilon_{33}^T R_L / \Delta} R_L, \tag{5}$$

where R_L is the external load resistance. The time-average of the output power of the PVDF film dissipated in the resistive load can be expressed by multiplication of the voltage and current flow (i.e., $I = \omega Q$) as

$$P_{av} = \frac{1}{2} \left(\frac{\omega b_t l_t d_{31} Y \overline{\varepsilon}_x}{1 + \omega C_p R_L} \right)^2 R_L. \tag{6}$$

Clearly, the output power is a function of the external load resistance R_L, the excitation frequency ω, the average dynamic strain $\overline{\varepsilon}_x$ of the mismatched segment, the piezoelectric constants, and dimensions of the energy harvester. The maximum value of the output power can be obtained by selecting the external load resistance as [16].

$$R_L^* = 1/\omega C_p, \tag{7}$$

which is related to the excitation frequency and the capacitance of the PVDF film.

In our modeling for energy harvesting, PVDF is assumed to have weak electromechanical coupling, in which converse coupling is neglected. The generated power is a quadratic function of the dynamic strain. Thus, the optimum placement of the PVDF film is related to the distribution of the dynamic strain field. It should be noted that the optimal size of the PVDF is not the focus in this work since for the considered defect mode, it only vibrates in the first bending mode and the harvested power will not be averaged out.

As shown in Figure 1, a PC beam having a mismatched unit cell (i.e., containing a perfect and a geometrically mismatched segment) placed in three different locations is considered. Each unit cell has two segments made of 6061 aluminum (i.e., material (1)) and acrylic (PMMA, material (2)). Each of the two materials has a size of 0.08 m (length) × 0.015 m (width) × 0.015 m (height). The height of the point defect (i.e., marked as (3) with the same material as (2), PMMA) is 0.008 m, and the other dimensions of the point defect are the same as the other segments. The local mismatched segment with a lower flexural rigidity serves as a perturbation to a perfect PC beam. For convenience, we, respectively, name the three PC beams as EHPC1, EHPC2, and EHPC3, in which "E" denotes energy and "H" denotes harvesting. A PVDF film energy harvester is attached on the top surface of the mismatched segment near the boundary between it and another segment in the first, second, and third unit cells of EHPC1, 2, and 3. The attachment of the PVDF film on EHPC1 is illustrated in Figure 1.

To validate the defect mode-based wave localization in the three PC beams, we calculate the flexural displacement transmission (denoted as frequency response function (FRF) in the following figures) using the finite element method (FEM). The displacement transmission is obtained by calculating $T(\omega) = 20\log_{10}|W_{out}(\omega)/W_{in}(\omega)|$, where $W_{out}(\omega)$ and $W_{in}(\omega)$ are, respectively, the line-average (along the width of the beam) displacement amplitudes at the two ends of the PC beam. In FEM simulations, the elastic constants of the materials are ρ_1 = 2735 kg/m^3, E_1 = 74.7 GPa, ρ_2 = 1142 kg/m^3, and E_1 = 4.5 GPa. The Poisson's ratios of the two materials are both 0.33. The damping is modeled in terms of a structural loss factor of 0.001 for aluminum and 0.01 for PMMA. A 1.74 μm vertical excitation (along the width of the beam) is given in the simulation.

From Figure 2a we can see that, after arranging a mismatched segment, the defect mode occurs in the band gaps. The passband resonance modes are in general larger than 0 dB in the displacement transmission. Although the considered PC beam is also truncated from an ideally infinite PC beam, we note that the defect modes are different from the truncation modes or surface modes in that wave will be localized around the mismatched unit cell at the frequency of the defect mode [29–31]. The insets in Figure 2a show the predicted output voltage around the two band gaps when an external load resistance of 500 kΩ is applied in the FEM simulation using the module "Piezoelectric Devices" in COMSOL MULTIPHYSICS. The PVDF film has a size of 0.0147 m (length) × 0.01 m (width) × 28 μm (thickness). To understand the behavior of the PVDF, the material constants of the PVDF are taken from the COMSOL database. We can see that, even compared with the resonance frequencies at the edges of the band gaps, the harvested voltage reaches the maximum when the PC beam is excited at the defect mode. The most efficient energy harvesting is obtained using the EHPC1 beam, where, in addition to the defect-mode wave localization, the elastic wave is less attenuated. To further gain insights into the less-attenuated defect-mode based energy harvesting, full-field strain distribution at the two defect modes of the defect and perfect PC beam are shown in Figure 2b,c, respectively. A non-smooth strain field distribution is observed due to the material discontinuities (i.e., impedance mismatch). The strains in the softer PMMA segments are larger than those in the harder aluminum segments. We can see that the mismatched segment possesses the largest dynamic strain compared to the other segments when the PC beam is excited at the defect modes.

One might think that the introduced mismatch segment acts as a local damage, and the variation of strain distribution must be sensitive according to the concept of damage detection. It is true that the strain mode shapes are sensitive to local damages. However, in PC beams, the mismatched flexible segment possesses higher strain mainly due to the confinement of flexural waves at the defect modes and the conditions of Bragg scattering. As an illustration, the displacement vibration shape and the strain distribution of the three PC beams (i.e., EHPC1, 2, 3) and two homogeneous beams with a thinner segment as that of the mismatched segment in the PC beam are simulated using the FEM, and the results are shown in Figure 3. All the strain responses are obtained at the same point on the mismatched segment, which is one third the length of the PVDF film and close to the discontinuity boundary. Although the strain responses are all obtained at the same point, the strain responses obtained from the PC beam are much larger than those obtained from the homogeneous beams, as shown in Figure 3. In addition, as the mismatched segment moves away from the excitation point (i.e., EHPC2 and EHPC3), the strain magnitude and their difference between the first and the second defect modes become significantly smaller compared to that of EHPC1. The difference of the strain magnitude between the defect mode and the passband resonance modes also becomes less obvious when the mismatched segment moves away from the excitation point. The strain magnitude shown in Figure 3 agrees with the predictions of the output voltage shown in Figure 2a.

In Figure 3 we observe that the defect modes behave similarly (e.g., similar larger local displacement shapes and strain distribution) and the vibrations in the other end of the PC beam are strongly attenuated because the defect modes still lie in the band gaps. In addition, from Figure 3, we can see that the magnitude of the dynamic strain distribution in the second defect mode around the mismatched segment is much larger than that in the first defect mode. This is because the second defect mode lies in a band gap deeper and wider than the first band gap where the first defect mode exists, as shown in Figure 2a. Thus, under the same vibration excitation condition, the input energy is more highly localized in the second defect mode. Figure 3 also suggests higher harvested power from the second defect mode, which will be confirmed later by our experiment results on energy harvesting. Note that for higher-order defect modes with even shorter wavelengths, the harvested voltage or power on the PVDF film might not be high since the short-wavelength strain might be averaged out in the PVDF film.

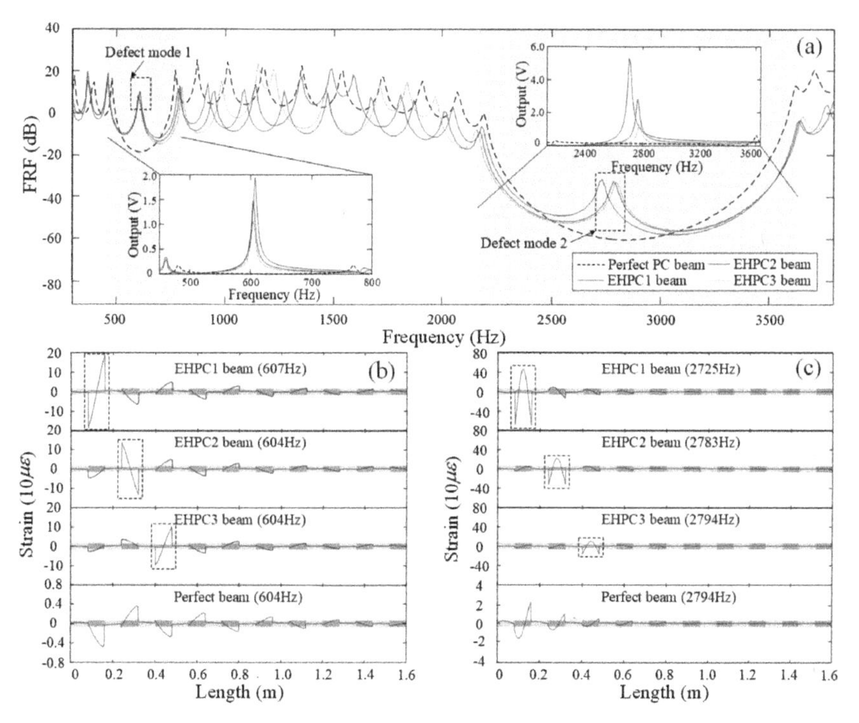

Figure 2. (**a**) Displacement transmission of the three PC beams and simulated output voltage in the band gaps. (**b**) Full-field strain distribution at the frequency of the first defect mode of the defect and perfect PC beam. (**c**) Full-field strain distribution at the frequency of the second defect mode of the defect and perfect PC beam.

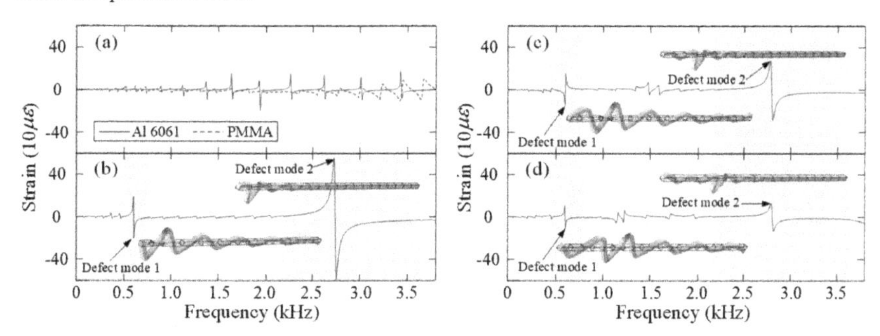

Figure 3. Strain responses in the (**a**): homogeneous beam with a mismatched segment, (**b**): EHPC1 beam, (**c**): EHPC2 beam, and (**d**): EHPC3 beam with the dynamic displacement vibration shapes and the strain field distribution at the defect modes (color online).

We further study the dependence of the harvested voltage. Figure 4a shows the simulated output voltage when the mismatched segment has different thicknesses. We see that the harvested voltage first increases then decreases when the thickness is varied from 12 mm to 6 mm. Thus, a moderate perturbation in the segment, instead of large impedance mismatching, is beneficial to the energy harvesting. The proposed PC beam has a thickness of 8 mm. In fact, a defect mode caused by a low-rigidity mismatched segment in a binary PC beam is in essential a shift of a resonance frequency from the upper (or lower for a high-rigidity mismatched segment) bandgap edge [32]. If the impedance difference of the constituent segments in the mismatched unit cell increases, the defect mode might largely shift to be close to the lower edge of the band gap and the wave localization at the defect mode might thus be minimized. Since the damping of the bonding material might affect the harvested output voltage, we further introduce damping at the bottom of the PVDF, and the damping is modeled as a loss factor ξ, as shown in Figure 4b. We see that the damping of the bonding layer will decrease the output voltage when the PC beam is excited at the defect mode. In Figure 4c, for EHPC1 beam, we change the external load resistance and observe the variation of the output power and voltage. The results indicate that the harvested voltage increases as the external load resistance increases. The harvested power first increases and then decreases after passing the optimized external load resistance 750 kΩ, where the trend and the optimized resistance agree with the theoretical predictions in Equations (6) and (7).

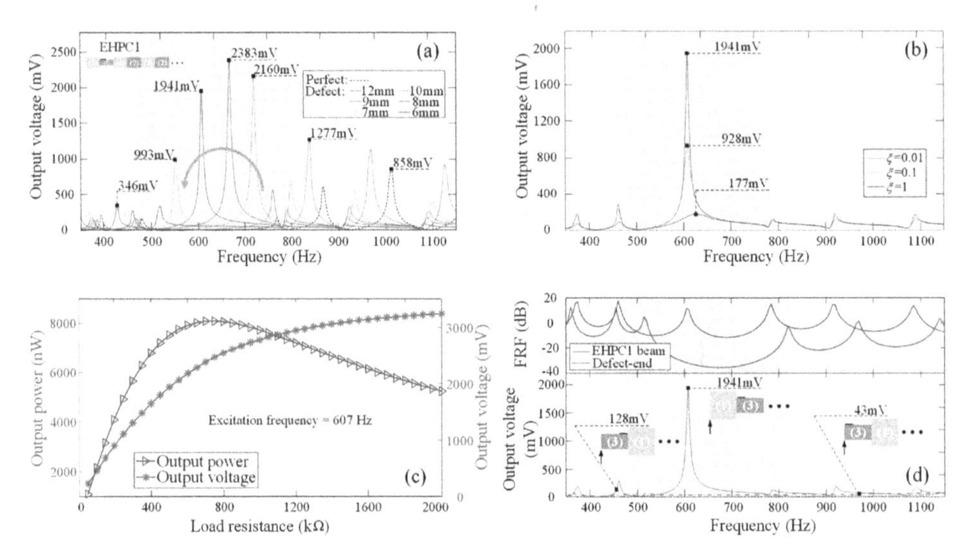

Figure 4. (**a**) Output voltage from the PC beams with different defect conditions. (**b**) Influences of the damping of the bonding layer. (**c**) Influences of the external load resistance to the output power and voltage. (**d**) Influences of the sequence of the defect segment in the first unit cell on displacement transmission and energy harvesting.

The possible combination of the wave confinement in the neighborhood of the excitation point and the wave localization around the point defect might cause the misunderstanding that the best placement of the geometrically mismatched segment is exactly at the excited point of the PC beam, where the excited flexural wave always reaches the largest magnitude without being strongly attenuated by Bragg scattering. However, if the mismatched segment is the first segment being excited by the excitation source, it only serves as a dissipative material that weakens the excitation magnitude instead of a part of the PC beam and does not contribute to the formation of the defect modes. To show the wrong idea of placing segment (3) adjacent to the excitation source, we compare the location of segment (3) in displacement transmission and energy harvesting as shown in Figure 4d. When segment (3) is located adjacent to the excitation, despite being less attenuated by Bragg scattering, segment (3) only serves as a part of the weakened excitation, and the PC beam behaves as defect-free because segment (3) is softer than the adjacent segment (1). Thus, when considering the placement of the point defect and the localization of flexural waves at the defect modes, the lower rigidity segment is suggested to be arranged as the second segment in the first unit cell that is connected to the excitation source.

3. Experimental Validations

Before validating the defect mode-based energy harvesting experimentally, we employed a high-sensitive fiber Bragg grating (FBG) displacement sensing system (as shown in Figure 5) to directly detect the defect modes in the displacement transmission [32]. The FBG displacement sensing system contains two FBG displacement sensors, located at the two extreme ends of the PC beam. White noise random signals, generated by the Simulink (The MathWorks, Natick, MA) and the dSPACE DS1104 system, are sent to a piezoelectric multilayered actuator through a power amplifier with a sampling frequency 50 kHz to excite flexural wave propagation. The displacement transmission is then obtained using a stochastic spectral estimation. The experimental transmissions of the three PC beams (i.e., EHPC1, 2, and 3), extracted from the responses of FBG2 (output sensor) and FBG1 (input sensor), are, respectively, shown in Figure 6a–c. The experimental results are compared with those obtained by the FEM simulations, where for clear observation the experimental transmission is shifted by −10 dB.

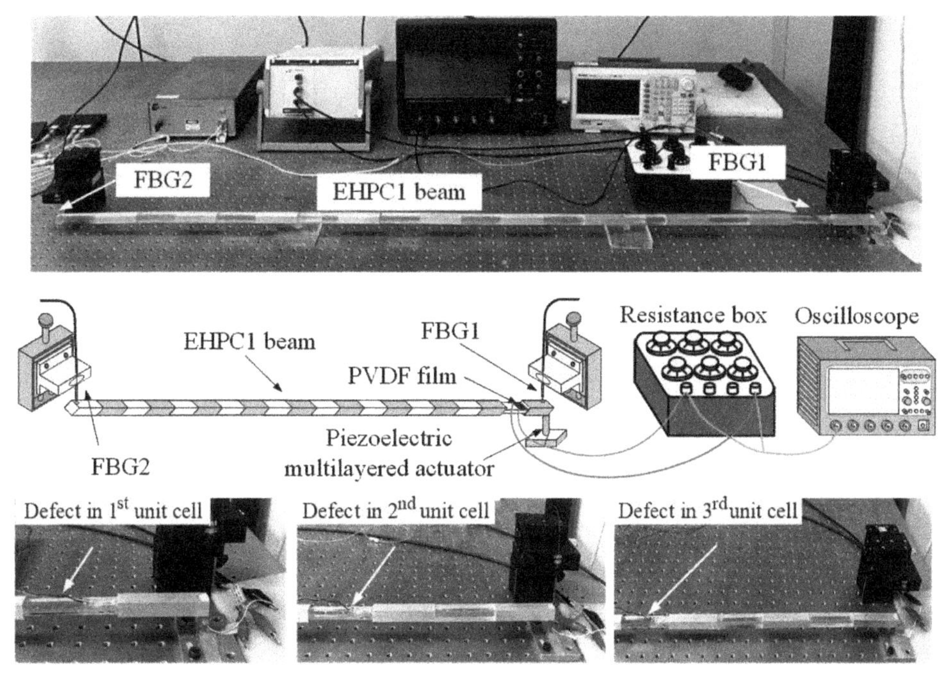

Figure 5. Experimental setup (including a fiber Bragg grating (FBG) displacement sensing system and the energy harvesting system) (color online).

From Figure 6, we can see that the regions of the band gaps and the pass bands are almost the same for the three PC beams with different placements of the mismatched segment. The simulated and measured band gaps and the defect modes in the three PC beams are listed in Table 1. The experimentally identified defect modes in the first two band gaps are indicated in the corresponding displacement transmission in Figure 6. From Figure 6 and Table 1, we can see that the displacement transmissions obtained from the FBG experiments and FEM simulations agree well with each other. Note that the good agreements between the trends of the resonance peaks in the pass bands, band gaps, and defect modes also indicate an excellent dynamic measurement capability of the FBG displacement sensing system.

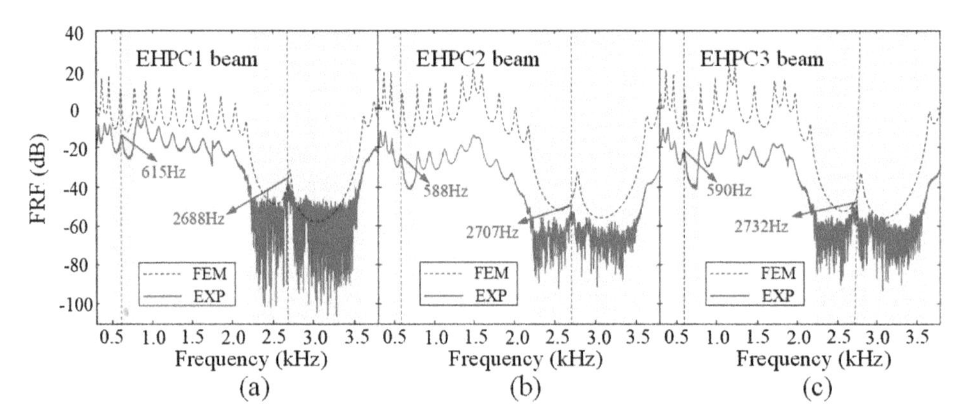

Figure 6. Displacement transmission of the PC beam: (**a**) EHPC1 beam, (**b**) EHPC2 beam, and (**c**) EHPC3 beam (color online).

Table 1. Comparisons of the band gaps and defect modes obtained by the finite element method (FEM) and fiber Bragg grating (FBG) [1].

Method	FEM	FBG
EHPC1 beam	Frequency	Frequency (Error %)
Band gap1	478–766	497–778
Band gap2	2180–3630	2148–3601
Defect mode1	607	615 (1.32)
Defect mode2	2725	2688 (−1.36)
EHPC2 beam	Frequency	Frequency (Error %)
Band gap1	480–775	493–780
Band gap2	2170–3650	2192–3653
Defect mode1	604	588 (−2.65)
Defect mode2	2783	2707 (−2.73)
EHPC3 beam	Frequency	Frequency (Error %)
Band gap1	478–788	492–771
Band gap2	2160–3660	2190–3655
Defect mode1	604	590 (−2.32)
Defect mode2	2794	2732 (−2.22)

[1] Unit: Hz

From Table 1 we see that, despite different locations of the mismatched segment, the frequency ranges of the band gaps and the frequencies of the defect modes are close to each other in the three PC beams. For example, in the experimental results, the first defect modes in the three PC beams are, respectively, 615 Hz, 588 Hz, and 590 Hz. As for the second defect modes, they are, respectively, 2688 Hz, 2707 Hz, and 2732 Hz. The discrepancies between the experimental drifts of the band gaps (or the defect modes) and the numerical results are within 3%.

Then, the defect mode-based wave localization was applied for energy harvesting using PVDFs (LDT0-028K, Measurement Specialties, Inc., Wayne, PA), having the same dimensions as those in the previous simulations. As illustrated in Figure 5, the PVDF film is attached to the mismatched segment near the edge. We will compare the harvested voltage and power using the three PC beams.

The harvested output power is related to the external load resistance. According to Equation (7), the optimal external load resistances for obtaining maximum output power using EHPC1 for the first defect mode (i.e., 615 Hz) is 517.6 kΩ. We experimentally compared the harvested output voltage and output power using the three designed PC beams by adjusting the optimized external load resistance through a resistance box (as shown in Figure 5) at different frequencies around the first defect mode. The results are shown in Figure 7a for output voltage and Figure 7b for output power, respectively. Since the highest output power can be obtained in EHPC1 compared to that in EHPC2 and EHPC3, the less-attenuated defect mode-based wave localization is validated.

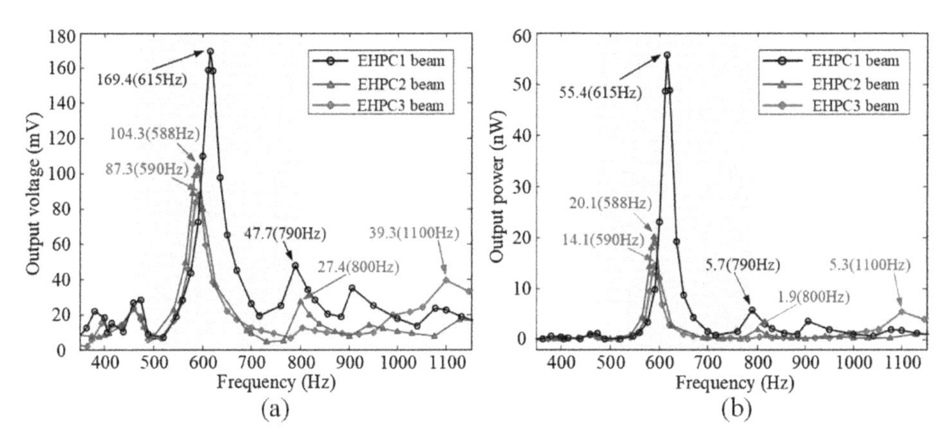

Figure 7. Experimental energy harvesting results around the first defect mode: (**a**) output voltage and (**b**) output power (color online).

According to Equation (6), under the same strain distribution, the harvested output power increases as the operational frequency increases. However, maximum output power (i.e., 55.4 nW) can be obtained in Figure 7 at the first defect mode at 615 Hz. It is interesting to point out that the defect mode-based wave localization might enable efficient energy harvesting even being compared to that operated at passband resonance frequencies (i.e., see resonance peaks outside the gray region in Figure 7). At the first defect mode, the maximum harvested power using EHPC2 and EHPC3 are, respectively, 20.1 nW and 14.1 nW. The decreasing of the harvested power at the defect modes, when the harvesting cell moves away from the excitation point, is expectable because the defect modes are inside the band gaps and waves are still attenuated as they propagate along each unit cell. They are, respectively, only 36.3% and 25.5% of the maximum harvested power using EHPC1. On the other hand, the harvested voltage or power at the resonance frequencies is less related to the location of the mismatched unit cell. From the experimental results in Figure 7, we can see that the defect modes behave quite differently from the passband resonance modes in that elastic waves are highly localized but globally and spatially decay along the unit cells. Thus, the defect-mode wave localization with the consideration of the placement of the mismatched segment in finite PC beams for energy harvesting is required. In this work, we only compare the optimal placement of the harvesting cell that contains the geometrically mismatched segment. The optimal placement and size of the energy harvester on the harvesting segment are not considered. Although not being considered, we note that at the considered two defect modes, the PVDF film only vibrates in its first bending mode, and thus the strain averaging effect needs not to be considered.

Finally, we validate the optimal load resistance for energy harvesting application. The harvested output power and voltage as a function of the external load resistance are shown in Figure 8. Figure 8a are the results obtained at the first defect mode (i.e., 615 Hz) and Figure 8b are those obtained at the second defect mode (i.e., 2688 Hz) using EHPC1. From Figure 8, we can see that the trends of the experimental results agree well with the theoretical predictions. It should be noted that the theoretical results derived through Equations (5) and (6), based on the experimental voltage with the corresponding calculated optimal load resistance, agreed with the theoretical predictions for the first defect mode at 615 Hz. The maximum output voltage and output power are, respectively, 169.4 mV and 55.4 nW. The corresponding experimental and theoretical (according to Equation (7)) optimal load resistance are, respectively, 465 kΩ and 517.6 kΩ. For the second defect mode at 2688 Hz, the maximum output voltage and output power are, respectively, 303.4mV and 777.5 nW. The corresponding experimental and theoretical optimal load resistance are, respectively, 110 kΩ and 118.4 kΩ.

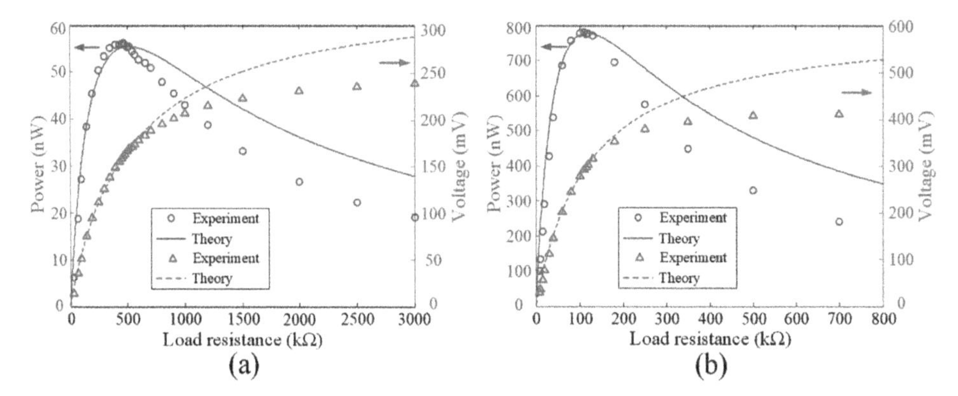

Figure 8. Experimental energy harvesting results at (**a**) the first defect mode (i.e., at 615 Hz) and (**b**) the second defect mode (i.e., 2688 Hz) using EHPC1 (color online).

Modeled as a single-degree-of-freedom (SDOF) harvester, the conversion efficiency of energy can be described as [28]

$$\eta_{\mathrm{me}} = \frac{R_{\mathrm{L}} k_{\mathrm{me}}}{R_{\mathrm{L}} k_{\mathrm{me}} + c + c\omega^2 C_p^2 R_{\mathrm{L}}^2} = \frac{k_{\mathrm{me}}}{k_{\mathrm{me}} + c\left(\frac{1}{R_{\mathrm{L}}} + \omega^2 C_p^2 R_{\mathrm{L}}\right)}, \tag{8}$$

where k_{me} is the modal piezoelectric coupling stiffness, and c is the modal mechanical damping. According to Equation (8), when the external load resistance is the optimal resistance R_{L}^*, the highest energy conversion efficiency can be obtained for weakly electromechanical coupling with the PVDF. The conversion efficiency at low resistance is smaller than the highest efficiency, which agrees well between the theoretical and experimental results. The cross-sensitivity between the drawn and transverse directions of the PVDF film due to the anisotropy of the piezoelectric effect might attribute to the discrepancies at high frequencies.

Our experimental investigation has validated that, for a normal PC beam, one can possibly perform energy harvesting by arranging a geometrically mismatched harvesting unit cell with a lower flexural rigidity between the PC beam and the excitation point. Although the PVDF is used in this work for demonstrating energy harvesting, higher power can be achieved in practice by using the piezoceramics (PZTs) with a higher value of d_{31} and with larger ambient vibration source.

4. Conclusions

By introducing a mismatched unit cell between the ambient vibration source and a PC beam operating in band gaps, we address a less-attenuated defect mode-based wave localization and apply it to energy harvesting. According to the wave confinement near the excitation point, the placement of the mismatched segment can be determined. From the wave localization at the defect mode, optimal ambient excitation frequencies can be decided.

A high-sensitive point-wise fiber Bragg grating (FBG) displacement sensing system is set up beforehand to obtain the displacement transmission of the PC beam for determining the frequencies of the defect modes. The harvested power at the defect mode is significant when the energy harvester is located on the mismatched segment in the first unit cell. In addition, the harvested voltage and power are higher when the PC beam is excited at the second defect mode that lies in the wider and deeper second band gap.

We have shown that energy harvesting in PC beams is feasible by introducing an imperfect segment to the original PC beams even when they are operated in suppressing unwanted ambient vibrations (i.e., operating inside the band gaps). It should be noted that the defect mode frequency, related to the dimensions of the mismatched segment, can be designed in practical PC beams to match the target known environmental, mechanical (or acoustic) excitation.

Author Contributions: Conceptualization, X.-F.L., Z.-Q.Z., and K.-C.C.; Formal analysis, X.-F.L., Z.-L.H., and K.-C.C.; Investigation, Z.-Q.Z. and K.-C.C.; Methodology, X.F. and K.-C.C.; Software, X.-F.L.; Supervision, K.-C.C.; Writing—original draft, X.-F.L., Z.-Q.Z., and K.-C.C.; Writing—review and editing, X.-F.L., X.F., and K.-C.C.

Funding: This research work was supported by the National Natural Science Foundation of China (Nos. 11672263 and 11532011).

Conflicts of Interest: The authors declare no conflict of interest. The funders had no role in the design of the study; in the collection, analyses, or interpretation of data; in the writing of the manuscript, or in the decision to publish the results.

References

1. Wu, Y.; Qiu, J.; Kojima, F.; Ji, H.; Xie, W.; Zhou, S. Design methodology of a frequency up-converting energy harvester based on dual-cantilever and pendulum structures. *AIP Adv.* **2019**, *9*, 045312. [CrossRef]
2. Saadon, S.; Sidek, O. A review of vibration-based MEMS piezoelectric energy harvesters. *Energy Convers. Manag.* **2011**, *52*, 500–504. [CrossRef]
3. Zhou, S.; Wang, J. Dual serial vortex-induced energy harvesting system for enhanced energy harvesting. *AIP Adv.* **2018**, *8*, 075221. [CrossRef]
4. Muthalif, A.G.; Nordin, N.D. Optimal piezoelectric beam shape for single and broadband vibration energy harvesting: Modeling, simulation and experimental results. *Mech. Syst. Signal Process.* **2015**, *54*, 417–426. [CrossRef]
5. Gonella, S.; To, A.C.; Liu, W.K. Interplay between phononic bandgaps and piezoelectric microstructures for energy harvesting. *J. Mech. Phys. Solids.* **2009**, *57*, 621–633. [CrossRef]
6. Chen, Z.; Yang, Y.; Lu, Z.; Luo, Y. Broadband characteristics of vibration energy harvesting using one-dimensional phononic piezoelectric cantilever beams. *Phys. B Condens. Matter* **2013**, *410*, 5–12. [CrossRef]
7. Chen, Z.; Guo, B.; Yang, Y.; Cheng, C. Metamaterials-based enhanced energy harvesting: A review. *Phys. B Condens. Matter* **2014**, *438*, 1–8. [CrossRef]
8. Hajhosseini, M.; Rafeeyan, M. Modeling and analysis of piezoelectric beam with periodically variable cross-sections for vibration energy harvesting. *Appl. Math. Mech.* **2016**, *37*, 1053–1066. [CrossRef]
9. Geng, Q.; Cai, T.; Li, Y. Flexural wave manipulation and energy harvesting characteristics of a defect phononic crystal beam with thermal effects. *J. Appl. Phys.* **2019**, *125*, 035103. [CrossRef]
10. Wang, Y.F.; Wang, Y.S.; Zhang, C. Bandgaps and directional propagation of elastic waves in 2D square zigzag lattice structures. *J. Phys. D Appl. Phys.* **2014**, *47*, 485102. [CrossRef]
11. Guo, Y.Q.; Fang, D.N. Formation of bending-wave band structures in bicoupled beam-type phononic crystals. *J. Appl. Mech.* **2014**, *81*, 011009. [CrossRef]
12. Huang, H.H.; Sun, C.T.; Huang, G.L. On the negative effective mass density in acoustic metamaterials. *Int. J. Eng. Sci.* **2009**, *47*, 610–617. [CrossRef]
13. Zhou, X.M.; Liu, X.N.; Hu, G.K. Elastic metamaterials with local resonances: An overview. *Theor. Appl. Mech. Lett.* **2012**, *2*, 041001. [CrossRef]
14. Liu, L.; Hussein, M.I. Wave motion in periodic flexural beams and characterization of the transition between Bragg scattering and local resonance. *J. Appl. Mech.* **2012**, *79*, 011003. [CrossRef]
15. Oudich, M.; Li, Y. Tunable sub-wavelength acoustic energy harvesting with a metamaterial plate. *J. Phys. D: Appl. Phys.* **2017**, *50*, 315104. [CrossRef]
16. Wang, W.C.; Wu, L.Y.; Chen, L.W.; Liu, C.M. Acoustic energy harvesting by piezoelectric curved beams in the cavity of a sonic crystal. *Smart Mater. Struct.* **2010**, *19*, 045016. [CrossRef]

17. Carrara, M.; Cacan, M.R.; Toussaint, J.; Leamy, M.J.; Ruzzene, M.; Erturk, A. Metamaterial-inspired structures and concepts for elastoacoustic wave energy harvesting. *Smart Mater. Struct.* **2013**, *22*, 065004. [CrossRef]
18. Lv, H.; Tian, X.; Wang, M.Y.; Li, D. Vibration energy harvesting using a phononic crystal with point defect states. *Appl. Phys. Lett.* **2013**, *102*, 034103. [CrossRef]
19. Sigalas, M.M. Defect states of acoustic waves in a two-dimensional lattice of solid cylinders. *J. Appl. Phys.* **1998**, *84*, 3026–3030. [CrossRef]
20. Khelif, A.; Djafari-Rouhani, B.; Vasseur, J.O.; Deymier, P.A. Transmission and dispersion relations of perfect and defect-containing waveguide structures in phononic band gap materials. *Phys. Rev. B* **2003**, *68*, 024302. [CrossRef]
21. Zhang, X.; Liu, Z.; Liu, Y.; Wu, F. Defect states in 2D acoustic band-gap materials with bend-shaped linear defects. *Solid State Commun.* **2004**, *130*, 67–71. [CrossRef]
22. Yao, Z.J.; Yu, G.L.; Wang, Y.S.; Shi, Z.F. Propagation of bending waves in phononic crystal thin plates with a point defect. *Int. J. Solids Struct.* **2009**, *46*, 2571–2576. [CrossRef]
23. Wang, G.; Wen, J.; Wen, X. Quasi-one-dimensional phononic crystals studied using the improved lumped-mass method: Application to locally resonant beams with flexural wave band gap. *Phys. Rev. B* **2005**, *71*, 104302. [CrossRef]
24. Wen, J.; Wang, G.; Yu, D.; Zhao, H.; Liu, Y. Theoretical and experimental investigation of flexural wave propagation in straight beams with periodic structures: Application to a vibration isolation structure. *J. Appl. Phys.* **2005**, *97*, 114907. [CrossRef]
25. Xiao, Y.; Wen, J.; Yu, D.; Wen, X. Flexural wave propagation in beams with periodically attached vibration absorbers: Band-gap behavior and band formation mechanisms. *J. Sound Vib.* **2013**, *332*, 867–893. [CrossRef]
26. Zhu, R.; Liu, X.N.; Hu, G.K.; Sun, C.T.; Huang, G.L. A chiral elastic metamaterial beam for broadband vibration suppression. *J. Sound Vib.* **2014**, *333*, 2759–2773. [CrossRef]
27. Zhang, H.; Xiao, Y.; Wen, J.; Yu, D.; Wen, X. Flexural wave band gaps in metamaterial beams with membrane-type resonators: Theory and experiment. *J. Phys. D Appl. Phys.* **2015**, *48*, 435305. [CrossRef]
28. Lu, F.; Lee, H.P.; Lim, S.P. Modeling and analysis of micro piezoelectric power generators for micro-electromechanical-systems applications. *Smart Mater. Struct.* **2003**, *13*, 57. [CrossRef]
29. Al Ba'ba'a, H.; Nouh, M.; Singh, T. Dispersion and topological characteristics of permutative polyatomic phononic crystals. *Proc. R. Soc. A* **2019**, *475*, 20190022. [CrossRef]
30. Hussein, M.I.; Biringen, S.; Bilal, O.R.; Kucala, A. Flow stabilization by subsurface phonons. *Proc. R. Soc. A* **2015**, *471*, 20140928. [CrossRef]
31. Davis, B.L.; Tomchek, A.S.; Flores, E.A.; Liu, L.; Hussein, M.I. Analysis of periodicity termination in phononic crystals. *ASME Paper No. IMECE* **2011**, *65666*, 973–977. [CrossRef]
32. Chuang, K.C.; Zhang, Z.Q.; Wang, H.X. Experimental study on slow flexural waves around the defect modes in a phononic crystal beam using fiber Bragg gratings. *Phys. Lett. A* **2016**, *380*, 3963–3969. [CrossRef]

Review

Recent Advances in Colloidal Photonic Crystal-Based Anti-Counterfeiting Materials

Mengyao Pan [1], Lebin Wang [1], Shuliang Dou [1], Jiupeng Zhao [2], Hongbo Xu [2], Bo Wang [1], Leipeng Zhang [1], Xiaobai Li [1], Lei Pan [2,*] and Yao Li [1,*]

[1] Center for Composite Material and Structure, Harbin Institute of Technology, Harbin 150001, China
[2] MIIT Key Laboratory of Critical Materials Technology for New Energy Conversion and Storage, School of Chemistry and Chemical Engineering, Harbin Institute of Technology, Harbin 150001, China
* Correspondence: panlei@hit.edu.cn (L.P.); yaoli@hit.edu.cn (Y.L.); Tel.: +86 0451 86403767 (Y.L.); Fax: +86 0451 86403767 (Y.L.)

Received: 27 July 2019; Accepted: 10 August 2019; Published: 12 August 2019

Abstract: Colloidal photonic crystal (PC)-based anti-counterfeiting materials have been widely studied due to their inimitable structural colors and tunable photonic band gaps (PBGs) as well as their convenient identification methods. In this review, we summarize recent developments of colloidal PCs in the field of anti-counterfeiting from aspects of security strategies, design, and fabrication principles, and identification means. Firstly, an overview of the strategies for constructing PC anti-counterfeiting materials composed of variable color PC patterns, invisible PC prints, and several other PC anti-counterfeiting materials is presented. Then, the synthesis methods, working principles, security level, and specific identification means of these three types of PC materials are discussed in detail. Finally, the summary of strengths and challenges, as well as development prospects in the attractive research field, are presented.

Keywords: colloidal photonic crystals; tunable photonic band gaps; anti-counterfeiting

1. Introduction

Photonic crystals (PCs) are a kind of artificial microstructure composed of periodic arrangements of materials with different dielectric constants [1–3]. Due to the periodic modulation of the dielectric constant, PCs exhibit photonic band gaps (PBGs) that can exclude the propagation of photons with a specific frequency range in a lattice [1,4,5]. When the energy of visible light is located in the PBG, the PCs show bright and iridescent reflected colors, which are well-known structural colors [6]. Based on their unique photonic band-gap properties and tunable structural colors, PCs have shown great application potential in developing optical waveguides, lasers, optical fibers [7–12], displays [13–16], sensors [17–19], and anti-counterfeiting materials [20–23]. When PCs are used as anti-counterfeiting materials, their stable and iridescent structural colors cannot be imitated by pigments or dyes [23]. Moreover, their distinctive reflection spectra are also useful for anti-counterfeiting [21,24,25]. Therefore, anti-counterfeiting materials based on PCs have attracted extensive attention in recent years.

In the past three decades, PCs have been prepared using various methods, such as mechanical drilling [26], layer-by-layer stacking technique [27], photolithography [28], and reactive ion beam etching [29]. The PCs prepared by these micro-fabricated approaches have precise structures but complex and time-consuming preparation processes. In addition, it is difficult to realize the band gap located in a visible light band for three-dimensional PCs fabricated by these methods [4]. Therefore, an alternative approach to overcome these limitations has been studied, which is to self-assemble colloidal nanoparticles into order colloidal PC structures. Common self-assembly methods include sedimentation [30], spray coating [31,32], spin-coating [33], electrodeposition [34], centrifugation [35], vertical deposition [36], and inkjet printing [6,16,37]. Compared to the micro-fabricated approach, the self-assembly approach is always simple and inexpensive [29]. Moreover, through careful selection of assembly methods and conditions, the colloidal PCs with different dimensions (1D, 2D, 3D) and geometry shapes (film, dots, lines, sphere, stars) can be easily realized [4,38]. These colloidal PCs show bright structural colors due to band gaps in the visible region [4]. Moreover, the structural colors and band gaps of colloidal crystals can be tuned by external stimuli, such as vapor [39,40], temperature [41], and electrical [42] and magnetic fields [23,43], which can be distinguished by the naked eye or spectrometers. These advantages of colloidal PC, such as sample preparation technology, easily tunable optical property, and flexible geometry structure, make it a more ideal candidate for anti-counterfeiting materials compared with micro-fabricated PC.

In this review, recent advances in colloidal PC materials for anti-counterfeiting purposes will be discussed from the aspects of security strategies, design and fabrication principles, and identification means. Firstly, we divide anti-counterfeiting colloidal PC materials into three categories: variable colored PC patterns, invisible PC prints, and a few other PC security materials, and discuss general strategies for creating these materials. Then, the synthesis methods, working principles, security level and specific identification means of these three kinds of colloidal PC materials are discussed in detail, respectively. Finally, a summary of anti-counterfeiting colloidal PC materials in terms of their opportunities and challenges is presented.

2. Strategies to Construct Anti-Counterfeiting Colloidal PC (ACPC) Materials

Most colloidal PCs can be used as anti-counterfeiting materials because of their tunable structural colors; however, a few are based on their other properties, such as fluorescence enhancement properties. According to this, we can divide colloidal PCs into two types: ACPC materials based on tunable structural colors and ACPC materials based on other properties. The strategies to construct the two types of ACPC materials are outlined below.

2.1. ACPC Materials Based on Tunable Structural Colors

ACPC materials based on tunable structural colors can be further classified into two types, visible ACPC patterns and invisible ACPC patterns, according to security strategies. A brief introduction to the strategies for constructing visible ACPC patterns and invisible ACPC patterns is presented in the following.

2.1.1. Principles for Constructing Visible ACPC Patterns

Visible colloidal PC patterns can be used for anti-counterfeiting purposes because their structural colors can be tuned. Therefore, it is necessary to briefly introduce the physical principles of the formation of structural colors in colloidal PCs.

Colloidal PCs are the spatially periodic structures assembled from colloidal nanoparticles, which can be divided into one-, two- or three-dimensional (1D, 2D, 3D) colloidal PCs according to their periodicity. The structural colors are originated from the diffraction of visible light from periodic microstructures. The diffraction properties can be described by Bragg's law. For example, 1D PCs, namely Bragg stacks or Bragg reflectors, obey Bragg–Snell's law and can be described by Equation (1) when incident light is perpendicular to the sample surface [44], where m is the order of diffraction, n_l and n_h represent the refractive indices of the low- and high-refractive-index components in the PC system, respectively, and d_l and d_h are the optical thicknesses of two components, respectively [4]:

$$m\lambda = 2(n_l d_l + n_h d_h) \quad (1)$$

The magnetically responsive PCs with chain-like structures are typical 1D colloidal PCs, and monolayer colloidal PCs are two-dimensional (2D) colloidal PCs [45]. The 3D colloidal PCs have opal structures and show bright structural colors due to diffraction of the 3D lattice [46]. The diffraction wavelength can be determined by Bragg–Snell's law (Equation (2)), where m is the diffraction order, λ is the diffraction wavelength, n_{eff} is the effective refractive index of the system, θ is the incident angle, and d is the interplanar spacing [4]:

$$m\lambda = 2n_{eff}d\cos\theta \tag{2}$$

$$m\lambda = \sqrt{\frac{8}{3}}D(n_{eff}^2 - \sin^2\theta)^{\frac{1}{2}} \tag{3}$$

$$n_{eff} = \left(n_p^2V_p + n_m^2(1 - V_p)\right)^{1/2} \tag{4}$$

Since the colloidal particles usually self-assemble into (111) planes of face-centered cubic (FCC) lattices in the direction parallel to the sample surface, the diffraction wavelength can be also calculated using distance D between the center of the nearest spheres in (111) planes [46]. Thus, the Bragg law can be expressed by Equation (3) [47]. The effective refractive index n_{eff} can be expressed as Equation (4), where n_p and n_m denote the refractive indices of particles and another medium, respectively, and V_p is the volume fraction of nanoparticles.

These equations provide guidance for the design of visible ACPC materials. In general, the changes in parameters of the equations that lead to the change in structural colors can be employed for constructing the visible ACPC materials, which can be summarized as follows:

1. Changing the colloidal PC plane spacing is a good strategy for designing visible ACPC materials. Generally, when colloidal PCs are embedded into the polymers or the polymers with 3D inverse opal structures are formed, the expansion of the polymer under external stimulus will lead to the increase of distance between the nearest nanoparticles or pores, thereby changing their structural color. This color change is apparent to the naked eye and very useful for anti-counterfeiting.
2. The change of the effective refractive index will lead to the change in the structural color. The absorption of new substances and the temperature-induced phase transition are two methods for changing the refractive index [4], where the former has been used to design visible ACPC materials. When the visible ACPC materials are composed of mesoporous nanoparticles, the hierarchical porous structure of particles is conducive to the absorption of gas molecules, thereby resulting in the change in the refractive index.
3. Changing the incident angle can also cause a change in structural color. The majority of visible colloidal PC patterns utilized in the anti-counterfeit field are based on their angle-dependent structural colors. When the incident angle changes, the abundant color state transition of the patterns can be observed by the naked eye, which makes them high-level security materials.
4. The transformation from disordered states to ordered states can also tune the structural color and can be applied in the field of anti-counterfeiting. A typical example here is magnetically responsive colloidal PC anti-counterfeit materials. In the absence of a magnetic field, the disordered superparamagnetic nanoparticles dispersed in solution show inherent chemical colors originating from light absorption, while these colloidal PCs exhibit bright structural colors under a magnetic field due to the ordered self-assembly of particles.

The above is a summary of the principles for constructing visible ACPC patterns. In the following section, preparation methods and working principles of different visible ACPC patterns designed using these four strategies will be discussed in detail.

2.1.2. Strategies to Create Invisible ACPC Patterns

Invisible photonic patterns are also known as invisible photonic prints. The patterns are invisible under normal circumstances and can only be displayed by applying corresponding an external stimulus [48], which makes them very useful for anti-counterfeiting. The key to constructing an invisible photonic print is to create the "pattern" and "background" with the same color and reflected wavelength ($\Delta\lambda_0 = 0$) at a normal state, but a different color and reflected wavelength ($\Delta\lambda \neq 0$) under the corresponding external stimulus [48–50]. In other words, the pattern and background must have very similar structures but different abilities to respond to the external stimulus, such as an electric field [14], magnetic field [50], deformation [48,51], or chemical stimuli [49,52]. According to this, many strategies to prepare invisible photonic prints have been proposed, which can be classified into four categories:

1. Selective immobilization is the most widely used method to prepare invisible prints that can be shown by a magnetic field or applied stress. For the invisible print displayed in a magnetic field, in the initial state, the superparamagnetic particles are randomly dispersed into a photocurable polymer to form a stable colloidal suspension, which displays a uniform brown color derived from the selective absorption of light by particles. By using selective ultraviolet (UV) irradiation of the suspension through a hollow patterned mask, the superparamagnetic particles are fixed into the polymer matrix in the irradiated region (background region), while the particles can move freely in the non-irradiated region (pattern region). When the external magnetic field is absent, both pattern and background regions show uniform brown colors, thereby hiding the pattern. After applying the external magnetic field, the background region maintains the brown color because of the fixation of superparamagnetic particles, while the pattern region shows a bright structural color because of the self-assembly of superparamagnetic particles. The color contrast makes the pattern visible. In addition to UV curing, selective drying of the solvent in the magnetically responsive PC can be also considered as another selective immobilization method, which is because the particles are fixed after the solvent is completely removed. The reason for hiding and displaying patterns is similar to the former pattern fabricated by selective UV irradiation. As for invisible prints shown by deformation, only the pattern region is cross-linked due to the selective UV exposure. The cross-linked region and non-cross-linked region have similar color and different elasticity under the normal state, which makes different changes in lattice constants when applying the external force, thereby further leading to the color contrast between the pattern and background, and the appearance of the pattern.
2. Selective modification is a simple and straightforward way of obtaining the pattern and background with different properties. Selective hydrophobic and hydrophilic modifications are common selective modification methods. The former approach is usually applied to the system in which colloidal PCs are embedded in hydrophilic polymers, where only the background region is modified by hydrophobic treatment. Thereby, the patterns on the prints are invisible in a dry state due to the similar structure and same color between pattern and background, while they can be shown by soaking in water, because the different water-absorbing swelling behaviors of the hydrophilic and hydrophobic regions result in a different lattice expansion and a large color contrast between them.
3. Using nanoparticles of the same intrinsic color but different sizes to prepare invisible prints is also a good method, especially for an invisible print shown under a magnetic field. The pattern is invisible at a normal state because of the uniform inherent color of the superparamagnetic particles. The pattern can be revealed under a magnetic field because self-assembly of superparamagnetic particles with different sizes leads to the structural color contrast between pattern and background.
4. In some cases, the patterned substrate will disturb the self-assembly of colloidal particles. Therefore, the introduction of the patterned substrate into the PC system can enable the formation of an invisible print. For example, the invisible print based on electrically responsive PCs has been successfully prepared in this way, where the substrate electrode was patterned by covering with an insulating polymer film. The pattern is invisible under normal circumstances due to the uniform color originating from the suspension of charged composite microspheres. The pattern can be displayed under the electric field because the pattern displays the original color derived from a random arrangement of charged microspheres, while the background showed the structural color.

In the following section, the working mechanisms, tuning methods and relative merits of invisible ACPC patterns created by these methods will be discussed in detail.

2.2. ACPC Materials Based on Their Other Properties

Almost all PC materials used for anti-counterfeiting are based on their tunable structural colors that originate from Bragg diffraction. In addition to the tunable structural colors, other important optical properties of PCs are also useful in the security field, including unique reflectivity properties and fluorescence enhancement effects. Based on these two properties, Song and co-workers designed a novel ACPC material composed of PC dots with three shapes, which were obtained by printing colloidal particles onto poly(dimethylsiloxane) (PDMS) substrates with different rheology states via an ink-jet printing method [53]. Due to the large optical differences in reflectivity property, and fluorescence enhancement of the three differently shaped PC dots, the ACPC material can show different images according to different optical conditions.

Tunable structural color generally means that the diffraction wavelength or PBG can be tuned by lattice parameters. In addition to tunable color, tunable diffraction intensity is also very useful for anti-counterfeiting. For example, Nam et al. [54] reported a monolayered PC anti-counterfeiting pattern fabricated by an inkjet-printing method, where the PC consisted of silica particles with a refractive index of 1.45 and voids with a refractive index of 1.0. The PC pattern showed an extremely weak structural color because the inkjet-printed colloidal PC possessed a short-range ordered structure and the monolayered structure further reduced the diffraction intensity of the PC [37]. This pattern was barely visible on a white background in daylight, while it became obvious by applying an external light source or changing the background. This is because the enhancement of diffraction color can be achieved by absorbing stray light through a black background or increasing the intensity of incident light.

Based on a similar security strategy, Shang et al. [55] further fabricated an anti-counterfeiting PC film by firstly constructing the magnetically responsive PC pattern and then fixing the PC pattern under a magnetic field, where the pattern was prepared by selective photo-polymerization of a photocurable colloidal suspension containing Fe_3O_4@C nanoparticles and a photocurable resin through a patterned mask. In the non-patterned region that was not protected by a mask, the monomer cross-links to form a polymer network and Fe_3O_4@C particles were steadily fixed into the network with a random arrangement, so that the non-patterned region lost its responsive ability to external magnetic fields and showed a constant brown color. In contrast, the suspension containing Fe_3O_4@C particles in the patterned region protected by a mask showed the structural color when a magnetic field was applied. Due to strong light scattering in the PC pattern, the pattern showed a very weak color and low diffraction intensity. The small color contrast between the patterned and non-patterned region makes it invisible when it was placed on a white background, while it was visible on a black background, because the black background can reduce array scattering and increase the signal-to-noise ratio [54].

These ACPC materials can easily realize the transition from invisible to visible state by changing light conditions or backgrounds without changing the intrinsic structure, which endows them with good durability and good usability.

3. Visible Colloidal PC Patterns for Anti-Counterfeiting

A visible colloidal PC anti-counterfeiting pattern is one that can alter the structural color when the incident angles or viewing angles change or external stimuli are applied, and remains in a visible state even in the absence of stimulations. The structural color can be tuned by angles that originate from the inherent PBG property of the PC, while tuning by the external stimulus is due to the introduction of a responsive PC. Common external stimuli include the magnetic field, temperature field, and chemical stimulus, which lead to changes of lattice parameters in the PC structure, such as the angle of incidence, the lattice constant, refractive index, and order of the photonic structure, thereby changing the structural color of the PC pattern. The color state transition can be easily observed by the naked eye, which means it has great application potential in the field of anti-counterfeiting.

3.1. Visible Single-Colored PC Patterns Based on Their Angle-Dependent Structural Colors for Anti-Counterfeiting

Most visible PC patterns utilized in the security field are based on their angle-dependent structural colors. Since the typical example was reported by Yang's group [20], a lot of PC security patterns have been developed, which have the same security strategy but different structures. Typical PC structures include non-close-packed and close-packed FCC structures and inverse opal (I-opal) structures. For example, Yang's group reported a security material that was composed of a SiO_2 colloidal PC with a non-close-packed FCC structure embedded into a photocurable ethoxylated trimethylolpropane triacrylate (ETPTA) matrix. The PC film can be easily patterned by using conventional photolithography techniques. The patterned PC film showed a very bright structural color and had four-color states transition when viewing angles changed between 0° and 55° (Figure 1a), which can be attributed to the low refractive index contrast between silica (n = 1.45) and ETPTA (n = 1.4689) [20].

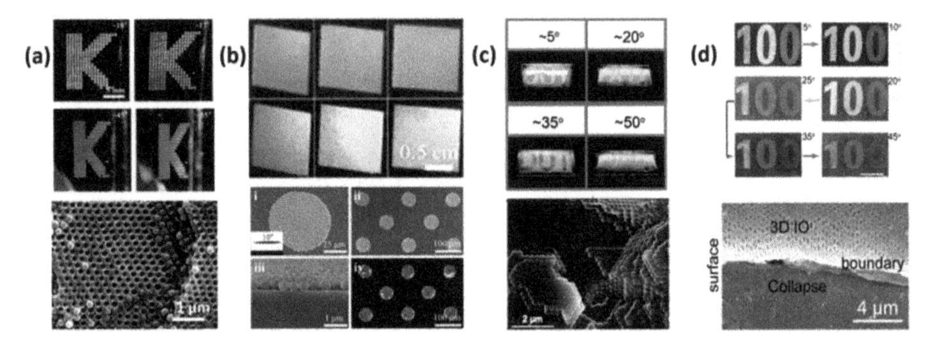

Figure 1. Colloidal photonic crystal (PC) patterns with different structures based on their angular-dependent structural colors for anti-counterfeiting. (**a**) Anti-counterfeiting colloidal PC (ACPC) patterns with non-close-packed FCC structure. Reused with permission from Reference [20], Copyright 2013, American Chemical Society. (**b**) ACPC patterns composed of mono-layered closest-packed nanoparticles. Reused with permission from Reference [54], Copyright 2016, Springer Nature. (**c**) ACPC patterns with close-packed FCC structures. Reused with permission from Reference [56], Copyright 2018, The Royal Society of Chemistry. (**d**) ACPC patterns with inverse opal structures. Reused with permission from Reference [22], Copyright 2018, Wiley-VCH.

The colloidal PC with close-packed FCC structure can be further divided into two types. One is only composed of colloidal nanoparticles with the closest packing. The typical example is an inkjet-printed colloidal PC, where colloidal nanoparticles are printed onto substrates and can arrange into highly ordered FCC arrays under the evaporation-driven self-assembly process. For example, Nam et al. [54] reported an ACPC pattern which was prepared via inkjet-printing method, which consisted of mono-layered closest-packed nanoparticles. The PC pattern can change its color as the viewing angle changes (Figure 1b). Wu and co-workers also reported an inkjet-printed PC pattern composed of CdS particles that also had a close-packed structure [37]. The PC pattern showed a bright structural color when the viewing angle was relatively close to the incident angle, which was due to a relatively large refractive index contrast between CdS particles (n = 2.5) and the air voids (n = 1). Moreover, the structural color can change at different viewing angles, which is useful for security purposes. It is also should be noted that PC patterns can display a uniform yellow color resulting from the inherent color of the CdS microspheres when the viewing angle is away from the incident angle. Therefore, the pattern can realize the conversion from visible to invisible states under varied viewing angles by using monodispersed CdS nanoparticles with different sizes to construct the pattern and background, respectively, which further improves the security level of this pattern.

The other type is a composite PC material formed by filling a polymer into the interstitial space of the FCC close-packed structure. Wu's group reported an ACPC material in which the uniform polystyrene (PS) particles self-assembled into a close-packed FCC lattice embed into the PDMS polymer matrix [56]. The anti-counterfeiting PC pattern fabricated by a spraying method showed different structural colors at different angles (Figure 1c). It is also worth mentioning that sandwiching an anti-counterfeiting pattern between two layers of PDMS can ensure its durability.

I-opal structure is also a common 3D PC structure formed by first permeating the polymers or gels into the interstitial space of an opal structure to form the composite PC structure and then removing the opal template via chemical methods. The I-opal structure always possesses low mechanical strength, thereby affecting its durability in the anti-forgery field. However, recently, Meng and coworkers reported a robust anti-counterfeiting patterned polyvinylidene fluoride (PVDF) film with I-opal structure, where the good durability of the pattern was due to intrinsically excellent mechanical properties of PVDF resin [22]. The patterned film obtained by the pressing method had a typical well-ordered I-opal structure, thereby producing the structural color with strong angle-dependency. Thus, six distinct color states transition of patterned PVDF film can be observed when viewing angles changed (Figure 1d), which makes it a good security material.

3.2. *Visible ACPC Patterns Based on Magnetically Responsive PCs*

3.2.1. Visible Magnetically Responsive PC Security Patterns Based on the Transition from Disorder to Order

Magnetically responsive PCs have been prepared by self-assembly of uniform colloidal superparamagnetic nanoparticles, where the nanoparticles are composed of tiny magnetite (Fe_3O_4) nanocrystals with an average size less than 10 nm [57]. The structures of magnetic-responsive PCs include 1D chain-like structures, 2D sheets, and 3D crystal structures, where the dimensionality of the ordered structure mainly depends on the nanoparticle concentration [58]. Magnetically responsive structures have attracted much attention in the field of anti-counterfeiting in recent years because they can achieve an instantaneous transition between assembled and disassembled states when applying or removing an external magnetic field, where the assembled and disassembled states respectively show a bright structural color and a brown color. The color contrast can be easily observed by the naked eye.

Hu et al. [23] developed a simple and cost-effective method to fabricate PC security materials based on magnetically induced self-assembly of superparamagnetic nanoparticle technology. They were patterned by firstly making letter-shaped cavities and then infiltrating the suspension containing magnetic particles into the cavities (Figure 2a). In these letter patterns, some contained carbon-capped superparamagnetic nanoparticles (Fe_3O_4@C) with uniform size, while others contained nanoparticles with two different sizes. All letters showed uniform pale brown colors in the absence of the magnetic field, while patterns showed bright structural colors under the magnetic field because of the formation of 1D chain-like PC structures parallel to the direction of the magnetic field (Figure 2b,c). It should be noted that the PCs containing the nanoparticles with two different sizes had double PBG heterostructures and displayed uniform structural colors (Figure 2b) due to the additive effect of monochrome. This large color contrast before and after applying magnetic fields can be easily identified by the naked eye, and special double diffraction peaks can be detected with a fiber optic spectrometer, which makes it difficult to be mimicked by conventional dyes and pigments.

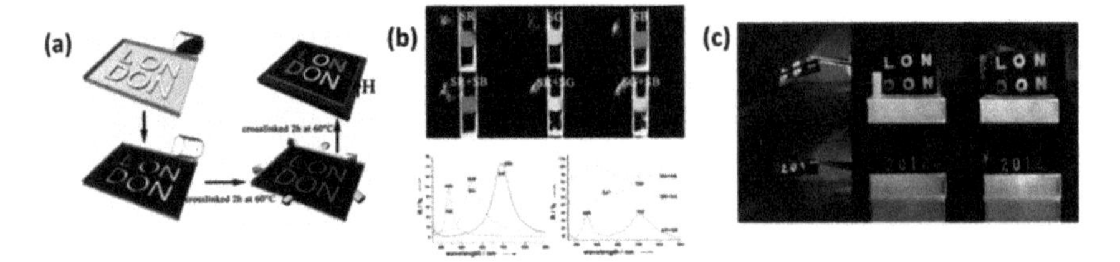

Figure 2. Visible magnetic-responsive Photonic crystals (PC) patterns for anti-counterfeiting based on the transition from disorder to order. (**a**) Schematic illustrations for fabricating anti-counterfeiting PC patterns. (**b**) Magnetic-responsive PC patterns containing uniform nanoparticles with one or two sizes had one or two photonic band gaps (PBGs), respectively, but displayed single colors. (**c**) The large color contrast of PC patterns appeared when applying the external magnetic field. Reused with permission from Reference [23], Copyright 2012, The Royal Society of Chemistry.

Another security material based on magnetic-responsive PCs was also reported by Hu's group, which was composed of uniform Fe_3O_4@C superparamagnetic nanoparticles in ethylene glycol (EG) emulsion droplets distributed in a cured resin matrix [59]. This lithography produced security pattern could realize a large color contrast when applying or removing a magnetic field, where the reason for the color change was the same as above. This pattern had a 1D chain-like structure and can be used as an anti-counterfeiting watermark on banknotes because of its flexibility, translucency, and ultra-thinness.

3.2.2. Visible Magnetic-Responsive PC Security Patterns Based on Their Photonic Bandgap and Birefringence Properties

In previous reports, magnetic-responsive PC security patterns always had 1D chain-like structures composed of spherical silica particles, which is because this type of PC had a lower volume fraction of superparamagnetic nanoparticles, so that they could easily realize instantaneous switching between ordered and completely disordered states [60].

Recently, Li and coworkers reported an anti-counterfeiting device based on a 3D colloidal PC structure self-assembled from uniform Fe_3O_4@SiO_2 nanorods in concentrated suspension [61]. Differently from the previously reported magnetic-responsive PC with 1D chain-like structures, these Fe_3O_4@SiO_2 nanorods could also form an ordered polycrystalline structure without an external magnetic field because the particle content was over the critical concentration in the suspension. Therefore, it showed a structural color instead of a uniform brown color even in the absence of a magnetic field. When applying a vertical magnetic field, the nanorods self-assembled into a 3D monoclinic crystal structure instead of a polycrystalline structure and showed a uniform structural color. In addition to the distinct structural color, the PC also had other important optical properties, namely, the optical birefringence property produced by the shape anisotropy of the nanorods. The anti-counterfeiting device based on a unique structural color and the birefringence property was prepared by sandwiching this colloidal PC containing spontaneously arranged nanorods between two cross polarizers. When the device was exposed to daylight or other external light sources, it worked in reflection mode and showed its bright structural color. The structural color of the device can be easily changed when applying a magnetic field or rotating the magnet (Figure 3a). Moreover, controlling the field distribution can realize the display of different patterns on the device due to the large color contrast between two regions with and without an external magnetic field (Figure 3b). In addition to reflection mode, it can also work in transmission mode when it is under the illumination of the backlight. In this mode, the device can achieve a transition between dark and bright states by changing the direction of the magnetic field in the horizontal plane (Figure 3c). Moreover, the device shows different patterns when applying different shaped magnets, which is caused by the large brightness contrast between two regions under magnetic fields with different directions (Figure 3d). These rapid and reversible transitions, including color states, light and dark states, and pattern shapes, can be easily recognized by the naked eye, thereby making it a higher-level security material compared to anti-fake materials based on the transition in color states [61].

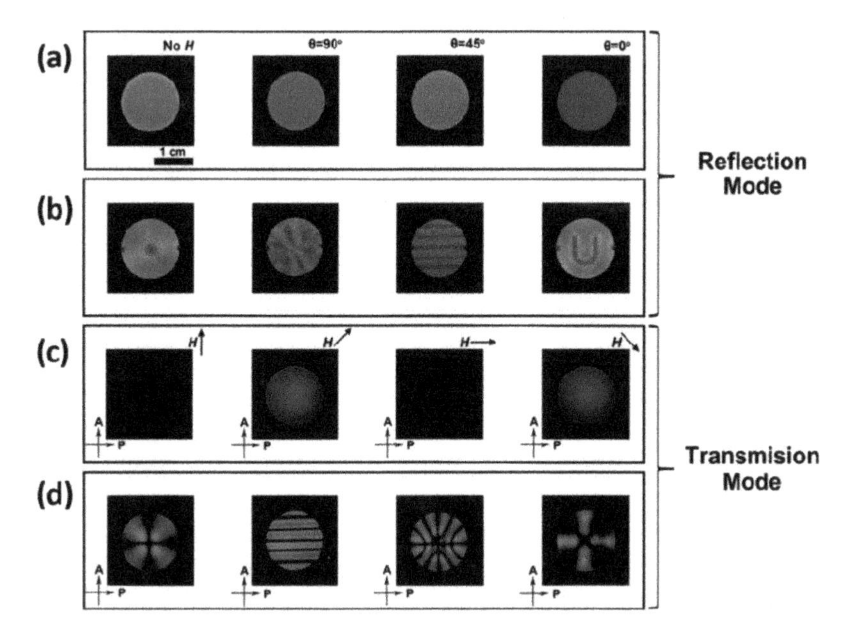

Figure 3. Magnetic-responsive PC patterns based on their photonic band-gap and birefringence properties for anti-counterfeiting. (**a**) Digital photos showing that the structural color of the anti-counterfeiting device based on magnetic-responsive PC strongly depends on the magnetic field and the direction of the magnetic field. (**b**) Digital photos of the anti-counterfeiting device showing different patterns under different non-uniform magnetic fields. (**c**) The transition between dark and bright states through changing the direction of the magnetic field in the horizontal plane. (**d**) The device showing different patterns when applying a different non-uniform magnetic field. Reused with permission from Reference [61], Copyright 2019, Wiley-VCH.

3.3. Visible Thermoresponsive PC Security Patterns Based on the Change in Lattice Spacing

Poly(N-isopropyl acrylamide) (PNIPAM) is a temperature-sensitive polymer that has a low volume phase transition temperature of about 32 °C. When the temperature increases, the PNIPAM gel will shrink, so that integrating a temperature-sensitive PNIPAM gel into PC systems is a good strategy to prepare thermoresponsive PCs. For example, Asher et al. [62] successfully fabricated a thermoresponsive PC material [62] by embedding the colloidal PC structure into a temperature-sensitive PNIPAM gel matrix [41], where the diffraction wavelength of this thermoresponsive PC can be tuned between 704 and 460 nm when the temperature was switched between 10 and 35 °C [62]. In addition to the composite PC structure, an ordered porous PNIPAM gel with an I-opal structure can also achieve a sharp and reversible response to temperature by removing the colloidal PC framework, which has been reported by Takeoka and Watanabe et al. [63,64].

Based on the thermoresponsive PC material, Zhao's group developed a near-infrared (NIR) light-responsive striped hydrogel fiber with an I-opal structure for anti-counterfeiting, which was composed of thermoresponsive PNIPAM and NIR light-sensitive reduced graphene oxide (rGO) [65]. The stripe-patterned PNIPAM/rGO fiber can be obtained by firstly fabricating the stripe-patterned colloidal PC on the inner surface of a capillary tube, then permeating hydrogel solutions consisting of NIPAM monomers, cross-linkers, initiators, and graphene oxide into the interstices of a striped colloidal PC and completely solidifying the solution mixtures, and finally etching the silica colloidal PC arrays and the glass capillary by HF solution (Figure 4a). It is also worth mentioning that the formation of a stripe-patterned colloidal PC instead of a homogeneous colloidal PC is caused by a lower colloid concentration of the suspension. Moreover, the width, spacing, and color of the striped pattern can be precisely controlled by adjusting the concentration of suspension and the self-assembly parameters. In the PNIPAM/rGO composite fiber, rGO had strong infrared absorption properties, high photothermal conversion efficiency, and high thermal conductivity, so that PNIPAM absorbed heat and displayed shrinking behavior when exposed to NIR light, leading to the change in lattice spacing and reflected color. Because the NIR light was only irradiated from one side of the fiber and irradiation was concentrated in the middle of the sample, the fiber showed inconsistent shrinking behavior (Figure 4b), resulting in an uneven change in lattice spacing. This further led to different degrees of bending behavior and different blue-shift values in different stripes of the fiber (Figure 4c), which can be easily observed by the naked eye. These properties have not been reported by previous materials with tunable structural colors, thereby making the material ideal for use in anti-counterfeit barcodes.

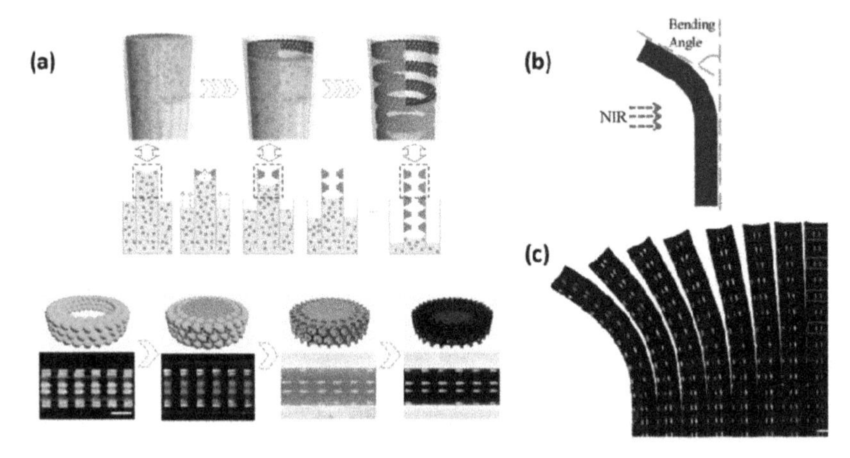

Figure 4. Thermo-responsive PC patterns based on the change in lattice spacing for anti-counterfeiting. (**a**) Formation of a near-infrared (NIR) light-responsive striped Poly(N-isopropyl acrylamide)/ reduced graphene oxide (PNIPAM/rGO) fiber with I-opal structure. (**b**) PNIPAM/rGO fiber bending toward the direction of light. (**c**) A group of digital photos showing that the fiber has a varying colored striped pattern at different bending angles. Reused with permission from Reference [65], Copyright 2017, Wiley-VCH.

3.4. *Visible Vapor Responsive PC Security Patterns Based on the Change in Refractive Index*

In addition to introducing responsive materials or changing the incident angle of the light source, changing the effective refractive index is also a common way to tune the structural colors in PC systems, where the change of refractive index can be achieved through infiltrating chemical vapors and solvents into the porous PC structures.

Bai and coworkers reported a security pattern based on vapor-responsive PCs, where the PCs were formed by self-assembly of mesoporous silica nanoparticles composed of SiO_2 solid cores and mesoporous shells [39]. The vapor-responsive PCs possessed special hierarchical porous structures so that the vapor could be adsorbed on the structures, leading to the increase of the mean refractive index and change of structural color. It is worth mentioning that the mean refractive index of the mesoporous PC depends on the respective refractive index and volume rate of the SiO_2 solid core, SiO_2 in the shells, mesopores in the shells, and the void spaces between the order colloidal arrays. In fact, when the PC was exposed to different vapor conditions, only the refractive index of the mesopores changed, and the refractive index of the void spaces remained the same. Therefore, when PCs are composed of solid silica particles, they can hardly change their colors because of no response to vapor. Furthermore, the color response ranges of the vapor-responsive PCs can be adjusted through controlling the core and shell thicknesses. This security pattern was created by integrating three types of PCs with different vapor-responsive capacities into a whole via an inkjet-printing method. As shown in Figure 5, the anti-counterfeiting tree pattern was composed of tree trunks, tree leaves and fruit, respectively formed by self-assembly of solid silica particles, and mesoporous silica particles with different core and shell thicknesses, where the trunk, leaf, and fruit images were composed of PCs with very close structures but different responses to vapor. Therefore, the tree pattern showed a uniform green color under an N_2 atmosphere, while in ethanol (EtOH) vapor, the tree trunk still displayed the green color, and the tree leaves and fruit turned a yellow and red color, respectively. This multicolored pattern can be easily revealed by non-toxic ethanol vapor and recognized by the naked eye, so is favorable for anti-counterfeiting.

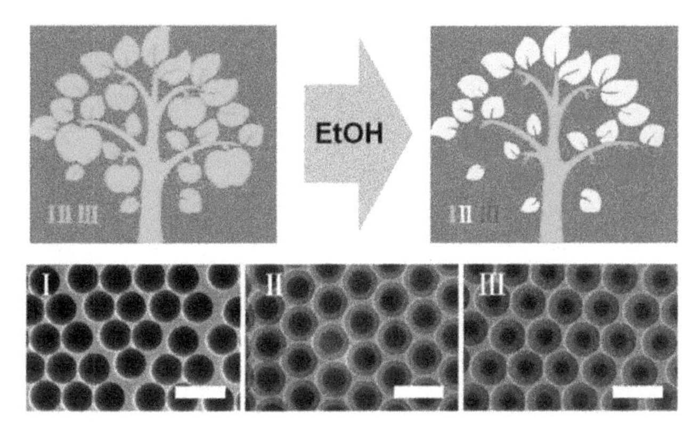

Figure 5. The structural color of the inkjet-printed tree pattern under N_2 and saturated EtOH atmospheres, where a pattern composed of tree trunks, tree leaves and fruit, respectively, is formed by self-assembly of solid silica particles, and mesoporous silica particles with different core and shell thicknesses. Reused with permission from Reference [39], Copyright 2014, American Chemical Society.

4. Invisible Colloidal PC Prints for Anti-Counterfeiting

The invisible photonic print consists of the pattern that was hidden under normal circumstances but revealed under a specific stimulus (e.g., chemical or mechanical stimulus, and electrical or magnetic field stimulus). That transition between invisible state and visible state is instantaneous, reversible and easily recognizable to the naked eye, which makes it well-suited for an anti-counterfeiting device. To build invisible photonic prints, two kinds of PCs with the same color but different response capability to external stimuli need to be integrated into a print, where the integration strategies include selective immobilization, cross-linking and modification of prints, and selective modification of substrates and self-assembly of nanoparticles with different sizes. A good invisible photonic print for anti-counterfeiting is usually characterized by a simple and nontoxic identification method, good reversibility, fast decryption speed, good stability, and durability.

4.1. Invisible Colloidal PC Patterns Shown by Chemical Stimulus

4.1.1. Invisible Colloidal PC Patterns Shown by Water Fabricated by Selective Modification or Selective Cross-Linking

Invisible colloidal PC patterns shown by water have been fabricated by selectively cross-linking or modifying the water-responsive PC films, where the classic system of water-responsive PCs was constructed by introducing the water-swellable polymers into periodic colloidal PC structures.

For example, Ge's group reported an invisible photonic print displayed by soaking in water, which was prepared by permeating NaOH solution or 3-aminopropyl-trimethoxysilane (APS) solution into the unshielded region (background region) on a pre-made PC film composed of a chain-like colloidal PC structure embedded in a polymer matrix [52], where the pre-made PC film was composed of $Fe_3O_4@SiO_2$ colloids and the polymer matrix formed by the cross-link of three monomers (poly(ethylene glycol) diacrylate, poly(ethylene glycol) methacrylate, and 3-trimethoxysilyl-propyl-methacrylate). The NaOH or APS solution can further cross-link the polymer matrix in the background region, which makes the background region display higher levels of cross-linking than the pattern region protected by the mask. The cross-linking process rarely changed the structure of the PC, so that pattern and background regions have a low diffraction wavelength contrast, resulting in the hiding of the photonic print. Because a different cross-linking degree leads to a different swelling speed of the pattern and background, the pattern will be easily shown by soaking in water (Figure 6a). The print has good reversibility but slower decryption speed (about several minutes).

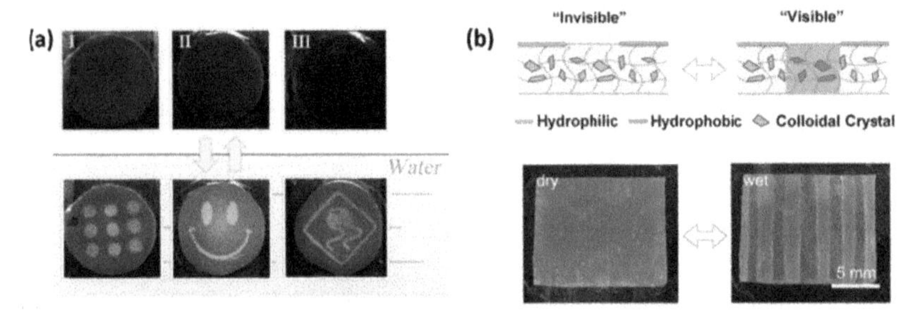

Figure 6. Invisible colloidal PC patterns shown by water. (**a**) The invisible prints prepared by selective cross-linking method revealed by soaking in water for 5 min. Reused with permission from Reference [52], Copyright 2012, The Royal Society of Chemistry. (**b**) The invisible print prepared by selective hydrophobic modification revealed by soaking in water within 10 s. Reused with permission from Reference [49], Copyright 2015, The Royal Society of Chemistry.

Based on similar design strategies, this group further fabricated an invisible colloidal PC pattern shown by water with a fast display speed [49]. In this case, the invisible print can be obtained by filling fluoroalkylsilane vapor into the unshielded background region on the hydrophilic PC film, where the PC consisted of a SiO_2 colloidal crystal framework embedded into a poly(ethylene glycol) methacrylate (PEGMA) matrix, and the refractive indices of the two components were 1.45 and 1.46, respectively. This process of selective modification made the unshielded background region hydrophobic, while the shielded pattern region remained hydrophilic. Generally, the pattern was invisible under normal circumstances because the modification process hardly changed the lattice constant. When it was soaked in water, the pattern could appear within 10 s (Figure 6b). Its decryption speed was much faster than the former invisible photonic print shown by water, because the uncross-linked characteristic gave it superior swelling ability. The pattern could achieve reversible transformation between visible and invisible states by soaking in water or evaporating the water.

4.1.2. Invisible Colloidal PC Patterns Shown by Vapor

In addition to the introduction of responsive materials into the PC pattern, changing the refractive index in the pattern region by infiltrating chemicals, such as vapors or solvents, is another effective method to tune the structural color. Zhong et al. [66] reported the synthesis of invisible prints through the hydrophilic/hydrophobic modification of hollow SiO_2 colloidal PC (HSCPC) film (Figure 7a). The HSPC film was composed of hollow SiO_2 nanoparticles and voids, where nanoparticles were arranged into a non-packed FCC structure (Figure 7b). When the water vapor was filled into the voids between hollow particles, the reflection wavelength of the HSCPC film had a red-shift due to the increase of average reflective index caused by the exchange of water vapor, with a high reflective index (1.34), and air, with a low reflective index (1.0) in the voids. This effect can be well explained by Bragg–Snell's law, as expressed in Equation [4] mentioned in Section 2. After the HSCPC film underwent selective hydrophilic (oxygen plasma etching) and hydrophobic (octadecyltrichlorosilane treating) modification, the hydrophilic pattern was invisible under normal conditions (Figure 7c), which is because the surface hydrophobization/hydrophilization neither changed the reflective index nor the PC structure between the pattern and the background region. The invisible "KUL" pattern can be revealed by water vapor flow, because the water vapor only diffuses into the hydrophilic pattern region, leading to an increase of the reflective index and a change of the structural color from blue to green (Figure 7d–f). In contrast, the structural color of the hydrophobic background cannot change, which makes a color contrast with the hydrophilic region. Thereby, an obvious pattern can be easily observed by the naked eye. The print can be reversibly shown and hidden by water vapor flow and evaporation. In addition, the print shows an ultrafast display speed (≈100 ms), multiple reversible cycles (at least 200 cycles) and good hiding effect ($\Delta\lambda_0 = 0$), which favor its application in identification marking, encoding, and anti-counterfeiting.

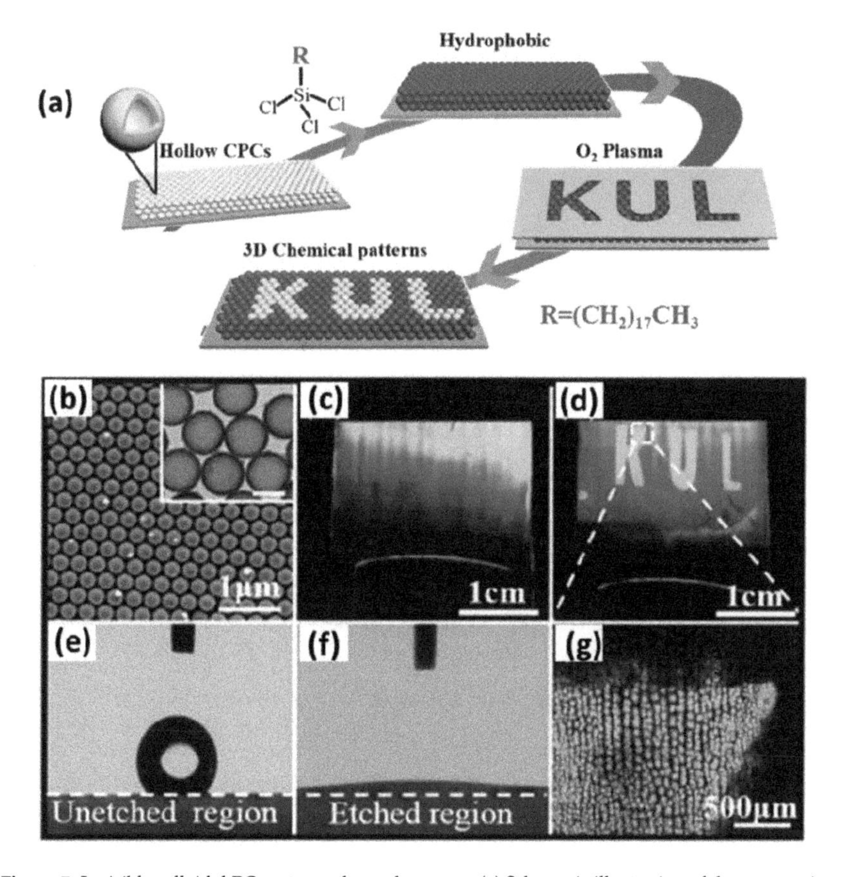

Figure 7. Invisible colloidal PC patterns shown by vapor. (**a**) Schematic illustration of the preparation procedure of the invisible PC pattern. (**b**) The SEM image of hollow PC. (**c**) The optical image of the invisible PC pattern under normal circumstances. (**d**) The pattern revealed by water vapor flow. (**g**) The microscopic image of the pattern that has been revealed. (**e–f**) Microscopic images of the water drop deposited on the unetched region (**e**) and O_2 plasma-etched regions (**f**) of the photonic print. Reused with permission from Reference [66], Copyright 2018, Wiley-VCH.

4.2. Invisible Colloidal PC Patterns Revealed by Mechanical Stretching Fabricated by Selective Immobilization

Compared with chemical stimulus, mechanical stretching or pressing is a more straightforward and readily available way for decryption of invisible patterns. To achieve an invisible pattern shown by mechanical stretching, a pattern and background region on photonic prints with the same color but different deformation capabilities should be designed.

For example, Ye and co-workers reported this type of invisible colloidal PC print, which was prepared by infiltrating a cross-linking agent (PEGDA) into pre-prepared mechanochromic PC film and selectively cross-linking the soaked mechanochromic PC film through selective UV irradiation (Figure 8a), where the mechanochromic PC film was composed of the silica colloidal crystalline array (CCA) fixed into the mixed matrix of EG and PEGMA [48]. The UV irradiation process can hardly cause the change in lattice constant, so that the irradiated region (pattern region) unprotected by a mask and un-irradiated region (background region) protected by the mask show the same color, resulting in the invisible image. The UV irradiation process can greatly improve the level of cross-linking in the pattern region, which gives it a higher hardness than the background region. When the invisible print was stretched or squeezed, the soft background region had larger deformation and color shift than the stiff pattern region, therefore revealing the pattern (Figure 8b). It is worth mentioning that the display of invisible patterns is instant once applying the external force and the pattern can be reversibly shown and hidden. Fast encryption speed and good reversibility are due to good elasticity of the background region caused by a large amount of EG solution in the mechanochromic PC system, and the elasticity of packaging material (PDMS). However, due to the inevitable evaporation of EG solution in the system, the print will lose elasticity over time and fail to display the pattern region, which means the invisible print has low durability.

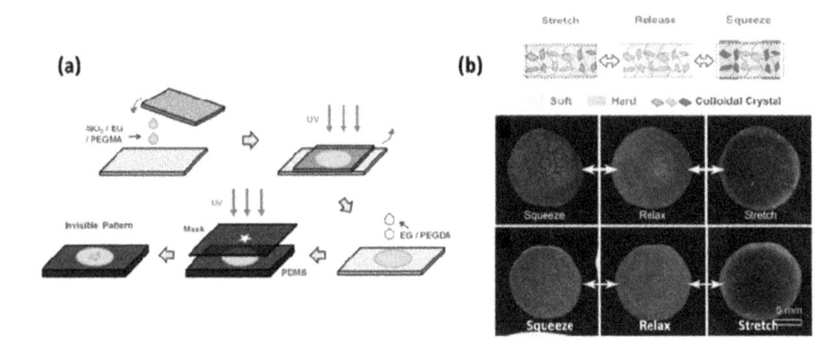

Figure 8. (**a**) Schematic illustration of the preparation process of the invisible colloidal PC pattern shown by deformation. (**b**) Sunlight and rabbit patterns on the photonic prints hidden in a relaxed state and shown by deformation. Reused with permission from Reference [66], Copyright 2014, Wiley-VCH.

To overcome the problem of low durability caused by the presence of solvent, Ding's group reported an all-solid-state invisible photonic print shown by deformation for anti-counterfeiting, which was prepared by selectively irradiating the polymer PC film (POF) composed of regularly arranged arrays of polystyrene core, poly(methylmethacrylate) interlayer, and poly(ethyl-acrylate) shell (PS@PMMA@PEA) nanospheres [51]. When the POF was selectively irradiated by UV light through the mask, the irradiated pattern region was hard due to the cross-linking reaction of PEA, while the un-irradiated background region was uncross-linked and showed good softness due to the good elasticity of the PEA shell. The as-prepared photonic print was invisible when the sample was not stretched (Figure 9a) because the diffraction wavelength in the pattern region (580 nm) was close to the background region (560 nm) (Figure 9b,c). The difference in diffraction wavelength was caused by the change of refractive index when cross-linking. The invisible "PhC" pattern became progressively clearer as the strain increased. When strains changed from 0% to 12%, the reflection wavelength of the soft background region changed from 560 to 490 nm (Figure 9b), showing good mechanochromic capability, and its reflection wavelength reverted to 550 nm when the strain was released. The reflection wavelength in the pattern region remained at 580 nm whether stretched or relaxed (Figure 9d). Therefore, the print had good reversibility and repeatability (Figure 9e). In addition, the print also showed good durability due to the absence of solvent in this system. When it was stored in ambient conditions for 10 months, such an invisible pattern could be reversibly shown and hidden.

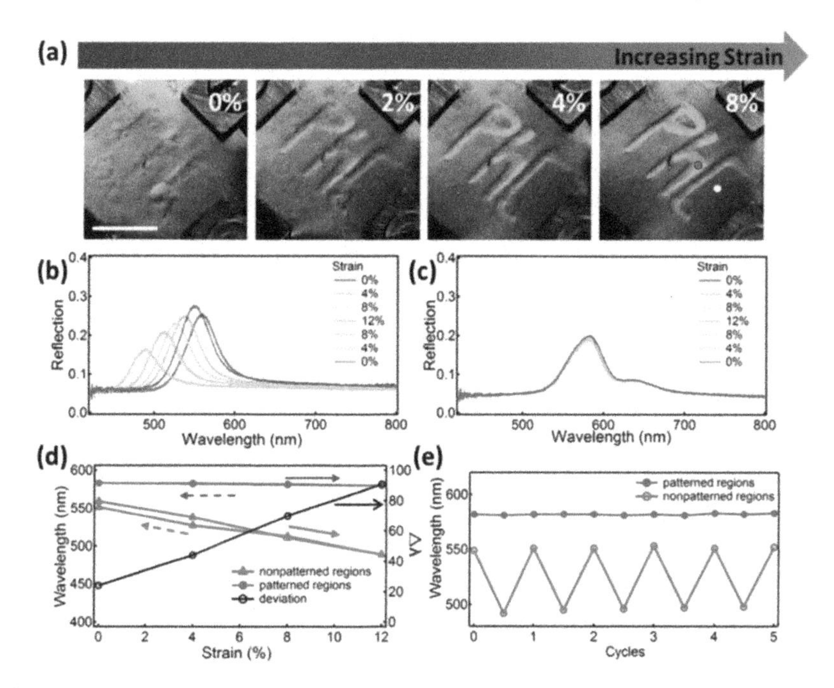

Figure 9. The invisible photonic print shown by mechanical stretching. (**a**) Invisible photonic print hidden in a relaxed state and shown by tensile state. Scale bar is 1 cm. (**b**,**c**) Reflection spectra in (**b**) background region and (**c**) pattern region under a stretch–release cycle. (**d**) The relationship between elongation and the reflection wavelength in the background region (red line) and the pattern region (blue line). The black line shows the relationship between elongation and the reflection wavelength difference. (**e**) Reflection wavelength change in the pattern region (blue line) and background region (red line) in five cycles of the stretching–releasing test. Reused with permission from Reference [51], Copyright 2015, American Chemical Society.

4.3. *Invisible Colloidal PC Patterns Displayed by Magnetic Field*

The photonic prints as mentioned above can be hidden under normal conditions because the pattern and background region have the same structural color derived from the ordered PC structure. Differently from the invisible prints shown by chemical stimulation or mechanical stretching, photonic prints displayed by a magnetic field can achieve invisibility under normal conditions due to the same chemical color in the pattern and background region caused by the random arrangement of superparamagnetic particles in the solvents. The invisible prints shown by the magnetic field can be obtained by selectively immobilizing the background region or constructing pattern and background regions using nanoparticles with different sizes. The invisible prints have a very fast decryption speed due to fast response speed (time <1 s) of the magnetically responsive PCs [67].

4.3.1. Invisible Colloidal PC Patterns Displayed by Magnetic Field Fabricated by Selective Immobilization

The key to designing an invisible colloidal PC print displayed by a magnetic field is to create the pattern and background with the same color but different magnetochromic capabilities.

Hu's group reported an invisible print shown by a magnetic field prepared by selectively irradiating a certain region of the photonic paper using UV light through a mask [50]. The magnetically responsive photonic paper was composed of EG droplets containing Fe_3O_4@C super-paramagnetic nanoparticles dispersed in the solidified PDMS matrix (Figure 10a). The UV-irradiation caused PDMS aging and resulted in the rupture of the PDMS network so that the EG solvent in the unshielded region (background region) leaked out through the ruptured PDMS, which fixed Fe_3O_4@C nanoparticles in the ruptured solid PDMS so that they could not respond to an applied magnetic field. Therefore, the background region remained a brown color. In contrast, in the unexposed region (pattern region) covered by the mask, the PDMS network was intact, so that the nanoparticles dispersed in EG droplets were randomly arranged without a magnetic field and arranged into an ordered 1D chain-like structure when a magnetic field was applied. The pattern respectively showed a brown color and a bright structural color when the magnetic field was withdrawn and applied (Figure 10b). The same brown color in the two regions means the print was well-hidden in the absence of a magnetic field. When applying a magnetic field, the background showed a brown chemical color and the pattern showed a bright green structural color, making the print highly visible. In addition, the state transition between the invisible and the visible state was instantaneous (responsive time usually is less than 1 s) and completely reversible, which can be attributed to the inherent advantages of magnetically responsive PCs. However, the print may exhibit low durability, because the properties of superparamagnetic nanoparticles in a ruptured solid PDMS could be easily affected by the external environment. Furthermore, the UV aging process caused the leakage of EG organic liquids, which was not environmentally friendly.

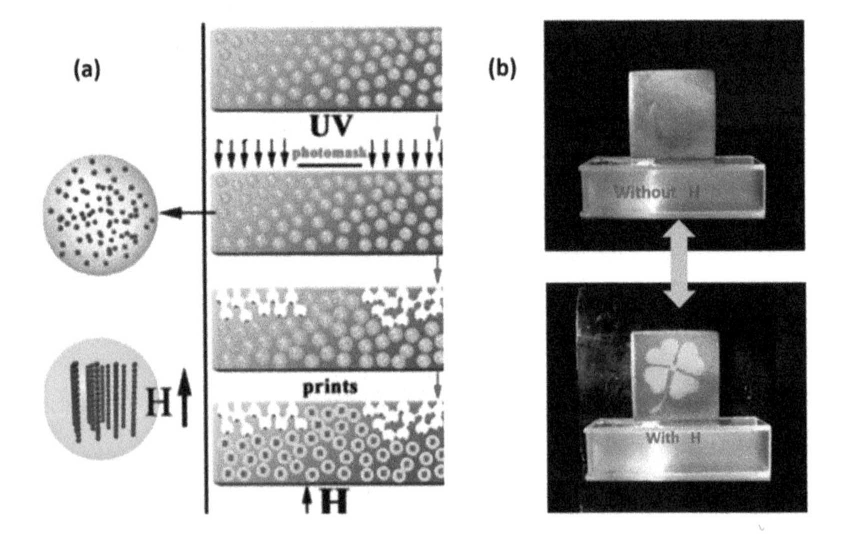

Figure 10. Invisible print shown by a magnetic field was fabricated by selective immobilization. (**a**) Schematic illustration of the preparation process of the invisible colloidal PC pattern shown by magnetic field. (**b**) The pattern was invisible in the absence of a magnetic field but visible when applying a magnetic field. Reused with permission from Reference [50], Copyright 2012, Springer Nature.

4.3.2. Invisible Colloidal PC Patterns Displayed by Magnetic Field Fabricated by Self-Assembly of Nanoparticles with Different Sizes

Another strategy for fabricating the invisible photonic print was to use $Fe_3O_4@SiO_2$ colloidal nanoparticles of different sizes to construct the pattern and background. For example, Yin's group fabricated the print displayed by magnetic field using this strategy, where the pattern and background region were composed of EG/PDMS mixtures containing $Fe_3O_4@SiO_2$ superparamagnetic particles with a core/shell size of 110/16 nm and 110/28 nm, respectively (Figure 11a,b) [68]. When the magnetic field was absent, the pattern and background both showed a brown color derived from the inherent color of the particles, thereby hiding the pattern. In contrast, the blue/green contrast could be easily observed by the naked eye when applying a strong magnetic field due to different diffraction colors of the chain-like structures self-assembled by particles with different sizes (Figure 11c). This print showed good flexibility due to the flexibility of the PDMS matrix, so it was suitable for pasting on target surfaces with various curvatures for anti-counterfeiting purposes. In addition, due to no leakage of EG in the preparation process, it was environmentally friendly. The limitation of this anti-counterfeiting print is that the pattern is not completely invisible under normal circumstances because of a deep impression on the joint between the pattern and background, which is caused by the letter molds with the same thickness as the print.

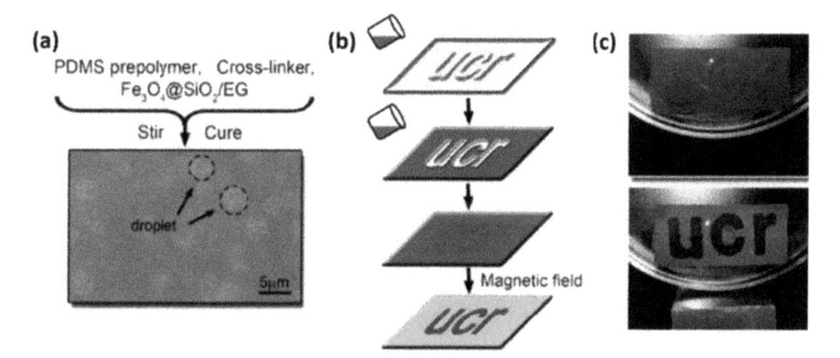

Figure 11. Invisible colloidal PC pattern shown by the magnetic field fabricated by self-assembly of super-paramagnetic nanoparticles with different sizes. (**a**) The optical microscope image shows that EG solvent containing $Fe_3O_4@SiO_2$ superparamagnetic particles was dispersed as circular droplets into the PDMS matrix. (**b**) Schematic diagram of a two-step procedure for fabricating a patterned print shown by a magnetic field. (**c**) The blue/green color contrast of the prints can appear with the application of an applied magnetic field. Reused with permission from Reference [67], Copyright 2008, Wiley-VCH.

4.4. Invisible Colloidal PC Pattern Shown by Electric Field Fabricated by Selective Modification of Substrates

Invisible patterns can be usually obtained by selectively modifying or immobilizing a certain portion of the responsive PC. Recently, however, Chen and co-workers [14] created an invisible pattern shown by an electric field, which was realized by using a selective modified substrate obtained by taping a patterned insulating polymer onto the conductive substrate, thereby inducing the selective deposition of charged nanoparticles only on the non-patterned conductive region (background region) under an electric field. The structure of the invisible patterned photonic anti-counterfeiting device is shown in Figure 12a, where the photonic suspension containing PS-co-G3Vi/Ag composite microspheres with high surface charge density was sandwiched between two fluorine-doped tin oxide (FTO) substrates and one of the FTO glasses was glued to a "maple leaf" shaped insulated polymer film. In the absence of an electric field, the device showed a very uniform yellow color derived from the PS-co-G3Vi/Ag particle, so that the pattern was invisible (Figure 12b). Once the voltage was applied, the maple leaf-patterned region also showed a yellow color due to the disordered arrangement of particles, while the background region gradually changed from yellow to green because of the formation of an ordered PC structure composed of PS-co-G3Vi/Ag particles with a higher refractive index (n = 1.58) and deionized water with a lower refractive index (n = 1.34), resulting in the display of the maple leaf pattern. When the field was withdrawn, the pattern gradually disappeared again. Therefore, the pattern shown by the electric field possessed good hiding and display abilities, and good reversibility. However, the brittle and rigid nature of FTO substrates greatly limits their further application to flexible target surfaces.

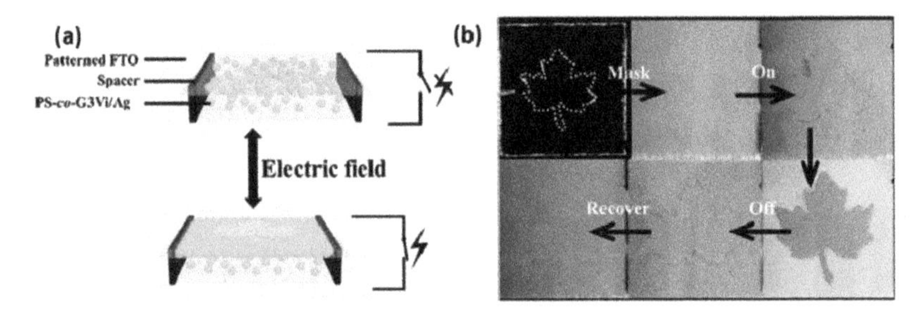

Figure 12. Invisible colloidal PC pattern shown by an electric field. (**a**) Schematic diagram showing the structure of an invisible patterned photonic anti-counterfeiting device. (**b**) The optical images showing that the anti-counterfeiting device can be reversibly displayed and hidden with the application and removal of an electric field. Reused with permission from Reference [14], Copyright 2016, The Royal Society of Chemistry.

5. Conclusions and Perspectives

In this review, recent developments of colloidal PC based anti-counterfeiting materials composed of variable colored PC patterns, invisible PC prints, and a few other PC security materials have been summarized from the aspects of security strategies, design and fabrication principles, and identification means. These anti-counterfeiting PC materials have multiple security features, which can achieve distinct color state transitions as viewing angles change, and can realize a reversible transition between invisible state and visible state when removing or applying specific external stimuli or lighting conditions. In addition, the micro-patterned PC material can be easily obtained through a photolithography technique or inkjet-printing method. Thus, the anti-counterfeiting materials based on PCs possess almost all anti-counterfeiting features of optical anti-counterfeiting techniques in existence, such as invisible watermarks, optical variable inks, laser anti-counterfeiting markings, and miniature printing. In addition, the structural colors of PCs originate from periodic structures, not fluorescence or dye molecules, thereby providing them with higher stability than variable inks. As a result, colloidal PCs provide a new avenue to develop anti-counterfeiting systems.

However, until now, an advanced anti-counterfeiting material with fast decryption, good hiding and display effects, as well as high durability, has not been realized in reality. For example, many invisible prints fabricated by selective cross-linking methods have good display effects when applying a specific external stimulation because of the larger color contrast between the pattern and the background region. However, solvents in the PC system will evaporate over time, leading to failure in the response to the external stimuli, resulting in low durability. In addition, these prints cannot realize true (optical) invisibility because cross-linked and uncross-linked regions have different reflected wavelengths. The prints fabricated by selective hydrophilic/hydrophobic modification can be classified into two types according to their structure, with both having good hiding effects because the modification process can hardly change the structure parameters of the PC. However, one type of invisible photonic print containing a responsive polymer exhibits a slow decryption speed (several minutes). The other type, composed of well-ordered hollow nanoparticles, has an ultrafast showing time (about 200 ms) but an indistinctive display effect. To develop the performance of existing PC-based anti-counterfeiting materials, we need to study the surface modification techniques of PCs or introduce a porous structure into the PC system.

Some challenges in PC-based anti-counterfeiting materials lie beyond the PC itself. For instance, the invisible photonic print shown by an electric field has fast speed, and good hiding and revealing effects, but its rigid and brittle conductive substrates limit its further application to flexible target surfaces. Therefore, using new flexible electrodes to replace FTO, such as conductive polymer electrodes [69], will broaden their application fields.

Based on this impressive research progress, a promising future can be expected for colloidal PC-based anti-counterfeiting materials.

Author Contributions: Y.L. and L.P. designed the study; M.P. wrote the paper; L.W., S.D. and J.Z. collected the literatures; H.X., B.W. and L.Z. revised the manuscript; X.L., Y.L. and L.P. edited and proofread the manuscript.

Funding: This research was funded by National Natural Science Foundation of China (No. 51572058, 51502057, 51761135123), the National Key Research & Development Program (2016YFB0303903, 2016YFE0201600), the International Science & Technology Cooperation Program of China (2013DFR10630, 2015DFE52770), and the Foundation of Equipment Development Department (6220914010901). Heilongjiang Postdoctoral Fund (LBH-Z15078, LBH-Z16080).

Conflicts of Interest: The author declares no conflict of interest.

References

1. Xia, Y.N.; Gates, B.; Li, Z.Y. Self-assembly approaches to three-dimensional photonic crystals. *Adv. Mater.* **2001**, *13*, 409–413. [CrossRef]
2. Yablonovitch, E. Photonic band-gap structures. *J. Opt. Soc. Am. B* **1993**, *10*, 283–295. [CrossRef]
3. Yablonovitch, E. Photonic band-gap crystals. *J. Phys-Condens. Mat.* **1993**, *5*, 2443–2460. [CrossRef]
4. Ge, J.; Yin, Y. Responsive photonic crystals. *Angew. Chem. Int. Ed.* **2011**, *50*, 1492–1522. [CrossRef] [PubMed]
5. John, S. Localization of light. *Phys. Today* **1991**, *44*, 32–40. [CrossRef]
6. Hou, J.; Li, M.Z.; Song, Y.L. Recent advances in colloidal photonic crystal sensors: materials, structures and analysis methods. *Nano Today* **2018**, *22*, 132–144. [CrossRef]
7. Rinne, S.A.; García-Santamaría, F.; Braun, P.V. Embedded cavities and waveguides in three-dimensional silicon photonic crystals. *Nat. Photonics* **2007**, *2*, 52–56. [CrossRef]
8. Kim, S.-H.; Kim, S.-H.; Jeong, W.C.; Yang, S.-M. Low-threshold lasing in 3D dye-doped photonic crystals derived from colloidal self-assemblies. *Chem. Mater.* **2009**, *21*, 4993–4999. [CrossRef]
9. Knight, J.C. Photonic crystal fibres. *Nature* **2003**, *424*, 847. [CrossRef]
10. Zhang, H.; Zhang, X.; Li, H.; Deng, Y.; Xi, L.; Tang, X.; Zhang, W. The orbital angular momentum modes supporting fibers based on the photonic crystal fiber structure. *Crystals* **2017**, *7*, 286. [CrossRef]
11. Du, C.; Wang, Q.; Zhao, Y.; Hu, S. Ultrasensitive long-period gratings sensor works near dispersion turning point and mode transition region by optimally designing a photonic crystal fiber. *Opt. Laser Technol.* **2019**, *112*, 261–268. [CrossRef]
12. Wang, S.; Chen, H.; Chen, M.; Xie, S. Ultrawideband pulse generation and bipolar coding based on optical cross-polarization modulation in highly nonlinear photonic crystal fiber. *Opt. Eng.* **2009**, *48*, 105006. [CrossRef]
13. Lee, S.Y.; Kim, S.-H.; Hwang, H.; Sim, J.Y.; Yang, S.-M. Controlled pixelation of inverse opaline structures towards reflection-mode displays. *Adv. Mater.* **2014**, *26*, 2391–2397. [CrossRef] [PubMed]
14. Chen, M.; Tian, Y.; Zhang, J.; Hong, R.; Chen, L.; Chen, S.; Son, D.Y. Fabrication of crack-free photonic crystal films via coordination of microsphere terminated dendrimers and their performance in invisible patterned photonic displays. *J. Mater. Chem. C* **2016**, *4*, 8765–8771. [CrossRef]
15. Yang, D.P.; Ye, S.Y.; Ge, J.P. From metastable colloidal crystalline arrays to fast responsive mechanochromic photonic gels: an organic gel for deformation-based display panels. *Adv. Funct. Mater.* **2014**, *24*, 3197–3205. [CrossRef]
16. Kuang, M.; Wang, J.; Bao, B.; Li, F.; Wang, L.; Jiang, L.; Song, Y. Inkjet printing patterned photonic crystal domes for wide viewing-angle displays by controlling the sliding three phase contact line. *Adv. Opt. Mater.* **2014**, *2*, 34–38. [CrossRef]
17. Zhou, J.; Zhou, T.J.; Li, J.G.; He, K.B.; Qiu, Z.R.; Qiu, B.C.; Zhang, Z.Y. Proposal and numerical study of a flexible visible photonic crystal defect cavity for nanoscale strain sensors. *Opt. Express* **2017**, *25*, 23645–23653. [CrossRef] [PubMed]

18. Cai, Z.; Smith, N.L.; Zhang, J.T.; Asher, S.A. Two-dimensional photonic crystal chemical and biomolecular sensors. *Anal. Chem.* **2015**, *87*, 5013–5025. [CrossRef]
19. Yang, D.Q.; Wang, B.; Chen, X.; Wang, C.; Ji, Y.F. Ultracompact on-chip multiplexed sensor array based on dense integration of flexible 1D photonic crystal nanobeam cavity with large free spectral range and high Q-factor. *IEEE Photonics J.* **2017**, *9*, 1–12.
20. Lee, H.S.; Shim, T.S.; Hwang, H.; Yang, S.-M.; Kim, S.-H. Colloidal photonic crystals toward structural color palettes for security materials. *Chem. Mater.* **2013**, *25*, 2684–2690. [CrossRef]
21. Heo, Y.; Kang, H.; Lee, J.S.; Oh, Y.K.; Kim, S.H. Lithographically encrypted inverse opals for anti-counterfeiting applications. *Small* **2016**, *12*, 3819–3826. [CrossRef] [PubMed]
22. Meng, Y.; Liu, F.F.; Umair, M.M.; Ju, B.Z.; Zhang, S.F.; Tang, B.T. Patterned and iridescent plastics with 3D inverse opal structure for anticounterfeiting of the banknotes. *Adv. Opt. Mater.* **2018**, *6*, 1701351. [CrossRef]
23. Hu, H.; Chen, Q.-W.; Tang, J.; Hu, X.-Y.; Zhou, X.-H. Photonic anti-counterfeiting using structural colors derived from magnetic-responsive photonic crystals with double photonic bandgap heterostructures. *J. Mater. Chem.* **2012**, *22*, 11048–11053. [CrossRef]
24. Zhang, J.; Yang, S.; Tian, Y.; Wang, C.F.; Chen, S. Dual photonic-bandgap optical films towards the generation of photonic crystal-derived 2-dimensional chemical codes. *Chem. Commun.* **2015**, *51*, 10528–10531. [CrossRef] [PubMed]
25. Yang, D.; Qin, Y.; Ye, S.; Ge, J. Polymerization-induced colloidal assembly and photonic crystal multilayer for coding and decoding. *Adv. Funct. Mater.* **2014**, *24*, 817–825. [CrossRef]
26. Yablonovitch, E.; Gmitter, T.J.; Leung, K.M. Photonic band structure: The face-centered-cubic case employing nonspherical atoms. *Phys. Rev. Lett.* **1991**, *67*, 2295–2298. [CrossRef] [PubMed]
27. Roundy, D.; Joannopoulos, J. Photonic crystal structure with square symmetry within each layer and a three-dimensional band gap. *Appl. Phys. Lett.* **2003**, *82*, 3835–3837. [CrossRef]
28. Král, Z.; Ferré-Borrull, J.; Trifonov, T.; Marsal, L.F.; Rodriguez, A.; Pallarès, J.; Alcubilla, R. Mid-IR characterization of photonic bands in 2D photonic crystals on silicon. *Thin Solid Films* **2008**, *516*, 8059–8063. [CrossRef]
29. Schaffner, M.; England, G.; Kolle, M.; Aizenberg, J.; Vogel, N. Combining bottom-up self-assembly with top-down microfabrication to create hierarchical inverse opals with high structural order. *Small* **2015**, *11*, 4334–4340. [CrossRef]
30. Meseguer, F.; Blanco, A.; Miguez, H.; Garcia-Santamaria, F.; Ibisate, M.; Lopez, C. Synthesis of inverse opals. *Colloid. Surface. A* **2002**, *202*, 281–290. [CrossRef]
31. Cui, L.; Zhang, Y.; Wang, J.; Ren, Y.; Song, Y.; Jiang, L. Ultra-fast fabrication of colloidal photonic crystals by spray coating. *Macromol. Rapid Commun.* **2009**, *30*, 598–603. [CrossRef]
32. Yang, H.W.; Pan, L.; Han, Y.P.; Ma, L.H.; Li, Y.; Xu, H.B.; Zhao, J.P. A visual water vapor photonic crystal sensor with PVA/SiO_2 opal structure. *Appl. Surf. Sci.* **2017**, *423*, 421–425. [CrossRef]
33. Jiang, P.; McFarland, M.J. Large-scale fabrication of wafer-size colloidal crystals, macroporous polymers and nanocomposites by spin-coating. *J. Am. Chem. Soc.* **2004**, *126*, 13778–13786. [CrossRef]
34. Liu, X.; Zhao, J.P.; Hao, J.; Su, B.L.; Li, Y. 3D ordered macroporous germanium fabricated by electrodeposition from an ionic liquid and its lithium storage properties. *J. Mater. Chem. A* **2013**, *1*, 15076–15081. [CrossRef]
35. Johnson, N.P.; McComb, D.W.; Richel, A.; Treble, B.M.; De la Rue, R.M. Synthesis and optical properties of opal and inverse opal photonic crystals. *Synthetic Met.* **2001**, *116*, 469–473. [CrossRef]
36. Ye, Y.H.; LeBlanc, F.; Hache, A.; Truong, V.V. Self-assembling three-dimensional colloidal photonic crystal structure with high crystalline quality. *Appl. Phys. Lett.* **2001**, *78*, 52–54. [CrossRef]
37. Wu, S.; Liu, B.; Su, X.; Zhang, S. Structural color patterns on paper fabricated by inkjet printer and their application in anticounterfeiting. *J. Phys. Chem. Lett.* **2017**, *8*, 2835–2841. [CrossRef]
38. Hou, J.; Li, M.; Song, Y. Patterned colloidal photonic crystals. *Angew. Chem. Int. Ed.* **2018**, *57*, 2544–2553. [CrossRef]
39. Bai, L.; Xie, Z.; Wang, W.; Yuan, C.; Zhao, Y.; Mu, Z.; Zhong, Q.; Gu, Z. Bio-inspired vapor-responsive colloidal photonic crystal patterns by inkjet printing. *Acs Nano* **2014**, *8*, 11094–11100. [CrossRef]
40. Xiong, C.; Zhao, J.; Wang, L.; Geng, H.; Xu, H.; Li, Y. Trace detection of homologues and isomers based on hollow mesoporous silica sphere photonic crystals. *Mater. Horiz.* **2017**, *4*, 862–868. [CrossRef]
41. Ma, H.R.; Zhu, M.X.; Luo, W.; Li, W.; Fang, K.; Mou, F.Z.; Guan, J.G. Free-standing, flexible thermochromic films based on one-dimensional magnetic photonic crystals. *J. Mater. Chem. C* **2015**, *3*, 2848–2855. [CrossRef]

42. Chen, K.; Fu, Q.; Ye, S.; Ge, J. Multicolor printing Using electric-field-responsive and photocurable photonic crystals. *Adv. Funct. Mater.* **2017**, *27*, 1702825. [CrossRef]
43. Ge, J.; Hu, Y.; Yin, Y. Highly tunable superparamagnetic colloidal photonic crystals. *Angew. Chem. Int. Ed.* **2007**, *46*, 7428–7431. [CrossRef] [PubMed]
44. Johnson, S.G.; Mekis, A.; Fan, S.H.; Joannopoulos, J.D. Molding the flow of light. *Comput. Sci. Eng.* **2001**, *3*, 38–47. [CrossRef]
45. Ye, X.Z.; Qi, L.M. Recent advances in fabrication of monolayer colloidal crystals and their inverse replicas. *Sci.China Chem.* **2014**, *57*, 58–69. [CrossRef]
46. Fenzl, C.; Hirsch, T.; Wolfbeis, O.S. Photonic crystals for chemical sensing and biosensing. *Angew. Chem. Int. Ed.* **2014**, *53*, 3318–3335. [CrossRef] [PubMed]
47. Fudouzi, H. Fabricating high-quality opal films with uniform structure over a large area. *J. Colloid Interf. Sci.* **2004**, *275*, 277–283. [CrossRef]
48. Ye, S.; Fu, Q.; Ge, J. Invisible photonic prints shown by deformation. *Adv. Funct. Mater.* **2014**, *24*, 6430–6438. [CrossRef]
49. Ye, S.; Ge, J. Soaking based invisible photonic print with a fast response and high resolution. *J. Mater. Chem. C* **2015**, *3*, 8097–8103. [CrossRef]
50. Hu, H.; Tang, J.; Zhong, H.; Xi, Z.; Chen, C.; Chen, Q. Invisible photonic printing: Computer designing graphics, UV printing and shown by a magnetic field. *Sci. Rep.* **2013**, *3*, 1484. [CrossRef]
51. Ding, T.; Cao, G.; Schaefer, C.G.; Zhao, Q.; Gallei, M.; Smoukov, S.K.; Baumberg, J.J. Revealing invisible photonic inscriptions: images from strain. *Acs Appl. Mater. Inter.* **2015**, *7*, 13497–13502. [CrossRef]
52. Xuan, R.; Ge, J. Invisible photonic prints shown by water. *J. Mater. Chem. C* **2012**, *22*, 367–372. [CrossRef]
53. Hou, J.; Zhang, H.; Su, B.; Li, M.; Yang, Q.; Jiang, L.; Song, Y. Four-dimensional screening anti-counterfeiting pattern by inkjet printed photonic crystals. *Chem. Asian J.* **2016**, *11*, 2680–2685. [CrossRef]
54. Nam, H.; Song, K.; Ha, D.; Kim, T. Inkjet printing based mono-layered photonic crystal patterning for anti-counterfeiting structural colors. *Sci. Rep.* **2016**, *6*, 30885. [CrossRef]
55. Shang, S.; Zhang, Q.; Wang, H.; Li, Y. Fabrication of magnetic field induced structural colored films with tunable colors and its application on security materials. *J. Colloid Interf. Sci.* **2017**, *485*, 18–24. [CrossRef]
56. Meng, Z.; Wu, S.; Tang, B.; Ma, W.; Zhang, S. Structurally colored polymer films with narrow stop band, high angle-dependence and good mechanical robustness for trademark anti-counterfeiting. *Nanoscale* **2018**, *10*, 14755–14762. [CrossRef]
57. Wang, M.; Yin, Y. Magnetically responsive nanostructures with tunable optical properties. *J. Am. Chem. Soc.* **2016**, *138*, 6315–6323. [CrossRef]
58. Hu, H.; Chen, C.; Chen, Q. Magnetically controllable colloidal photonic crystals: Unique features and intriguing applications. *J. Mater. Chem. C* **2013**, *1*, 6013–6030. [CrossRef]
59. Hu, H.B.; Zhong, H.; Chen, C.L.; Chen, Q.W. Magnetically responsive photonic watermarks on banknotes. *J. Mater. Chem. C* **2014**, *2*, 3695–3702. [CrossRef]
60. Ge, J.P.; Yin, Y.D. Magnetically responsive colloidal photonic crystals. *J. Mater. Chem.* **2008**, *18*, 5041–5045. [CrossRef]
61. Li, H.; Li, C.; Sun, W.; Wang, Y.; Hua, W.; Liu, J.; Zhang, S.; Chen, Z.; Wang, S.; Wu, Z.; et al. Single-stimulus-induced modulation of multiple optical properties. *Adv. Mater.* **2019**, *31*, 1900388. [CrossRef]
62. Weissman, J.M.; Sunkara, H.B.; Tse, A.S.; Asher, S.A. Thermally switchable periodicities and diffraction from mesoscopically ordered materials. *Science* **1996**, *274*, 959. [CrossRef]
63. Takeoka, Y.; Watanabe, M. Tuning structural color changes of porous thermosensitive gels through quantitative adjustment of the cross-linker in pre-gel solutions. *Langmuir* **2003**, *19*, 9104–9106. [CrossRef]
64. Kumoda, M.; Watanabe, M.; Takeoka, Y. Preparations and optical properties of ordered arrays of submicron gel particles: Interconnected state and trapped state. *Langmuir* **2006**, *22*, 4403–4407. [CrossRef]
65. Zhao, Z.; Wang, H.; Shang, L.; Yu, Y.; Fu, F.; Zhao, Y.; Gu, Z. Bioinspired heterogeneous structural color stripes from capillaries. *Adv. Mater.* **2017**, *29*, 1704569. [CrossRef]
66. Zhong, K.; Li, J.; Liu, L.; Van Cleuvenbergen, S.; Song, K.; Clays, K. Instantaneous, simple, and reversible revealing of invisible patterns encrypted in robust hollow sphere colloidal photonic crystals. *Adv. Mater.* **2018**, *30*. [CrossRef]
67. Jia, X.; Wang, K.; Wang, J.; Hu, Y.; Shen, L.; Zhu, J. Full-color photonic hydrogels for pH and ionic strength sensing. *Eur Polym J.* **2016**, *83*, 60–66. [CrossRef]

68. Ge, J.; Yin, Y. Magnetically tunable colloidal photonic structures in alkanol solutions. *Adv. Mater.* **2008**, *20*, 3485–3491. [CrossRef]
69. Zhou, L.; Yu, M.; Chen, X.; Nie, S.; Lai, W.-Y.; Su, W.; Cui, Z.; Huang, W. Screen-printed poly(3,4-ethylenedioxythiophene):poly(styrenesulfonate) grids as ITO-free anodes for flexible organic light-emitting diodes. *Adv. Funct. Mater.* **2018**, *28*, 1705955. [CrossRef]

Article

Band Tunability of Coupled Elastic Waves along Thickness in Laminated Anisotropic Piezoelectric Phononic Crystals

Qiangqiang Li [1], Yongqiang Guo [2,*], Yajun Wang [2] and Haibo Zhang [1]

1 College of Urban and Rural Construction, Shanxi Agricultural University, Jinzhong 030801, China
2 School of Civil Engineering and Mechanics, Lanzhou University, and Key Laboratory of Mechanics on Disaster and Environment in Western China, Ministry of Education, Lanzhou 730000, China
* Correspondence: guoyq@lzu.edu.cn; Tel.: +86-139-1905-1830

Received: 22 May 2019; Accepted: 13 August 2019; Published: 16 August 2019

Abstract: Although the passively adjusting and actively tuning of pure longitudinal (primary (P-)) and pure transverse (secondary or shear (S-)) waves band structures in periodically laminated piezoelectric composites have been studied, the actively tuning of coupled elastic waves (such as P-SV, P-SH, SV-SH, and P-SV-SH waves), particularly as the coupling of wave modes is attributed to the material anisotropy, in these phononic crystals remains an untouched topic. This paper presents the analytical matrix method for solving the dispersion characteristics of coupled elastic waves along the thickness direction in periodically multilayered piezoelectric composites consisting of arbitrarily anisotropic materials and applied by four kinds of electrical boundaries. By switching among these four electrical boundaries—the electric-open, the external capacitance, the electric-short, and the external feedback control—and by altering the capacitance/gain coefficient in cases of the external capacitance/feedback-voltage boundaries, the tunability of the band properties of the coupled elastic waves along layering thickness in the concerned phononic multilayered crystals are investigated. First, the state space formalism is introduced to describe the three-dimensional elastodynamics of arbitrarily anisotropic elastic and piezoelectric layers. Second, based on the traveling wave solutions to the state vectors of all constituent layers in the unit cell, the transfer matrix method is used to derive the dispersion equation of characteristic coupled elastic waves in the whole periodically laminated anisotropic piezoelectric composites. Finally, the numerical examples are provided to demonstrate the dispersion properties of the coupled elastic waves, with their dependence on the anisotropy of piezoelectric constituent layers being emphasized. The influences of the electrical boundaries and the electrode thickness on the band structures of various kinds of coupled elastic waves are also studied through numerical examples. One main finding is that the frequencies corresponding to $qH = n\pi$ (with qH the dimensionless characteristic wavenumber) are not always the demarcation between pass-bands and stop-bands for coupled elastic waves, although they are definitely the demarcation for pure P- and S-waves. The other main finding is that the coupled elastic waves are more sensitive to, if they are affected by, the electrical boundaries than the pure P- and S-wave modes, so that higher tunability efficiency should be achieved if coupled elastic waves instead of pure waves are exploited.

Keywords: coupled elastic waves; laminated piezoelectric phononic crystals; arbitrarily anisotropic materials; band tunability; electrical boundaries; dispersion curves

---

1. Introduction

Periodically multilayered composite structures [1–3] are constituted by periodically arranged unit cell with multilayered configuration. The different constituent layers in the unit cell have disparate material parameters such as material density and elastic stiffness constants. When elastic waves propagate in the periodically multilayered composites, their abovementioned particular construction form leads to the frequency band property resulting from the periodic scattering and further interference phenomenon of partial waves due to impedance mismatch at the interfaces between alternating constituent layers. This property, caused by the well-known Bragg scattering mechanism [2,3], refers to the elastic wave with frequency in pass-bands that propagates without attenuation, and the elastic wave with frequency in stop-bands that attenuates without propagation. During the recent three decades, the Bragg bands in periodically multilayered composites have been subsequently extended by the innovative concepts such as superlattices [4,5], phononic crystals [2,3,6,7], and metamaterials [2,6,8–10]. Besides the frequency bands, other novel elastic wave phenomena in these periodically multilayers such as the negative refraction, beam steering, and mode switching have also been revealed. On account of this progress, elastic wave (also referred to Floquet/Bloch waves) propagation in periodically multilayered composites (also known as laminated phononic crystals) becomes an even more attractive research topic.

Among the extensive studies on elastic waves in laminated phononic crystals, the very earliest object of study is the periodically multilayered composites made of only elastic materials. The incipient motivation for studying elastic wave propagation in periodically elastic multilayers is to develop nondestructive evaluation strategy for composite materials [2]. To achieve this goal, various elastic waves in the corresponding periodically elastic multilayers have been investigated, i.e., the bulk waves including the pure longitudinal (primary (P-)), pure transverse (secondary or shear (S-)), and their coupling modes in infinitely periodic multilayered elastic media [10–15], the surface/interface waves including the Rayleigh, Love, and surface transverse waves in semi-infinitely periodic multilayered elastic half-spaces [15–17], and the guided waves such as the Lamb and SH-type guided wave in periodically laminated elastic slab of finite thickness [10,15]. Besides the frequency bands of these wave modes, particular attentions have also been paid on the dispersion properties of various waves influenced by material anisotropy [12,13] as well as material elastic constants [13] and on the anisotropy of the characteristic waves like shear horizontal (abbr. SH) wave [14]. With the deepening understanding of the Floquet/Bloch waves in periodically multilayered elastic composites and the pressing demand for acoustic wave devices, researchers realized as early as 1980s that the piezoelectric materials can be introduced periodically into multilayers to form periodically laminated piezoelectric composites [4,5,18], and developed high-performed filters, guiders, and splitters, or delicately control mechanical waves through electricity. Since the piezoelectric material couples the mechanical field with the electrical field that is relatively easier to excite, detect and control as compared to other physical fields, the piezoelectric material is the most important and extensively-used intelligent material, especially in many functional devices like the bulk acoustic wave (BAW) and surface acoustic wave (SAW) devices [19]. Therefore, the combination of the piezoelectric effect and the elastic wave bands in the laminated piezoelectric phononic crystals is a very natural advance to improve the performance of and to add new function for acoustic wave devices [18]. It is also a promising strategy to control mechanical waves in a feasible and relatively easy way.

To push forward the development and design of these wave devices and the control strategy for various elastic waves, multifarious laminated piezoelectric periodic structures (phononic crystals) have been conceived. Moreover, the properties of the corresponding elastic waves thereof have been studied [4,5,18,20–71]. Nevertheless, the studies on elastic waves in laminated piezoelectric phononic crystals may be difficult on four counts. The first aspect, which appears definitely, is that the coupling between mechanical and electrical fields exists in piezoelectric phononic crystals, which represents the most essentially different property when compared to the purely elastic periodic composites. This electromechanical coupling causes the dependence of some or all the mechanical waves on the electrical field, besides results in the addition of a new wave mode (electric potential wave) mainly dominated by the electric field. Accordingly, these waves are sensitive to the electrical boundary conditions, which should be specified during the study of the wave properties and can be sorted into two main classes. One class is called as the passive-type electrical boundary, which remains fixed after applied like either the electric-open or the electric-short one as the most common boundary condition. The other class is referred to as the active-type electrical boundary, which is able to be switched between diverse passive electrical boundaries or adjusted through connection to external electric circuits. The second aspect, which may possibly happen, is that the constituent materials in multilayered piezoelectric phononic crystals are anisotropic with enough low crystal symmetry to bring about the coupling among the original pure horizontally-transverse (shear horizontal (SH)), pure longitudinal (P), and pure vertically transverse (shear vertical (SV)) modes even for the normally-propagating elastic waves. The third aspect, which depends on the utilized intention of the multilayered piezoelectric phononic crystals, is that the propagation direction of the elastic waves may affect their properties, even as the constituent materials have weak anisotropy (i.e., high symmetry). The fourth aspect is that the multiple reflection and transmission together with possible mode conversion of elastic waves at the surfaces/interfaces of constituent layers give rise to the complex bulk wave modes in infinite composites, or the surface/interface waves including the surface SH, Rayleigh, Love, and Stoneley waves in semi-infinite half-spaces, or the guided waves such as Lamb wave in finite thickness slabs. Note that the first aspect is exclusive of piezoelectric multilayered phononic crystals, while the latter three aspects are all common to multilayered phononic crystals of any materials. In addition, the second aspect, i.e., the material anisotropy actually affects the coupling and the electric-field dependence of wave modes in combination with the third and fourth aspects, i.e., the propagation direction and surface/interface property. For example, as the constituent piezoelectric and elastic layers in the multilayered piezoelectric phononic crystals all have enough high crystal symmetries, the mechanical waves thereof will be pure P-, pure SV-, and pure SH-waves, among which only one is influenced by the electric field and boundary. Nonetheless, the uncoupling requirement to the crystal symmetry may be different for disparate propagation direction. When the constituent materials entail crystal symmetries that do not satisfy the uncoupling condition of mechanical waves in a specific direction, some or all wave modes propagating in that direction will be coupled. The uncoupling or coupling of mechanical waves also influences the complexity of bulk wave and the formation of surface, interface, and guided waves.

Taking the above four aspects into account, we can review the research progresses on elastic waves in laminated piezoelectric phononic crystals by classification in two levels including the electrical boundary and wave type as follows.

Firstly, the investigations considering passive-type electrical boundaries (actually nearly all literature so far, except those specified otherwise concerning the electric open condition although not clearly pointed out), will be surveyed according to the wave type studied. For all kinds of bulk waves, including the pure SH wave, the uncoupled P and SV waves, and the coupled P-SV wave, propagating both normally and obliquely in infinite media with transversely isotropic (hexagonal crystal) constituent materials: Sapriel & Djafari-Rouhani [4] and Nougaoui & Djafari-Rouhani [5] summarized some research achievements before 1990 involving the dispersion curves, the effective constants and the analysis methods. Li and Wang [20,22] and Li et al. [21,23] studied the localization factor and length in randomly disordered piezoceramic–polymer composites by the transfer matrix method (TMM) with particular attention on the disorders of the thicknesses and thickness ratio of constituent layers and the piezoelectric/elastic constants of the piezoelectric layer. The effects of the piezoelectricity, the piezocomposite sort, the propagation direction, the nondimensional wavenumber, and the electrical potential on wave band and localization were discussed. Guo and Wei [24], considering initial stresses and their effects on constitutive equations, governing equations and boundary conditions, analyzed by TMM the dispersion curves in phononic crystal composed of two piezoelectric materials, whose dependences on the normal and shear initial stresses and the corresponding boundary conditions were discussed based on the numerical results. Golub et al. [25] and Fomenko et al. [26] analyzed by TMM the dispersion properties (such as dispersion curves and transmission/reflection coefficients), the localization factor and the classification of pass-bands and band gaps in phononic crystals composed of a specific number of periodically arranged unit cells with homogenous or functionally-graded interlayers and elastic half-spaces on both sides, whose dependences on the incident angle and the gradation and geometrical properties of interlayers were also discussed. Very recently, this functionally-graded model was extended by Fomenko et al. [27] to two cases, i.e., the infinite layered phononic crystals and finite counterparts between two isotropic half-spaces. The unit cell of both cases was composed of four piezoelectric sublayers with two being homogeneous and two being functionally graded. A semi-analytical method based on the transfer matrix of a unit cell was proposed to analyze the dispersion curves (phase and attenuation coefficients), the energy transmission coefficients, the localization factor and the classification of pass-bands and band gaps, whose dependences on the number of unit cells, the angle and type of incident waves, the thickness and material properties of the functionally graded sublayers, and the geometrical and material properties of the homogeneous layers were also discussed. Chen et al. [28] and Yan et al. [29], based on the nonlocal piezoelectricity continuum theory, analyzed the dispersion curves and the localization factor by TMM and the transmission/reflection spectra by the stiffness matrix method (SMM), respectively, in nanoscale phononic crystals consisting of two piezoelectric materials. The dependences of these wave propagation behaviors together with the found cutoff frequency on the ratio of internal to external characteristic lengths (R), the piezoelectric constant, the impedance ratio, and the incident angle were discussed with referring to numerical results. The influence of R on the mode conversion and the influence of the mode conversion on band gaps were also analyzed. Only concerning the pure SH waves in infinite media with transversely isotropic (hexagonal crystal) constituent materials: Zinchuk et al. [30] proposed a matrizant method to analyze the dispersion characteristics with particular attention on the effects of the piezoelectricity, the unit cell configuration, and the relative thickness ratio in numerical examples. This analysis was later extended by Zinchuk et al. [31] to periodic medium made of alternating metal and piezoelectric layers. Alshits and Shuvalov [32] analyzed the reflection/transmission of inclined SH waves in periodic composites having finite number of unit cells consisting of identical piezoelectric layers with metallized interfaces of fixed electropotential at alternating distances and semi-infinite substrates on its both sides, in which the existence of Bragg resonances was revealed. Qian et al. [33] obtained the phase velocity equations of normally and parallelly propagating waves in piezoelectric-elastic composites, and discussed the basic wave properties such as the filter effect and the effects of thickness and shear modulus ratios of the piezoelectric layer to elastic layer on phase velocity. Lan and Wei [34] studied the influence of imperfect interfaces (modeled by the mass-spring parameters) on dispersion curves and band gaps of both normally and obliquely propagating waves in piezoceramic–polymer composites with incidental attention on the piezoelectricity and the thickness-ratio effects. Only regarding the pure P waves, Faidi and Nayfeh [35] developed an improved continuum mixture model for analyzing the dispersion curves (phase velocity spectra) of the parallelly propagating longitudinal waves in periodic bi-laminated orthotropic piezoelectric media. As far as only the coupled P-SV waves are concerned, Geng and Zhang [36] analyzed the dispersion curves of parallelly propagating coupled P-SV waves in periodic piezoceramic–polymer composites by the method of partial wave expansion with special attention on the effects of the volume fraction and the polymer properties, which are partially validated through the experimentally measured thickness resonance (with polarization parallel to interface) and lateral resonance (with polarization along periodic direction) spectra. Regarding the three-dimensional (3D) coupled waves, besides some occasional discussions in Fomenko et al. [27], Podlipenets [37] presented without validation a Hamiltonian system formalism to analyze the dispersion equations of bulk, surface, and plate waves in respectively the infinite, semi-infinite, and finite phononic crystals with constituent materials of *mm2* or higher symmetry crystal. With respect to the surface/interface and guided waves, the laminated semi-infinite and finite transversely isotropic piezoelectric phononic crystals, respectively, with bounding plane either parallel or perpendicular to the layering plane have all been considered. Initially in the parallel case, Zinchuk et al. [38] analyzed the dispersion curves of the SH-type surface, Stoneley, and guided waves and discussed the state variables distribution of corresponding modes in periodic piezoelectric and metal composites under metallized mechanically free condition. The distinctive/interrelation characteristics of dispersion spectra for guided (normal) wave in even-layered periodic finite thickness plate with those for the surface wave in semi-infinite half-space were also discussed. Yan et al. [39] investigated the dispersion curves and the mechanical displacements and electrical potential variations of the symmetrical Lamb waves in nanoscale periodic piezoelectric composites based on the nonlocal piezoelectricity continuum theory. The influences of the piezoelectric effect, the ratio of internal to external characteristic lengths (i.e., R representing nanoscale size-effect), and the volume fractions on these wave behaviors particularly like the found mode conversion and cut-off frequency were analyzed in details. Later in the perpendicular case, Wang et al. [40] analyzed using TMM the localization factor of Rayleigh waves in periodic piezoceramic–polymer half-space due to disorders in polymer thickness or piezoceramic material constant, and discussed its influence by the thickness ratios of constituent layers. Alippi et al. [41] proposed an approximate theoretical model by neglecting the mode coupling to analyze the dispersion curves of Lamb wave below the frequency of thickness resonance. By experimentally exciting the band edge resonances, the analysis was validated and the fractional volume dependence of stop-bands was discussed. This work was later improved by Craciun et al. [42] who experimentally and theoretically studied the resonance spectra of the lateral and thickness resonances with polarizations along and perpendicular to the periodic direction, respectively, based on the Kronig-Penney and effective medium models, when considering the anisotropy, piezoelectricity, and volume fraction factors, which together with the coupling of two resonant modes were verified by the corresponding dispersion curves of pure elastic P and electroelastic SV waves. Note from the above literatures considering passive electrical boundaries that although the passive-type electrical boundaries are easy to realize and already provide the laminated piezoelectric phononic crystals with abundant elastic wave properties such as the frequency bands, but these wave properties are fixed or work at fixed frequency ranges once the laminated piezoelectric phononic crystals are fabricated. This passive feature obviously limits the application of laminated piezoelectric phononic crystals with passive electrical boundaries in working occasions where the external dynamic excitations have varying frequencies. Therefore, active-type electrical boundaries are proposed to tune the elastic wave properties such as the frequency bands anytime for adapting them to the external excitations. Two approaches have been proposed so far to actively adjust the electrical boundaries: One is to switch between diverse passive electrical boundaries and the other is to connect to external electrical network with alterable parameters such as capacitance, inductance, resistance, voltage, etc. In what follows, the studies related to the two adjusting manners will be reviewed successively.

Secondly, the investigations simultaneously considering many passive electrical boundaries for possibly switching between will be surveyed according to the wave type studied. For all kinds of bulk waves, including the pure SH wave, the uncoupled P and SV waves, and the coupled P-SV wave, propagating both normally and obliquely, Guo et al. [43] used TMM to analyze the dispersion curves of the phononic crystals composed of two transversely isotropic piezoelectric materials, considering four kinds of dielectric imperfect interfaces including weak conducting, high conducting, low dielectric, and grounded metallized interfaces together with four kinds of mechanical imperfect interfaces such as normal compliant, tangent compliant, tangent fixed, and tangent slippery interfaces. The influences of the piezoelectricity and these mechanically and dielectrically imperfect interfaces on the dispersion curves were discussed based on the numerical results. As far as only the pure SH waves in infinitely laminated piezoelectric phononic crystals with transversely isotropic (hexagonal symmetric) constituent materials are concerned, Ghazaryan and Piliposyan [44] investigated the effects of three kinds of interfaces, including electrically shorted with mechanically continuous condition, magnetically closed with mechanically continuous condition, and electrically open with mechanically smooth contacts, on dispersion properties such as band structures and bandgap width of obliquely-propagating Bloch–Floquet waves, with special attention on the factors like piezoelectricity, incident wavenumber, and filling fraction. In cases of electric-open and electric-short conditions, Zhao et al. [45] computed the dispersion curves, the transmission coefficients, and the reflective spectrum by the global transfer matrix method with attention on the effects of the piezoelectricity, the volume fraction and the propagation direction. With respect to coupled P-SV waves in infinite media, Zhao et al. [46], also in cases of electric-open and electric-short conditions and by the modified transfer matrix recursive algorithm, theoretically studied the dispersion curves (band gaps) and the transmission/reflection spectra in periodic structures containing fluid layer and orthotropic piezoelectric layer with special attention on the influences of piezoelectricity and propagation direction. As regard to the surface/interface or guided waves, the usually considered laminated piezoelectric semi-infinite or finite phononic crystals have bounding plane parallel to the layering plane. For this kind of phononic crystals, Sapriel & Djafari-Rouhani [4] and Nougaoui & Djafari-Rouhani [5] summarized research results on the dispersion curve of SH-type surface wave in cases of metallized or non-metallized surfaces before 1990s, with special attention on the influence of the nature of the film at the surface. Considering the same electrical boundaries, Zinchuk et al. [47] and Zinchuk & Podlipenets [48] analyzed the dispersion curves of SH-type surface and guided waves, the corresponding mode distributions of state variables by a matrizant method based on periodic Hamiltonian system, as the constituent materials belong to *6mm* crystal class. The effects of the electrical boundaries, the unit cell configuration and relative thickness ratio on the dispersion properties were studied by numerical examples. Zinchuk and Podlipenets [49] introduced their previous method to the analysis of dispersion curves of Rayleigh waves also with the *6mm* materials assumption for three kinds of electrical boundaries including electric-short, non-metalized contact with a vacuum and electric-open conditions, whose influence by the piezoelectricity was examined in the case of free mechanical boundary. Alami et al. [50], on basis of Green's function method, derived and validated the dispersion relation and state density of the SH-type surface waves in a semi-infinite superlattice composed of *6mm* class piezoelectric–metallic layers and capped with a piezoelectric layer with open- or short-circuited surfaces. The interaction between this SH-type surface wave and the possibly-existing interface, guided and pseudo-guided (leaky) waves induced by the cap layer and the dependences of electromechanical coupling coefficient on the cap layer thickness and the metallic layer property were also investigated. In addition, the laminated piezoelectric finite phononic crystals with bounding planes perpendicular to the layering plane have also been considered. The model studied so far is the plane-strain plate consisting of piezoceramic–polymer or two piezoelectric layers with hexagonal (usually *6mm*) symmetry and with the length in the periodic direction and the thickness parallel to the interfaces. In cases of electric open and short boundaries, Otero et al. [51] computed the dispersion curves of Lamb waves along periodic direction using the effective coefficients approximated by the first order asymptotic homogenization method considering four piezoelectric volume fractions. The behaviors of the mechanical displacement and the electric potential for different wave modes were illustrated. Considering the same electric boundaries and Lamb wave, Zou et al. [52] studied, using the plane wave expansion (PWE) method, the band properties such as the dispersion curves, widths and starting frequencies of the first bandgaps and their influences by the filling fraction, the thickness to lattice pitch ratio, and the polarizations directions. In cases of non-piezoelectricity, open-circuit, and short-circuit conditions, Zhu et al. [53] analyzed and compared the dispersion curves, transmission spectra, eigenmode displacements of different modes by the finite element method (FEM) on COMSOL Multiphysics software, based on which the piezoelectric-sensitive mode was defined and its physical mechanism such as its dependences on the piezoelectric constants, the filling ratio, and the ratio of thickness to lattice pitch were revealed. Considering both metallized and non-metallized interfaces, Piliposyan et al. [54] analyzed by TMM the dispersion curves and the reflection/transmission of inclined SH-type guided waves in the infinite periodic composites or in the finite counterparts with a defect layer as the mirror symmetry center and two piezoelectric half-spaces on both sides as substrates. The dependences of these wave behaviors on the ratio of the unit cell's length to the waveguide's height, the piezoelectric material properties, the boundary condition distribution on the lower and upper walls, and the presence of defect layer were discussed, with special attention on the Bragg resonances and the presence of trapped modes and slow waves. Notice from the above surveyed research works that the switching method between diverse passive electrical boundaries can adjust the frequency bands and other wave properties of laminated piezoelectric phononic crystals among several discrete states. However, in practical applications there is a trend toward the flexible usage of the frequency spectrum since the external dynamic excitations in working environments usually have continuously-varying frequency components rather than discrete-varying ones. This trend requires continuously-tunable frequency bands and other wave properties, which can be realized via installing external electrical circuits with alterable electrical parameters on the laminated piezoelectric phononic crystals that serves as the other approach to actively adjust the electrical boundaries. Next, the studies on elastic waves in laminated piezoelectric phononic crystals with connecting to external electrical network will be reviewed.

Thirdly, by the category of studied wave type and the alterable parameters such as capacitance, inductance, resistance, and voltage in the connected electrical circuits, the investigations on laminated piezoelectric phononic crystals possessing continuously-varying wave behaviors will be surveyed. For all kinds of uncoupled waves including pure P and S (SH or SV) modes, Li et al. [55] studied the dispersion curves of waves along the thickness direction in infinite media with orthotropic materials (the lowest symmetry crystal guaranteeing the decoupling of three wave modes) in cases of four electrical boundaries including the electric-open, applied electric capacitance, electric-short, and applied feedback voltage that are capable of both discretely and continuously tuning the frequency bands. Particular attentions were also on the influence of electrode thickness and on the characteristics of wave dispersion. Regarding to pure P waves in infinite phononic crystals with the faces of the piezoelectric layers being electroded and connected to electrical capacitor: Ponge et al. [56,57] briefly reviewed the dispersion properties induced by the periodic electrical boundary conditions (with open-circuit and short-circuit as reference) in homogeneous piezoelectric material, and then accordingly designed a Fabry-Perot cavity whose length might be tuned by a spatial shift of electrical boundaries along with the position of the transducer driven electrically by a voltage. The optimum performance of the cavity device achieved through compromise among the series/parallel resonance frequencies, the band gaps of phononic crystal (influenced by number and length of unit cells), and the coupling coefficient was corroborated by the 1D analytical analysis using TMM, the numerical simulation using finite element method (FEM), and the experiment. Kutsenko et al. [58,59] analyzed using TMM the dispersion spectra of normally-propagating wave in periodic structure of identical transversely isotropic piezoelectric layers or of alternating elastic and piezoelectric layers. Special attentions were on the unusual features of dispersion spectra in the case of negative capacitance value, such as the quasistatic stop-band, the poles of attenuation constant spectra corresponding to jumps of phase constant from 0 to π, and the occurrence of quasiflat dispersion branches and dispersion curves with infinitely growing group velocity [58,59]. Kutsenko et al. [60] extended the model to a general case where a unit cell may consist of several piezoelectric or elastic–piezoelectric multilayers (possibly functionally graded), and studied the effective properties characterizing the homogenized medium such as effective density, elastic constant, as well as Willis coupling constant, and their tunability by the capacitance. Recently, Kutsenko et al. [61] further extended the electrical boundary to 2D semi-infinite periodic network whose unit cell contains two capacitors in parallel and in series, and studied the dispersion spectra and wave fields by deriving explicit equations. The control of the dispersion spectrum, which can come out either as a discrete set of curves, or as a continuous band, or as a superposition of both, by the signs and values of the two capacitances in the unit cell was studied with considering examples and the limiting cases of open and short circuits. As far as pure P waves in infinite media with electrical circuits other than simple capacitor are concerned, Mansoura et al. [62] investigated theoretically the dispersion curves of normally-propagating waves in phononic piezoelectric-elastic crystals, paying special attention to the newly-opening hybridization gap due to a coupling of electrical resonance with the wave propagation as the electrical impendence was inductances. The results were verified experimentally by the measured transmission. Zhu et al. [63] proposed an active acoustic metamaterial consisting of periodic layers of steel, polyuria, and piezoelectric ceramic transducer (PZT) with the PZT layer shunted by an inductor. Its band structure and transmission coefficient were calculated by the TMM, and its effective material parameters were computed by homogenization method. The extremely narrow stop band induced by the resonance circuit was able to be controlled through the inductor and the corresponding effective parameters behaved negative. Parra et al. [64] calculated using TMM the dispersions, impedances, and displacements in phononic crystals with periodically distributed local and interconnected LC (inductance-capacitance) shunts, which were experimentally validated. The ability of local or interconnected shunts to control the width or depth of bandgap was also discussed. As regard to the coupled P-SV (or identical P-SH) waves, Fomenko et al. [65] analyzed using TMM the dispersion curves, transmission coefficients, and localization factor of the obliquely-propagating waves in infinite phononic crystals with the unit cell consisting of two piezoelectric interlayers and with or without electric potentials at the metallic interfaces of two kinds of configurations, i.e., an infinite periodic structure and a periodic structure surrounded by two half-spaces. The influences of the electrical potential and the incident angle on these wave properties were also investigated. As for the pure longitudinal waves and vibrations along the thickness in finite thickness plates have bounding plane parallel to the layering plane: Mansoura et al. [66,67] theoretically and experimentally investigated the electrical impedance of one piezoelectric layer within phononic crystals made of alternating piezoelectric and elastic layers while the other piezoelectric layers were connected to the external circuit including electric-open, electric-short, and external capacitance conditions, which is related to the band structures and transmissions together with the electromechanical coupling. Mansoura et al. [68] further verified, by comparing the experimentally measured transmissions with theoretically analyzed dispersion curves, the wave control in phononic piezoelectric elastic crystals through the negative capacitance connected on the electroded faces of piezoelectric layers that broadening band gaps. Darinskii et al. [69] theoretically studied using TMM the reflection/transmission along the 6-fold symmetry axis in periodic piezoelectric structure constructed by inserting infinitesimally thin metallic electrodes into a homogeneous piezoelectric material of *6mm* symmetry class and then interconnecting each two successive electrodes by an external capacitor (with electric-open and electric-short conditions corresponds to zero and infinity capacitance). Allam et al. [70] proposed an active acoustic metamaterial (AMM) with the unit cell consisting of transversely isotropic piezoelectric and isotropic brass layers clamped in air, whose effective density in real-time might be controlled by the closed feedback control loop connected between the piezoelectric layers. The stability, characterization, and behavior of the AMM were predicted by vibro-acoustic analytic model and were verified experimentally. The adaptive and programmable control of AMM's effective density through three types of controllers was also studied. Wang et al. [71] theoretically studied using TMM the transmission and the pass-band in piezoelectric laminated phononic crystals inserted with a 0.2 mol% Fe-doped relaxor-based ferroelectric 0.62 $Pb(Mg_{1/3}Nb_{2/3})O_3$–0.38$PbTiO_3$ (PMN–0.38PT) single crystal defect layer, whose dependences on the thickness/strain adjusted nonlinearly by the external electrical voltage and on the acoustic impedance of constituent materials were analyzed. As for the vibrations associated with coupled P-SV waves propagation in finite thickness plates that have bounding plane perpendicular to the layering plane, Geng and Zhang [36] studied theoretically and experimentally the vibration properties of thickness and lateral resonances (such as the surface vibration profile, the electric impedance distribution, and the electromechanical coupling factor) under an external driving AC electric field in both air and water media and under external incident pressure wave from the water, respectively, with particular attention on the influence of the aspect ratio.

In summary of the above literature review on various types of elastic waves in laminated piezoelectric phononic crystals with both passive and active electrical boundary conditions, we propose the following five insufficiencies that exist in the investigations so far:

1. Although the electric-open, the electric-short, and the applied electrical capacitance boundaries have all received plenty attentions, but the feedback control condition, which plays an important role in vibration control of structures in many engineering fields [72,73], is nearly not concerned except that in Allam et al. [70] the feedback strategy was adopted to control the real-time effective density of an unit cell of the proposed active acoustic metamaterial involving pure P wave propagation in laminated piezoelectric phononic crystals of finite thickness. Hence, the authors Li et al. [55] extended the external feedback control voltage boundary into the infinite media

and investigated its contrast with the electric-open, the electric-short, and the applied electrical capacitance boundaries when considering the pure P- and S-waves propagation. However, when the coupled mechanical waves in laminated piezoelectric phononic crystals are investigated, the applied feedback control electrical condition has never been considered so far.

2. Throughout the studies on coupled elastic waves in cases of all electrical boundaries, the coupling of the wave modes is due to the obliquely-propagating direction rather than the material anisotropy, since the piezoelectric and elastic constituent layers in all these investigations are assumed as transversely isotropic and isotropic, respectively, with hexagonal or higher symmetries. Notice from Li et al. [55] that as long as the interlayers all have orthorhombic or higher symmetries, the normally-propagating elastic waves are decoupled. That is to say, the wave coupling caused by material anisotropy actually has nearly not been investigated in literatures except the following two [74,75] to the best of authors' knowledge. In Honein et al. [74] and Minagawa [75], the analysis of the 2D coupled P-SV waves and the 3D coupled waves along arbitrary (and occasionally along thickness) direction were considered, respectively, both for periodically multilayered piezoelectric media with general anisotropic constituent materials. Concretely, Honein et al. [74] introduced without validation a surface impedance matrix method based on state vector formalism. Minagawa [75] proposed the TMM and evaluated the influence of piezoelectric effect on the phase velocity–wavenumber relation. Therefore, the dependence of various coupled waves on the material anisotropy alone is yet to know.
3. In cases of electrical boundaries with alterable parameters, the dominantly studied wave type is the pure P wave. Although Li et al. [55] concerned the pure S wave in cases of applied electric capacitance and feedback voltage conditions, and Fomenko et al. [65] and Geng & Zhang [36] considered the coupled P-SV mode in respectively infinite and finite laminated piezoelectric phononic crystals with applied electric potential boundary, but the coupled SV-SH, P-SH and P-SV-SH waves in cases of diverse electrical boundaries have not been comprehensively investigated yet.
4. In all the previous studies, the mechanical effect of electrodes is omitted by neglecting their thickness. However, the feasibility of this manner to treat the electrode has not been validated as far.
5. In the previous studies, except that the phase velocity spectrum is concerned occasionally in literatures [33,35,75], the dispersion curve is dominantly represented as the frequency–wavenumber relation. The properties of other forms of dispersion curves such as the frequency-wavelength spectra have not been discussed up to now.

In order to make up for the above deficiencies, this paper analyzes, by introducing the TMM [76] based on the state space formalism [77], the coupled elastic waves, including P-SH, P-SV, SV-SH, and P-SV-SH modes, along thickness direction in infinitely laminated piezoelectric phononic crystals with the unit cell consisting of any number of arbitrarily anisotropic piezoelectric and elastic layers and having four electrical boundaries, such as the electric-open, the applied electrical capacitance, the electric-short, and the applied feedback control conditions. The electrodes on the surfaces of the piezoelectric interlayers are all modeled as elastic layers along with the inserted elastic materials themselves, whose mechanical and electrical functions are both considered in the analysis. The configuration of the analysis model hereof renders the presence of coupled waves without introducing the effect of inclined propagation direction. Consequently, the forming of coupled waves is solely caused by the material anisotropy, the individual effect of which on the multifarious frequency related dispersion curves in cases of four electrical boundaries is the main motivation of our research. Compared with our previous similar work [55], this paper innovatively considers the coupled-mode waves due to material anisotropy as an extension to the previously concerned pure P and pure S waves and as a remedy to deficiencies (2) and (3), in spite that a similar analysis process is adopted.

This paper is organized as follows. Following the research background and motivation in Section 1, the basic model of the general periodically laminated arbitrarily-anisotropic piezoelectric composites is described in Section 2. Section 3 provides the state space formalism for describing the elastodynamics of the constituent layers. Based on the wave solutions to the state variables of constituent layers, the formulation of transfer matrix method (TMM) is presented in Section 4 to establish the governing relations of the unit cell, which are combined with the Floquet–Bloch theorem [2,78,79] to bring out the dispersion equation of the whole system. The numerical examples for phononic crystals with various layouts of piezoelectric materials and four kinds of electrical boundaries are given in Section 5, with special attention on the general features of diversified frequency related dispersion curves and their dependences on the electrode thickness and electrical boundaries. Some findings from this research are drawn as conclusions in Section 6.

2. Basic Model

Consider the elastic waves propagating along the thickness direction, i.e., perpendicular to the interface, in infinitely periodically laminated piezoelectric composite structures whose unit cell, as shown in Figure 1, consists of any number of arbitrarily anisotropic (triclinic) piezoelectric and elastic layers with the piezoelectric interlayers having anyone of four electrical boundaries such as the electric-open, the applied electrical capacitance, the electric-short, and the applied feedback control conditions. These electrical boundaries are applied through the electrodes coated on the surfaces of piezoelectric layers. Here in this paper, the electrodes are also modeled as elastic layers along with the inserted elastic materials themselves, whose mechanical and electrical functions are both considered in the analysis, so that their effect on the wave propagation can be revealed. According to the Floquet–Bloch theorem [2,78,79], the unit cell model in Figure 1 together with the periodic boundary condition is adequate for the analysis of characteristic waves with wavenumber q and circular frequency ω. Since arbitrarily anisotropic (triclinic) materials are considered in the model, whose crystal symmetry is far lower than orthotropy, the lowest crystal symmetry requirement for decoupling the elastic waves along thickness direction [55]. Therefore, coupled-mode waves are actually analyzed via the model. In the unit cell shown by Figure 1, altogether any number (m) of interlayers, involving the shaded elastic layers, gray piezoelectric layers, and black electrode layers, and correspondingly N ($N = m + 1$) interfaces are assumed for the sake of modeling all sorts of structural configurations.

As shown in Figure 1, the constituent layers, with finite thicknesses and being unbounded on their layering planes, in the unit cell are labeled in sequence from top to bottom as $(1), (2), \cdots, (j), \cdots, (m)$, while correspondingly those surfaces/interfaces are denoted as $1, 2, \cdots, J, \cdots, N$, for the convenience of description. The thickness of a typical layer, say (j), is denoted as $h_{(j)}$, shown both in Figure 1 and in the enlarged view of layer (j) in Figure 2b, while the thickness of the unit cell is $H = \sum_{(j)=(1)}^{(m)} h_{(j)}$. For the piezoelectric layer (j) as depicted by Figure 2b, a local coordinate system $\left(x_{(j)}, y_{(j)}, z_{(j)}\right)$ is established with its origin on the top surface of the layer for facilitating the description of physical quantities. $u_{x(j)}$, $u_{y(j)}$, and $u_{z(j)}$ signify the displacements along x, y, and z axes at any position on a plane within layer (j) parallel to the layer surfaces, while $\tau_{zx(j)}$, $\tau_{zy(j)}$, and $\sigma_{z(j)}$ denote the corresponding stresses on that plane. These mechanical quantities, if the linear theory of piezoelectricity [80–82] are resorted, are known to be related to the six partial mechanical waves whose wavenumbers are the components of vector $\mathbf{\Lambda}_{(j)}$. The six partial waves can be divided into two groups propagating respectively in downward and upward directions and into three pairs representing respectively the three modes. Because of the piezoelectric effect in the piezoelectric layer, these partial waves can also be tuned by the electrical boundaries applied on the electrode layers, which are coated on the piezoelectric layer surfaces and are connected to the four external circuits including the electric-open, the applied electric capacitance, the electric-short, and the applied feedback control conditions. Both switching between the four electrical boundaries and adjusting the applied capacitance $C_{(j)}$ in the case of connecting external capacitor or the gain coefficient $K_{g(j)}$ in the connecting feedback control condition can actively change the electric charge on the electrode $Q_{(j)}$ and the voltage (electric potential difference) between electrodes $V_{(j)}$, and further alter the thickness distribution of the electric potential

$\phi_{(j)}$, the electrical displacement $D_{z(j)}$ along the thickness direction, and the intensity of electrical field $E_{z(j)}$ along the thickness direction. Note that in the case of applied feedback control boundary, $V_{(j)}$ is related to $K_{g(j)}$ through $V_{(j)} = -K_{g(j)}[u_{z(j)}(h_{(j)}) - u_{z(j)}(0)]$, where $u_{z(j)}(0)$ and $u_{z(j)}(h_{(j)})$ represent the displacements along the thickness direction on the top and bottom surfaces of layer (j) at the sampling position, respectively.

The descriptions of the local coordinates and the pertaining physical quantities of other piezoelectric layers and of elastic layers denoting either the electrodes or the inserted elastic materials are exactly the same as those for piezoelectric layer (j), except that the quantities related to electrical field of elastic layers are not provided as can be noticed from Figure 2a for elastic layer (i), because they will not be involved in later formulation. Moreover, no electrical boundaries are applied on the elastic layers, as also noted from Figure 2a. Here and after, all the quantities pertaining to a layer are denoted by a subscript of layer label, while those pertaining to a surface/interface will be signified by a superscript of surface/interface label. All the vector-type physical variables are deemed as positive when their directions are coincident with the positive coordinate axes, while the scalar electric quantities are deemed as positive when they are corresponding to the positive electric charge, and vice versa.

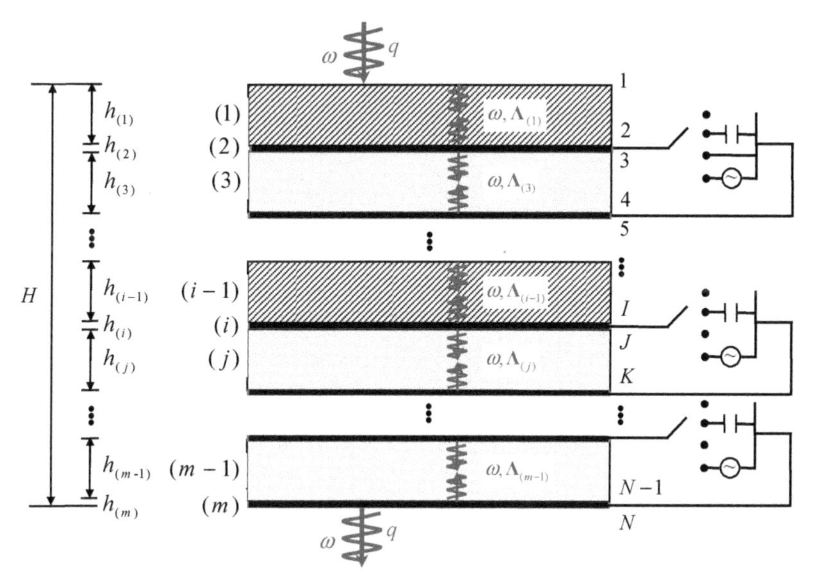

Figure 1. The unit cell of the general periodically laminated piezoelectric composite structures with arbitrarily anisotropic constituent interlayers and the description about the propagation characteristics of coupled elastic waves along thickness direction.

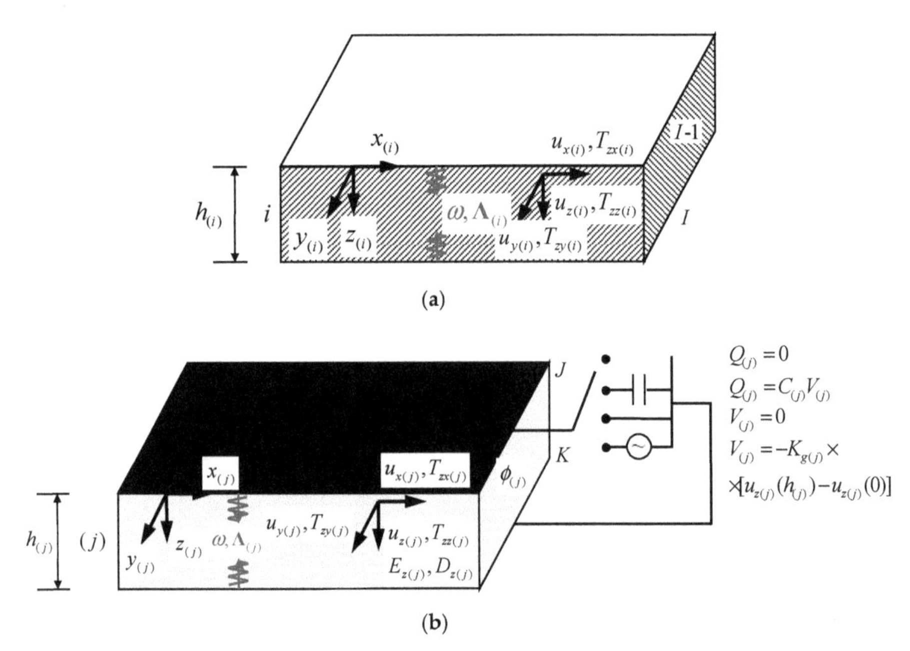

Figure 2. The descriptions of the local coordinate systems and the physical quantities of typical elastic/piezoelectric layers in the unit cell. (**a**) A typical elastic layer. (**b**) A typical piezoelectric layer covered by the electrode layers with four electrical boundaries: electric-open, applied capacitance, electric-short, and applied feedback control.

3. State Space Formalism

Since the considered piezoelectric phononic crystals and their unit cells are laminated, the state space formalism [77], which is essentially a displacement–stress hybrid method suitable for structures with unidirectional configuration, is introduced to describe the elastodynamics of the constituent layers. For any elastic or piezoelectric constituent layer, by using the method of separation of variables and by choosing the displacements and stresses on the plane parallel to the surfaces as the components in the state vector, the state equation governing the spectral state vector in frequency–wavenumber domain can be derived from all the fundamental equations of three-dimensional (3D) elasticity and piezoelectricity, respectively. The solution to the state equation can then be obtained by virtue of the theory of first order ordinary differential equations. It should be pointed out that all the equations and their solutions in this section are given in the local coordinate system, but the subscript indicating the pertinent layer and coordinates is omitted for brevity.

3.1. *The State Equation of a Layer Derived from the 3D Elastodynamics*

The derivation of the state equation for elastic and piezoelectric layers will be provided successively, although their route is generally similar except for the introduction of electrical boundaries to the piezoelectric layer. For the piezoelectric layer, the differences of the deriving process as compared to that for the elastic layer will be emphasized, and the similarities will be abbreviated, as indicated in Section 3.1.2.

3.1.1. The State Equation of An Elastic Layer

According to the 3D elasticity, the fundamental equations governing the elastodynamics of a typical arbitrarily anisotropic elastic layer (i) as seen in Figure 2a can be written based on mechanical and electrical considerations [82]. The mechanical equations include the equation of motion (without body force here), the strain–displacement relation, and the elastic constitutive equation, which are written in fully matrix notation as

$$(\nabla_s)^{\mathrm{T}}\mathbf{T} = \rho\frac{\partial^2 \mathbf{u}}{\partial t^2},\ \mathbf{S} = \nabla_s \mathbf{u},\ \mathbf{T} = \mathbf{cS}, \tag{1}$$

where the superscript "T" denotes the transposition of a matrix or a vector here and after; $\mathbf{u} = [u_x, u_y, u_z]^{\mathrm{T}}$, $\mathbf{T} = [T_{xx}, T_{yy}, T_{zz}, T_{yz}, T_{zx}, T_{xy}]^{\mathrm{T}}$, and $\mathbf{S} = [S_{xx}, S_{yy}, S_{zz}, 2S_{yz}, 2S_{zx}, 2S_{xy}]^{\mathrm{T}}$ are the displacement, stress, and strain vectors, respectively; ρ denotes the mass density; and ∇_s and $\mathbf{c}$ are the 6×3 operator matrix and the 6×6 symmetric elastic stiffness constant matrix, respectively, with the following components

$$\nabla_s = \begin{bmatrix} \partial/\partial x & 0 & 0 \\ 0 & \partial/\partial y & 0 \\ 0 & 0 & \partial/\partial z \\ 0 & \partial/\partial z & \partial/\partial y \\ \partial/\partial z & 0 & \partial/\partial x \\ \partial/\partial y & \partial/\partial x & 0 \end{bmatrix},\ \mathbf{c} = \begin{bmatrix} c_{11} & c_{12} & c_{13} & c_{14} & c_{15} & c_{16} \\ c_{21} & c_{22} & c_{23} & c_{24} & c_{25} & c_{26} \\ c_{31} & c_{32} & c_{33} & c_{34} & c_{35} & c_{36} \\ c_{41} & c_{42} & c_{43} & c_{44} & c_{45} & c_{46} \\ c_{51} & c_{52} & c_{53} & c_{54} & c_{55} & c_{56} \\ c_{61} & c_{62} & c_{63} & c_{64} & c_{65} & c_{66} \end{bmatrix}. \tag{2}$$

The electrical equations under the quasi-static assumption to electric field include the charge equation (Gauss's law) of electrostatics (with free charge here), the electric field-electric potential relation, and the electric constitutive equation, which are written in fully matrix notation as

$$(\nabla)^{\mathrm{T}}\mathbf{D} = 0,\ \mathbf{E} = -\nabla\phi,\ \mathbf{D} = \boldsymbol{\varepsilon}\mathbf{E}, \tag{3}$$

where ϕ is the scalar electric potential; $\mathbf{D} = [D_x, D_y, D_z]^{\mathrm{T}}$ and $\mathbf{E} = [E_x, E_y, E_z]^{\mathrm{T}}$ denote the electric displacement and electric field intensity vectors, respectively; and ∇ and $\boldsymbol{\varepsilon}$ are the 3×1 Hamilton operator vector and the 3×3 dielectric constant matrix, respectively, having the following components

$$\nabla = \begin{bmatrix} \partial/\partial x \\ \partial/\partial y \\ \partial/\partial z \end{bmatrix},\ \boldsymbol{\varepsilon} = \begin{bmatrix} \varepsilon_{11} & \varepsilon_{12} & \varepsilon_{13} \\ \varepsilon_{21} & \varepsilon_{22} & \varepsilon_{23} \\ \varepsilon_{31} & \varepsilon_{32} & \varepsilon_{33} \end{bmatrix}. \tag{4}$$

According to the method of separation of variables, the harmonic solutions to the physical quantities can be given as

$$\Gamma(x, y, z, t) = \hat{\Gamma}(k_x; k_y; z; \omega)\mathrm{e}^{\mathrm{i}(\omega t - k_x x - k_y y)} = \hat{\Gamma}(z; \omega)\mathrm{e}^{\mathrm{i}\omega t}(\Gamma = \mathbf{u}, \mathbf{T}, \mathbf{S}, \phi, \mathbf{D}, \mathbf{E}), \tag{5}$$

where $\mathrm{i} = \sqrt{-1}$ is the imaginary unit; ω is the circular frequency; k_x and k_y are the wavenumbers along the local coordinates x and y, respectively; and the superscript "$\hat{}$" means corresponding physical quantities in the $k_x - k_y - \omega$ domain. In Equation (5), the vanishing of wavenumbers in the layering plane $k_x = k_y = 0$ has been taken into account, because only the elastic waves propagating along the thickness direction are considered in this paper. Besides, since the considered layer is infinite in the layering plane and has finite thickness, then only the displacements and stresses on the surfaces will appear in the mechanical boundaries. Consequently, we choose the displacements and stresses on the plane parallel to the surfaces [77] as the components of the state vector $\hat{\mathbf{v}} = [(\hat{\mathbf{v}}_u)^{\mathrm{T}}, (\hat{\mathbf{v}}_T)^{\mathrm{T}}]^{\mathrm{T}} = [\hat{u}_x, \hat{u}_y, \hat{u}_z, \hat{T}_{zx}, \hat{T}_{zy}, \hat{T}_{zz}]^{\mathrm{T}}$, with $\hat{\mathbf{v}}_u = \hat{\mathbf{u}} = [\hat{u}_x, \hat{u}_y, \hat{u}_z]^{\mathrm{T}}$ and $\hat{\mathbf{v}}_T = [\hat{T}_{zx}, \hat{T}_{zy}, \hat{T}_{zz}]^{\mathrm{T}}$ being referred to as the displacements and stresses state vectors, respectively.

Note from Equations (1) and (3) that the mechanical and electrical fields in the elastic layer are mutually independent each other. Since here in this paper the mechanical field rather than electrical field is continuous across the adjacent layers, only Equation (1) is used to derive the state equation of the elastic layer. Substitution of Equation (5) into the first formula in Equation (1) leads to

$$\frac{\mathrm{d}\hat{\mathbf{v}}_T}{\mathrm{d}z} = -\rho\omega^2\hat{\mathbf{v}}_u. \tag{6}$$

Substituting the second formula into the third formula in Equation (1), then introducing Equation (5) into the resulting equation $\mathbf{T} = \mathbf{c}\nabla_s\mathbf{u}$, and finally selecting from the consequent equation out those pertaining to the stress state vector, we can obtain

$$\hat{\mathbf{v}}_T = \mathbf{G}\frac{\mathrm{d}\hat{\mathbf{v}}_u}{\mathrm{d}z}, \tag{7}$$

which further leads, through matrix inversion, to

$$\frac{\mathrm{d}\hat{\mathbf{v}}_u}{\mathrm{d}z} = \mathbf{G}^{-1}\hat{\mathbf{v}}_T, \tag{8}$$

The 3×3 coefficient matrix $\mathbf{G}$ is composed of

$$\mathbf{G} = \begin{bmatrix} c_{55} & c_{54} & c_{53} \\ c_{45} & c_{44} & c_{43} \\ c_{35} & c_{34} & c_{33} \end{bmatrix}. \tag{9}$$

The combination of Equations (6) and (8) results in the state equation of the elastic layer

$$\frac{\mathrm{d}\hat{\mathbf{v}}}{\mathrm{d}z} = \mathbf{M}\hat{\mathbf{v}}, \tag{10}$$

where the 6×6 coefficient matrix $\mathbf{M}$ of the state equation are formed as

$$\mathbf{M} = \begin{pmatrix} 0 & \mathbf{G}^{-1} \\ -\rho\omega^2\mathbf{I}_3 & 0 \end{pmatrix}, \tag{11}$$

with $\mathbf{I}_3$ the third order identity matrix.

3.1.2. The State Equation of a Piezoelectric Layer

According to the 3D piezoelectricity, the fundamental elastodynamic equations of a typical arbitrarily anisotropic piezoelectric layer (j), as seen in Figure 2b, can be given. From both mechanical and electrical considerations [80–83], all the equations are identical to those of elastic layer except that the constitutive equation currently should be written in fully matrix notation as

$$\mathbf{T} = \mathbf{cS} - \mathbf{e}^{\mathrm{T}}\mathbf{E},\ \mathbf{D} = \mathbf{eS} + \varepsilon\mathbf{E}, \tag{12}$$

where $\mathbf{e}$ is the 3×6 piezoelectric constant matrix composed of

$$\mathbf{e} = \begin{bmatrix} e_{11} & e_{12} & e_{13} & e_{14} & e_{15} & e_{16} \\ e_{21} & e_{22} & e_{23} & e_{24} & e_{25} & e_{26} \\ e_{31} & e_{32} & e_{33} & e_{34} & e_{35} & e_{36} \end{bmatrix}. \tag{13}$$

Equation (12) indicates that the mechanical and electrical fields in piezoelectric layer are coupled through the constitutive relation. Consequently, although here in this paper only the mechanical field is continuous across the adjacent layers, the effect of the electrical field on the mechanical field needs to be considered during the derivation of the state equation for the piezoelectric layer.

Bear in mind that the solutions to the mechanical and electrical quantities of the piezoelectric layer can be still expressed as Equation (5), and the displacements, stresses, and whole state vectors of the piezoelectric layer are formed identically as those of the elastic layer shown before. Therefore, Equation (6) is still right for piezoelectric layer when Equation (5) is substituted into the equation of motion without body force. In addition, by substituting the strain–displacement relation as the second formula in Equation (1), the electric field-electric potential relation as the second formula in Equation (3), and Equation (5) into the constitutive Equation (12), through the similar process as before in Section 3.1.1, we gain

$$\hat{\mathbf{v}}_T = \mathbf{G}\frac{\mathrm{d}\hat{\mathbf{v}}_u}{\mathrm{d}z} + \mathbf{F}^{\mathrm{T}}\frac{\mathrm{d}\hat{\phi}}{\mathrm{d}z},\ \hat{D}_z = \mathbf{F}\frac{\mathrm{d}\hat{\mathbf{v}}_u}{\mathrm{d}z} - \varepsilon_{33}\frac{\mathrm{d}\hat{\phi}}{\mathrm{d}z}, \tag{14}$$

where $\mathbf{G}$ is the same as Equation (9)

$$\mathbf{F} = \begin{bmatrix} e_{35} & e_{34} & e_{33} \end{bmatrix}, \tag{15}$$

and only the expression of $\hat{D}_z$ among the three components of electric displacement vector is sorted out since only it can be used, after the electrical boundaries of the piezoelectric layer have been introduced, to represent $\mathrm{d}\hat{\phi}/\mathrm{d}z$ by $\mathrm{d}\hat{\mathbf{v}}_u/\mathrm{d}z$ as

$$\frac{\mathrm{d}\hat{\phi}}{\mathrm{d}z} = \frac{\mathbf{F}}{\varepsilon_{33}}\frac{\mathrm{d}\hat{\mathbf{v}}_u}{\mathrm{d}z} - \mathbf{R}[\hat{\mathbf{v}}_u(h) - \hat{\mathbf{v}}_u(0)]. \tag{16}$$

The 1×3 coefficient matrix $\mathbf{R}$ is related to $\mathbf{F}$ diversely in cases of four different electrical boundaries as shown from Table A1 in Appendix A, where the derivation of Equation (16) is provided in detail. Substituting Equation (16) into the first formula of Equation (14), it is obtained by further simplification that

$$\frac{\mathrm{d}\hat{\mathbf{v}}_u}{\mathrm{d}z} = \left(\mathbf{G} + \frac{\mathbf{F}^{\mathrm{T}}\mathbf{F}}{\varepsilon_{33}}\right)^{-1}\hat{\mathbf{v}}_T + \left(\mathbf{G} + \frac{\mathbf{F}^{\mathrm{T}}\mathbf{F}}{\varepsilon_{33}}\right)^{-1}\mathbf{F}^{\mathrm{T}}\mathbf{R}[\hat{\mathbf{v}}_u(h) - \hat{\mathbf{v}}_u(0)]. \tag{17}$$

The combination of Equations (6) and (17) gives the state equation of the piezoelectric layer as

$$\frac{\mathrm{d}\hat{\mathbf{v}}}{\mathrm{d}z} = \mathbf{M}\hat{\mathbf{v}} + \mathbf{N}[\hat{\mathbf{v}}(h) - \hat{\mathbf{v}}(0)], \tag{18}$$

where the 6×6 coefficient matrix $\mathbf{M}$ and the 6×6 inhomogeneous term matrix $\mathbf{N}$ are given by

$$\mathbf{M} = \begin{pmatrix} 0 & (\mathbf{G} + \mathbf{F}^{\mathrm{T}}\mathbf{F}/\varepsilon_{33})^{-1} \\ -\rho\omega^2\mathbf{I}_3 & 0 \end{pmatrix},\ \mathbf{N} = \begin{pmatrix} (\mathbf{G} + \mathbf{F}^{\mathrm{T}}\mathbf{F}/\varepsilon_{33})^{-1}\mathbf{F}^{\mathrm{T}}\mathbf{R} & 0 \\ 0 & 0 \end{pmatrix}. \tag{19}$$

Notice that the state Equation (18) of the piezoelectric layer is naturally degenerated to that of the elastic layer in Equation (10), when the piezoelectric constants are set to zero, i.e., $\mathbf{N}$ is a matrix of zeros due to $\mathbf{F} = 0_{1\times3}$, with 1×3 signifying the numbers of row and column of the zero matrix 0 here and after.

3.2. The Traveling Wave Solution to the State Equation of a Layer

The state equation of an elastic layer is a set of one order homogeneous linear ordinary differential equations as shown in Equation (10), whose general solution can be directly written according to the mathematical theory of this kind of equations as

$$\hat{\mathbf{v}} = \mathbf{\Phi}\mathrm{e}^{\mathbf{\Lambda}z}\mathbf{w}, \tag{20}$$

where $\mathbf{w}$ is the 6 order column vector composed of undetermined amplitudes; and $\mathbf{\Lambda}$ and $\mathbf{\Phi}$ are the 6×6 matrices composed of eigenvalues and eigenvectors of coefficient matrix $\mathbf{M}$, respectively. The

6×6 diagonal matrix $\mathrm{e}^{\mathbf{\Lambda} z}$ is formed by placing $\mathrm{e}^{\gamma_i z}$ on the main diagonal with γ_i the ith $(i = 1,2,3,4,5,6)$ eigenvalue.

Nevertheless, the state equation of a piezoelectric layer is a set of one order *inhomogeneous* linear ordinary differential equations as shown in Equation (18), whose complete solution is comprised of the general solution to the corresponding homogeneous equations and the particular solution to the inhomogeneous equations and is expressed as

$$\hat{\mathbf{v}} = \mathbf{\Phi}(\mathrm{e}^{\mathbf{\Lambda} z} - \mathbf{P})\mathbf{w}, \tag{21}$$

where $\mathbf{P}$ is the 6×6 matrix related to matrix $\mathbf{N}$ via $\mathbf{P} = \mathbf{\Lambda}^{-1}\mathbf{\Phi}^{-1}\mathbf{N}\mathbf{\Phi}(\mathrm{e}^{\mathbf{\Lambda} h} - \mathbf{I}_6)$ with $\mathbf{I}_6$ denoting the sixth order identity matrix here and after.

Notice that the solution to the state equation of the elastic layer in Equation (20) can also be achieved through degeneration to that of the piezoelectric layer when the piezoelectric coefficients equal to zero, i.e., $\mathbf{P}$ is a matrix of zeros due to $\mathbf{N} = 0_{6\times 6}$.

4. Transfer Matrix Method

After the state equations and their solutions determining the state vectors of the elastic and piezoelectric constituent layers have been obtained, the classical transfer matrix method (TMM) [76] is further introduced to derive the equation governing the elastodynamics of the whole unit cell. This equation is represented by a relation between the state vector on the top surface and that on the bottom surface in the TMM formulation. The derived transfer relation of the unit cell will then be combined with the Floquet–Bloch theorem [2,78,79] to bring out the dispersion equation governing the dispersion characteristics of coupled elastic waves in the laminated arbitrarily anisotropic piezoelectric phononic crystals.

4.1. Transfer Relation of an Elastic Layer

Take a typical elastic layer (i) (as depicted in Figure 2a for illustration), the state vectors of its top and bottom surfaces, when referring to Equation (20), should be expressed respectively as

$$\hat{\mathbf{v}}_{(i)}(0) = \mathbf{\Phi}_{(i)}\mathbf{w}_{(i)}, \ \hat{\mathbf{v}}_{(i)}(h_{(i)}) = \mathbf{\Phi}_{(i)}\mathrm{e}^{\mathbf{\Lambda}_{(i)}h_{(i)}}\mathbf{w}_{(i)}. \tag{22}$$

It is obtained from the former formula in Equation (22) that $\mathbf{w}_{(i)} = \mathbf{\Phi}_{(i)}^{-1}\hat{\mathbf{v}}_{(i)}(0)$, which is substituted into the latter formula in Equation (22) to provide the transfer relation of elastic layer (i)

$$\hat{\mathbf{v}}_{(i)}(h_{(i)}) = \mathbf{\Phi}_{(i)}\mathrm{e}^{\mathbf{\Lambda}_{(i)}h_{(i)}}\mathbf{\Phi}_{(i)}^{-1}\hat{\mathbf{v}}_{(i)}(0) = \mathbf{B}_{(i)}\hat{\mathbf{v}}_{(i)}(0), \tag{23}$$

where $\mathbf{B}_{(i)} = \mathbf{\Phi}_{(i)}\mathrm{e}^{\mathbf{\Lambda}_{(i)}h_{(i)}}\mathbf{\Phi}_{(i)}^{-1}$ is referred to as the transfer matrix of elastic layer (i).

4.2. Transfer Relation of a Piezoelectric Layer

A typical piezoelectric layer (j) is accounted for demonstration. By referring to Equation (21) one can express the state vectors at the top and bottom surfaces of layer (j) as

$$\hat{\mathbf{v}}_{(j)}(0) = \mathbf{\Phi}_{(j)}(\mathbf{I}_6 - \mathbf{P}_{(j)})\mathbf{w}_{(j)}, \ \hat{\mathbf{v}}_{(j)}(h_{(j)}) = \mathbf{\Phi}_{(j)}(\mathrm{e}^{\mathbf{\Lambda}_{(j)}h_{(j)}} - \mathbf{P}_{(j)})\mathbf{w}_{(j)}. \tag{24}$$

When the expression $\mathbf{w}_{(j)} = (\mathbf{I}_6 - \mathbf{P}_{(j)})^{-1}\mathbf{\Phi}_{(j)}^{-1}\hat{\mathbf{v}}_{(j)}(0)$ gotten from the former formula in Equation (24) is introduced into its latter formula, the transfer relation is easily obtained by this process of eliminating $\mathbf{w}_{(j)}$ as

$$\hat{\mathbf{v}}_{(j)}(h_{(j)}) = \mathbf{\Phi}_{(j)}(\mathrm{e}^{\mathbf{\Lambda}_{(j)}h_{(j)}} - \mathbf{P}_{(j)})(\mathbf{I}_6 - \mathbf{P}_{(j)})^{-1}\mathbf{\Phi}_{(j)}^{-1}\hat{\mathbf{v}}_{(j)}(0) = \mathbf{B}_{(j)}\hat{\mathbf{v}}_{(j)}(0), \tag{25}$$

where $\mathbf{B}_{(j)} = \mathbf{\Phi}_{(j)}(\mathrm{e}^{\mathbf{\Lambda}_{(j)}h_{(j)}} - \mathbf{P}_{(j)})(\mathbf{I}_6 - \mathbf{P}_{(j)})^{-1}\mathbf{\Phi}_{(j)}^{-1}$ is referred to as the transfer matrix of piezoelectric layer (j).

4.3. Transfer Relation of an Interface

Consider a typical interface J between adjacent layers (i) and (j). According to the compatibility of displacements and equilibrium of stresses, the state vector at the top surface of layer (j) is related to that at the bottom surface of layer (i) by

$$\hat{\mathbf{v}}_{(j)}(0) = \hat{\mathbf{v}}_{(i)}(h_{(i)}) = \mathbf{B}^J\hat{\mathbf{v}}_{(i)}(h_{(i)}), \tag{26}$$

where $\mathbf{B}^J = \mathbf{I}_6$ is referred to as the transfer matrix of interface J.

4.4. Transfer Relation of the Unit Cell

For the whole unit cell, by recurring the transfer relations of layers and interfaces alternately from bottom to top, the state vector at the bottom surface of the bottommost layer (m) can be expressed by the state vector at the top surface of the topmost layer (1) as

$$\begin{aligned}\hat{\mathbf{v}}_{(m)}(h_{(m)}) &= \mathbf{B}_{(m)}\hat{\mathbf{v}}_{(m)}(0) = \mathbf{B}_{(m)}\mathbf{B}^M\hat{\mathbf{v}}_{(m-1)}(h_{(m-1)}) = \cdots \\ &= \mathbf{B}_{(m)}\mathbf{B}^M\mathbf{B}_{(m-1)}\mathbf{B}^{(M-1)}\cdots\mathbf{B}_{(j)}\mathbf{B}^J\cdots\mathbf{B}_{(2)}\mathbf{B}^2\mathbf{B}_{(1)}\hat{\mathbf{v}}_{(1)}(0) = \mathbf{B}\hat{\mathbf{v}}_1(0)\end{aligned} \tag{27}$$

which is referred to as the transfer relation of the unit cell. The 6×6 matrix $\mathbf{B} = \mathbf{B}_{(m)}\mathbf{B}^M\mathbf{B}_{(m-1)}\mathbf{B}^{(M-1)}\cdots\mathbf{B}_{(j)}\mathbf{B}^J\cdots\mathbf{B}_{(2)}\mathbf{B}^2\mathbf{B}_{(1)}$ is the transfer matrix of the unit cell.

4.5. Dispersion Relation of the Laminated Arbitrarily anisotropic Piezoelectric Phononic Crystals

Since the unit cells with arbitrarily anisotropic elastic and piezoelectric constituent layers are periodically arranged in the considered laminated phononic crystals, in view of the Floquet–Bloch theorem [2,78,79] for periodic structures, the state vector $\hat{\mathbf{v}}_{(m)}(h_{(m)})$ should also be related to $\hat{\mathbf{v}}_{(1)}(0)$ through

$$\hat{\mathbf{v}}_{(m)}(h_{(m)}) = \mathrm{e}^{\mathrm{i}qH}\hat{\mathbf{v}}_{(1)}(0), \tag{28}$$

where q and H are the wavenumber of the characteristic coupled elastic waves in and the height of the unit cell, respectively, as can be seen in Figure 1.

Combining Equation (28) with Equation (27) and eliminating the state vector $\hat{\mathbf{v}}_{(m)}(h_{(m)})$, one gets

$$\mathbf{B}\hat{\mathbf{v}}_{(1)}(0) = \mathrm{e}^{\mathrm{i}qH}\hat{\mathbf{v}}_{(1)}(0) \text{ or } (\mathbf{B} - \mathrm{e}^{\mathrm{i}qH}\mathbf{I}_6)\hat{\mathbf{v}}_{(1)}(0) = 0. \tag{29}$$

The former formula in Equation (29) comes down to the matrix eigenvalue problem, while the latter indicates that the determinant of the coefficient matrix $(\mathbf{B} - \mathrm{e}^{\mathrm{i}qH}\mathbf{I}_6)$ should be zero to give nontrivial solution to $\hat{\mathbf{v}}_1(0)$, i.e.,

$$\mathrm{e}^{iqH} = \text{Eigenvalues}(\mathbf{B}) = \mu \text{ or } \det(\mathbf{B} - \mathrm{e}^{\mathrm{i}qH}\mathbf{I}_6) = 0. \tag{30}$$

These formulas are the dispersion equations governing the characteristics of the coupled elastic waves along the thickness direction in the laminated arbitrarily anisotropic piezoelectric phononic crystals, which actually specify the relation between the wavenumber q and the frequency ω. Notice that the phase velocity and the wavelength are expressed as $v = \omega/q$ and $\lambda = 2\pi/q$, respectively. When the frequency ω is given within a range at specified increment, the frequency-related dispersion curves including the eigenvalue, wavenumber, wavelength and phase velocity spectra, can all be obtained after the former formula in Equation (30) has been solved numerically by direct eigenvalue operation. Otherwise, as anyone among ω, q (or λ) and v is specified, the other two can be obtained by numerically solving the latter formula in Equation (30), so as to provide the comprehensive dispersion curves including the frequency-related, wavenumber-related, wavelength-related, and phase velocity-related dispersion curves. Regardless, in the following section, only the frequency-related dispersion curves are illustrated since they can already describe the dispersion characteristics of coupled elastic waves in a relatively complete way.

5. Numerical Examples

The above derived analysis method will be utilized in this section to calculate the comprehensive frequency-related dispersion curves of elastic waves along thickness direction, including the eigenvalue amplitude spectra, the wavenumber spectra, the wavelength spectra, and the phase velocity spectra in laminated piezoelectric phononic crystals with four exemplified unit cell configurations and with four kinds of electrical boundaries as stated in Section 2. The four unit cell configurations guarantee the presence of four representative patterns about the elastic wave coupling and decoupling, which are composed of the Glass as the inserted elastic layer, the Brass as the electrode layer, and anyone or two layers among the three types piezoelectric layers labeled as "$LiNbO_3{}^A$", "$LiNbO_3{}^B$", and "$LiNbO_3{}^C$", respectively. These three piezoelectric layers are actually all acquired from the piezoelectric material $LiNbO_3$ crystal with *3m* symmetry but are arranged in three different directions. The layers "$LiNbO_3{}^A$", "$LiNbO_3{}^B$", and "$LiNbO_3{}^C$" are formed, respectively, when the threefold symmetry axis *3* are parallel to the local coordinate axes *z*, *y*, and *x*, and correspondingly the three layers are interpreted as *3//z*, *3//y*, and *3//x*, respectively. Because "$LiNbO_3{}^B$" and "$LiNbO_3{}^C$" have lower crystal symmetry in the layering plane than the orthotropic crystal discussed by the authors of [55], the coupling of normally-propagating elastic waves exists when they are the constituent layers of the unit cell. By contrast, "$LiNbO_3{}^A$" layer still has high crystal symmetry in the layering plane. Thus, when the unit cell is Glass-Brass-$LiNbO_3{}^A$-Brass, which is the first analysis model, all the elastic waves along thickness direction are decoupled. The computed dispersion curves of the pure mode waves in this configuration will be later taken as the reference to show the differences of dispersion characteristics of the coupled mode waves in the other three configurations, i.e., Glass-Brass-$LiNbO_3{}^B$-Brass, Glass-Brass-$LiNbO_3{}^C$-Brass, and Glass-Brass-$LiNbO_3{}^B$-Brass-$LiNbO_3{}^C$-Brass. In all the four considered configurations, the thicknesses of the inserted elastic Glass layer and the piezoelectric layers are all 10 mm, and the thickness of the Brass electrodes is 0.025 mm, except that it is varied as stated in Section 5.3 where the effect of the electrode thickness on the dispersion curves is discussed. The material parameters of these constituent layers are listed in Table 1. Note that the parameters for "$LiNbO_3{}^A$" are exactly excerpted from Auld [82], while those for "$LiNbO_3{}^B$" and "$LiNbO_3{}^C$" are obtained from further coordinate transformation. The $(c_{11})_{Glass} = (c_{22})_{Glass} = (c_{33})_{Glass} = 83.34$ GPa and $(\rho)_{Glass} = 2540\ kg/m^3$ for Glass are quoted from Kutsenko et al. [60] and the other parameters are computed from the Poisson's ratio 0.2163 and the Young's modulus 73.39 GPa determined by rough average of the many values in material handbook [84]. The stiffness constants of Brass are all computed from the Poisson's ratio 0.337 and the Young's modulus 106.80 GPa that are roughly determined together with the mass density $(\rho)_{Brass} = 8320\ kg/m^3$ by referring material handbook [84]. The material parameters of Glass and Brass have also been provided in our previous studies [55,85].

For the convenience of drawing the dispersion curves, the dimensionless frequency $\omega H/(\pi \overline{v})$, the dimensionless wavenumber qH/π, the dimensionless wavelength λ/H, and the dimensionless phase velocity $v/\overline{v}$ are employed with $\overline{v} = \sqrt{(c_{55})_{LiNbO_3{}^A}/(\rho)_{LiNbO_3}}$ the velocity of shear wave in $LiNbO_3{}^A$.

Since in previous literatures the coupled waves and the electrode thickness are not considered, in order to validate our proposed analysis method, these factors have to be left out. Thus, we further calculated the band structures (frequency–wavenumber spectra) of pure P and S waves along thickness in periodically multilayered Glass-$LiNbO_3{}^A$ composites with electric-open boundary, which can be compared with the results computed by analytical formula in Galich et al. [86] or identically in Shen & Cao [87]. All the results are provided in Appendix B.

Table 1. Material parameters of the constituent layers in the unit cell of exemplified laminated phononic crystals.

Materials		$LiNbO_3{}^A$ [82] (3//z)		$LiNbO_3{}^B$ (3//y)		$LiNbO_3{}^C$ (3//x)		Glass [55,60,84]	Brass [55,84,85]
Elastic constants ($\times 10^{10}$ $N \cdot m^{-2}$)	c_{11}		20.300		20.300		24.500	8.334	16.246
	c_{22}		20.300		24.500		20.300	8.334	16.246
	c_{33}		24.500		20.300		20.300	8.334	16.246
	c_{44}		6.000		6.000		7.500	3.017	3.994
	c_{55}		6.000		7.500		6.000	3.017	3.994
	c_{66}		7.500		6.000		6.000	3.017	3.994
	c_{12}		5.300		7.500		7.500	2.300	8.258
	c_{13}		7.500		5.300		7.500	2.300	8.258
	c_{23}		7.500		7.500		5.300	2.300	8.258
		c_{14}	0.900	c_{14}	−0.900	c_{36}	−0.900	-	
		c_{24}	−0.900	c_{34}	0.900	c_{26}	0.900		
		c_{56}	0.900	c_{56}	−0.900	c_{45}	−0.900		
Piezoelectric constants ($C \cdot m^{-2}$)		e_{15}	3.700	e_{15}	−2.500	e_{11}	−1.300	- or $e_{\alpha i} = 0.000$ ($\alpha = 1,2,3, i = 1,\cdots,6$)	
		e_{16}	−2.500	e_{16}	−3.700	e_{12}	−0.200		
		e_{21}	−2.500	e_{21}	−0.200	e_{13}	−0.200		
		e_{22}	2.500	e_{22}	−1.300	e_{22}	2.500		
		e_{24}	3.700	e_{23}	−0.200	e_{23}	−2.500		
		e_{31}	0.200	e_{31}	−2.500	e_{26}	−3.700		
		e_{32}	0.200	e_{33}	2.500	e_{34}	−2.500		
		e_{33}	1.300	e_{34}	−3.700	e_{35}	−3.700		
Dielectric constants ($\times 10^{-10}$ $F \cdot m^{-1}$)	ε_{11}		3.89576		3.89576		2.56766	-	-
	ε_{22}		3.89576		2.56766		3.89576	-	-
	ε_{33}		2.56766		3.89576		3.89576	-	-
Mass density ($kg \cdot m^{-3}$)	ρ		4700		4700		4700	2540	8320

5.1. Dispersion Properties of Coupled Elastic Waves

To search the general features of the comprehensive frequency-related dispersion curves, the electric-short boundary is adopted in the calculation, because the spectra associated with this electrical boundary serve as the limiting curves for both the applied capacitance and the applied feedback control boundaries as will be shown in Section 5.2. Figures 3–6 give the results of these comprehensive frequency-related dispersion curves for the periodic structures consisting of Glass-Brass-$LiNbO_3{}^A$-Brass, Glass-Brass-$LiNbO_3{}^B$-Brass, Glass-Brass-$LiNbO_3{}^C$-Brass, and Glass-Brass-$LiNbO_3{}^B$-Brass-$LiNbO_3{}^C$-Brass configurations, respectively; subfigures (a–d) in these figures represent the relations between the frequency and the eigenvalue amplitude ($|\mu|$), the wavenumber (qH/π), the wavelength (λ/H), and the phase velocity ($v/\overline{v}$), respectively. The corresponding wave modes of all the spectra in these figures are marked out.

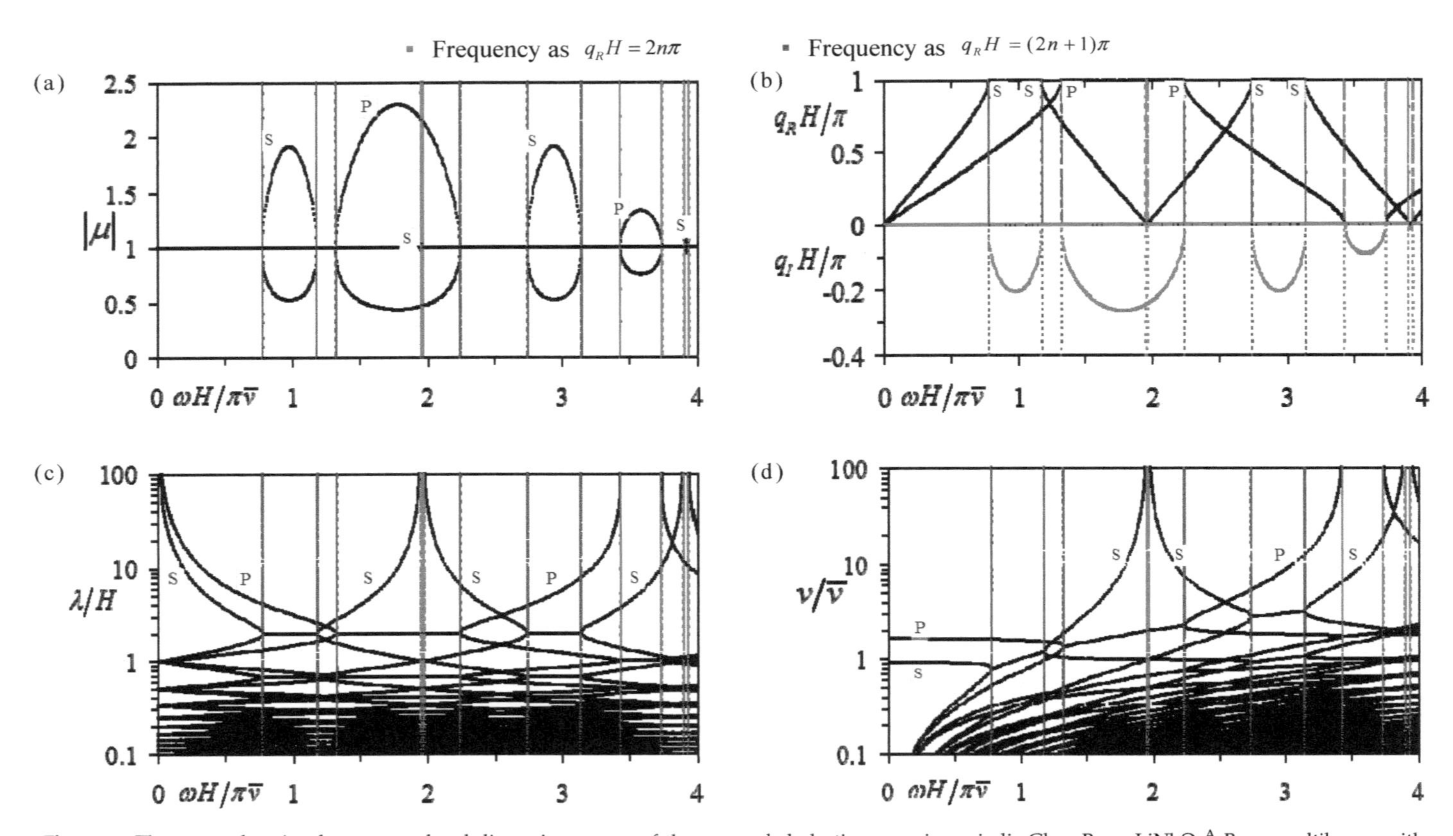

Figure 3. The comprehensive frequency-related dispersion curves of the uncoupled elastic waves in periodic Glass-Brass-LiNbO$_3$^{A}-Brass multilayers with electric-short boundary: (**a**) eigenvalue amplitude $|\mu|$ spectra; (**b**) wavenumber spectra; (**c**) wavelength spectra; (**d**) phase velocity spectra.

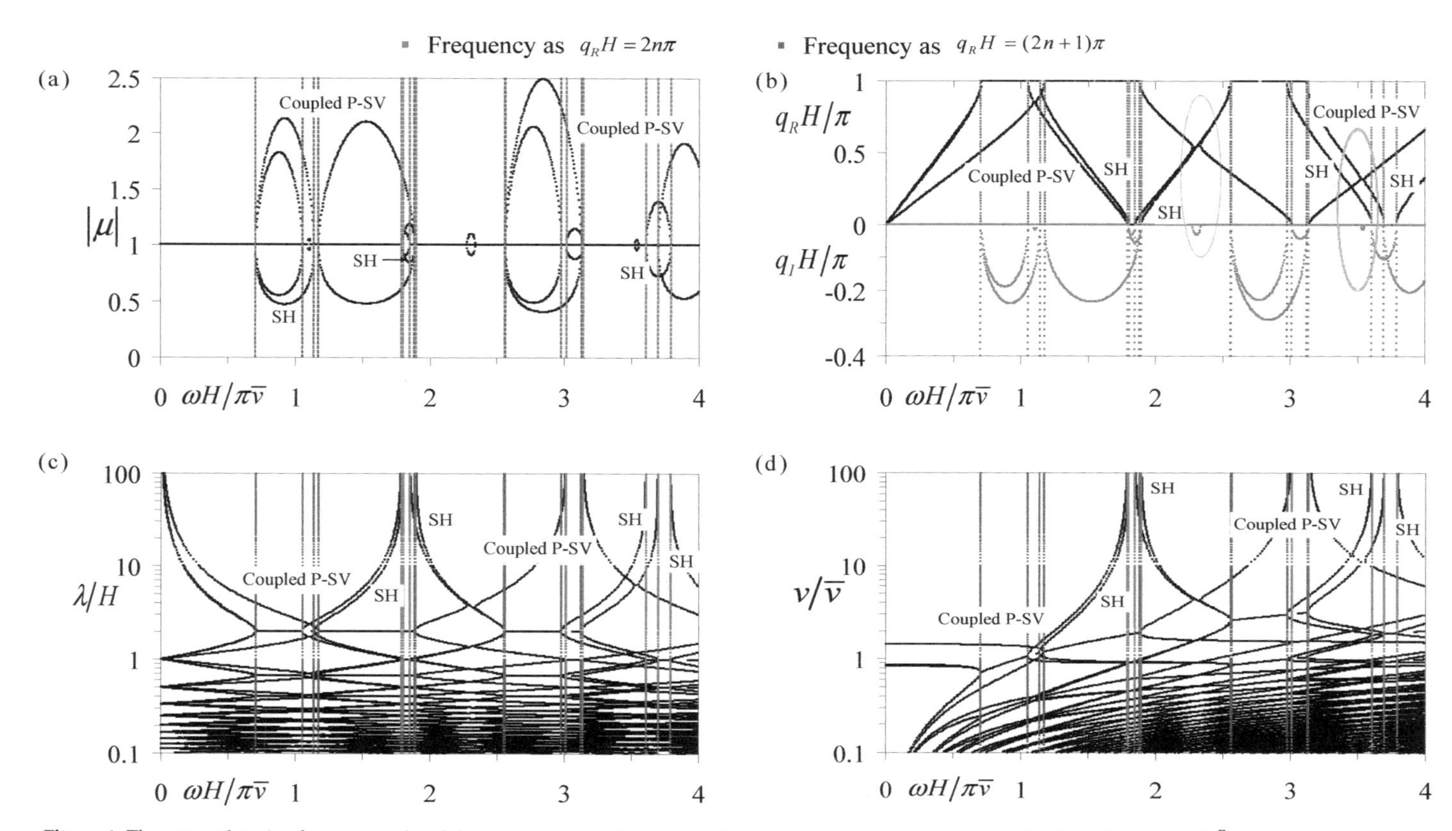

Figure 4. The comprehensive frequency-related dispersion curves of the coupled P-SV and pure SH waves in periodic Glass-Brass-$LiNbO_3{}^B$-Brass multilayers with electric-short boundary: (**a**) eigenvalue amplitude $|\mu|$ spectra; (**b**) wavenumber spectra; (**c**) wavelength spectra; (**d**) phase velocity spectra.

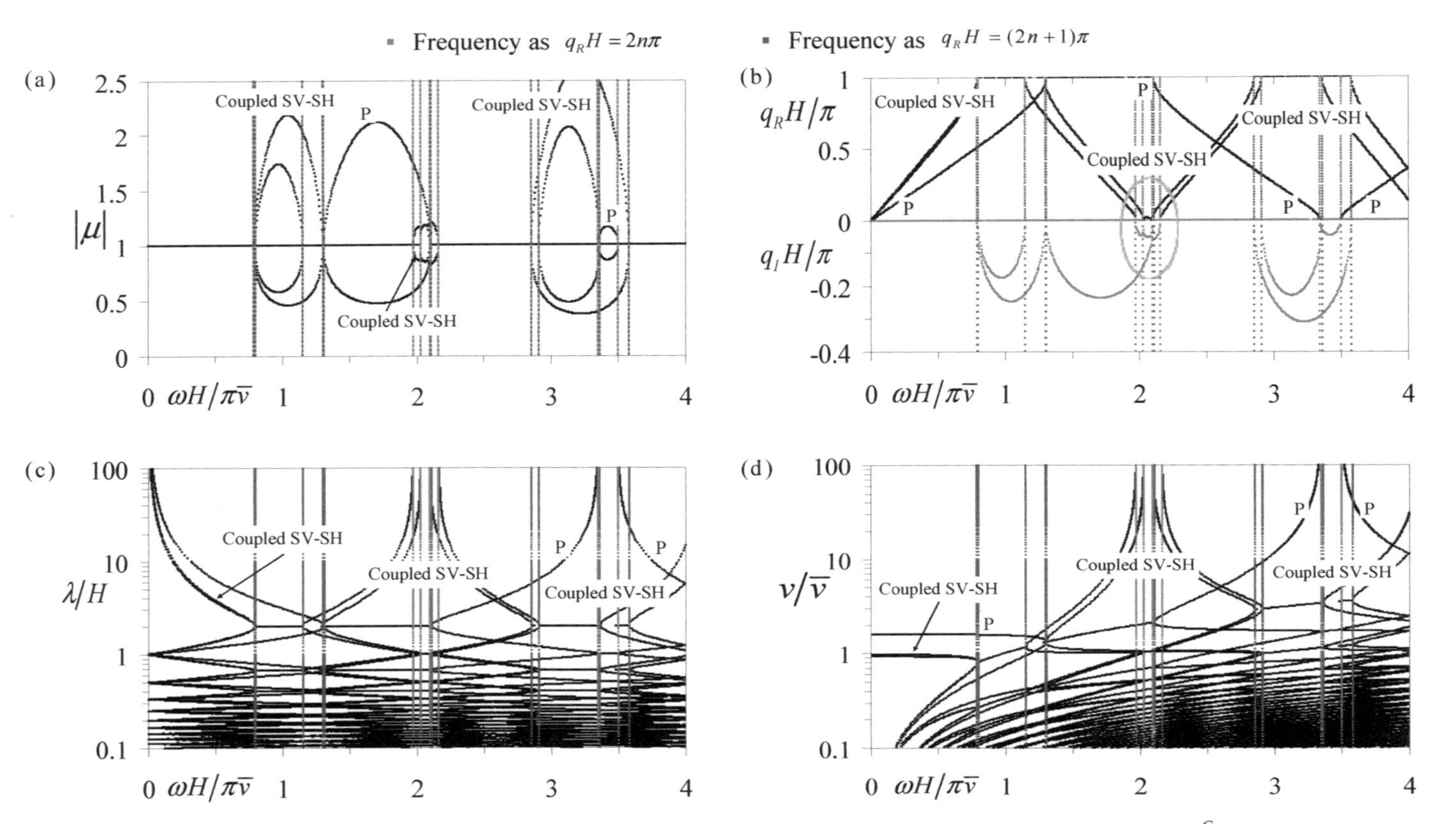

Figure 5. The comprehensive frequency-related dispersion curves of the coupled SV-SH and pure P waves in periodic Glass-Brass-$LiNbO_3{}^C$-Brass multilayers with electric-short boundary: **(a)** The eigenvalue amplitude $|\mu|$ spectra; **(b)** The wavenumber spectra; **(c)** The wavelength spectra; **(d)** The phase velocity spectra.

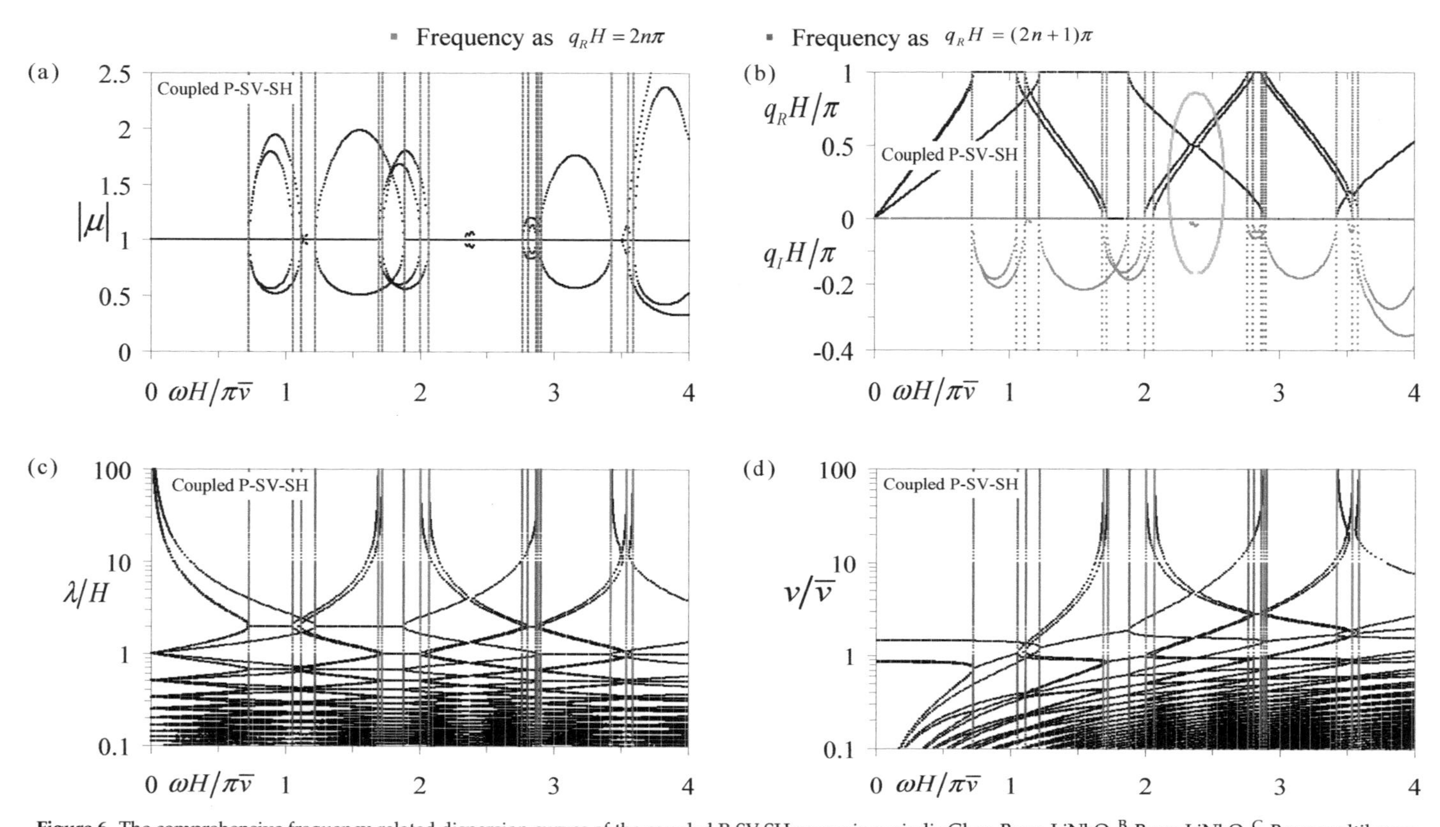

Figure 6. The comprehensive frequency-related dispersion curves of the coupled P-SV-SH waves in periodic Glass-Brass-LiNbO$_3^B$-Brass-LiNbO$_3^C$-Brass multilayers with electric-short boundary: (**a**) eigenvalue amplitude $|\mu|$ spectra; (**b**) wavenumber spectra; (**c**) wavelength spectra; (**d**) phase velocity spectra.

The subfigures in Figures 4–6 indicate, as compared with their counterparts in Figure 3, that the coupled mechanical waves in laminated phononic crystals have some dispersion features identical to those of the uncoupling pure waves, which can be found in [55] but some of which will mention briefly as follows. For example, (1) any kind of frequency-related dispersion curve can show the frequency bands clearly. In frequency ranges where the spectra associated with the real wavenumber (or wavelength/phase velocity) appear, the corresponding wave modes propagate without attenuation, while in frequency ranges where the spectra associated with the imaginary wavenumber (or non-unit eigenvalue amplitude) emerge, the corresponding wave modes attenuate without propagation. These frequency ranges are called as the pass-bands and stop-bands, respectively. (2) When the wavelength tends to infinity, the first order spectrum of each mode asymptotically tend to zero frequency, and the corresponding phase velocity gradually tends to a cut-off value; the higher order spectra of each mode asymptotically tend to the bounding frequencies with phase $2n\pi$ (n is an integer), and the corresponding phase velocity also asymptotically tends to infinity. These features are actually common to all kinds of waves in diverse periodic structures such as the longitudinal waves in periodic elastic rods [88] and periodic piezoelectric rods [85]. In addition, as a complement to the summarization in previous studies, it is also found that the eigenvalue amplitude spectra appear in reciprocal pairs.

Beside these same features, the subfigures in Figures 4–6 also exhibit three dispersion features of coupled elastic waves that are different from those of the pure mode waves. The first remarkable one is that a narrow stop-band can be observed near the intersections of the dispersion curves of different coupled and/or uncoupled waves, which are shown in the green ellipse of Figure 4b, Figure 5b, and Figure 6b. The reason for forming these bands may involve strong coupling and interaction between different wave modes near specific frequency as a result of the material anisotropy. The second feature is that the pass-bands and stop-bands of the coupled wave of one mode no longer appear alternately, which is exactly opposite to those of the pure wave of one mode shown in Figure 3b. The third feature is that for some pass-bands/stop-bands, the frequency lines corresponding to phase $q_R H = 2n\pi$ and $q_R H = (2n+1)\pi$, i.e., the Brillouin zone boundaries, are no longer the demarcation line between a pass-band and a stop-band. Thus the dispersion curves of these pass-bands/stop-bands are not entirely constrained in-between the Brillouin zone boundaries.

5.2. Tuning the Dispersion Characteristics of Coupled Elastic Waves by the Electrical Boundaries

In order to show the influence of the electrical boundaries on the dispersion characteristics of coupled elastic waves, the band structures (wavenumber spectra) of the periodic anisotropic piezoelectric composites with the four configurations of the unit cell are computed in cases of four electrical boundaries, including the electric-open, the applied electrical capacitance, the electric-short, and the applied feedback control conditions. The wavenumber spectra are selected from the four kinds of frequency-related dispersion curves because they contain the most information and thus they are the representative dispersion curves. For the sake of expressing the influences more clearly, we always take the electric-open and electric-short boundaries as reference and individually consider the applied positive capacitance, the applied negative capacitance, and the applied feedback control conditions. Figure 7 shows the band structures in cases of electric-open, electric-short, and applied capacitances $C/A = 1.0 \times 10^{-8}$ F/m^2 and $C/A = 1.0 \times 10^{-7}$ F/m^2, with the subfigures (a), (b), (c), and (d) denoting the results for the unit cell configurations of Glass-Brass-$LiNbO_3{}^A$-Brass, Glass-Brass-$LiNbO_3{}^B$-Brass, Glass-Brass-$LiNbO_3{}^C$-Brass, and Glass-Brass-$LiNbO_3{}^B$-Brass-$LiNbO_3{}^C$-Brass, respectively. Figure 8a–d provides the corresponding wavenumber spectra in cases of the applied capacitance with the values $C/A = -2.5 \times 10^{-8}$ F/m^2 and $C/A = -2.6 \times 10^{-8}$ F/m^2 for periodic Glass-Brass-$LiNbO_3{}^A$-Brass composite as well as $C/A = -3.8 \times 10^{-8}$ F/m^2 and $C/A = -4.0 \times 10^{-8}$ F/m^2 for the other configurations, together with the electric-open ($C/A = 0$) and electric-short ($C/A = \infty$) condition, while Figure 9a–d provides the corresponding ones in the case of applied feedback control with the gain coefficient being $K_g = 5.0 \times 10^{10}$ V/m and $K_g = 1.0 \times 10^{12}$ V/m together with the electric-open and electric-short ($K_g = 0.0$ V/m) conditions. Note that the values of the positive capacitance and the gain coefficient are chosen so as to most clearly show their influences on the band structures and their correlation with the electric-open and electric-short conditions. The negative capacitance values are determined so as to closely near but not at the singular point specified by $\varepsilon_{33}^2 S/C + \varepsilon_{33} h = 0$ that can be obtained through the vanishing of the denominator of the expression for vector $\mathbf{R}$ as given in Table A1. In addition, notice also from the expression for vector $\mathbf{R}$ in Table A1 that $C/A = +\infty$ and $C/A = -\infty$ actually lead to the same $\mathbf{R}$ and accordingly the same computed band structures. In Figure 10, the band structures associated with the negative capacitances are compared with those associated with a very big gain coefficient $K_g = 1.0 \times 10^{12}$ V/m, so that the correlation between applied negative capacitance and applied feedback control can be revealed. In Figures 7–10, the band structures corresponding to the pure wave that is independent on the electric field are marked by solid arrows pointing to the direction of increasing frequency, these are trivial curves for our investigation. In the following interpretations, we only focus on the elastic waves that are dependent on electric field and thus the electrical boundary. Each unit cell configuration exactly corresponds to one elastic wave. As also noticed in Section 5.1, these waves are pure P, coupled P-SV, coupled SV-SH, and coupled P-SV-SH modes in subfigures (a–d), respectively. Common in this numerical example Section, the pure P wave in the periodic Glass-Brass-$LiNbO_3{}^A$-Brass structure is used as the object of comparison, so that the differences between the coupled waves and the pure modes waves can be clearly exhibited.

From all the subfigures in Figure 7, it is seen that as the positive capacitance varies from $C/A = 0$ to $C/A = \infty$, the band structures gradually changes from those associated with the electric-open condition to those associated with the electric-short condition. Generally, the central frequency of bandgaps moves downward, and the bandgap widths becomes narrower. Nevertheless, the entire alteration is not very significant, since the band structures associated with the electric-short condition are actually relatively near the corresponding curves associated with the electric-open condition. Comparing Figure 7b–d with Figure 7a, it is seen that the dispersion curves of coupled waves are more obviously influenced than those of the pure wave, so that the coupled waves are more sensitive to the applied positive capacitance.

From all the subfigures in Figure 8, it is seen that for any waves related to the electric field, either uncoupled or coupled modes, sharp variation of the attenuation constants (imaginary wavenumbers) to infinity exists at some frequency ranges, where the phase constants also sharply and corresponding changed. These phenomena are similarly found in Refs. [55,58] for pure P and S waves. As the negative capacitance varies from $C/A = 0$ to $C/A = -\infty$, the band structures does not change in a monotonous way, which is attributed to the singular property as the applied capacitance corresponds to the zero denominator of the expression for vector $\mathbf{R}$. The comparison between Figure 8a–d further indicates that the coupled P-SV-SH wave has more complex variation pattern of the attenuation and phase constants spectra near the singular range.

It is seen from Figure 9 that the gain coefficient of the applied feedback control has very obvious influence on the band structures of both uncoupled and coupled waves except those of the coupled SV-SH wave. The insensitivity of coupled SV-SH wave to the applied feedback control condition is resulting from the $e_{33} = 0$ for $LiNbO_3{}^C$ as shown in Table 1, which further leads to the vanishing of the last component of vector $\mathbf{F}$ and accordingly the vanishing of the components at the last row of $\mathbf{F}^T\mathbf{R}$. As the gain coefficient K_g increases from zero, the band structures of coupled P-SV and P-SV-SH waves associated with the applied feedback control condition, as compared to those associated with the electric-short condition, changes substantially including forming new phase/attenuation constants spectra as shown in subfigures (b,d). However, those of pure P wave simply move downwards to the smaller frequency side.

Figure 10 indicates, except the coupled SV-SH wave, all other waves can realize the singular pattern of dispersion by either applying proper negative capacitance boundary or applying feedback control condition with relatively big gain coefficient. For coupled SV-SH wave in periodic Glass-Brass-$LiNbO_3{}^C$-Brass composites discussed here, only the proper negative capacitance can be applied.

Moreover, from all the above discussion it is known that the electric-short boundary is both the upper value limiting of the applied positive/negative capacitance boundary and the lower value limiting of the applied feedback control boundary. This is the reason that the electric-short boundaries are utilized in the previous Section 5.1 and the following Section 5.3.

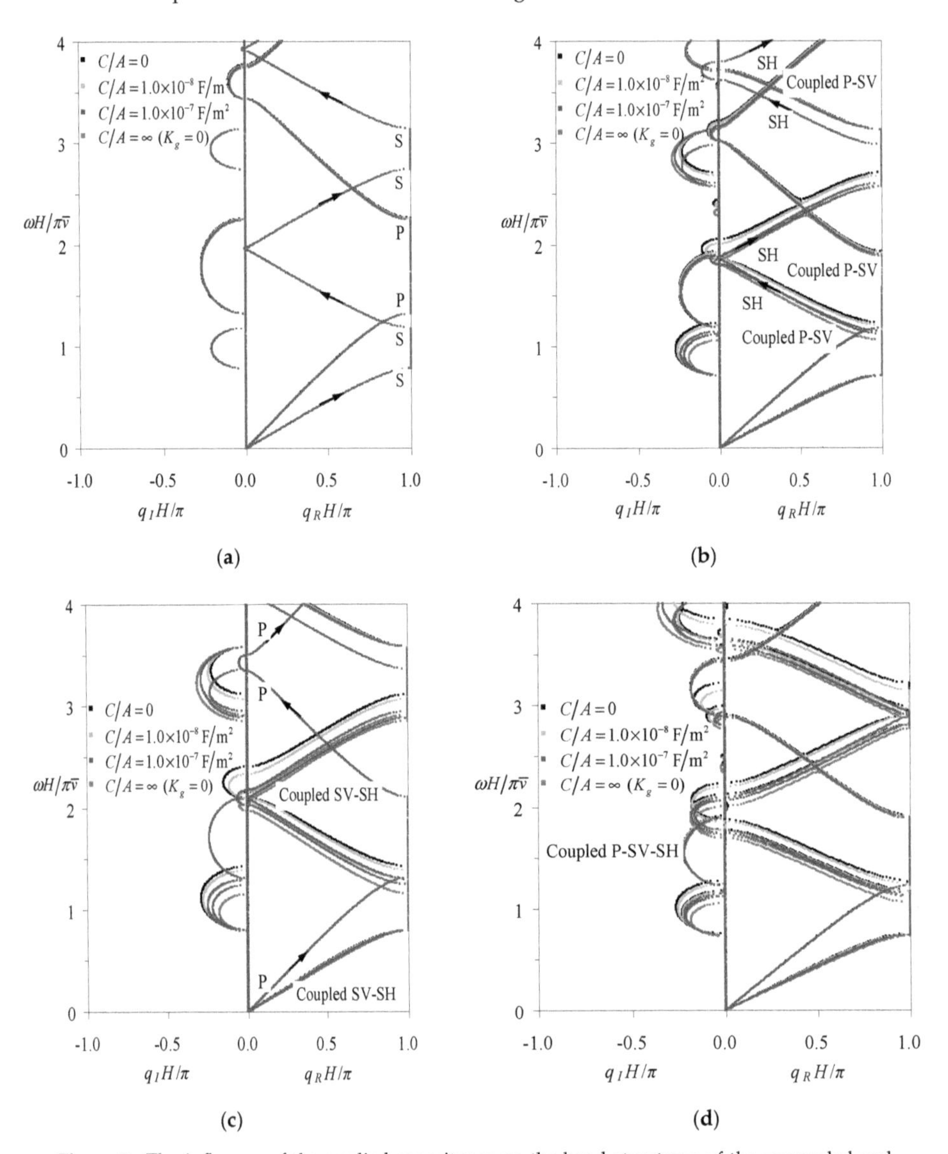

Figure 7. The influence of the applied capacitance on the band structures of the uncoupled and coupled elastic waves in periodic anisotropic piezoelectric composites with the unit cell consisting of (**a**) Glass-Brass-$LiNbO_3{}^A$-Brass, (**b**) Glass-Brass-$LiNbO_3{}^B$-Brass, (**c**) Glass-Brass-$LiNbO_3{}^C$-Brass, and (**d**) Glass-Brass-$LiNbO_3{}^B$-Brass-$LiNbO_3{}^C$-Brass.

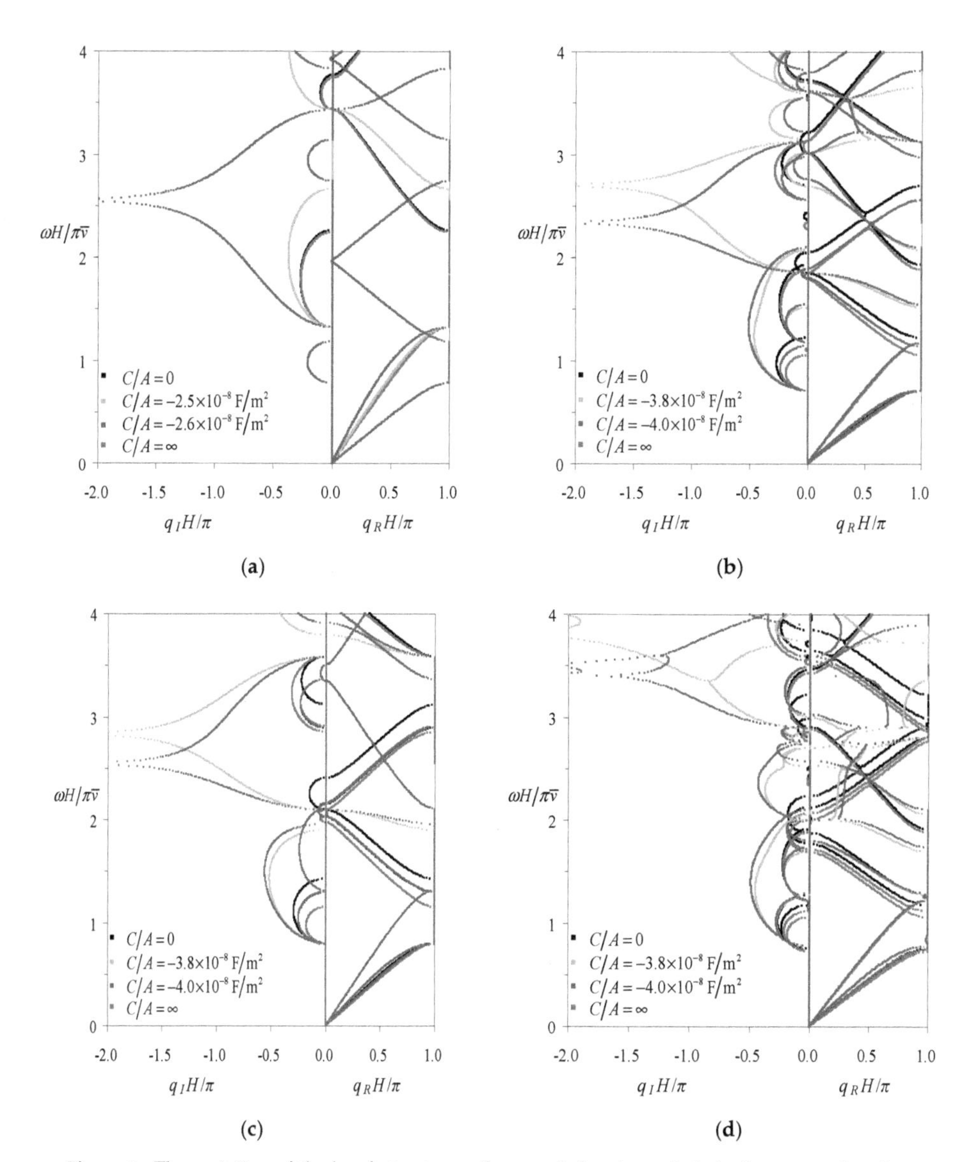

Figure 8. The variation of the band structures of uncoupled and coupled elastic waves when the applied negative capacitance is at the vicinity of the singular point in periodic anisotropic piezoelectric composites with the unit cell consisting of (**a**) Glass-Brass-$LiNbO_3{}^A$-Brass, (**b**) Glass-Brass-$LiNbO_3{}^B$-Brass, (**c**) Glass-Brass-$LiNbO_3{}^C$-Brass, and (**d**) Glass-Brass-$LiNbO_3{}^B$-Brass-$LiNbO_3{}^C$-Brass.

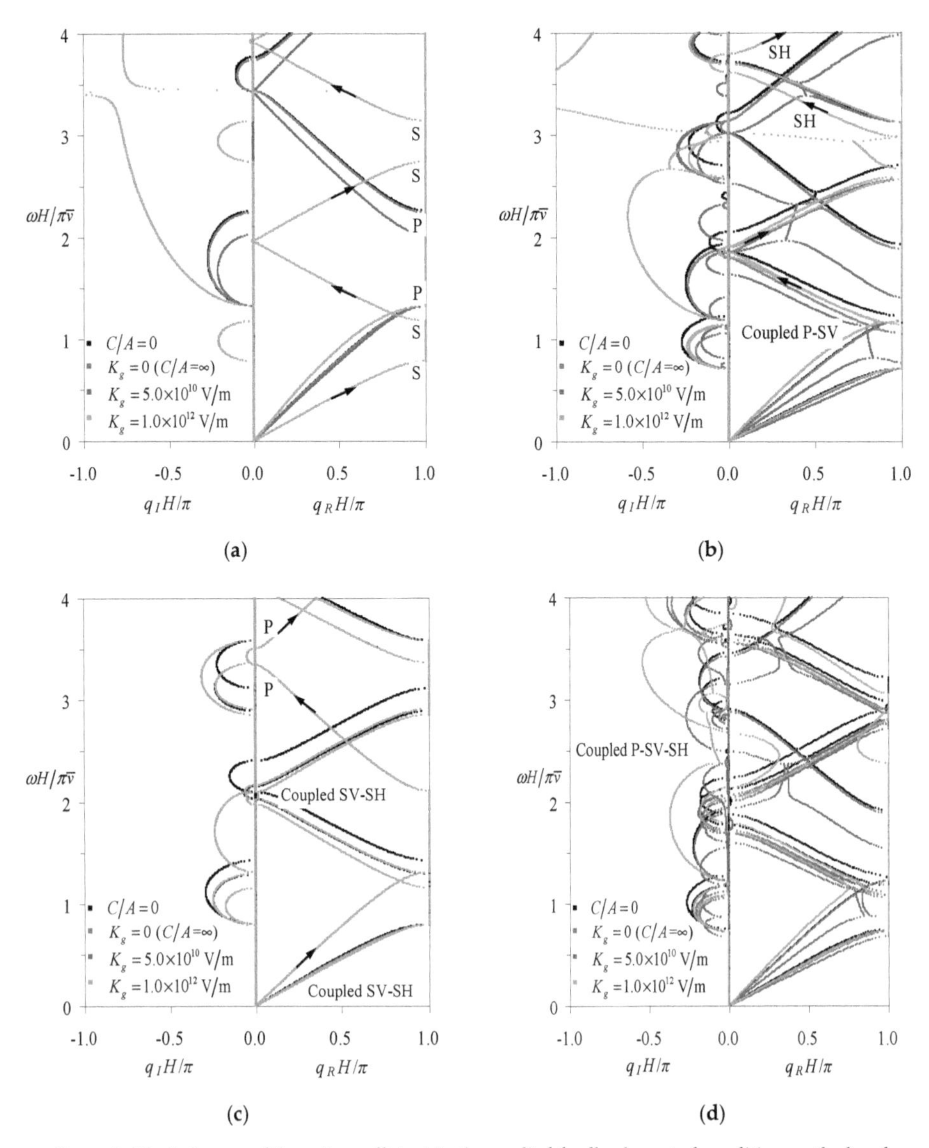

Figure 9. The influence of the gain coefficient in the applied feedback control condition on the band structures of the uncoupled and coupled elastic waves in periodic anisotropic piezoelectric composites with the unit cell consisting of (**a**) Glass-Brass-$LiNbO_3{}^A$-Brass, (**b**) Glass-Brass-$LiNbO_3{}^B$-Brass, (**c**) Glass-Brass-$LiNbO_3{}^C$-Brass, and (**d**) Glass-Brass-$LiNbO_3{}^B$-Brass-$LiNbO_3{}^C$-Brass.

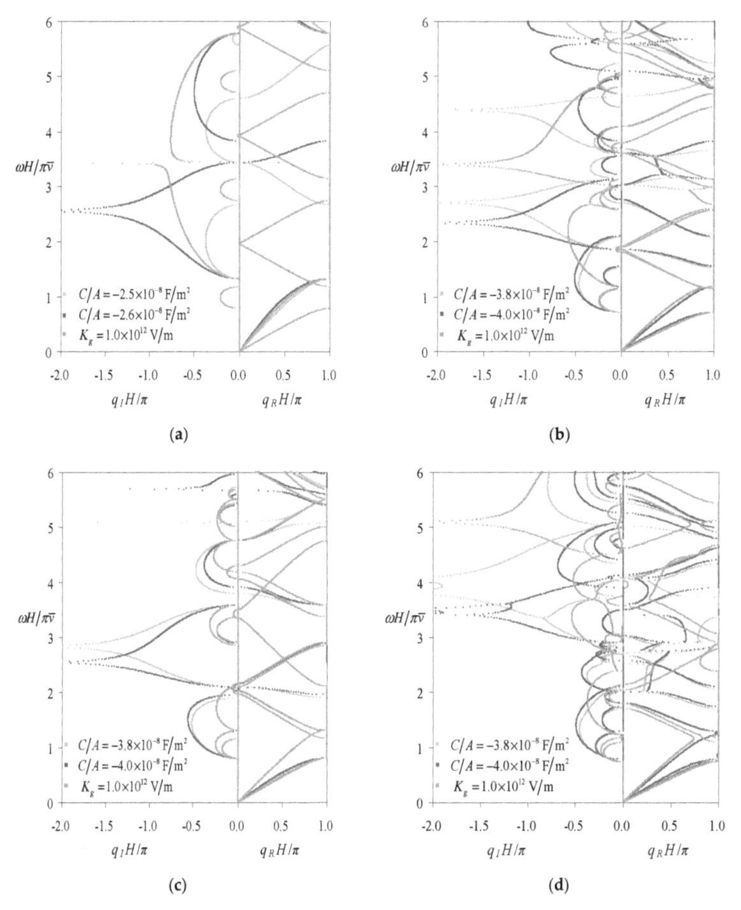

Figure 10. The comparison of the band structures associated with the applied feedback control condition of big gain coefficient ($K_g = 1.0 \times 10^{12}$ V/m) and those associated with the applied negative capacitance near the singular point in periodic anisotropic piezoelectric composites as the unit cell consisting of (**a**) Glass-Brass-$LiNbO_3{}^A$-Brass, (**b**) Glass-Brass-$LiNbO_3{}^B$-Brass, (**c**) Glass-Brass-$LiNbO_3{}^C$-Brass, and (**d**) Glass-Brass-$LiNbO_3{}^B$-Brass-$LiNbO_3{}^C$-Brass.

5.3. Electrode Thickness Influence on the Dispersion Characteristics of Coupled Elastic Waves

For the sake of knowing the mechanical effect of electrode layers on coupled elastic waves besides their electrical function, the band structures (wavenumber spectra) of the periodic anisotropic piezoelectric composites with the above-mentioned four configurations of unit cell are computed by specifying the electric-short boundary and four values of the electrode thickness, i.e., 0 mm, 0.025 mm, 0.25 mm, and 2.5 mm. The results are provided in Figure 11, with subfigures (a–d) corresponding to the configuration of Glass-Brass-$LiNbO_3{}^A$-Brass, Glass-Brass-$LiNbO_3{}^B$-Brass, Glass-Brass-$LiNbO_3{}^C$-Brass, and Glass-Brass-$LiNbO_3{}^B$-Brass-$LiNbO_3{}^C$-Brass, respectively. Adopting the electric-short boundary is because the associated band structures serve exactly as the limiting curves for both the applied capacitance and the applied feedback control boundaries as obtained from Section 5.2, and thus they are typical. In Figure 11a–d, for any kind of wave, when the ratio of the electrode thickness to the thickness of the piezoelectric layer (called as the relative electrode thickness) is less than $0.25\%(= 0.025/10)$, the mechanical effect of the electrode is too weak to affect the band structures. When it reaches to $2.5\%(= 0.25/10)$, the band structures of the relatively high order modes are distinguishably affected by the mechanical effect of the electrode, though the band structures of low order modes are nearly not influenced. Also inferred from these figures that, when the relative electrode thickness is higher than 2.5%, with the increasing of electrode thickness, the band structures will be more and more obviously affected by the mechanical effect of the electrode. The variation is characterized by the decreasing of the central frequencies of the pass-bands/stop-bands and the changes of the bandgap widths without uniform rule. From the comparisons of Figure 11b–d with Figure 11a, it can be noticed that the band structures of coupled wave modes associated with the narrow bandgaps discussed in Section 5.1 are more sensitive to the electrode thickness than those of the other coupled wave modes and uncoupled waves. With the increasing of electrode thickness, these sensitively-influenced bandgaps gradually move down, but their widths remain almost unchanged.

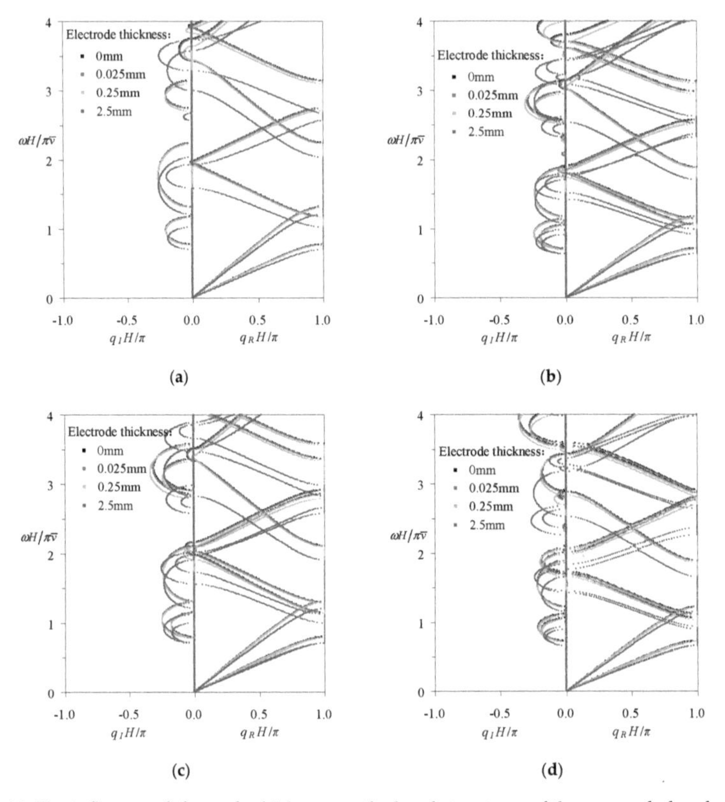

Figure 11. The influence of electrode thickness on the band structures of the uncoupled and coupled elastic waves in periodic anisotropic piezoelectric composites with electric-short boundary and with the unit cell consisting of (**a**) Glass-Brass-$LiNbO_3{}^A$-Brass, (**b**) Glass-Brass-$LiNbO_3{}^B$-Brass, (**c**) Glass-Brass-$LiNbO_3{}^C$-Brass, and (**d**) Glass-Brass-$LiNbO_3{}^B$-Brass-$LiNbO_3{}^C$-Brass.

6. Conclusions

This paper studies the active tuning of coupled elastic waves propagating perpendicular to the interface in general laminated piezoelectric phononic crystals composed of arbitrarily anisotropic piezoelectric and elastic materials, where the wave coupling is attributed only to the material anisotropy. The active tuning is realized through the continuous adjustment of the electrical parameters in the external electric circuit connected to the piezoelectric layers or the switching between different electric circuits. The considered electrical boundaries include the electric-open, the electric-short, the applied capacitance, and the applied feedback control conditions. The analytical matrix method is proposed, on the bases of the state space formalism for modeling the elastodynamics of constituent layers, the transfer matrix method for modeling the relations between constituent layers, and the Floquet–Bloch theorem for modeling the periodic conditions of the unit cell, to derive the dispersion equation of general periodic laminated piezoelectric composites. By validating and adopting the analysis method in numerical examples, the dispersion characteristics of coupled elastic waves as well as the influences of electrical boundaries and electrode thickness on the band structures of coupled waves were investigated. The following conclusions may be drawn from these studies for the coupled waves induced by only material anisotropy other than inclined propagation.

1. The coupled elastic waves may have innovative dispersion phenomena such as the presence of narrow band gaps that are not bounded between frequency lines associated with phase $2n\pi$ or $(2n+1)\pi$ (n is an integer). That is to say, the dispersion curves in some pass-bands/stop-bands of coupled waves are not entirely constrained in between the Brillouin zone boundaries.
2. The coupled elastic waves are generally more sensitive than pure waves to the electrical boundaries if they are related to electric field. The electric-short boundary is both the upper value limiting of the applied positive/negative capacitance boundary and the lower value limiting of the applied feedback control boundary. The applied negative capacitance and the applied feedback control generally have much bigger extent of tunability on the band structures of couple waves than the applied positive capacitance has. The negative capacitance and the gain coefficient do not monotonously affect the band structures of coupled waves, but the positive capacitance does.
3. For most coupled waves, the singular pattern of dispersion curves characterized as infinity attenuation constant (imaginary wavenumber) may be realized by either applying proper negative capacitance boundary or applying feedback control condition with relatively big gain coefficient.
4. The mechanical effect of the electrode should be considered, as its relative thickness (defined as the ratio of the electrode thickness to the thickness of the piezoelectric layer) reaches ~2.5% or above if the accurate band structures of the coupled waves are required.

All these findings will push forward the design and nondestructive evaluation of laminated anisotropic piezoelectric phononic crystals that may be applied in the smart structures with active wave control function.

Author Contributions: Conceptualization, Y.Q.G.; methodology, Y.Q.G.; software, Y.Q.G. and Q.Q.L.; validation, Q.Q.L., Y.J.W., and H.B.Z.; formal analysis, Q.Q.L., Y.J.W., and H.B.Z.; investigation, Y.Q.G.; resources, Y.Q.G.; data curation, Q.Q.L.; writing—original draft preparation, Q.Q.L., Y.J.W., and H.B.Z.; writing—review and editing, Y.Q.G. and Q.Q.L.; visualization, Q.Q.L.; supervision, Y.Q.G.; project administration, Y.Q.G. and Q.Q.L.; funding acquisition, Y.Q.G. and Q.Q.L.

Funding: This work was funded by the National Natural Science Foundation of China, grant number 11372119, and the Youth Innovation Funds in Shanxi Agricultural University, grant number 2018003.

Conflicts of Interest: The authors declare no conflicts of interest.

Appendix A

In this section, the derivation of the relation between $\mathrm{d}\hat{\phi}/\mathrm{dz}$ and $\mathrm{d}\hat{\mathbf{v}}_u/\mathrm{dz}$ will be given in detail, based on all the electrical considerations including the electrical boundary conditions and their influences on mechanical field.

First, substituting Equation (5) into the charge equation (Gauss's law) of electrostatics with free charge as the first formula in Equation (3), one gets $\mathrm{d}\hat{D}_z/\mathrm{dz} = 0$, which implies that $\hat{D}_z$ is a constant. The relation between $\hat{D}_z$ and the electric charge on electrodes $\hat{Q}$ is $\hat{D}_z = \hat{Q}/A$ with A the area of electrodes, which further indicates that $\hat{Q}$ is a constant. It is then substituted into the latter formula of Equation (14) to express $\mathrm{d}\hat{\phi}/\mathrm{dz}$ as

$$\frac{\mathrm{d}\hat{\phi}}{\mathrm{dz}} = \frac{\mathbf{F}}{\varepsilon_{33}}\frac{\mathrm{d}\hat{\mathbf{v}}_u}{\mathrm{dz}} - \frac{\hat{Q}}{\varepsilon_{33}A} \tag{A1}$$

Second, the voltage (potential difference) $\hat{V}$ between the electrodes can be computed from the electric potential by

$$\begin{aligned}\hat{V} &= \hat{\phi}(h) - \hat{\phi}(0) = \int_0^h \frac{\mathrm{d}\hat{\phi}}{\mathrm{d}z}\mathrm{d}z = \int_0^h \left(\frac{\mathbf{F}}{\varepsilon_{33}}\frac{\mathrm{d}\hat{\mathbf{v}}_u}{\mathrm{d}z} - \frac{\hat{Q}}{\varepsilon_{33}A}\right)\mathrm{d}z \\ &= \frac{\mathbf{F}}{\varepsilon_{33}}[\hat{\mathbf{v}}_u(h) - \hat{\mathbf{v}}_u(0)] - \frac{\hat{Q}h}{\varepsilon_{33}A}\end{aligned} \tag{A2}$$

where Equation (A1) has been involved. Equation (A2) provides the relation with respect to three quantities, i.e., the voltage $\hat{V}$, the electric charge $\hat{Q}$, and the displacement difference $[\hat{\mathbf{v}}_u(h) - \hat{\mathbf{v}}_u(0)]$. On the basis of Equation (A2) and the mathematical relation specifying either $\hat{V}$ or $\hat{Q}$ or a relation between $\hat{V}$ and $\hat{Q}$ corresponding to the four considered electrical boundaries, the expressions of $\hat{V}$ and $\hat{Q}$ can be further obtained as listed in Table A1.

Table A1. Expressions of relative electrical quantities in cases of four electrical boundaries.

Electrical Boundaries	Electric-Open	Applied Capacitance	Electric-Short	Applied Feedback Control [1]
Mathematical relation	$\hat{Q} = 0$	$\hat{Q} = C\hat{V}$	$\hat{V} = 0$	$\hat{V} = -K_g\times [\hat{u}_z(h) - \hat{u}_z(0)]$
Expressions of $\hat{V}$	$\frac{\mathbf{F}}{\varepsilon_{33}}[\hat{\mathbf{v}}_u(h) - \hat{\mathbf{v}}_u(0)]$	$\frac{\mathbf{F}}{\varepsilon_{33}+Ch/A}\times [\hat{\mathbf{v}}_u(h) - \hat{\mathbf{v}}_u(0)]$	0	$-\mathbf{K}[\hat{\mathbf{v}}_u(h) - \hat{\mathbf{v}}_u(0)]$
Expressions of $\hat{Q}$	0	$\frac{\mathbf{F}}{\varepsilon_{33}/C+h/A}\times [\hat{\mathbf{v}}_u(h) - \hat{\mathbf{v}}_u(0)]$	$\frac{\mathbf{F}}{h/A}[\hat{\mathbf{v}}_u(h) - \hat{\mathbf{v}}_u(0)]$	$\frac{\mathbf{F}+\varepsilon_{33}\mathbf{K}}{h/A}\times [\hat{\mathbf{v}}_u(h) - \hat{\mathbf{v}}_u(0)]$
Expressions of $\mathbf{R}$	$0_{1\times 3}$	$\frac{\mathbf{F}}{[\varepsilon_{33}A/(Ch)+1]\varepsilon_{33}h}$	$\frac{\mathbf{F}}{\varepsilon_{33}h}$	$\frac{\mathbf{F}+\varepsilon_{33}\mathbf{K}}{\varepsilon_{33}h}$

[1] $\mathbf{K} = [0\ \ 0\ \ K_g]$.

Finally, substituting the expressions of $\hat{Q}$ into Equation (A1) gives,

$$\frac{\mathrm{d}\hat{\phi}}{\mathrm{d}z} = \frac{\mathbf{F}}{\varepsilon_{33}}\frac{\mathrm{d}\hat{\mathbf{v}}_u}{\mathrm{d}z} - \mathbf{R}[\hat{\mathbf{v}}_u(h) - \hat{\mathbf{v}}_u(0)] \tag{A3}$$

i.e., Equation (16), where $\mathbf{R}$ is naturally obtained from $\hat{Q}$ and represented by $\mathbf{F}$ in cases of four electrical boundaries as also listed in Table A1.

Appendix B

In this section, the frequency–wavenumber spectra (band structures) of elastic waves along thickness in laminated piezoelectric phononic crystals with the unit cell consisting of Glass-$LiNbO_3^A$ layers and the piezoelectric layer being in a state of electric open are analyzed by the formulation described above, so that the proposed method and the corresponding computer program can be validated in this degeneration situation. The elastic waves in this case are decoupled into pure SH, SV and P waves, so the computed results can be compared with their counterparts obtained by the explicit formula in Galich et al. [86] or identically in Shen & Cao [87], as provided in Figure A1. It is worth noting that the equivalent elastic constant $c_{33} + e_{33}^2/\varepsilon_{33}$ is used for the $LiNbO_3^A$ layer in the explicit formula so as to take its piezoelectric effect into account. From Figure A1, it is seen that the band structures gained by the two methods are exactly coincident, which proves our proposed method in this paper. Besides, note that the wavenumber spectra of the SH wave are superposed exactly on those of the SV wave due to the same wave speed.

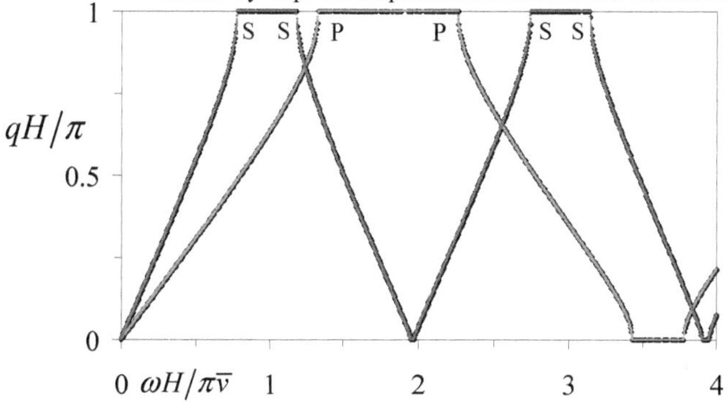

Figure A1. Comparisons of the wavenumber spectra for the periodic Glass-$LiNbO_3{}^A$ multilayers with electric-open boundary computed by the proposed method and the explicit dispersion relation from Galich et al. [86].

References

1. Lee, E.H.; Yang, W.H. On waves in composite materials with periodic structure. *SIAM J. Appl. Math.* **1973**, *25*, 492–499. [CrossRef]
2. Hussein, M.I.; Leamy, M.J.; Ruzzene, M. Dynamics of phononic materials and structures: Historical origins, recent progress, and future outlook. *ASME Appl. Mech. Rev.* **2014**, *66*, 040802. [CrossRef]
3. Banerjee, A.; Das, R.; Calius, E.P. Waves in structured mediums or metamaterials: A review. *Arch. Computat. Methods Eng.* **2018**, 1–30. [CrossRef]
4. Sapriel, J.; Djafari-Rouhani, B. Vibrations in superlattices. *Surf. Sci. Rep.* **1989**, *10*, 189–275. [CrossRef]
5. Nougaoui, A.; Djafari-Rouhani, B. Vibrations in elastic and piezoelectric superlattices. *J. Electron. Spectrosc. Relat. Phenom.* **1987**, *45*, 197–206. [CrossRef]
6. Deymier, P.A. *Acoustic Metamaterials and Phononic Crystals*; Springer: Berlin, Germany, 2013; pp. 1–12.
7. Gomopoulos, N.; Maschke, D.; Koh, C.Y.; Thomas, E.L.; Tremel, W.; Butt, H.-J.; Fytas, G. One-dimensional hypersonic phononic crystals. *Nano Lett.* **2010**, *10*, 980–984. [CrossRef] [PubMed]
8. Nemat-Nasser, S.; Sadeghi, H.; Amirkhizi, A.V.; Srivastava, A. Phononic layered composites for stress-wave attenuation. *Mech. Res. Commun.* **2015**, *68*, 65–69. [CrossRef]
9. Srivastava, A. Metamaterial properties of periodic laminates. *J. Mech. Phys. Solids* **2016**, *96*, 252–263. [CrossRef]
10. Datta, S.K.; Shah, A.H. *Elastic Waves in Composite Media and Structures: With Applications to Ultrasonic Nondestructive Evaluation*; CRC Press: Boca Raton, FL, USA, 2009; pp. 37–164.
11. Shen, M.R.; Cao, W.W. Acoustic band-gap engineering using finite-size layered structures of multiple periodicity. *Appl. Phys. Lett.* **1999**, *75*, 3713–3715. [CrossRef]
12. Nayfeh, A.H. The general problem of elastic wave propagation in multilayered anisotropic media. *J. Acoust. Soc. Am.* **1991**, *89*, 1521–1531. [CrossRef]
13. Aly, A.H.; Mehaney, A.; Abdel-Rahman, E. Study of physical parameters on the properties of phononic band gaps. *Int. J. Mod. Phys. B* **2013**, *27*, 1350047. [CrossRef]
14. Helbig, K. Anisotropy and dispersion in periodically layered media. *Geophysics* **1984**, *49*, 364–373. [CrossRef]
15. Camley, R.E.; Djafari-Rouhani, B.; Dobrzynski, L.; Maradudin, A.A. Transverse elastic waves in periodically layered infinite, semi-infinite, and slab media. *J. Vac. Sci. Technol. B* **1983**, *1*, 371–375. [CrossRef]
16. Auld, B.A.; Beaupre, G.S.; Herrmann, G. Horizontal shear surface waves on a laminated composite. *Electron. Lett.* **1977**, *13*, 525–527. [CrossRef]
17. Bulgakov, A.A. Surface acoustic oscillations in a periodically layered medium. *Solid State Commun.* **1985**, *55*, 869–872. [CrossRef]

18. Maréchal, P.; Haumesser, L.; Tran-Huu-Hue, L.P.; Holc, J.; Kuščer, D.; Lethiecq, M.; Feuillard, G. Modeling of a high frequency ultrasonic transducer using periodic structures. *Ultrasonics* **2008**, *48*, 141–149. [CrossRef] [PubMed]
19. Royer, D.; Dieulesaint, E. *Elastic Waves in Solids II: Generation, Acousto-optic Interaction, Applications*; Springer: Berlin, Germany, 2000; ISBN 3-540-65931-5.
20. Li, F.-M.; Wang, Y.-S. Study on wave localization in disordered periodic layered piezoelectric composite structures. *Int. J. Solids Struct.* **2005**, *42*, 6457–6474. [CrossRef]
21. Li, F.-M.; Xu, M.-Q.; Wang, Y.-S. Frequency-dependent localization length of SH-wave in randomly disordered piezoelectric phononic crystals. *Solid State Commun.* **2007**, *141*, 296–301. [CrossRef]
22. Li, F.-M.; Wang, Y.-S. Study on localization of plane elastic waves in disordered periodic 2-2 piezoelectric composite structures. *J. Sound Vib.* **2006**, *296*, 554–566. [CrossRef]
23. Li, F.-M.; Wang, Y.-Z.; Fang, B.; Wang, Y.-S. Propagation and localization of two-dimensional in-plane elastic waves in randomly disordered layered piezoelectric phononic crystals. *Int. J. Solids Struct.* **2007**, *44*, 7444–7456. [CrossRef]
24. Guo, X.; Wei, P.J. Dispersion relations of elastic waves in one-dimensional piezoelectric phononic crystal with initial stresses. *Int. J. Mech. Sci.* **2016**, *106*, 231–244. [CrossRef]
25. Golub, M.V.; Fomenko, S.I.; Alexandrov, A.A. Simulation of plane 3D wave propagation in layered piezoelectric phononic crystals. In Proceedings of the International Conference Days on Diffraction (DD) 2014, St. Petersburg, Russia, 26–30 May 2014; IEEE: Piscataway, NJ, USA, 2014; pp. 83–88. [CrossRef]
26. Fomenko, S.I.; Golub, M.V.; Alexandrov, A.A.; Chen, A.L.; Wang, Y.S.; Zhang, Ch. Band-gaps and low transmission pass-bands in layered piezoelectric phononic crystals. In Proceedings of the International Conference Days on Diffraction (DD) 2016, St. Petersburg, Russia, 27 June–1 July 2016; IEEE: Piscataway, NJ, USA, 2016; pp. 149–154. [CrossRef]
27. Fomenko, S.I.; Golub, M.V.; Chen, A.L.; Wang, Y.S.; Zhang, C.Z. Band-gap and pass-band classification for oblique waves propagating in a three-dimensional layered functionally graded piezoelectric phononic crystal. *J. Sound Vib.* **2019**, *439*, 219–240. [CrossRef]
28. Chen, A.-L.; Yan, D.-J.; Wang, Y.-S.; Zhang, C.Z. Anti-plane transverse waves propagation in nanoscale periodic layered piezoelectric structures. *Ultrasonics* **2016**, *65*, 154–164. [CrossRef] [PubMed]
29. Yan, D.-J.; Chen, A.-L.; Wang, Y.-S.; Zhang, C.Z.; Golub, M. In-plane elastic wave propagation in nanoscale periodic layered piezoelectric structures. *Int. J. Mech. Sci.* **2018**, *142*, 276–288. [CrossRef]
30. Zinchuk, L.P.; Podlipenets, A.N.; Shul'ga, N.A. Electroelastic shear waves in a stratified periodic medium. *Sov. Appl. Mech.* **1988**, *24*, 245–250. [CrossRef]
31. Zinchuk, L.P.; Levchenko, V.V.; Shul'ga, N.A. Propagation of three-dimensional electroelastic shear waves in a regularly layered medium of metal-piezoelectric type. *J. Math. Sci.* **1993**, *63*, 298–302. [CrossRef]
32. Alshits, V.I.; Shuvalov, A.L. Resonance reflection and transmission of shear elastic waves in multilayered piezoelectric structures. *J. Appl. Phys.* **1995**, *77*, 2659–2665. [CrossRef]
33. Qian, Z.H.; Jin, F.; Wang, Z.K.; Kishimoto, K. Dispersion relations for SH-wave propagation in periodic piezoelectric composite layered structures. *Int. J. Eng. Sci.* **2004**, *42*, 673–689. [CrossRef]
34. Lan, M.; Wei, P.J. Laminated piezoelectric phononic crystal with imperfect interfaces. *J. Appl. Phys.* **2012**, *111*, 013505. [CrossRef]
35. Faidi, W.I.; Nayfeh, A.H. An improved model for wave propagation in laminated piezoelectric composites. *Mech. Mater.* **2000**, *32*, 235–241. [CrossRef]
36. Geng, X.C.; Zhang, Q.M. Evaluation of piezocomposites for ultrasonic transducer applications—Influence of the unit cell dimensions and the properties of constituents on the performance of 2-2 piezocomposites. *IEEE Trans. Ultrason. Ferroelectr. Freq. Control* **1997**, *44*, 857–872. [CrossRef]
37. Podlipenets, A.N. Wave propagation in periodically layered elastic and electroelastic media. In *IUTAM Symposium on Anisotropy, Inhomogeneity and Nonlinearity in Solid Mechanics, Nottingham, UK, 30 August–3 September 1994*; Solid Mechanics and Its Applications book series (SMIA, volume 39); Parker, D.F., England, A.H., Eds.; Springer: Dordrecht, The Netherlands, 1995; pp. 469–474.
38. Zinchuk, L.P.; Podlipenets, A.N.; Shul'ga, N.A. Surface and normal electroelastic shear waves in even-layered structures. *J. Math. Sci.* **1995**, *74*, 1157–1162. [CrossRef]
39. Yan, D.-J.; Chen, A.-L.; Wang, Y.-S.; Zhang, C.Z.; Golub, M. Propagation of guided elastic waves in nanoscale layered periodic piezoelectric composites. *Eur. J. Mech. A Solids* **2017**, *66*, 158–167. [CrossRef]

40. Wang, Y.-Z.; Li, F.-M.; Huang, W.-H.; Wang, Y.-S. The propagation and localization of Rayleigh waves in disordered piezoelectric phononic crystals. *J. Mech. Phys. Solids* **2008**, *56*, 1578–1590. [CrossRef]
41. Alippi, A.; Craciun, F.; Molinari, E. Stopband edges in the dispersion curves of Lamb waves propagating in piezoelectric periodical structures. *Appl. Phys. Lett.* **1988**, *53*, 1806–1808. [CrossRef]
42. Craciun, F.; Sorba, L.; Molinari, E.; Pappalardo, M. A coupled-mode theory for periodic piezoelectric composites. *IEEE Trans. Ultrason. Ferroelectr. Freq. Control* **1989**, *36*, 50–56. [CrossRef] [PubMed]
43. Guo, X.; Wei, P.J.; Li, L. Dispersion relations of elastic waves in one-dimensional piezoelectric phononic crystal with mechanically and dielectrically imperfect interfaces. *Mech. Mater.* **2016**, *93*, 168–183. [CrossRef]
44. Ghazaryan, K.B.; Piliposyan, D.G. Interfacial effects for shear waves in one dimensional periodic piezoelectric structure. *J. Sound Vib.* **2011**, *330*, 6456–6466. [CrossRef]
45. Zhao, J.; Pan, Y.; Zhong, Z. Theoretical study of shear horizontal wave propagation in periodically layered piezoelectric structure. *J. Appl. Phys.* **2012**, *111*, 064906. [CrossRef]
46. Zhao, J.; Pan, Y.; Zhong, Z. A study of pressure-shear vertical wave propagation in periodically layered fluid and piezoelectric structure. *J. Appl. Phys.* **2013**, *113*, 054903. [CrossRef]
47. Zinchuk, L.P.; Podlipenets, A.N.; Shul'ga, N.A. Dispersion relations for electroelastic shear waves in a periodically layered medium. *Sov. Appl. Mech.* **1990**, *26*, 1092–1099. [CrossRef]
48. Zinchuk, L.P.; Podlipenets, A.N. Vibrational modes in a surface shear wave propagating in a regularly layered electroelastic half-space. *Sov. Appl. Mech.* **1991**, *27*, 775–779. [CrossRef]
49. Zinchuk, L.P.; Podlipenets, A.N. Dispersion equations for Rayleigh waves in a piezoelectric periodically layered structure. *J. Math. Sci.* **2001**, *103*, 398–403. [CrossRef]
50. Alami, M.; El Boudouti, E.H.; Djafari-Rouhani, B.; El Hassouani, Y.; Talbi, A. Surface acoustic waves in one-dimensional piezoelectric-metallic phononic crystal: Effect of a cap layer. *Ultrasonics* **2018**, *90*, 80–97. [CrossRef] [PubMed]
51. Otero, J.A.; Rodríguez-Ramos, R.; Monsivais, G.; Pérez-Alvarez, R. Dynamical behavior of a layered piezocomposite using the asymptotic homogenization method. *Mech. Mater.* **2005**, *37*, 33–44. [CrossRef]
52. Zou, X.-Y.; Chen, Q.; Cheng, J.-C. The band gaps of plate-mode waves in one-dimensional piezoelectric composite plates: Polarizations and boundary conditions. *IEEE Trans. Ultrason. Ferroelectr. Freq. Control* **2007**, *54*, 1430–1436. [CrossRef] [PubMed]
53. Zhu, Y.-F.; Yuan, Y.; Zou, X.-Y.; Cheng, J.-C. Piezoelectric-sensitive mode of lamb wave in one-dimensional piezoelectric phononic crystal plate. *Wave Motion* **2015**, *54*, 66–75. [CrossRef]
54. Piliposyan, D.G.; Piliposian, G.T.; Ghazaryan, K.B. Propagation and control of shear waves in piezoelectric composite waveguides with metallized interfaces. *Int. J. Solids Struct.* **2017**, *106*, 119–128. [CrossRef]
55. Li, Q.Q.; Guo, Y.Q.; Wang, J.Y.; Chen, W. Band structures analysis of elastic waves propagating along thickness direction in periodically laminated piezoelectric composites. *Crystals* **2018**, *8*, 351. [CrossRef]
56. Ponge, M.-F.; Dubus, B.; Granger, C.; Vasseur, J.O.; Thi, M.P.; Hladky-Hennion, A.-C. Theoretical and experimental analyses of tunable Fabry-Perot resonators using piezoelectric phononic crystals. *IEEE Trans. Ultrason. Ferroelectr. Freq. Control* **2015**, *62*, 1114–1121. [CrossRef]
57. Ponge, M.F.; Dubus, B.; Granger, C.; Vasseur, J.; Thi, M.P.; Hladky-Hennion, A.-C. Optimization of a tunable piezoelectric resonator using phononic crystals with periodic electrical boundary conditions. *Phys. Proc.* **2015**, *70*, 258–261. [CrossRef]
58. Kutsenko, A.A.; Shuvalov, A.L.; Poncelet, O.; Darinskii, A.N. Quasistatic stopband and other unusual features of the spectrum of a one-dimensional piezoelectric phononic crystal controlled by negative capacitance. *C. R. Mecanique* **2015**, *343*, 680–688. [CrossRef]
59. Kutsenko, A.A.; Shuvalov, A.L.; Poncelet, O.; Darinskii, A.N. Quasistatic stopband in the spectrum of one-dimensional piezoelectric phononic crystal. *arXiv* **2015**, arXiv:1410.7215v2.
60. Kutsenko, A.A.; Shuvalov, A.L.; Poncelet, O.; Darinskii, A.N. Tunable effective constants of the one-dimensional piezoelectric phononic crystal with internal connected electrodes. *J. Acoust. Soc. Am.* **2015**, *137*, 606–616. [CrossRef] [PubMed]
61. Kutsenko, A.A.; Shuvalov, A.L.; Poncelet, O. Dispersion spectrum of acoustoelectric waves in 1D piezoelectric crystal coupled with 2D infinite network of capacitors. *J. Appl. Phys.* **2018**, *123*, 044902. [CrossRef]
62. Mansoura, S.A.; Morvan, B.; Maréchal, P.; Hladky-Hennion, A.-C.; Dubus, B. Study of an hybridization gap in a one dimensional piezoelectric phononic crystal. *Phys. Procedia* **2015**, *70*, 279–282. [CrossRef]

63. Zhu, X.H.; Qiao, J.; Zhang, G.Y.; Zhou, Q.; Wu, Y.D.; Li, L.Q. Tunable acoustic metamaterial based on piezoelectric ceramic transducer. In Proceedings of the Active and Passive Smart Structures and Integrated Systems 2017, SPIE 10164, SPIE Smart Structures and Materials+Nondestructive Evaluation and Health Monitoring 2017, Portland, OR, USA, 25–29 March; Park, G., Erturk, A., Han, J.-H., Eds.; SPIE: Bellingham, WA, USA, 2017; p. 1016411.
64. Parra, E.A.F.; Bergamini, A.; Lossouarn, B.; Damme, B.V.; Cenedese, M.; Ermanni, P. Bandgap control with local and interconnected LC piezoelectric shunts. *Appl. Phys. Lett.* **2017**, *111*, 111902. [CrossRef]
65. Fomenko, S.I.; Golub, M.V.; Doroshenko, O.V.; Chen, A.-L.; Wang, Y.-S.; Zhang, C.Z. Wave motion in piezoelectric layered phononic crystals with and without electroded surfaces. In Proceedings of the International Conference Days on Diffraction (DD) 2017, St. Petersburg, Russia, 19–23 June 2017; IEEE: Piscataway, NJ, USA, 2017; pp. 116–121. [CrossRef]
66. Mansoura, S.A.; Maréchal, P.; Morvan, B.; Hladky-Hennion, A.C.; Dubus, B. Active control of a piezoelectric phononic crystal using electrical impedance. In Proceedings of the 2014 IEEE International Ultrasonics Symposium (IUS), Chicago, IL, USA, 3–6 September 2014; IEEE: Piscataway, NJ, USA, 2014; pp. 951–954. [CrossRef]
67. Mansoura, S.A.; Maréchal, P.; Morvan, B.; Dubus, B. Analysis of a phononic crystal constituted of piezoelectric layers using electrical impedance measurement. *Phys. Procedia* **2015**, *70*, 283–286. [CrossRef]
68. Mansoura, S.A.; Morvan, B.; Maréchal, P.; Benard, P.; Lhadky-Hennion, A.-C.; Dubus, B. Tunability of the band structure of a piezoelectric phononic crystal using electrical negative capacitance. In Proceedings of the 2015 IEEE International Ultrasonics Symposium (IUS), Taipei, Taiwan, 21–24 October 2015; IEEE: Piscataway, NJ, USA, 2015; pp. 1–4. [CrossRef]
69. Darinskii, A.N.; Shuvalov, A.L.; Poncelet, O.; Kutsenko, A.A. Bulk longitudinal wave reflection/transmission in periodic piezoelectric structures with metallized interfaces. *Ultrasonics* **2015**, *63*, 118–125. [CrossRef]
70. Allam, A.; Elsabbagh, A.; Akl, W. Experimental demonstration of one-dimensional active plate-type acoustic metamaterial with adaptive programmable density. *J. Appl. Phys.* **2017**, *121*, 125106. [CrossRef]
71. Wang, Y.L.; Song, W.; Sun, E.W.; Zhang, R.; Cao, W.W. Tunable passband in one-dimensional phononic crystal containing a piezoelectric $0.62Pb(Mg_{1/3}Nb_{2/3})O_3$-$0.38PbTiO_3$ single crystal defect layer. *Physica E* **2014**, *60*, 37–41. [CrossRef]
72. Chopra, I. Review of state of art of smart structures and integrated systems. *AIAA J.* **2002**, *40*, 2145–2187. [CrossRef]
73. Song, G.; Sethi, V.; Li, H.-N. Vibration control of civil structures using piezoceramic smart materials: A review. *Eng. Struct.* **2006**, *28*, 1513–1524. [CrossRef]
74. Honein, B.; Braga, A.M.B.; Barbone, P.; Herrmann, G. Wave propagation in piezoelectric layered media with some applications. *J. Intell. Mater. Syst. Struct.* **1991**, *2*, 542–557. [CrossRef]
75. Minagawa, S. Propagation of harmonic waves in a layered elasto-piezoelectric composite. *Mech. Mater.* **1995**, *19*, 165–170. [CrossRef]
76. Pestel, E.C.; Leckie, F.A. *Matrix Methods in Elasto Mechanics*; McGraw-Hill: New York, NY, USA, 1963; ISBN 0-070-49520-3.
77. Tarn, J.Q. A state space formalism for piezothermoelasticity. *Int. J. Solids Struct.* **2002**, *39*, 5173–5184. [CrossRef]
78. Mead, D.J. Wave propagation in continuous periodic structures: Research contributions from Southampton, 1964–1995. *J. Sound Vib.* **1996**, *190*, 495–524. [CrossRef]
79. Brillouin, L. *Wave Propagation in Periodic Structures: Electric Filters and Crystal Lattices*, 2nd ed.; Dover Publications: Mineola, NY, USA, 1953.
80. ANSI/IEEE Std 176-1987. *IEEE Standard on Piezoelectricity*; IEEE: New York, NY, USA, 1988; pp. 3–30. ISBN 0-7381-2411-7.
81. Tiersten, H.F. *Linear Piezoelectric Plate Vibrations*; Plenum Press: New York, NY, USA, 1969; pp. 33–39. ISBN 978-1-4899-6221-8.
82. Auld, B.A. *Acoustic Fields and Waves in Solids*; John Wiley & Sons: New York, NY, USA, 1973; Volume 1, pp. 101–134. ISBN 0-471-03700-1.
83. Kochervinskiĭ, V.V. Piezoelectricity in crystallizing ferroelectric polymers: Poly(vinylidene fluoride) and its copolymers (a review). *Crystallogr. Rep.* **2003**, *48*, 649–675. [CrossRef]

84. Cardarelli, F. *Materials Handbook: A Concise Desktop Reference*, 2nd ed.; Springer: London, UK, 2008; pp. 184–186.
85. Li, L.F.; Guo, Y.Q. Analysis of longitudinal waves in rod-type piezoelectric phononic crystals. *Crystals* **2016**, *6*, 45. [CrossRef]
86. Galich, P.I.; Fang, N.X.; Boyce, M.C.; Rudykh, S. Elastic wave propagation in finitely deformed layered materials. *J. Mech. Phys. Solids* **2017**, *98*, 390–410. [CrossRef]
87. Shen, M.R.; Cao, W.W. Acoustic bandgap formation in a periodic structure with multilayer unit cells. *J. Phys. D Appl. Phys.* **2000**, *33*, 1150–1154. [CrossRef]
88. Guo, Y.Q.; Fang, D.N. Analysis and interpretation of longitudinal waves in periodic multiphase rods using the method of reverberation-ray matrix combined with the Floquet-Bloch theorem. *J. Vib. Acoust.* **2014**, *136*, 011006. [CrossRef]

Article

An Optimization of Two-Dimensional Photonic Crystals at Low Refractive Index Material

Thanh-Phuong Nguyen [1], Tran Quoc Tien [2] and Quang Cong Tong [2,*] and Ngoc Diep Lai [3,*]

1 School of Engineering Physics, Hanoi University of Science and Technology, No. 1 Dai Co Viet, Hai Ba Trung, Hanoi 100000, Vietnam
2 Institute of Materials Science, Vietnam Academy of Science and Technology, 18 Hoang Quoc Viet, Cau Giay, Hanoi 100000, Vietnam
3 Laboratoire de Photonique Quantique et Moléculaire, UMR 8537, École Normale Supérieure Paris-Saclay, Centrale Supélec, CNRS, Université Paris-Saclay, 61 avenue de Président Wilson, 94235 Cachan, France
* Correspondence: congtq2004@gmail.com (Q.C.T.); ngoc-diep.lai@ens-paris-saclay.fr (N.D.L.)

Received: 1 August 2019; Accepted: 21 August 2019; Published: 24 August 2019

Abstract: Photonic crystal (PC) is usually realized in materials with high refractive indices contrast to achieve a photonic bandgap (PBG). In this work, we demonstrated an optimization of two-dimensional PCs using a low refractive index polymer material. An original idea of assembly of polymeric multiple rings in a hexagonal configuration allowed us to obtain a circular-like structure with higher symmetry, resulting in a larger PBG at a low refractive index of 1.6. The optical properties of such newly proposed structure are numerically calculated by using finite-difference time-domain (FDTD) method. The proposed structures were realized experimentally by using a direct laser writing technique based on low one-photon absorption method.

Keywords: photonic crystals; photonic bandgaps; polymer materials; direct laser writing

---

1. Introduction

Photonic crystal (PC), an artificial material in which the refractive index is modulated at wavelength scale, offers presently many interesting applications in different domains [1,2]. The most important property of the PCs is the existence of a so-called photonic bandgap (PBG). In case of one-dimensional (1D) PCs, the PBG can be easily obtained by assembling multiple layers of two different materials with very low refractive indices contrast [3–6]. However, in case of two- and three-dimensional (2D and 3D) PCs, it requires that the refractive indices contrast should be as high as possible. Some 2D and 3D PCs are made by polystyrenes or SiO_2 nanoparticles [7–10], which are fabricated by a simple fabrication method, but these PCs do not have a true PBG, or in other word, they have a so-called pseudo or partial PBG, due to their weak refractive index. Therefore, most 2D and 3D PCs are made by semiconductor materials with high refractive indices, which however makes the fabrication procedure complicated and expensive. Different optimizations have been proposed to obtain the PBG as large as possible and with a material with a smaller refractive index. By increasing the symmetry of the PCs, for example in the case of 2D PCs, from square (four-fold) to hexagon (six-fold), and to quasi-periodic PCs, such as Penrose structure [11,12], the PBG can be opened at lower refractive index materials, such as polymers [13,14].

The most difficult task of PCs investigations is the fabrication, which should satisfy several conditions, such as simple, rapid, inexpensive, and particularly flexible. Today, many fabrication techniques have been proposed to realize PCs, for example, e-beam lithography, self-assembly of opals, laser interference, and direct laser writing. The choice of fabrication technique depends on desired applications. Among available fabrication technologies, the direct laser writing (DLW) is the most powerful one allowing fabrication of any desired PCs, in multi-dimension ((1D, 2D, 3D) PCs) [15,16].

The working principle of the DLW is to focus tightly a laser beam into a small region of a photoresist to induce a local polymerization or depolymerization effect. By moving the focusing spot following a well-designed structure, a corresponding polymeric pattern can be obtained.

In this work, we demonstrated an optimization of the 2D PCs to obtain a large PBG even with low refractive index materials, such as polymer. We proposed an original idea of high symmetry 2D PCs, and a new experimental way to realize these PCs by the DLW technique. The PBGs of a standard honeycomb structure and of a newly proposed structure are calculated and compared by using a Finite-difference time-domain (FDTD) method. These structures are then fabricated by a new scanning method using the one-photon absorption-based DLW technique [17].

2. Theoretical Calculation

For calculation of the PBG of different PCs, we have used the FDTD method (Lumerical Software) with Bloch boundary conditions. We assumed that the refractive index of material is 1.6, a typical value of polymer without considering the dispersion effect of the material. This choice is justified by the values experimentally measured in our case, i.e.; the refractive index of the used polymer is in between 1.55 and 1.62 for the wavelengths ranging from 400 nm to 800 nm. For simulations of 2D PCs, we distinguished two different modes, TM and TE, which are the transverse magnetic or transverse electric modes where the magnetic or electric field is parallel to the PC plane, respectively.

It is well-known that the honeycomb lattice is the best 2D PC, which provides a largest PBG, as compared to the square or hexagonal lattice. We first calculated the PBG of this structure at low refractive index material, by considering two types: one is the honeycomb lattice comprising of polymeric cylinders (n = 1.6) in an air background (n = 1) and the another is an inverse geometry consisting of air-holes in the polymeric background. Figure 1a,b show a honeycomb lattice and its reciprocal space, where a is lattice constant, and r is the radius of polymeric cylinders or air-holes. For the case of polymeric cylinders in the air background, the PBG only reveals a small gap for TE mode. Figure 2a shows a band structure for TE mode, obtained with an optimum radius of r/a = 0.3. Figure 2b shows gap map obtained with various filling factors (different ratios of r/a). We see that the largest PBG for TE mode is about 0.02 at r = 0.3a. When considering a honeycomb lattice made of air cylinders in the polymeric background, we found that the PBG only reveals for TM mode. A band structure for TM mode (r/a = 0.35) is shown in Figure 3a. The PBG (for TM mode) as a function of the r/a ratio was also calculated, and a gap map is presented in the Figure 3b. We can see that the PBG is increased as compared to the one obtained in previous case, and reaches 0.05 at r = 0.35a. These results are somehow similar to results obtained with higher refractive index [1,2]. But at low refractive index material, the PBG exists but not complete, i.e.; no common PBG for both TE and TM modes. Table 1 summarizes the PBG of the honeycomb lattice at low refractive index material for both configurations.

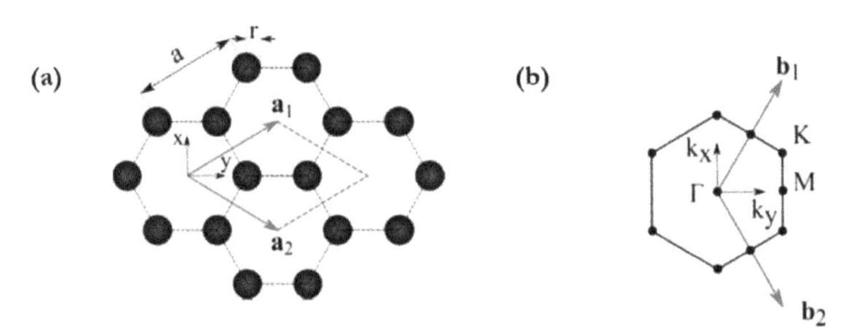

Figure 1. (**a**) A two-dimensional honeycomb lattice with two main lattice vectors: $\mathbf{a}_1 = [\sqrt{3}a/2, a/2]$, $\mathbf{a}_2 = [\sqrt{3}a/2, -a/2]$. The artificial atome has a cylindrical form with a radius r. (**b**) The reciprocal space of corresponding two-dimensional honeycomb lattice with the primitive vectors: $\mathbf{b}_1 = 2\pi/a\ (1/3\sqrt{3}, 1/3)$ and $\mathbf{b}_2 = 2\pi/a\ (1/3\sqrt{3}, -1)/3)$. The reduced first Brillouin zone is identified by Γ, K and M points.

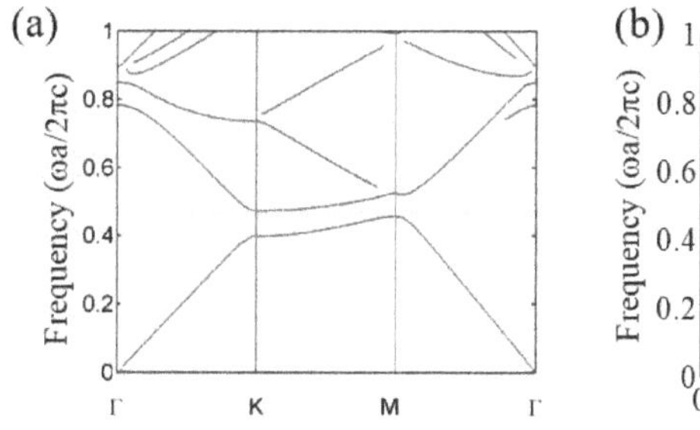

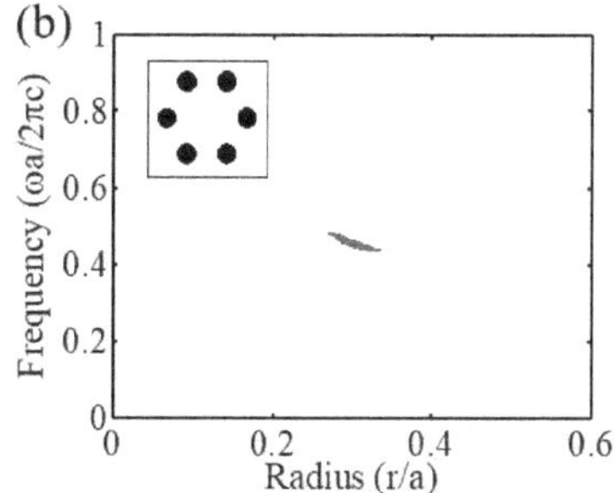

Figure 2. Photonic band structureof a 2D honeycomb lattice of polymeric cylinders in an air background (see insert of b). (**a**) Photonic bandgap for TE mode with r/a = 0.3. (**b**) Map of photonic bandgap as a function of r/a (shaded blue region).

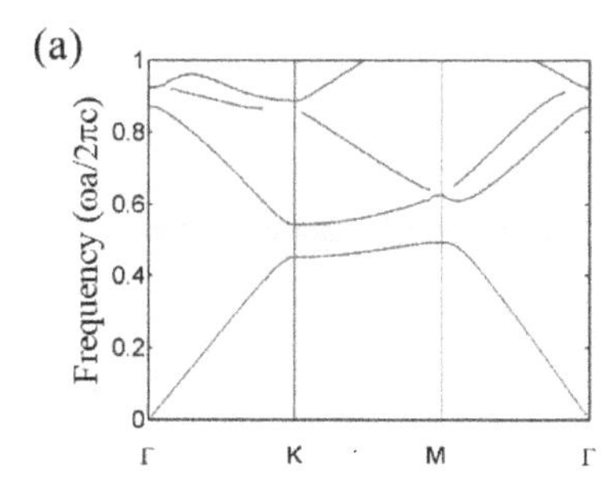

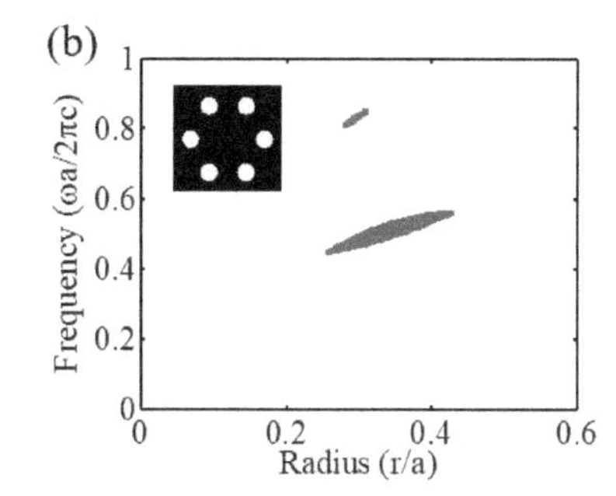

Figure 3. Photonic band structure of a 2D honeycomb lattice of air cylinders in a polymeric background (see insert of b). (**a**) Photonic bandgap for TM mode with r/a =0.35. (**b**) Map of photonic bandgap as a function of r/a (shaded blue region).

Table 1. Summary of the best photonic bandgaps of honeycomb lattices obtained with a polymer material with a refractive index of n = 1.6.

Polarization Mode	Type of Cylinders	The Existence of PBG	The Largest PBG
TM	Air	Yes	0.05
TM	Polymer	No	NA
TE	Air	No	NA
TE	Polymer	Yes	0.01

It has been well-known that the larger PBG can be obtained by increasing the structure symmetry [18]. Different structures have been proposed [11–14], and larger PBG are obtained, even at low refractive index materials. In particular, it is important to emphasis on the most symmetric one, called circular photonic crystal (CPC) [19,20], which possesses a perfect symmetry in a 2D plane. This CPC structure enables particularly an isotropic PBG, and consequently a PBG at low refractive index. However, the CPC is a non-periodic structure and usually developed around a symmetric point. Therefore, its photonic property is only characterized by a transmission calculation or measurement but not its PBG.

In this work, we proposed a novel kind of equivalent 2D PC by arranging multi-rings in a periodic configuration, such as a hexagonal lattice. This novel structure allowed us to combine the advantages of both CPC structure, thanks to the perfect symmetry of the rings (dielectric or air), and the periodic organization of these rings in a well-known hexagonal lattice, which can be numerically simulated to obtain a true PBG. As it will be shown in the experimental part, the proposed structure is also matched to the capacity of the fabrication technology.

Figure 4a presents an assembled multi-rings lattice and Figure 4b illustrates its corresponding reciprocal space, where a is the lattice constant, r is the radius of the rings, and d is the width of rings. Similar to honeycomb PC, the PBG of the assembled multi-rings lattice can be calculated for two possible configurations: the lattice containing multi-rings of air ($n = 1$) in the polymeric background ($n = 1.6$) and the another is an inverse geometry consisting of multi-rings of polymer in the air background. In practices, these two configurations can be optically realized by using a negative photoresist or a positive photoresist, or simply by using a positive photoresist with controllable laser power [17]. We demonstrated that the multi-rings assembly allowed obtaining different 2D PCs with various filling factors. That can be done by changing the r/a ratio. It is clear that if a > 2r, all rings will be separated, and the structure is a simple hexagonal PC with a ring as an "artificial atom". However, if a < 2r, multiple rings will overlap to each other, resulting in a complicated 2D PC, with a perfect symmetry determined by the ring.

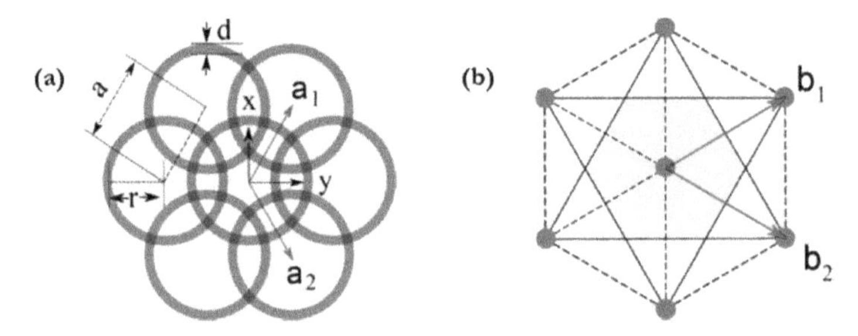

Figure 4. (**a**) Proposal of a 2D assembled-multirings structure arranged in a hexagonal configuration. The main lattice vectors are $\mathbf{a}_1 = a\left(1/2, \sqrt{3}/2\right)$ and $\mathbf{a}_2 = a\left(1/2, -\sqrt{3}/2\right)$; r is the radius of rings; d is the width of ring-walls; a is the lattice constant. (**b**) The corresponding 2D reciprocal space off the 2D assembled-multirings structure. $\mathbf{b}_1 = 2\pi/a\left(1, 1/\sqrt{3}\right)$ and $\mathbf{b}_2 = 2\pi/a\left(1, -1/\sqrt{3}\right)$: the primitive vectors of the reciprocal lattice. The Brillouin zone is identified by a yellow hexagon.

Figure 5 shows the simulation results for the case of assembled polymeric multi-rings ($n = 1.6$) in the air background. Similar to the case of the honeycomb, it has seen that the PBG reveals only for TE mode. For the simulation shown in Figure 5a, we assumed that the ring width is 300 nm, the ring separation a = 1.5 μm, and the ring radius r/a = 0.51, which is suitable to the fabrication capacity. A larger PBG is obtained, as compared to similar case of the honeycomb lattice shown in Figure 2a. By changing the radius of rings, a gap map (shaded blue region) as a function of r/a for TE mode has been carried out, and shown in Figure 5b. It can be seen that the largest PBG is obtained with a value of $a\Delta\omega/2\pi c = 0.035$ with r = 0.51a.

For the case of a 2D lattice consisting of air rings ($n = 1$) in the polymeric background, in contrast, the PBG reveals only for TM mode. Figure 6a shows its PBG obtained with a radius ratio of r/a = 0.78 with a = 1.5 μm. Again, the PBG is larger, as compared to that obtained by the honeycomb lattice shown in Figure 3a. A map of PBGs for TM mode versus the ratio of r/a is shown in Figure 6b. It can be seen that multiple gaps can be obtained with this structure, even with low refractive index contrast. The PBG can reach a value of $a\Delta\omega/2\pi c = 0.089$ at r = 0.78a, which is larger than 0.05, obtained by the honeycomb PC. A summary of PBGs of multi-rings lattice is shown in Table 2. This novel PC still does not produce a complete PBG, but the individual PBG (for TE or TM mode) is better than those obtained by a best standard 2D PC (e.g., honeycomb).

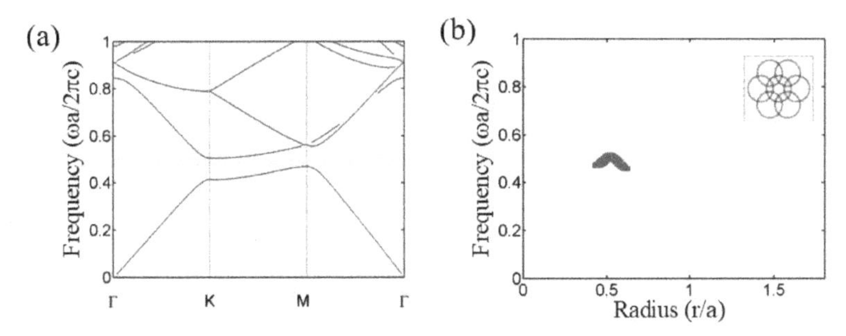

Figure 5. (**a**) Photonic band structure (TE mode) of an assembled-multirings lattice consisted of polymer-rings in an air background. (**b**) Map of photonic bandgap (TE mode) as a function of r/a (shaded blue region). The simulations are realized with r/a = 0.51, a = 1.5 μm, d = 300 nm, and n = 1.6.

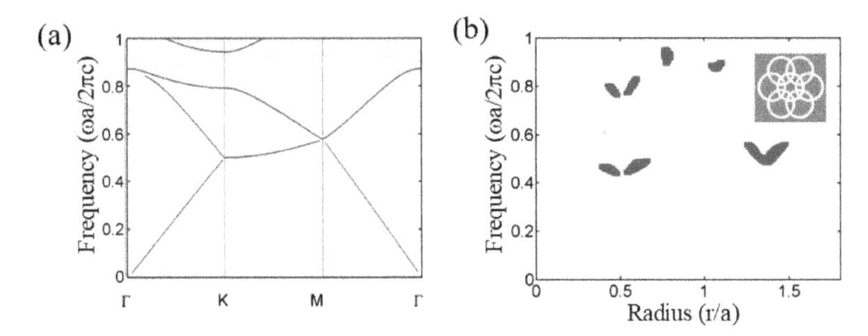

Figure 6. (**a**) Photonic band structure (TM mode) of an assembled-multirings lattice consisted of air rings in the polymer background. (**b**) Map of photonic bandgap (TM mode) as a function of r/a (shaded blue region). The simulations are realized with r/a = 0.78, a = 1.5 μm, d = 300 nm, and n = 1.6.

Table 2. Summary of photonic bandgaps of assembled multi-rings structures. The calculations were realized with a ring width of d = 300 nm and the polymer material has a refractive index of n = 1.6.

Polarization Mode	Assembled Multi-Rings of	The Existence of PBG	The Largest PBG
TM	Air	Yes	0.089
TM	Polymer	No	NA
TE	Air	No	NA
TE	Polymer	Yes	0.035

We note that in a standard case of square or hexagonal 2D PCs, the rod-type has a TM-bandgap and the hole-type has a TE-bandgap, respectively [1,2]. However, when working with particular structures, such as honeycomb or multi-ring PCs, the situation changes. In these cases, the existence of TM or TE bandgaps varies as a function of the ratio between the rod size and the lattice periodicity. With low refractive index, we can see that the rod-type favors a TE-bandgap while the hole-type favors a TM-bandgap, respectively.

3. Fabrication of Proposed Structure by Direct Laser Writing Method

To fabricate these proposed PCs, two possible methods can be used, such as e-beam lithography and DLW method. In this work, we have demonstrated the fabrication by using one-photon absorption-based DLW method, which is simple, low cost, and it can even produce 3D PCs [16,17,21]. The experimental setup is shown in Figure 7 and can be explained as follows. A simple and low-cost continuous-wave (cw) 532 nm laser was used to write the desired structures. The laser power was adjusted by using a combination of a half-wave plate and a polarizing beam splitter. The laser beam is sent through an electronic shutter (S), which allows controlling the exposure time. The quarter-wave plate was used to obtain a circular polarization of the excitation beam, just before focusing. The laser beam was sent on a 50/50 beam-splitting cube (BS), after which one part goes to a powermeter and the other is directed to an oil-immersion objective lens (OL) with a numerical aperture (NA) of 1.3, which focuses the laser beam on the photoresist sample. The last one was placed on a piezoelectric translation stage (PZT), which allows moving the sample in 3D with a resolution of 0.1 nm. Thanks to the high NA OL, the laser is tightly focused onto the sample with a focusing spot of about 300 nm. To determine the focusing position, the same OL is used to collect the fluorescence photons, emitted from the photoresist sample, which then propagate in opposite direction with respect to the excitation beam. The fluorescence beam is passed through a long-pass filter (F) with a cut-off wavelength of 580 nm, to remove the excitation laser light, and directed towards an avalanche photodiode (APD). It has been demonstrated that this technique can fabricate structures with a thickness as large as 100 micrometers, depending on the absorption coefficient of the photoresist [16,21], with a periodicity down to 400 nm and the smallest size of pattern can be as small as 57 nm [17].

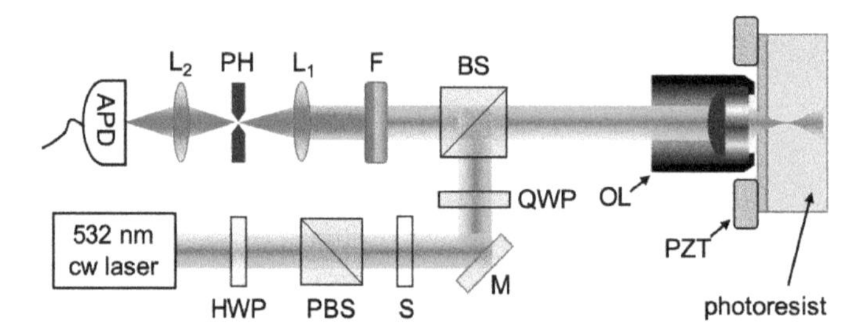

Figure 7. Experimental setup of the low one-photon absorption-based direct laser writing method used to fabricate desired polymeric photonic crystals. HWP: half-wave plate, PBS: polarizing beam splitter, S: electronic shutter, M: mirror, QWP: quarter-wave plate, BS: beam splitter, OL: objective lens, PZT: piezoelectric translator, F: long-pass filter, $L_{1,2}$: lenses, PH: pinhole, APD: avalanche photodiode.

In this work, we proposed a new way to realize desired structures. Indeed, in e-beam lithography or DLW, the structures are usually created by moving the focusing spot following a design. We proposed to assembly multiple lines or rings to achieve desired structures. Various structures can be created with the same lines or rings assembly, which simplifies a lot for fabrication procedure. For demonstration, we used a positive S1818 photoresist with a thickness of 1 μm. With low excitation laser power, the focusing spot produced an air-hole in polymer thin film (after development), and by moving this focusing, we obtained an assembly of air multi-rings lattice. Please note that with the same photoresist and with higher excitation laser power, we can obtain a polymeric multi-ring thanks to the optically induced thermal effect [17].

Figure 8a shows a new proposed trajectory of the focusing laser beam to draw a honeycomb lattice. By this method, we can obtain different structures, starting from a standard air-hole hexagonal structure to a honeycomb structure with controllable filling factors and dots shape. Figure 8b–e show scanning electron microscope (SEM) images of different structures (period of a = 1.5 μm) fabricated by a constant laser power of 10 μW and by different moving velocities from 9 μm/s to 6 μm/s. We can see that with velocities of 9 μm/s and 8 μm/s, typical hexagons with air-holes are obtained, while with velocities of 7 μm/s and 6 μm/s (higher dose), honeycombs with different filling factors are demonstrated. Thus, by using the simple scanning scheme shown in Figure 8a, and by controlling the scanning speed, various types of honeycomb lattices can be obtained.

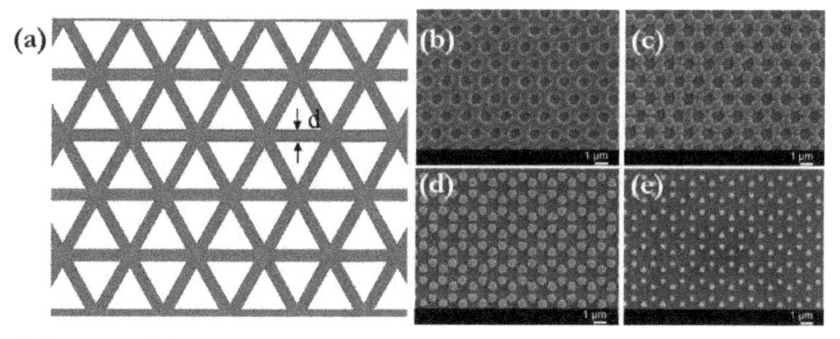

Figure 8. Fabrication of a honeycomb lattice by a simple direct laser writing method. (**a**) Illustration of a writing model in which the focused laser beam is scanned line by line with a period a, following a triangular configuration. The line width, d, is controlled by the writing parameters, such as laser power, writing speed, and focusing lens. The final structures vary from air-holes hexagonal structure to polymeric-dots honeycomb, depending on the writing doses. (**b**–**e**) SEM images of fabricated 2D structures (period, a = 1.5 μm), obtained with different doses. For these structures, the laser power was fixed at 10 μW and the scanning speeds were 9 μm/s (**b**), 8 μm/s (**c**), 7 μm/s (**d**), and 6 μm/s (**e**), respectively.

By the same method, assembled multi-rings lattices were fabricated by scanning the focusing laser beam following the trajectory shown in Figure 9a. Different structures have been demonstrated by changing the structures parameters, such as the ring width, d, the lattice constant, a, as well as the ring radius, r. Figure 9b–e show SEM images of four structures realized by keeping the same laser power, 10 μW while changing the moving velocities of laser beam from 9 μm/s to 6 μm/s. The lattice constant and the ring radius were chosen as 2 μm and 1.51 μm, respectively. When varying the scanning velocities, the ring width changes from 150 nm (Figure 9b), 200 nm (Figure 9c), 300 nm (Figure 9d), to 400 nm (Figure 9e), resulting in different 2D structures with different filling factors and structure compositions. This perfectly agrees with the theoretical prediction, and their PBGs are also demonstrated to be better than those obtained by the standard 2D PCs.

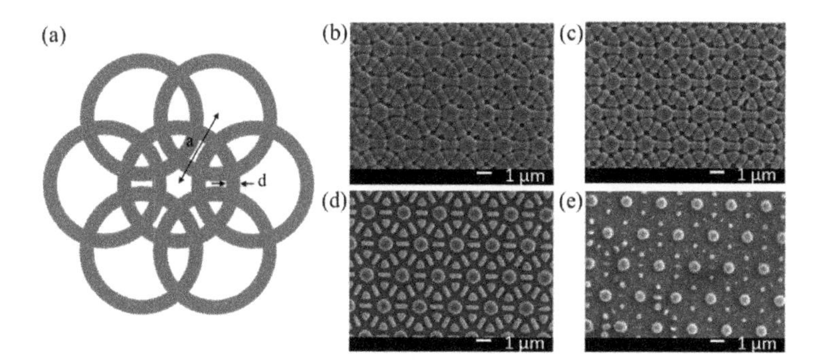

Figure 9. Fabrication of various high symmetry 2D photonic crystals by assembling multiple rings. (**a**) Illustration of the assembly of multi-rings, created by moving the focused laser beam, following a hexagonal configuration. The ring width, d, is controlled by the writing dose, and the rings are separated, from center to center, by a lattice constant a. By changing the exposure doses (laser power and writing speed) as well as the lattice constant, a, the final structures can be varied from a Penrose-like structure (**b**) to a hexagonal structure (**e**) consisting of different polymeric dots. (**b**–**e**) SEM images of 2D assembled-multi-rings lattices with a lattice constant a = 2 μm and fabricated by the same laser power of 10 μW. Different structures were obtained, as predicted by the simulations, with different ring widths d = 150 nm (**b**), 200 nm (**c**), 300 nm (**d**), and 400 nm (**e**), realized by different scanning speeds, 9 μm/s (**b**), 8 μm/s (**c**), 7 μm/s (**d**), and 6 μm/s (**e**), respectively.

Please note that it is not possible at this moment to experimentally characterize the PBGs of these fabricated structures, because the structures sizes are limited to 100 μm × 100 μm due to the scanning system of the DLW technique. Also, with a lattice constant of 2 μm, an assembled multi-rings PC produces a PBG in the range of 2.85–3.15 μm, which is in the mid-infrared domain. The limited PC thickness (1 μm) also makes the direct PBG characterization difficult. Further reduction of the lattice constant (a) and of the ring radius (r), and increase of the PCs thickness are therefore necessary to obtain a PBG in visible or near IR range, which can be very promising for many applications. A direct application of such fabricated structures is to make PC-based organic laser, with controllable emission wavelength and desired beam shape.

4. Conclusions

In conclusion, we have demonstrated an optimized two-dimensional photonic crystal with a better photonic bandgap at low refractive index contrast. The numerical calculations of the photonic bandgaps of honeycombs and assembled multi-rings lattices at low refractive index (n = 1.6) have been done by using FDTD method. The results showed that the TM mode has larger PBGs, as compared to the TE mode, and the largest PBG is approximately 0.089 (= $a\Delta\omega/2\pi c$) at r = 0.78a, obtained by the newly proposed assembled multi-rings photonic crystal. This is due to the perfect symmetry at local area of the photonic crystal, and the PBG calculation is possible thanks to the organization of multiple rings in a periodic structure. Both types of lattices, honeycomb and assembled multi-rings, have then been fabricated by using a low one-photon absorption-based direct laser writing technique, in a simple way. By controlling the exposure doses (laser power and exposure time), the lattice parameters, such as structure filling factor and lattice periodicity, were well controlled, allowing an optimization of the photonic bandgap.

Author Contributions: Q.C.T., T.Q.T. and N.D.L. proposed the idea and designed the experiments. T.-P.N. performed simulations and Q.C.T. realized experiments and analyzed the data. All authors wrote, reviewed and approved the final version of manuscript.

Funding: This research is supported by a public grant of Ministry of Science and Technology of Vietnam under the project [ĐTĐLCN.01/2017] and by a public grant overseen by the French National Research Agency in the frame of GRATEOM project [ANR-17-CE09-0047-01].

Conflicts of Interest: The authors declare no conflict of interest.

References

1. Joannopoulos, J.D.; Meade R.D.; Winn, J.N. *Photonic Crystals: Molding the Flow of Light*; Princeton University Press: Princeton, NJ, USA, 1995.
2. Akoda, K. *Optical Properties of Photonic Crystals*, 2nd ed.; Springer Science & Business Media: Berlin, Germany, 2004; ISBN 978-3-662-14324-7.
3. Gao, S.; Tang, X.; Langner, S.; Osvet, A.; Harrei, C.; Barr, M.K.S.; Spiecker, E.; Bachmann, J.; Brabec, C.J. Forberich, K. Time-resolved analysis of dielectric mirrors for vapor sensing. *ACS Appl. Mater. Interfaces* **2018**, *10*, 36398–36406 [CrossRef] [PubMed]
4. Lova, P. Selective polymer distributed bragg reflector vapor sensors. *Polymers* **2018**, *10*, 1161. [CrossRef] [PubMed]
5. Lova, P.; Manfredi, G.; Comoretto, D. Advances in functional solution processed planar 1D photonic crystals. *Adv. Opt. Mater.* **2018**, *6*, 1800730. [CrossRef]
6. Kang, H.S.; Lee, J.; Cho, S.M.; Park, T.H.; Kim, M.J.; Park, C.; Lee, S.W.; Kim, K.L.; Ryu, D.Y.; Huh, J. Printable and rewritable full block copolymer structural color. *Adv. Mater.* **2017**, *29*, 1700084. [CrossRef] [PubMed]
7. Lee, W.; Kim, S.; Kim, S.; Kim, J.-H.; Lee, H. Hierarchical opal grating films prepared by slide coating of colloidal dispersions in binary liquid media. *J. Colloid Interface Sci.* **2015**, *440*, 229–235. [CrossRef] [PubMed]
8. Kuo, W.-K.; Hsu J.-J.; Nien, C.-K.; Yu, H.H. Moth-eye inspired biophotonic surfaces with antireflective and hydrophobic characteristics. *ACS Appl. Mater. Interfaces* **2016**, *8*, 32021–32030 . [CrossRef] [PubMed]

9. Nucara, L.; Piazza, V.; Greco, F.; Robbiano, V.; Cappello, V.; Gemmi, M.; Cacialli, F.; Mattoli, V. Ionic strength responsive sulfonated polystyrene opals. *ACS Appl. Mater. Interfaces* **2017**, *95*, 4818–4827. [CrossRef] [PubMed]
10. Nien, C.-K.; Yu, H.H. The applications of biomimetic cicada-wing structure on the organic light-emitting diodes. *Mater. Chem. Phys.* **2019**, *227*, 191–199. [CrossRef]
11. Zoorob, M.E.; Charlton, M.D.B.; Parker, G.J.; Baumberg, J.J.; Netti, M.C. Complete photonic bandgaps in 12-fold symmetric quasicrystals. *Nature* **2000**, *404*, 740–743. [CrossRef] [PubMed]
12. Zhang, X.; Zhang, Z.Q.; Chan, C.T. Absolute photonic bandgaps in 12-fold symmetric photonic quasicrystals. *Phys. Rev. B* **2001**, *63*, 081105R . [CrossRef]
13. Lai, N.D.; Lin, J.H.; Huang, Y.Y.; Hsu, C.C. Fabrication of two- and three-dimensional quasi-periodic structures with 12-fold symmetry by interference technique. *Opt. Express* **2006**, *14*, 10746–10752. [CrossRef] [PubMed]
14. Jia, L.; Bita, I.; Thomas, E.L. Level set photonic quasicrystals with phase parameters. *Adv. Funct. Mater.* **2012**, *22*, 1150–1157. [CrossRef]
15. Deubel, M.; von Freymann, G.; Wegener, M.; Pereira, S.; Busch, K.; Soukoulis, C.M. Direct laser writing of three-dimensional photonic-crystal templates for telecommunications. *Nat. Mater.* **2004**, *3*, 444–447. [CrossRef] [PubMed]
16. Nguyen, T.; Do, M.T.; Li, Q.; Tong, Q.C.; Au, T.H.; Lai, N.D. One-photon absorption-based direct laser writing of three-dimensional photonic crystals. In *Theoretical Foundations and Application of Photonic Crystals;* Vakhrushev, A., Ed.; INTech: Rijeka, Croatia, 2018; pp. 133–157. ISBN 978-953-51-3962-1.
17. Tong, Q.C.; Nguyen, D.T.T.; Do, M.T.; Luong, M.H.; Journet, B.; Ledoux-Rak, I.; Lai, N.D. Direct laser writing of polymeric nanostructures via optically induced local thermal effect. *Appl. Phys. Lett.* **2016**, *108*, 183104. [CrossRef]
18. Anderson, C.M.; Giapis, K.P. Larger two-dimensional photonic band gaps. *Phys. Rev. Lett.* **1996**, *77*, 2949. [CrossRef] [PubMed]
19. Horiuchi, N.; Segawa, Y.; Nozokido, T.; Mizuno, K.; Miyazaki, H. Isotropic photonic gaps in a circular photonic crystal. *Opt. Lett.* **2004**, *29*, 1084. [CrossRef] [PubMed]
20. Chang, D.; Scheuer, J.; Yariv, A. Optimization of circular photonic crystal cavities—Beyond coupled mode theory. *Opt. Express* **2005**, *13*, 9272. [CrossRef] [PubMed]
21. Do, M.T.; Nguyen, T.T.N.; Li, Q.; Benisty, H.; Ledoux-Rak, I.; Lai, N.D. Submicrometer 3d structures fabrication enabled by one-photon absorption direct laser writing. *Opt. Express* **2013**, *21*, 20964–20969. [CrossRef] [PubMed]

Article

Bubbly Water as a Natural Metamaterial of Negative Bulk-Modulus

Pi-Gang Luan

Department of Optics and Photonics, National Central University, Jhongli District, Taoyuan City 32001, Taiwan; pgluan@dop.ncu.edu.tw

Received: 18 August 2019; Accepted: 30 August 2019; Published: 1 September 2019

Abstract: In this study, an oscillator model of bubble-in-water is proposed to analyze the effective modulus of low-concentration bubbly water. We show that in a wide range of wave frequency the bubbly water acquires a negative effective modulus, while the effective density of the medium is still positive. These two properties imply the existence of a wide acoustic gap in which the propagation of acoustic waves in this medium is prohibited. The dispersion relation for the acoustic modes in this medium follows Lorentz type dispersion, which is of the same form as that of the phonon-polariton in an ionic crystal. Numerical results of the gap edge frequencies and the dispersion relation in the long-wavelength regime based on this effective theory are consistent with the sonic band results calculated with the plane-wave expansion method (PWEM). Our theory provides a simple mechanism for explaining the long-wavelength behavior of the bubbly water medium. Therefore, phenomena such as the high attenuation rate of sound or acoustic Anderson localization in bubbly water can be understood more intuitively. The effects of damping are also briefly discussed. This effective modulus theory may be generalized and applied to other bubble-in-soft-medium type sonic systems.

Keywords: acoustic metamaterial; effective medium; bubble resonance; negative modulus

1. Introduction

Metamaterials are usually manmade structures designed to have some exotic properties that are rarely or never seen in natural materials. The split-ring resonators (SRRs) array is a well-studied structure that responds to incident electromagnetic waves in a range of frequencies like a negative-permeability material [1]. Similarly, the structure of periodically arranged rubber-shelled metal spheres embedded in an epoxy background is an acoustic metamaterial with negative mass density for low frequency sound [2]. In both cases, the exotic wave properties are caused by the coupling of the propagating wave in the background medium to the local resonance mechanism in each cell. The negativity of the material parameters in these two cases also implies the existence of a frequency gap [3]. No wave with frequency in the gap can propagate in this metamaterial if it occupies the whole space. If the metamaterial is distributed in a finite region, waves get attenuated before they penetrate through the medium.

Frequency gap and wave attenuation effects can also be achieved using the Bragg scattering [3] mechanism instead of local resonance. The bandgap of a usual photonic [4] or sonic crystal [5,6] is caused by the destructive interference between different 'component waves' in the Fourier series expansion of the Bloch mode. To make such a bandgap large, a high contrast of material parameters between the periodically distributed inclusions and the background material is required [7]. In addition, the center of the gap usually corresponds to a wavelength of about two to three times the lattice constant (spatial period) of the structure. This fact means that in order to isolate waves using non-resonant structures, the entire structure has to be bulky. Therefore, resonant structure is the better choice in this respect.

Recently, some researchers have noticed that bubble phononic crystals can attenuate sound very effectively, so they designed bubble phononic screens to isolate acoustic waves [8–11]. Similar to the negative-density metamaterial mentioned above, the high attenuation ability of the bubble phononic crystal is caused by the resonance coupling of elastic waves and bubble pulsation. In fact, a similar effect of bubble pulsation resonance in water is a well-known phenomenon and has been studied for decades [12]. The sound attenuation ability of bubbly water has also been discovered theoretically for about two decades [13,14]. To the best of our knowledge, the sound attenuation effect has been explained so far by the multiple scattering theory of waves when the bubble positions and bubble sizes are randomly distributed, whereas the concept of phononic bandgap was used if the bubbles were periodically arranged. We will show in this paper that bubbly water can be treated as a natural metamaterial, and the resonance coupling of the pressure wave with the bubble pulsation results in a negative modulus of the medium, giving a low frequency gap.

Acoustic gaps and dispersion effects in a short wavelength regime for sound waves propagating in liquid have also been studied recently and attracted much attention [15,16]. The phononic gaps at elevated frequencies emerge due to symmetry breaking in phonon interactions. These theoretical predictions bear a general character shared among soft and biological (non-crystalline) materials/metamaterials, as have been verified through inelastic X-ray scattering (IXS) experiments.

The metamaterial theory of bubbly water we developed in this paper is based on an oscillator model for a pulsating bubble in water. According to this model, the pulsating bubble can be regarded as a radially vibrating oscillator. The spherical bubble (the air region) acts as a spring, and a shell of water around the bubble is the mass connected to the spring. To derive the effective bulk modulus of the bubbly water, it is assumed that all the bubbles have the same size and are periodically distributed in the water background. For the sake of simplicity, we only consider simple cubic (SC) structures. Therefore, a cubic unit cell contains three regions: the air sphere, the mass shell (water), and the water outside the shell. We treat the material in each region as a lump material. We then derive the dynamical modulus by considering the radial oscillator motion and the pressure–volume (P–V) relation of the air region.

The purpose of this paper is to provide a simple theory to study the acoustic properties of bubbly water at low filling fraction and long wavelengths. The scattering function of a single air bubble in water and the corresponding oscillator model are derived in Section 2. We then use this model to derive the bulk modulus of the bubbly water in Section 3 and briefly discuss the effects of absorption and the possibility of generalizing this theory. Numerical results are presented in Section 4. Finally, we conclude this paper in Section 5.

2. Scattering Function and the Oscillator Model for a Single Bubble

In this section, we derive the acoustic scattering function f_s of a small air bubble for the incident plane wave. The resonance frequency ω_0 of the radial vibrating motion of the bubble can be read out directly from the scattering function. On the other hand, the radial stiffness constant C of the bubble can be derived by analyze the relation between the inner pressure and the radial displacement of the bubble. We then use the relation $C = m\omega^2$ to derive the effective mass m of the oscillator. The result obtained in this section will be used in the next section for constructing the dynamic model we mentioned in the last section.

2.1. Scattering Function of a Bubble

Suppose a spherical air bubble of radius r_a is surrounded by water, and the pulsation of the bubble is driven by a plane incident wave p_{inc} of wavelength much larger than the bubble size ($kr_a \ll 1$); here $k = 2\pi/\lambda$ is the wave number. In this limit the bubble scatters the incident wave isotropically, leading to the scattering wave of the form

$$p_{scat} = f_s p_{inc} G(r). \quad (1)$$

Here, $G(r) = e^{ikr}/r$ represents the spherical wave solution of the Helmholtz equation (with source) $\left(\nabla^2 + k^2\right)G(\mathbf{r}) = -4\pi\delta(\mathbf{r})$, and f_s is the scattering function to be determined later.

Although the bubble is assumed to be small, we also assume that the acoustic process is fast enough to avoid heat conduction between the bubble and the environment, thus the pulsation of the bubble can be approximated as an adiabatic process in the thermodynamics sense, satisfying the adiabatic P–V relation $PV^\gamma =$ constant. Here P is the interior pressure of the bubble, V is the bubble volume, and $\gamma = 1.4$ is the specific heat ratio for air. Under this assumption, the radial motion of the bubble satisfies

$$\frac{\Delta P}{P} = -\frac{\gamma \Delta V}{V} = -\frac{3\gamma \Delta R}{R}, \tag{2}$$

where ΔP, ΔV, and ΔR are the variation of the pressure, the volume, and the radius of the bubble, respectively, and R is the instantaneous radius of the bubble, respectively. Denote ΔP as p, ΔR as ξ, and assume both of them are small comparing to the static value of P and R. We can therefore rewrite Equation (2) as

$$p = -\frac{3\gamma P_0}{r_a}\xi = -\frac{3B_a}{r_a}\xi. \tag{3}$$

Here P_0 is the hydrostatic ambient pressure when the incident wave is absent, and $B_a = \gamma P_0$ is the bulk modulus of the air.

The pressure across the bubble surface must be continuous, therefore $p = p_{inc} + p_{scat}$. In addition, both p and ξ get a time factor $e^{-i\omega t}$ under the influence of the incident wave. Differentiate Equation (3) twice with respect to time, we get

$$\omega^2(p_{inc} + p_{scat}) = \frac{3\gamma P_0}{r_a}\ddot{\xi}. \tag{4}$$

However, the radial acceleration $\ddot{\xi}$ is related to the pressure gradient via

$$\rho_w \ddot{\xi} = -\frac{\partial p}{\partial R} = -\frac{\partial p_{scat}}{\partial R}. \tag{5}$$

Here ρ_w is density of the water, and the second equality is approximately correct under the long wavelength assumption. Combine Equations (1), (4), and (5) and use the fact $kr_a \ll 1$, we find

$$f_s = \frac{r_a}{\omega_0^2/\omega^2 - 1 - ikr_a}. \tag{6}$$

Here ω_0 is the Minnaert frequency, which is the resonance frequency of the 'bubble oscillator', given by

$$\omega_0 = \frac{1}{r_a}\sqrt{\frac{3\gamma P_0}{\rho_w}}. \tag{7}$$

As can be easily observed, the strongest scattering happens at a frequency very close to the resonance frequency ω_0. Therefore, the effective medium properties might be modified by the resonance. We will derive the effective bulk modulus in the next section.

2.2. The Stiffness Constant and the Mass of the Oscillator Model

The mechanism of this low frequency resonance can be understood intuitively. When the wavelength is much longer than the bubble radius, the pressure difference across the bubble itself is negligible, thus we can think of the bubble as a perfect spring. However, even under the long wavelength condition, the scattered wave rapidly drops in the radial direction, so there is a pressure difference between the bubble surface and an imaginary "outer surface" that surrounds the bubble and a water shell around the bubble. These two effects together cause the bubble to vibrate radially like an oscillator that is driven by a driving force. The bubble itself acts as a spring, the water shell is the mass, and the radial pressure difference is the driving force. The detailed derivation of this oscillator model can be found in [12]. Here we just derive the stiffness constant and the mass of the water shell for later use.

We denote the surface area $4\pi r_a^2$ of the bubble as S_i, and rewrite the bubble pressure p in Equation (3) as p_a, then according to Equation (3), the radial restoring force $F_r = p_a S_i$ is

$$F_r = -\frac{3B_a S_i}{r_a}\xi = -C\xi. \tag{8}$$

The radial stiffness constant C is related to the oscillator mass m through the relation $C = m\omega_0^2$, so we have

$$m = \frac{3B_a S_i}{\omega_0^2 r_a} = 4\pi r_a^3 \rho_w. \tag{9}$$

3. Effective Bulk Modulus and Dispersion Relation

In this section, we derive the effective bulk modulus of the bubbly water. For the sake of simplicity, we assume all bubbles are of the same size, and their centers are located periodically on the cites of a simple cubic lattice. We denote the bubble volume and the cell volume as V_a and V_c, and the water volume $V_c - V_a$ in one unit cell as V_w. The volume filling fraction of the bubbles is denoted as f (do not confuse it with the scattering function f_s). Thus, we have $f = V_a/V_c$ and $1 - f = V_w/V_c$. We also denote the bulk modulus for air and water as B_a and B_w, respectively. Similar notation rules also apply to the densities ρ_a, ρ_w, the vibration displacements $\boldsymbol{u}_a$, $\boldsymbol{u}_w$, and the vibration velocities $\boldsymbol{v}_a = \dot{\boldsymbol{u}}_a$, $\boldsymbol{v}_w = \dot{\boldsymbol{u}}_w$. In addition, the bubble surface is denoted either as S_a or S_i, and the outer surface of the 'mass shell' is denoted as S_o. Since the volume enclosed by S_o is four times of the bubble volume, and the 'outer radius' r_o corresponding to S_o is related to the bubble radius r_a as $r_0 = 2^{2/3} r_a$. Our goal is to derive the effective density of the medium ρ_{eff} and the effective bulk modulus B_{eff} for use in the following two equations

$$\frac{\partial}{\partial t}\left(\rho_{eff}\boldsymbol{v}\right) = -\nabla p, \quad \nabla \cdot \boldsymbol{v} = -\frac{\partial}{\partial t}\left(\frac{p}{B_{eff}}\right). \tag{10}$$

The left equation is Newton's second law expressed in density form. Here $\rho_{eff}\boldsymbol{v}$ is the effective momentum density of the vibrating medium, and $-\nabla p$ plays the role of the force density. Any component of ∇p is calculated from the pressure difference between the corresponding two boundaries of the cell, divided by the lattice constant. The right equation is in fact the time derivative of the equation $p = -B_{eff}\nabla \cdot \boldsymbol{u}$, which describes the relation between the volume variation rate $\frac{\Delta V_c}{V_c} = \nabla \cdot \boldsymbol{u}$ and the corresponding pressure variation p (see Figure 1). With these two equations we can derive the wave equation for monochromatic (single frequency) waves in this sound-bubble coupling system and the dispersion relations of the propagating modes. The results provide simple explanation of some dramatic wave phenomena such as the very high attenuation rate of sound and Anderson localization, both of which have not yet been explained using the simple concept of frequency-dependent effective bulk modulus.

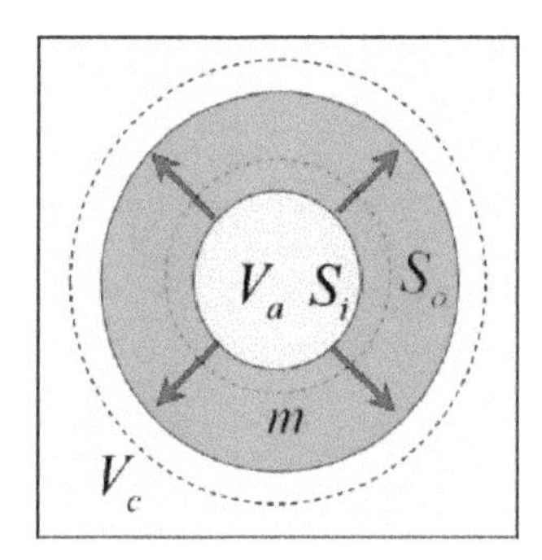

Figure 1. Oscillator model for deriving the effective bulk modulus and effective density. The meaning of the notations are explained in Section 3.

3.1. The Pressure–Volume Relations and the Effective Bulk Modulus

We begin with the following relation about the volumes:

$$\frac{\Delta V_c}{V_c} = \frac{\Delta V_a + \Delta V_w}{V_c} = f\frac{\Delta V_a}{V_a} + (1-f)\frac{\Delta V_w}{V_w}. \quad (11)$$

Now replace the volume variation rates by the corresponding divergence of the displacements, and apply the relations $\nabla \cdot \boldsymbol{u} = -p/B$, for them, we find

$$\nabla{\cdot}\mathbf{v} = -\frac{\partial}{\partial t}\left[f\frac{p_a}{B_a} + (1-f)\frac{p_w}{B_w}\right]. \quad (12)$$

Comparing Equation (12) with Equation (10), and making use of the equality $p_w = p$, we get

$$\frac{1}{B_{eff}} = \frac{f}{B_a}\left(\frac{p_a}{p}\right) + \frac{1-f}{B_w}. \quad (13)$$

The ratio p_a/p will be derived in the next subsection using the oscillator model.

3.2. The Dynamics Equation of the Radial Vibrating Oscillator

We now study the radial motion of the bubble. The mass of the oscillator can be thought of as a water shell of constant density ρ_w between the inner surface S_i and the outer surface S_o. The total radial force $p_aS_i - pS_o$ drives the radial motion of the mass shell, satisfying the equation of motion:

$$p_aS_i - pS_o = m\ddot{\xi}. \quad (14)$$

Applying the sinusoidal condition to ξ, (ξ has a time factor $e^{-i\omega t}$), we get

$$m\ddot{\xi} = -m\omega^2 r_a\left(\frac{\xi}{r_a}\right) = -\frac{m\omega^2 r_a}{3}\left(\frac{\Delta V_a}{V_a}\right) = \frac{m\omega^2 r_a}{3}\left(\frac{p_a}{B_a}\right). \quad (15)$$

Substitute Equation (15) into Equation (14), we find

$$\frac{p_a}{p} = \frac{3B_aS_o}{mr_a\left(\omega_0^2 - \omega^2\right)}; \quad (16)$$

here we have used the alternative expression of the Minnaert frequency

$$\omega_0 = \sqrt{\frac{3B_aS_i}{mr_a}}, \quad (17)$$

which can be derived from Equation (7), Equation (9), and the relation $B_a = \gamma P_0$.

3.3. The Effective Bulk Modulus as a Function of Frequency

Now it is straightforward to derive the frequency-dependent bulk modulus. Substitute Equation (16) into Equation (13), we get

$$\frac{1}{B_{eff}} = \frac{1-f}{B_w} + \frac{3fS_o}{mr_a\left(\omega_0^2 - \omega^2\right)}, \tag{18}$$

which can be written as

$$\frac{1}{B_{eff}} = \frac{1}{B_\infty}\left(\frac{\omega^2 - \omega_1^2}{\omega^2 - \omega_0^2}\right) = \frac{1}{B_\infty}\left(1 - \frac{F\omega_0^2}{\omega^2 - \omega_0^2}\right), \tag{19}$$

where the parameters B_∞, F, and ω_1^2 are given by

$$B_\infty = \frac{B_w}{1-f}, \quad F = \frac{\omega_1^2 - \omega_0^2}{\omega_0^2}, \quad \omega_1^2 = \omega_0^2 + \frac{3f}{1-f}\left(\frac{S_0 B_w}{mr_a}\right). \tag{20}$$

It is clear from Equation (19) that the inverse bulk modulus B_{eff}^{-1} as a function of frequency has the same form as the dielectric function in the Lorentz model, or that for the polariton wave in the polar medium. According to this observation we learn immediately that the effective bulk modulus for the bubbly water medium becomes negative in the frequency range $\omega_0 < \omega < \omega_1$. This frequency interval is also the frequency gap or forbidden band of the propagating modes.

The bulk modulus of air is given by $B_a = \rho_a c_a^2$, where c_a is the sound speed in air. Similarly, $B_w = \rho_w c_w^2$ gives the bulk modulus of the water. Substitute these two relations and $S_o/m = 2^{4/3}/\rho_w = 2.52/\rho_w$ into Equation (20), we get

$$B_\infty = \frac{\rho_w c_w^2}{1-f}, \quad F = \frac{2.52f}{1-f}\left(\frac{\rho_w c_w^2}{\rho_a c_a^2}\right), \quad \omega_1 = \sqrt{1+F}\omega_0. \tag{21}$$

The bulk modulus ratio between water and air is about 1.5×10^4, thus according to Equation (21) even a low filling fraction f will give a large F, corresponding to a wide gap. We will derive in the next section the dispersion relation of the propagating modes and discuss the effect of the frequency gap.

3.4. Dispersion Relation of Acoustic Modes

To determine the dispersion relation, besides the bulk modulus, we should know the effective mass density. Since we assume the wavelength is long enough, it is reasonable to assume that the bubble moves together (in phase) with the surrounding water in the same cell. This consideration leads us to the simple average result

$$\rho_{eff} = f\rho_a + (1-f)\rho_w. \tag{22}$$

The correctness of this expression will be tested later.

The wave equation for a monochromatic (single frequency) wave can be derived by substituting Equation (19) and Equation (22) into Equation (10). The result is

$$\nabla^2 p - \frac{\rho_{eff}}{B_{eff}}\frac{\partial^2 p}{\partial t^2} = 0. \tag{23}$$

Here the ratio ρ_{eff}/B_{eff} before the double time derivative is the inverse square speed of sound in the bubbly water medium. Adopting water as the standard medium or reference 'wave vacuum', we can define the refractive index of this medium as

$$n(\omega) = c_w\sqrt{\frac{\rho_{eff}}{B_{eff}}} = n_\infty\sqrt{\frac{\omega^2-\omega_1^2}{\omega^2-\omega_0^2}}. \tag{24}$$

Here $n_\infty = c_w\sqrt{\rho_{eff}/B_\infty}$ is a reference index at high frequency. This refractive index becomes imaginary in the frequency gap, which indicates that waves cannot propagate in the medium if its frequency is within the gap.

We can deepen our understanding of the refractive index function by studying the dispersion relation of the propagating modes. Substitute the plane wave form $p = p_0 e^{i(\boldsymbol{K}\cdot\boldsymbol{r}-\omega t)}$ into Equation (23), we get

$$K^2 = \frac{n^2(\omega)\omega^2}{c_\omega^2} = \frac{n_\infty^2\omega^2}{C_\omega^2}\left(\frac{\omega^2-\omega_1^2}{\omega^2-\omega_0^2}\right). \tag{25}$$

This dispersion relation is of the same form as that for the phonon-polariton waves. According to Equation (25), propagating modes can exist with frequency below ω_0 or above ω_1. Within the interval $\omega_0 < \omega < \omega_1$, the wave vector becomes imaginary, thus the wave amplitude decays along the wave vector direction. If the bubbly water medium is filled in a slab region of finite thickness, an incident wave having a frequency within the gap would get attenuated before it penetrates through the slab. Furthermore, if a pulsating source of sound surrounded by the 'bubble cloud' radiates sound waves of frequency in the gap, the sound wave would be trapped by the bubble cloud, leading to the phenomena such like the Anderson localization of acoustic waves.

The negative modulus of the bubbly water is caused by the dynamical coupling of the sound field in water and the local resonance of every bubble oscillator. Similar mechanism for polar crystal also leads to the negative dielectric function in the polariton gap. The main difference between these two systems is that in the bubbly water case the oscillators vibrate radially so they couple with the scalar pressure field, whereas in the polar crystal case the local dipoles (for example, the $Na^+ - Cl^-$ pairs) are coupled with the electric vector field.

3.5. *Absorption Effect and the Possibility of Generalization*

In the derivation of the effective bulk modulus, we never considered any absorption or loss effects. For real bubbles, the damping effect caused by thermal conductivity (non-adiabatic correction), shear viscosity absorption at the bubble surface, and the radiation loss due to scattering must be considered. These corrections will cause the resonance frequency of the bubble to change slightly, reducing the peak of the scattering function and widening its width. These corrections will also introduce a damping force on the radial motion of the bubble oscillator and change the bulk modulus such that it does not acquire an unphysical infinite value. Since the absorption effect is a very cumbersome effect, and our aim in this study is not to study this effect, we will not further address this issue. We follow the spirit of [13,14] to choose the proper bubble size so that we can avoid the effects of the absorption.

Another question is how to generalize our theory to elastic solid media containing bubbles. As the authors have demonstrated in [8–11], for elastic media with very low shear stress, the primary mechanism of the first band gap is still caused by the resonance of the bubble. Therefore, we believe that our theory can be generalized and applied to such a system without much difficulty.

4. Numerical Results

In this section we discuss some numerical results based on our theory. Hereafter we assume $\rho_a = 1.2\ \text{kg/m}^3$, $\rho_w = 1000\ \text{kg/m}^3$, $c_a = 343\ \text{m/s}$, and $c_w = 1490\ \text{m/s}$. These parameters give us the bulk modulus $B_a = \rho_a c_a^2 = 1.412 \times 10^5$ Pa and $B_w = \rho_w c_w^2 = 2.22 \times 10^9$ Pa. If the radius of a bubble is measured in mm, the Minnaert frequency measured in kHz and we get a simple formula $\omega_0 = 20.58/r_a$ or $\nu_0 = \omega_0/2\pi = 3.28/r_a$. To avoid high absorption, we assume the bubbles have radius larger than 100 μm [12,13].

The scattering function f_s as a function of dimensionless frequency $kr_a = \omega r_a/c_w$ is plotted in Figure 2. Here we replaced the f_s function by its dimensionless correspondence f_s/r_a because here the information of the size is irrelevant. According to Figure 1, the absolute value as well as the imaginary part of f_s goes to the maximum value $72.5r_a$ at $kr_a = 0.0138$, and behaves like an even function with respect to this peak location. The real part of f_s, on the other hand, first raises to $36.6r_a$, and then drops to zero at this place, and continually goes downwards to $-36.5r_a$, then approaches the horizontal axis without crossing the zero value. Resonance of this kind belongs to the Lorentz dispersion, which is responsible for the negative bulk modulus and the low frequency gap of this bubbly water medium.

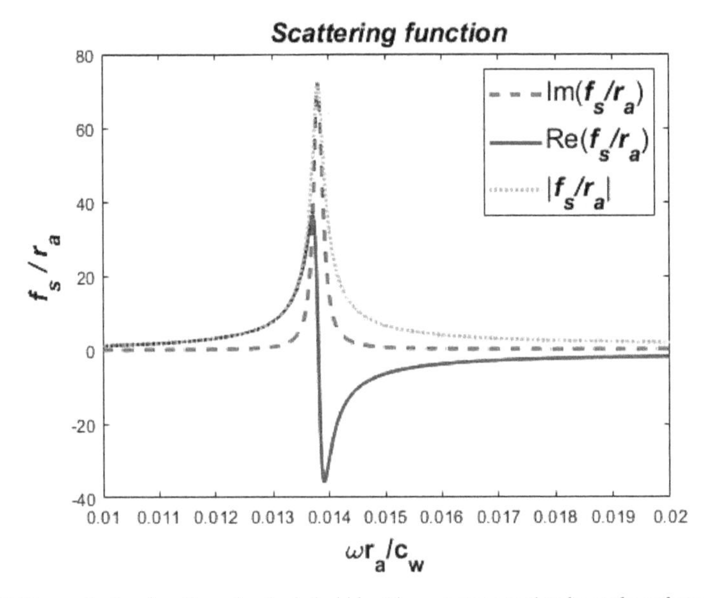

Figure 2. The scattering function of a single bubble. The notations in this figure have been defined in Section 2.

We now study the characteristics of the acoustic band structures of the bubbly-water sonic crystals. Plane wave expansion method (PWEM) were implemented for the calculations. Results for filling fraction $f = 0.001$ and $f = 0.01$ are shown in Figure 3. These results indicate that bubbly-water sonic crystal is an ideal structure to open large gap at low frequency. For the case with filling fraction $f = 0.001$, the gap is opened between $\omega_L r_a/c_w = 0.0185$ (at X) and $\omega_U r_a/c_w = 0.085$ (at Γ). Comparing them with the gap boundaries $\omega_0 r_a/c_w = 0.0138$ and $\omega_1 r_a/c_w = 0.0881$ predicted by the effective metamaterial theory, we find excellent agreement between them. The low frequency limit of the gap in Figure 3 is a little higher than the prediction. We believe this discrepancy is due to the bad convergence of the PWEM at low filling fraction. Similarly for the $f = 0.01$ case, we have the gap opened from $\omega_L r_a/c_w = 0.0173$ (at X) to $\omega_U r_a/c_w = 0.271$ (at ¡). The gap boundaries provided by the effective theory are $\omega_0 r_a/c_w = 0.0138$ and $\omega_1 r_a/c_w = 0.277$, also in very good agreement with the band structure result. We noticed that the agreement between ω_L and ω_0 in this case is better than in the former case. This fact is consistent with our judgement about the PWEM.

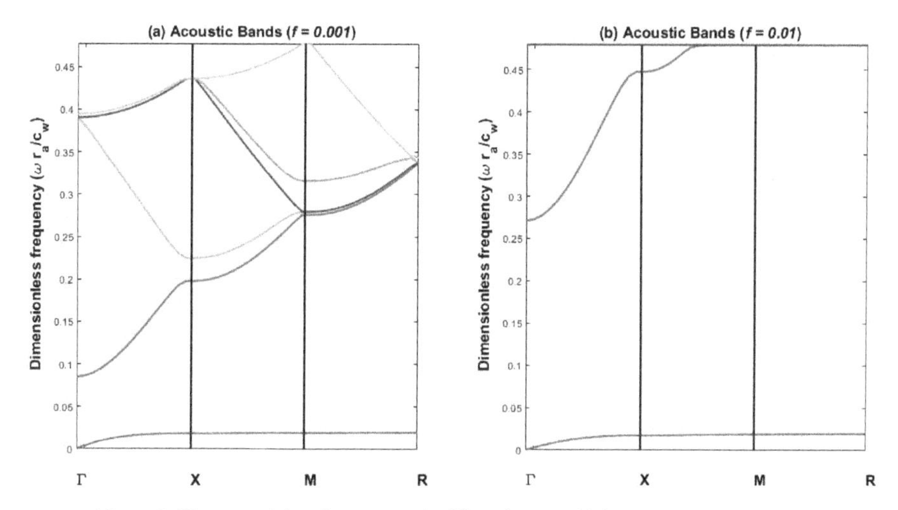

Figure 3. The acoustic band structures for filling fraction (**a**) f = 0.001, and (**b**) f = 0.01.

The metamaterial theory developed in this paper not only can provide the range of the gap, but also can give us the attenuation rate of the wave, which is directly related to the imaginary part of the effective wave number that determined by Equation (25). In addition, absorption or loss effect can be easily included by a simple replacement of the frequency: $\omega \rightarrow \omega(1 + i\delta)$ [12]. Here δ is a positive number smaller than 0.1 if we assume the that the bubble radius is larger then 100^1 m. For the purpose of demonstration, in the simulation of lossy media, we take $\delta = 0.05$. Figure 4a,b is the results for the lossless media, while Figure 4c,d is the results for the lossy media. Here the dimensionless variable *Ka* is the product of the wavenumber *K* in Equation (25) times the lattice constant *a* of the SC structure. the lattice constant in the SC structure. The real part of the wave number (wave vector) of the mode waves are represented by the blue solid curves, and the red broken lines represent the imaginary part of the wave vector. These dispersion relations are similar to the dispersion relations of the phonon-polariton that mentioned in the previous section. We noticed that the presence of loss or absorption does not destroy the attenuation effect because it is so large.

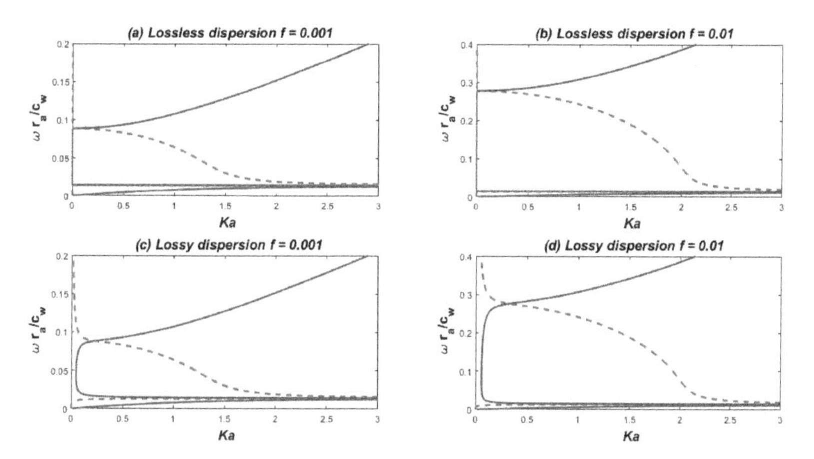

Figure 4. Dispersion relations for f = 0.001 and f = 0.01, lossless media are shown in (**a**) and (**b**). The corresponding results for lossy media are shown in (**c**) f = 0.001 and (**d**) f = 0.01.

Now we discuss the effective bulk modulus according to Equation (19) and its modified form when loss cannot be ignored. As before, four cases are studied, and their results (B_{eff}/B_w) are shown in Figure 5. The parameters used in the calculations are the same as those used in Figure 4. The small circles in the four subplots indicate the starting frequencies (changes from positive to negative) of the negative bulk modulus. For the lossless cases, the starting points and the resonance points (infinite effective modulus) are the same as the gap boundaries of the metamaterial. The loss effect seemed not to change the effective bulk modulus much if the frequency is not very close to the resonance point. However, the effective modulus at the resonance frequency no longer goes to the unphysical value of infinity, as expected.

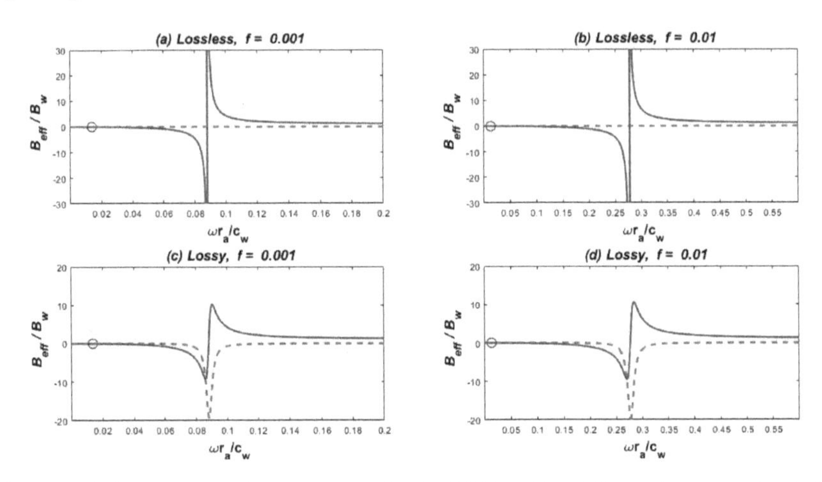

Figure 5. Bulk modulus for lossless cases with f = 0.001 and f = 0.01 are shown in (**a**) and (**b**). The corresponding lossy cases are shown in (**c**) f = 0.001 and (**d**) f = 0.01.

Finally, we discuss the valid range of the theory developed in this paper. Based on our numerical simulations, we found that the gap range given by the effective theory is in good agreement with that obtained from the band structure calculation (the discrepancy is less than 3%) if the filling fraction f of the bubbles is within the range 0.0005 to 0.02. Beyond this range, we are not sure about the accuracy, and need further investigation. In addition, the bubble radius must not be less than 100 µm, otherwise high absorption effects must be considered [12].

5. Conclusions

In this paper, we apply the oscillator model of air-bubble-in-water to the bubbly water medium and derive the effective bulk modulus of the medium. We show that in a wide range of frequency the effective modulus becomes negative, while the effective density of the medium is still positive. These properties imply the existence of a wide acoustic gap, consistent with band structure calculations. The dispersion relation for the acoustic modes in this medium has Lorentz type dispersion, like that of the phonon-polariton. The theory proposed in this paper provides a simple mechanism for explaining the long-wavelength behavior of the bubbly water medium. Using this theory, phenomena such as the high attenuation rate through a bubble screen or the acoustic Anderson localization in bubbly water can be understood more intuitively. The effects of damping are also discussed. This effective modulus theory may be generalized and applied to other similar structural sonic medium containing periodically arranged bubbles.

Funding: This research received no external funding.

Conflicts of Interest: The author declares no conflict of interest.

References

1. Smith, D.R.; Willie, J.P.; Vier, D.C.; Nemat-Nasser, S.C.; Schultz, S. Composite Medium with Simultaneously Negative Permeability and Permittivity. *Phys. Rev. Lett.* **2000**, *84*, 4184. [CrossRef] [PubMed]
2. Liu, Z.; Zhang, X.; Mao, Y.; Zhu, Y.Y.; Yang, Z.; Chan, T.C.; Sheng, P. Locally Resonant Sonic Materials. *Science* **2000**, *289*, 1734. [CrossRef] [PubMed]
3. Kittel, C. *Introduction to Solid State Physics*, 8th ed.; John Wiley &Sons Inc.: Hoboken, NJ, USA, 2005.
4. Joannopoulos, J.; Johnson, S.; Winn, J.; Meade, R. Photonic Crystals: Molding the Flow of Light. Princeton University Press, 2008. Available online: http://ab-initio.mit.edu/book/photonic-crystals-book.pdf (accessed on 31 August 2019).
5. Sánchez-Pérez, J.V.; Caballero, D.; Mártinez-Sala, R.; Rubio, C.; Sánchez-Dehesa, J.; Meseguer, F.; Llinares, J.; Gálvez, F. Sound Attenuation by a Two-Dimensional Array of Rigid Cylinders. *Phys. Rev. Lett.* **1998**, *80*, 5325. [CrossRef]
6. Kushwaha, M.S. Stop-bands for periodic metallic rods: Sculptures that can filter the noise. *Appl. Phys. Lett.* **1997**, *70*, 3218. [CrossRef]
7. Kushwaha, M.S.; Halevi, P. Giant acoustic stop bands in two-dimensional periodic arrays of liquid cylinders. *Appl. Phys. Lett.* **1996**, *69*, 31. [CrossRef]
8. Leroy, V.; Bretagne, A.; Fink, M.; Willaime, H.; Tabeling, P.; Tourin, A. Design and characterization of bubble phononic crystals. *Appl. Phys. Lett.* **2009**, *95*, 171904. [CrossRef]
9. Bretagne, A.; Tourin, A.; Leroy, V. Enhanced and reduced transmission of acoustic waves with bubble meta-screens. *Appl. Phys. Lett.* **2011**, *99*, 221906. [CrossRef]
10. Leroy, V.; Bretagne, A.; Lanoy, M.; Tourin, A. Band gaps in bubble phononic crystals. *AIP Adv.* **2016**, *6*, 121604. [CrossRef]
11. Leroy, V.; Chastrette, N.; Thieury, M.; Lombard, O.; Tourin, A. Acoustics of Bubble Arrays: Role Played by the Dipole Response of Bubbles. *Fluids* **2018**, *3*, 95. [CrossRef]
12. Clay, C.S.; Medwin, H. *Acoustical Oceanography: Principles and Applications*; John Wiley & Sons: Hoboken, NJ, USA, 1977.
13. Ye, Z.; Hsu, H. Phase transition and acoustic localization in arrays of air bubbles in water. *Appl. Phys. Lett.* **2001**, *79*, 1724. [CrossRef]
14. Kafesaki, M.; Penciu, R.S.; Economou, E.N. Air Bubbles in Water: A Strongly Multiple Scattering Medium for Acoustic Waves. *Phys. Rev. Lett.* **2000**, *84*, 6050. [CrossRef] [PubMed]
15. Bolmatov, D.; Zhernenkov, M.; Zav'yalov, D.; Stoupin, S.; Cunsolo, A.; Cai, Y.Q. Thermally triggered phononic gaps in liquids at THz scale. *Sci. Rep.* **2016**, *6*, 19469. [CrossRef] [PubMed]
16. Bolmatov, D.; Zhernenkov, M.; Zav'yalov, D.; Stoupin, S.; Cai, Y.Q.; Cunsolo, A. Revealing the mechanism of the viscous-to-elastic crossover in liquids. *J. Phys. Chem. Lett.* **2015**, *5*, 2785–2790. [CrossRef] [PubMed]

Article

All-Optical Ultra-Fast Graphene-Photonic Crystal Switch

Mohammad Reza Jalali Azizpour [1], Mohammad Soroosh [1,*], Narges Dalvand [2] and Yousef Seifi-Kavian [1]

[1] Department of Electrical Engineering, Shahid Chamran University of Ahvaz, Ahvaz 61357-83135, Iran
[2] Electrical Engineering Department, K. N. Toosi University of Technology, Tehran 19697-64499, Iran
* Correspondence: m.soroosh@scu.ac.ir

Received: 30 June 2019; Accepted: 28 August 2019; Published: 3 September 2019

Abstract: In this paper, an all-optical photonic crystal-based switch containing a graphene resonant ring has been presented. The structure has been composed of 15 × 15 silicon rods for a fundamental lattice. Then, a resonant ring including 9 thick silicon rods and 24 graphene-SiO_2 rods was placed between two waveguides. The thick rods with a radius of 0.41a in the form of a 3 × 3 lattice were placed at the center of the ring. Graphene-SiO_2 rods with a radius of 0.2a were assumed around the thick rods. These rods were made of the graphene monolayers which were separated by SiO_2 disks. The size of the structure was about 70 μm^2 that was more compact than other works. Furthermore, the rise and fall times were obtained by 0.3 ps and 0.4 ps, respectively, which were less than other reports. Besides, the amount of the contrast ratio (the difference between the margin values for logics 1 and 0) for the proposed structure was calculated by about 82%. The correct switching operation, compactness, and ultra-fast response, as well as the high contrast ratio, make the presented switch for optical integrated circuits.

Keywords: graphene; kerr effect; optical switch; photonic band gap; photonic crystal

---

1. Introduction

In recent years, increased demand for ultra-fast processing systems has necessitated the development of optical devices. Many attempts have been accomplished to introduce all-optical structures that allow circuits operating at a frequency of terahertz. Achieving a high data transferring rate with a possibility for integration is among the important issues in designing all-optical circuits [1,2].

Photonic crystals (PCs) were initially and simultaneously proposed by Yablonocvitch [3] and John [4] in 1987. PCs are structures composed of periodic layers with different dielectric constants and demonstrate useful characteristics such as photonic band gap (PBG) [3], slow light regime [5–7], and self-collimation [8–11]. Because of their compact size and the aforementioned properties, they are promising candidates in realizing all-optical devices such as optical waveguides [10–12], filters [13–18], demultiplexers [19–25], switches [26–30], logic gates [31–36], encoders [37–42], decoders [43–48], and analog-to-digital converters [49–54].

In photonic integrated circuits (PICs), optical switches play a crucial role in guiding light to the desired directions. Generally, optical switches include ring resonators, input and output ports that drop light with a particular wavelength. PCs-based switches have been accomplished by using the Kerr nonlinear effect [28,55]. Hache et al. developed the first concept using the Kerr nonlinearity effect in PIC. The proposed structure included two materials, Si and SiO_2, with a 1.5 μm stop band and 18 Gw/cm^2 functional power. Although their structure shed light in the all-optical circuits area, the level of power was not applicable for PICs [56]. Another optical switch suggested by Alipour-Banaei et al. consisted of the chalcogenide material with the nonlinear coefficient 9×10^{-17} m^2/W. The threshold power level for switching operation and the footprint were 2 $KW/\mu m^2$ and 259 μm^2, respectively [57].

Serajmohammadi et al. proposed a PC-based NAND gate using an optical switch. The structure was made of one bias signal, two input ports, and one output port. Using the bias signal, the Kerr effect was approached when both input signals were introduced into the switch. They also used the chalcogenide glass to gain the nonlinear properties for a resonant ring. Even though the threshold power and the footprint reduced to 1.5 KW/μm^2 and 230 μm^2, respectively, the power was so high to employ in integrated circuits [58]. In this way, Alipour-Banaei et al. proposed another optical switch by a focus on the power issue [59]. Although they approached the power level 1.5 KW/μm^2, the footprint was notably increased to as large as 360 μm^2.

Recently, Daghooghi et al. developed a PC-based switch for a decoding operation which used a 31 × 31 lattice of silicon rods in the air [60]. They employed nanocrystal material in which the nonlinear coefficient was 10^{-16} m^2/W and succeeded in reducing the threshold power level to 13 W/μm^2, while the footprint was considerably increased.

As mentioned, many attempts have been done to reduce the threshold power level for switching, as well as the size of the structure. The possibility of fabrication for the proposed structures is an important challenge that restricts researchers in selecting the different materials and sizes. So, the compatibility with conventional technologies helps researchers achieve new materials with distinct characteristics. Graphene has a substantially more Kerr coefficient than other materials such as chalcogenide and nanocrystal [35,61,62]. Several experiments and calculations have been done for determination of Kerr coefficient of graphene [63–70]. Soh et al. have shown the Kerr coefficient of graphene at a wavelength of 1550 nm is equal to 10^{-15} m^2/W [62]. They presented a comprehensive analysis for nonlinear optical Kerr effect in graphene including two-photon absorption, Raman transition, self-coupling, and quadratic AC Stark effect. They calculated the absorption rate using the S-matrix element and converted it to nonlinear refractive index coefficients. Then, they obtained the rates of distinct nonlinear processes that contribute to the Kerr nonlinear refractive index coefficient. Due to this fact, using graphene in PICs results in decreasing the functional optical intensity in comparison to other materials such as chalcogenide and nanocrystal. This is the main reason why we are utilizing graphene. Moreover, graphene has an impact on two other factors in PICs fabrication: contrast ratio and footprint.

In this study, a new graphene-based switch is proposed to enhance the power and footprint difficulties. The switch is composed of 9 thick silicon rods at the center of the resonant ring and 24 graphene-SiO_2 stacks as the nonlinear rods around them. Each nonlinear rod is made of the graphene monolayers which are separated by SiO_2 disks. Using graphene monolayers enhances the light-material interaction and results in decreasing the threshold intensity for switching operation. Furthermore, the more Kerr coefficient of graphene than chalcogenide and nanocrystal used in the previous works assists with enhancing the power difficulty in this work. The obtained results of the simulation present 70 μm^2 and 0.235 W/μm^2 for the footprint and the threshold intensity level, respectively. Furthermore, the normalized output power margins, 4% and 86% for logic 0 and 1 illustrate that the structure is potentially a good candidate in PIC applications.

In this paper, three other sections have been organized as follows; in Section 2, the structure will be presented in detail. Then, the simulation results will be provided in Section 3. In Section 4, the evaluation of the device and discussions will be described, along with a comparison of the obtained results with other works. Finally, a conclusion of the work will be presented in Section 5.

2. Materials and Methods

The primary structure includes a lattice of 15 × 15 rods in air background where the lattice constant is 558 nm. So, the overall size of the structure is about 70 μm^2. The refractive index of rods is 3.46, and the radius of rods for the fundamental structure is 0.2a, where a is the lattice constant. To calculate the band structure, the plane wave expansion (PWE) method is used [71]. According to the band structure (as shown in Figure 1), two PBGs at TE mode are obtained for $0.29 \leq a/\lambda \leq 0.42$ and $0.725 \leq a/\lambda \leq 0.74$ where λ is the wavelength. So, for 1290 nm $\leq \lambda \leq$ 1990 nm optical waves could not be allowed for propagation throughout the structure. Including the third optical communication window, the last interval is used for the proposed device.

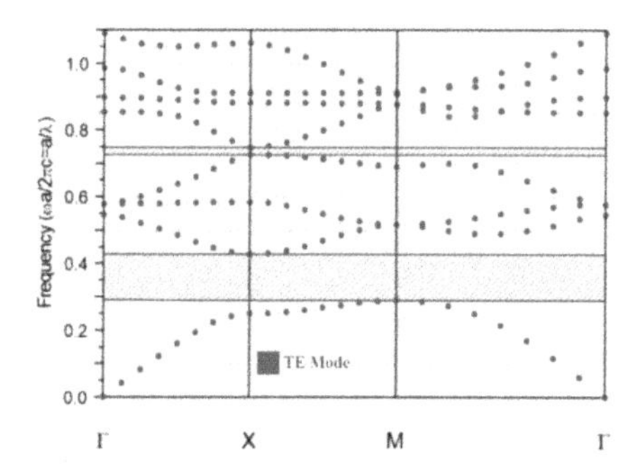

Figure 1. The band diagram of the proposed structure.

The formation of the switch was gained by removing and changing the radius of some rods in X and Z directions. One of the best features of the proposed structure is that there is not any bias or enable port. It can be seen in Figure 2 that there are two output ports (named as O1 and O2) and one input port (named as IN) that are connected via a resonant ring. A 3 × 3 lattice rod formed by changing in the radius 0.41a was placed in the center of the ring. Around the core, 24 graphene-SiO_2 rods (with blue color) were used in which the linear refractive index (n_1) and nonlinear coefficient (n_2) of graphene were 2.6 and 10^{-15} m^2/W, respectively [62]. These rods are made of the graphene monolayers which are sandwiched between SiO_2 disks. Berman et al. presented a photonic crystal structure including a periodic array of the graphene-SiO_2 stacks [72]. They calculated the transmission coefficient for different frequencies and showed that the structure can be used as frequency filters and waveguides for optical waves. According to their research, we used the graphene monolayers which were separated by SiO_2 disks for blue color rods. Using graphene in the ring resonator causes the possibility of achieving the nonlinear Kerr effect. The optical Kerr effect is generally known as the changing in the refractive index (n) in response to applied light intensity (I) and is defined by $n = n_1 + n_2 I$ [65,66]. The nonlinear coefficient of graphene is defined as [73]:

$$n_2 = \frac{3\eta}{n_1^2 \epsilon} \chi^{(3)}$$

where n_1 is the linear refractive index, η is the impedance, and $\chi^{(3)}$ is the nonlinear susceptibility. The nonlinear susceptibility has been described as follows [65]:

$$\chi^{(3)} = \frac{\sigma_g(\omega)}{\omega d_g}$$

where ω is the frequency and d_g is the thickness of the graphene monolayer. The dielectric constant is defined as [72]:

$$\varepsilon(\omega) = \varepsilon_0 + \frac{4\pi i \sigma_g(\omega)}{\omega d}$$

where ε_0 is the dielectric constant of SiO_2, d is the width of SiO_2 layer and equal to 1 nm. The dynamical conductivity of the graphene (σ_g) is defined by the following equation [72]:

$$\sigma_g(\omega) = \frac{e^2}{4\hbar}\left[\eta(\hbar\omega - 2\mu) + \frac{i}{2\pi}\left(\frac{16K_B T}{\hbar\omega} log\left[2cosh\left(\frac{\mu}{2K_B T}\right)\right] - log\frac{(\hbar\omega + 2\mu)^2}{(\hbar\omega - 2\mu)^2 + (2K_B T)^2}\right)\right]$$

where e is the electron charge, K_B is the Boltzmann constant, and μ is the chemical potential.

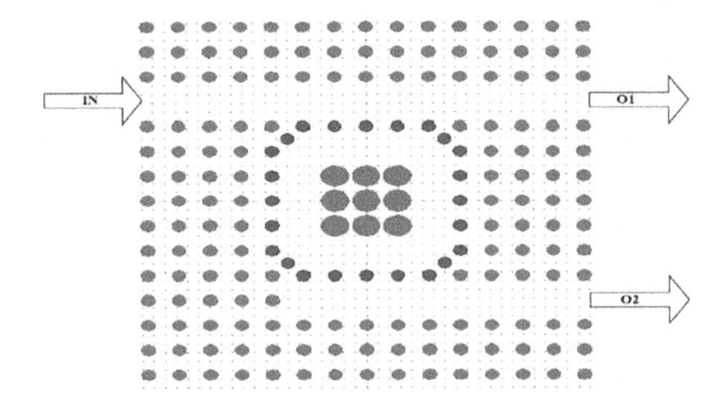

Figure 2. The proposed all-optical switch.

Changing the refractive index of the nonlinear rods results in changing the effective index of the ring resonator, and hence, the switching operation could be approached. The more nonlinear coefficient for graphene in comparison with chalcogenide and nanocrystal [60,62] assists in reducing the needed optical power for switching. In this study, with respect to the aforementioned issue, the graphene was used in the ring resonator to propose a low-threshold all-optical switch. Furthermore, the high contrast ratio for digital applications as the difference between the normalized power margins for logic 0 and 1 is another advantage of the presented structure.

3. Results

To simulate the optical wave propagation throughout the proposed structure, the finite difference time domain (FDTD) method was used [74]. According to the Courant condition, it was assumed as follows [74]:

$$c\Delta t < \frac{1}{\sqrt{(\frac{1}{\Delta x})^2 + (\frac{1}{\Delta z})^2}},$$

where c is the speed of light in vacuum, Δt is the time step, and Δx and Δz are mesh sizes in both X and Z directions. The second condition is about the grid spacing that should be less than $\lambda/10$. The time step was assumed $\Delta t = 0.2$ fs and the structure was discretized in which the length of unit cells was $\Delta x = \Delta z = 100$ nm. In the first stage, an optical pulse was applied to the input port and the pulse response was calculated at output ports. Figure 3 shows the frequency harmonic amplitude for O1 and O2 ports. It can be seen that the optical intensity between two output ports at $\lambda = 1547$ nm can be approached for switching applications.

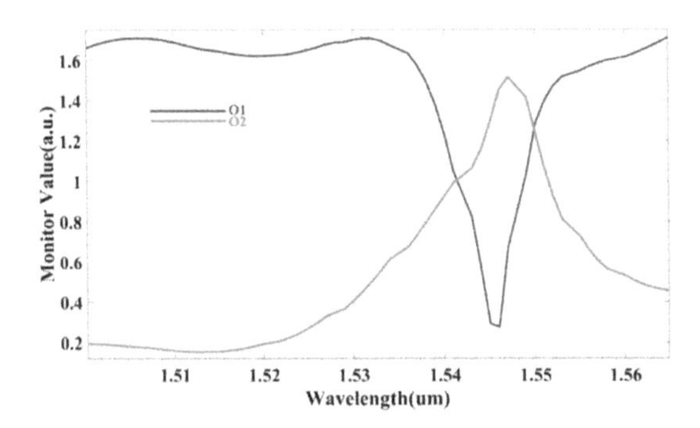

Figure 3. Response of the switch for different wavelengths.

To evaluate the resonance phenomenon, the transmission ratio for different input intensities was simulated (see in Figure 4). In this study, the intensity value at output port divides to input intensity is defined as the transmission ratio factor. When a signal is introduced to the input port, the output ports will be ON depending on the entrance optical intensity. If the input intensity is less than 0.235 W/μm^2, the resonance wavelength of the ring will not change. As a result, O1 and O2 ports will be ON and OFF, respectively. The resonance wavelength can be changed because of the Kerr effect, so the optical waves will considerably be coupled to the O2 port for amounts of more than 0.235 W/μm^2. In this case, O1 and O2 ports will be OFF and ON, respectively. It can be concluded that the threshold intensity for switching is around 0.235 W/μm^2. This value is less than one in other works [45,57,59,60], so it can be considered an advantage of the proposed switch.

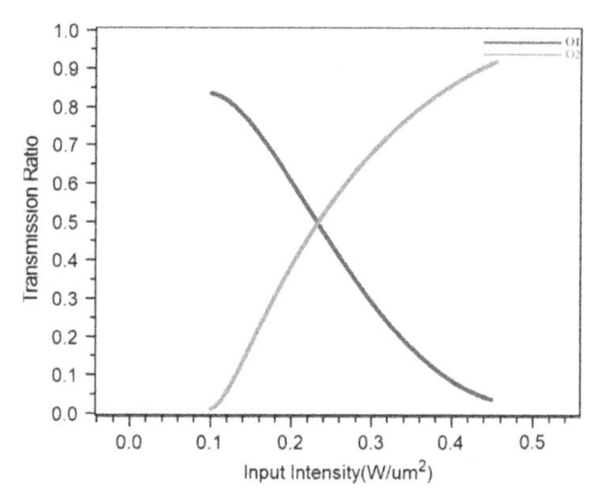

Figure 4. Transmission ratio of the structure for different input intensities.

Figure 5 shows the normalized electric field distribution for two optical intensities, 0.1 W/μm^2 and 0.45 W/μm^2. A color bar has been inserted on the right side in which the red and black colors are the response in +1 and −1 for the domain amplitude of the field. It can be seen that optical waves are dropped from the upper waveguide to the lower waveguide for I = 0.45 W/μm^2 while the dropping operation is not done for I = 0.1 W/μm^2.

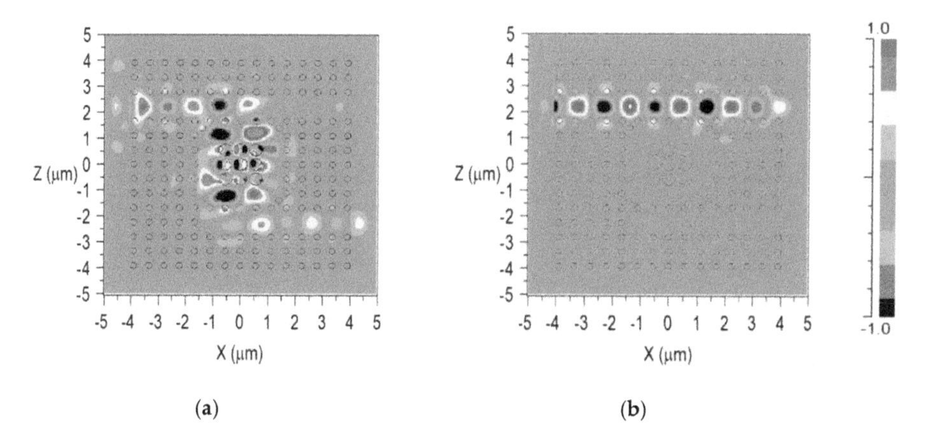

Figure 5. Electric field distribution into the structure for: (**a**) I = 0.45 W/μm^2; (**b**) I = 0.1 W/μm^2.

For more evaluation, the pulse response of the presented structure corresponding to Figure 5 has been calculated. As shown in Figure 6, the normalized powers at port O1 and O2 are obtained by 86%

and 2% for $I = 0.1$ W/μm^2, respectively, while ones are reached to 4% and 95% for $I = 0.45$ W/μm^2. Also, as Figure 7 presents, the rise and fall times of the structure are obtained by 0.3 ps and 0.4 ps, respectively. The rise time is defined as the needed time for coupling the optical waves from one waveguide to another. This time is recorded when the power at the output port reaches 90% steady-state value. The fall time is assumed as the during time for decreasing power from a steady-state value to 10% of it. Comparing the obtained times with the ones in other works [75] demonstrates that the proposed graphene-based switch is faster than them. This characteristic is an essential factor in optical switching applications.

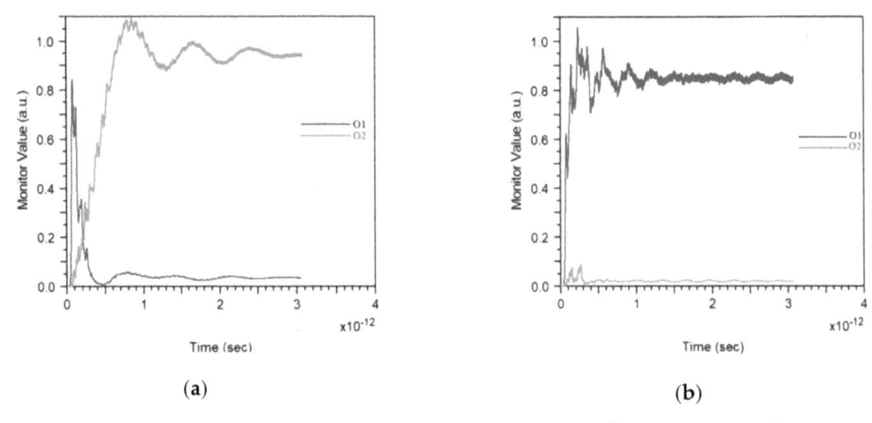

Figure 6. The normalized output power for: (**a**) $I = 0.45$ W/μm^2; (**b**) $I = 0.1$ W/μm^2.

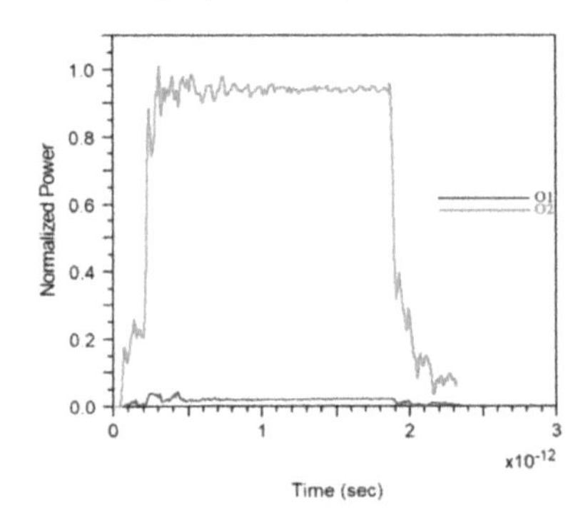

Figure 7. The pulse response for proposed optical switch.

4. Discussion

Approaching the fast response of a switch is one of the main challenges for designing different switched-based applications. Furthermore, the small size of the structure results in it being capable of being considered for integrated circuits. Also, the low power and high contrast ratio are other important parameters for designing different switches. Some attempts have been done to improve the characteristics of optical switches [61,75].

In this study, a graphene-based all-optical switch has been presented in Section 2. Figures 5 and 6 demonstrate the correct operation of the switch. To assess this work, the obtained results have been compared with references [57–60] in Table 1.

Table 1. Comparing the simulated results with ones in other works.

Works	Steady-State Time (ps)	Size(μm^2)	Contrast Ratio (%)	Functional Intensity ($W/\mu m^2$)
[57]	NA *	259	NA *	2000
[58]	NA *	230	NA *	1500
[59]	NA *	360	NA *	1500
[60]	4	250	82	13
This Work	2.75	70	82	0.235

* NA: Not Assigned.

As presented in Figure 6, the time analysis shows that the proposed switch is faster than the one in other works [60]. Approaching the amount of 0.3 ps and 0.4 ps for rise and fall times is the main advantage of this work. Because of the large nonlinear coefficient of graphene in comparison with materials such as chalcogenide, silicon, nanocrystals, and gallium arsenide, the needed threshold intensity for switching operation in this study has been reduced to 0.235 $W/\mu m^2$ in comparison to references [57–60]. This is another advantage of the graphene-based switch. Furthermore, the size of the proposed structure is less than the one in references [57–60]. It is obvious that the compactness of the switch is necessary for use in optical integrated circuits.

Contrast ratio of the output ports is calculated at 82%. The presented switch covers the high contrast ratio while this parameter has not been reported in references [57–60]. Besides all of these advantages, tuneability of some electrical and optical parameters for graphene is a principal characteristic for designing photonic crystal-based devices. Changing the refractive index of graphene in response to electric field affects optical power transmission toward the output ports. Also, the resonant frequency (or wavelength) of the switch could be changed to reach the desired value after the fabrication process.

5. Conclusions

In this study, a photonic crystal-based structure was proposed as an all-optical switch. Using the graphene-SiO_2 stacks as rods in the resonant ring, switching operation was correctly approached. The rise and fall times of the device were obtained by 0.3 ps and 0.4 ps, respectively, that were less than the ones in other reports. The size of the presented structure was just about 70 μm^2 and hence was smaller than the one in the aforementioned works of Section 1. The high nonlinear coefficient of the graphene resulted in a decreasing of the needed optical intensity to 0.235 $W/\mu m^2$ for switching operation in comparison with other works. Furthermore, simulation results showed the contrast ratio of the switch was about 82%. In respect of the obtained results, it can be concluded that the presented all-optical switch is capable of consideration for optical integrated circuits.

Author Contributions: Formal analysis, M.R.J.A. and N.D.; Methodology, M.S. and M.R.J.A.; Writing-Original Draft, M.R.J.A.; Writing-Review and editing, M.S. and Y.S.-K.

Funding: This research received no external funding.

Conflicts of Interest: The authors declare no competing financial interest.

References

1. Nagarajan, R.; Joyner, C.H.; Schneider, R.P.; Bostak, J.S.; Butrie, T.; Dentai, A.G.; Dominic, V.G.; Evans, P.W.; Kato, M.; Kauffman, M. Large-scale photonic integrated circuits. *IEEE J. Sel. Top. Quantum* **2005**, *11*, 50–65. [CrossRef]
2. Bogaerts, W.; Chrostowski, L. Silicon photonics circuit design: Methods, tools and challenges. *Laser Photonics Rev.* **2018**, *12*, 1700237. [CrossRef]
3. Yablonovitch, E. Inhibited spontaneous emission in solid-state physics and electronics. *Phys. Rev. Lett.* **1987**, *58*, 2059. [CrossRef] [PubMed]
4. John, S. Strong localization of photons in certain disordered dielectric superlattices. *Phys. Rev. Lett.* **1987**, *58*, 2486. [CrossRef] [PubMed]

5. Pu, S.; Dong, S.; Huang, J. Tunable slow light based on magnetic-fluid-infiltrated photonic crystal waveguides. *J. Opt.* **2014**, *16*, 045102. [CrossRef]
6. Soljačić, M.; Johnson, S.G.; Fan, S.; Ibanescu, M.; Ippen, E.; Joannopoulos, J. Photonic-crystal slow-light enhancement of nonlinear phase sensitivity. *JOSA B* **2002**, *19*, 2052–2059. [CrossRef]
7. Vlasov, Y.A.; O'boyle, M.; Hamann, H.F.; McNab, S.J. Active control of slow light on a chip with photonic crystal waveguides. *Nature* **2005**, *438*, 65. [CrossRef]
8. Noori, M.; Soroosh, M.; Baghban, H. Highly efficient self-collimation based waveguide for Mid-IR applications. *Photonics Nanostruct.* **2016**, *19*, 1–11. [CrossRef]
9. Noori, M.; Soroosh, M.; Baghban, H. All-angle self-collimation in two-dimensional square array photonic crystals based on index contrast tailoring. *Opt. Eng.* **2015**, *54*, 037111. [CrossRef]
10. Witzens, J.; Loncar, M.; Scherer, A. Self-collimation in planar photonic crystals. *IEEE J. Sel. Top. Quantum* **2002**, *8*, 1246–1257. [CrossRef]
11. Kosaka, H.; Kawashima, T.; Tomita, A.; Notomi, M.; Tamamura, T.; Sato, T.; Kawakami, S. Self-collimating phenomena in photonic crystals. *Appl. Phys. Lett.* **1999**, *74*, 1212–1214. [CrossRef]
12. Rani, P.; Kalra, Y.; Sinha, R. Realization of AND gate in Y shaped photonic crystal waveguide. *Opt. Commun.* **2013**, *298*, 227–231. [CrossRef]
13. Mansouri-Birjandi, M.A.; Tavousi, A.; Ghadrdan, M. Full-optical tunable add/drop filter based on nonlinear photonic crystal ring resonators. *Photonics Nanostruct.* **2016**, *21*, 44–51. [CrossRef]
14. Tavousi, A.; Mansouri-Birjandi, M.A.; Ghadrdan, M.; Ranjbar-Torkamani, M. Application of photonic crystal ring resonator nonlinear response for full-optical tunable add–drop filtering. *Photonic Netw. Commun.* **2017**, *34*, 131–139. [CrossRef]
15. Musavizadeh, S.M.; Soroosh, M.; Mehdizadeh, F. Optical filter based on photonic crystal. *Indian J. Pure Appl. Phys.* **2015**, *53*, 736–739.
16. Qiang, Z.; Zhou, W.; Soref, R.A. Optical add-drop filters based on photonic crystal ring resonators. *Opt. Express* **2007**, *15*, 1823–1831. [CrossRef] [PubMed]
17. Ying, C.; Jing, D.; Jia, S.; Qiguang, Z.; Weihong, B. Study on tunable filtering performance of compound defect photonic crystal with magnetic control. *Optik* **2015**, *126*, 5353–5356. [CrossRef]
18. Dideban, A.; Habibiyan, H.; Ghafoorifard, H. Photonic crystal channel drop filter based on ring-shaped defects for DWDM systems. *Phys. E* **2017**, *87*, 77–83. [CrossRef]
19. Mehdizadeh, F.; Soroosh, M. A new proposal for eight-channel optical demultiplexer based on photonic crystal resonant cavities. *Photonic Netw. Commun.* **2016**, *31*, 65–70. [CrossRef]
20. Mehdizadeh, F.; Soroosh, M.; Alipour-Banaei, H. An optical demultiplexer based on photonic crystal ring resonators. *Optik* **2016**, *127*, 8706–8709. [CrossRef]
21. Talebzadeh, R.; Soroosh, M.; Mehdizadeh, F. Improved low channel spacing high quality factor four-channel demultiplexer based on photonic crystal ring resonators. *Opt. Appl.* **2016**, *46*, 553–564.
22. Talebzadeh, R.; Soroosh, M.; Kavian, Y.S.; Mehdizadeh, F. Eight-channel all-optical demultiplexer based on photonic crystal resonant cavities. *Optik* **2017**, *140*, 331–337. [CrossRef]
23. Zavvari, M. Design of photonic crystal-based demultiplexer with high-quality factor for DWDM applications. *J. Opt. Commun.* **2019**, *40*, 135–138. [CrossRef]
24. Khorshidahmad, A.; Kirk, A.G. Composite superprism photonic crystal demultiplexer: Analysis and design. *Opt. Express* **2010**, *18*, 20518–20528. [CrossRef] [PubMed]
25. Jiu-Sheng, L.; Han, L.; Le, Z. Compact four-channel terahertz demultiplexer based on directional coupling photonic crystal. *Opt. Commun.* **2015**, *350*, 248–251. [CrossRef]
26. Mehdizadeh, F.; Soroosh, M.; Alipour-Banaei, H. A novel proposal for optical decoder switch based on photonic crystal ring resonators. *Opt. Quantum Electron.* **2016**, *48*, 20. [CrossRef]
27. Nozhat, N.; Taher Rahmati, A.; Granpayeh, N. An all optical switch based on nonlinear photonic crystal microcavities. In Proceedings of the Progress in Electromagnetics Research Symposium, Moscow, Russia, 18–21 August 2009.
28. Serajmohammadi, S.; Alipour-Banaei, H.; Mehdizadeh, F. All optical decoder switch based on photonic crystal ring resonators. *Opt. Quantum Electron.* **2015**, *47*, 1109–1115. [CrossRef]
29. Ouahab, I.; Naoum, R. A novel all optical 4 × 2 encoder switch based on photonic crystal ring resonators. *Optik* **2016**, *127*, 7835–7841. [CrossRef]

30. Teo, H.; Liu, A.; Singh, J.; Yu, M.; Bourouina, T. Design and simulation of MEMS optical switch using photonic bandgap crystal. *Microsyst. Technol.* **2004**, *10*, 400–406. [CrossRef]
31. Bai, J.; Wang, J.; Jiang, J.; Chen, X.; Li, H.; Qiu, Y.; Qiang, Z. Photonic not and nor gates based on a single compact photonic crystal ring resonator. *Appl. Opt.* **2009**, *48*, 6923–6927. [CrossRef]
32. Hussein, H.M.; Ali, T.A.; Rafat, N.H. A review on the techniques for building all-optical photonic crystal logic gates. *Opt. Laser Technol.* **2018**, *106*, 385–397. [CrossRef]
33. Saidani, N.; Belhadj, W.; AbdelMalek, F. Novel all-optical logic gates based photonic crystal waveguide using self-imaging phenomena. *Opt. Quantum Electron.* **2015**, *47*, 1829–1846. [CrossRef]
34. haq Shaik, E.; Rangaswamy, N. Improved design of all-optical photonic crystal logic gates using T-shaped waveguide. *Opt. Quantum Electron.* **2016**, *48*, 33. [CrossRef]
35. Andalib, P.; Granpayeh, N. All-optical ultracompact photonic crystal AND gate based on nonlinear ring resonators. *J. Opt. Soc. Am. Rev. B* **2009**, *26*, 10–16. [CrossRef]
36. Fasihi, K. Design and simulation of linear logic gates in the two-dimensional square-lattice photonic crystals. *Optik* **2016**, *127*, 4669–4674. [CrossRef]
37. Moniem, T.A. All-optical digital 4 × 2 encoder based on 2D photonic crystal ring resonators. *J. Mod. Opt.* **2016**, *63*, 735–741. [CrossRef]
38. Hassangholizadeh-Kashtiban, M.; Sabbaghi-Nadooshan, R.; Alipour-Banaei, H. A novel all optical reversible 4 × 2 encoder based on photonic crystals. *Optik* **2015**, *126*, 2368–2372. [CrossRef]
39. Haddadan, F.; Soroosh, M. Low-power all-optical 8-to-3 encoder using photonic crystal-based waveguides. *Photon. Net. Comm.* **2019**, *37*, 83–89. [CrossRef]
40. Alipour-Banaei, H.; Rabati, M.G.; Abdollahzadeh-Badelbou, P.; Mehdizadeh, F. Application of self-collimated beams to realization of all optical photonic crystal encoder. *Phys. E* **2016**, *75*, 77–85. [CrossRef]
41. Mehdizadeh, F.; Soroosh, M.; Alipour-Banaei, H. Proposal for 4-to-2 optical encoder based on photonic crystals. *IET Optoelectron.* **2016**, *11*, 29–35. [CrossRef]
42. Rajasekar, R.; Latha, R.; Robinson, S. Ultra-contrast ratio optical encoder using photonic crystal waveguides. *Matt. Lett.* **2019**, *251*, 144–147. [CrossRef]
43. Alipour-Banaei, H.; Ghorbanzadeh Rabati, M.; Abdollahzadeh-Badelbou, P.; Mehdizadeh, F. Effect of self-collimated beams on the operation of photonic crystal decoders. *J. Electromagn. Wave* **2016**, *30*, 1440–1448. [CrossRef]
44. Mehdizadeh, F.; Alipour-Banaei, H.; Serajmohammadi, S. Study the role of non-linear resonant cavities in photonic crystal-based decoder switches. *J. Mod. Opt.* **2017**, *64*, 1233–1239. [CrossRef]
45. Daghooghi, T.; Soroosh, M.; Ansari-Asl, K. A novel proposal for all-optical decoder based on photonic crystals. *Photonic Netw. Commun.* **2018**, *34*, 335–341. [CrossRef]
46. Chattopadhyay, T.; Roy, J.N. An all-optical technique for a binary-to-quaternary encoder and a quaternary-to-binary decoder. *J. Opt. A Pure Appl. Opt.* **2009**, *11*, 075501. [CrossRef]
47. Alipour-Banaei, H.; Mehdizadeh, F.; Serajmohammadi, S.; Hassangholizadeh-Kashtiban, M. A 2*4 all optical decoder switch based on photonic crystal ring resonators. *J. Mod. Opt.* **2015**, *62*, 430–434. [CrossRef]
48. Zhang, C.; Qiu, K. Design and analysis of coherent OCDM en/decoder based on photonic crystal. *Opt. Lasers Eng.* **2008**, *46*, 582–589. [CrossRef]
49. Tavousi, A.; Mansouri-Birjandi, M.A.; Saffari, M. Successive approximation-like 4-bit full-optical analog-to-digital converter based on Kerr-like nonlinear photonic crystal ring resonators. *Phys. E* **2016**, *83*, 101–106. [CrossRef]
50. Mehdizadeh, F.; Soroosh, M.; Alipour-Banaei, H.; Farshidi, E. Ultra-fast analog-to-digital converter based on a nonlinear triplexer and an optical coder with a photonic crystal structure. *Appl. Opt.* **2017**, *56*, 1799–1806. [CrossRef]
51. Mehdizadeh, F.; Soroosh, M.; Alipour-Banaei, H.; Farshidi, E. All optical 2-bit analog to digital converter using photonic crystal based cavities. *Opt. Quantum Electron.* **2017**, *49*, 38. [CrossRef]
52. Mehdizadeh, F.; Soroosh, M.; Alipour-Banaei, H.; Farshidi, E. A novel proposal for all optical analog-to-digital converter based on photonic crystal structures. *IEEE Photonics J.* **2017**, *9*, 1–11. [CrossRef]
53. Xu, C.; Liu, X. Photonic analog-to-digital converter using soliton self-frequency shift and interleaving spectral filters. *Opt. Lett.* **2003**, *28*, 986–988. [CrossRef] [PubMed]
54. Youssefi, B.; Moravvej-Farshi, M.K.; Granpayeh, N. Two-bit all-optical analog-to-digital converter based on nonlinear Kerr effect in 2D photonic crystals. *Opt. Commun.* **2012**, *285*, 3228–3233. [CrossRef]

55. Tajaldini, M.; Jafri, M.Z. An optimum multimode interference coupler as an all-optical switch based on nonlinear modal propagation analysis. *Optik* **2015**, *126*, 436–441. [CrossRef]
56. Haché, A.; Bourgeois, M. Ultrafast all-optical switching in a silicon-based photonic crystal. *Appl. Phys. Lett.* **2000**, *77*, 4089–4091. [CrossRef]
57. Alipour-Banaei, H.; Serajmohammadi, S.; Mehdizadeh, F. All optical NOR and NAND gate based on nonlinear photonic crystal ring resonators. *Optik* **2014**, *125*, 5701–5704. [CrossRef]
58. Serajmohammadi, S.; Absalan, H. All optical NAND gate based on nonlinear photonic crystal ring resonator. *IPA* **2016**, *3*, 119–123.
59. Alipour-Banaei, H.; Serajmohammadi, S.; Mehdizadeh, F. All optical NAND gate based on nonlinear photonic crystal ring resonators. *Optik* **2017**, *130*, 1214–1221. [CrossRef]
60. Daghooghi, T.; Soroosh, M.; Ansari-Asl, K. A low power all optical decoder based on photonic crystal nonlinear ring resonators. *Optik* **2018**, *174*, 400–408. [CrossRef]
61. Ogusu, K.; Yamasaki, J.; Maeda, S.; Kitao, M.; Minakata, M. Linear and nonlinear optical properties of Ag–As–Se chalcogenide glasses for all-optical switching. *Opt. Lett.* **2004**, *29*, 265–267. [CrossRef]
62. Soh, D.B.S.; Hamerly, R.; Mabuchi, H. Comprehensive analysis of the optical Kerr coefficient of graphene. *Phys. Rev. A* **2016**, *94*, 023845. [CrossRef]
63. Soavi, G.; Wang, G.; Rostami, H.; Purdie, D.G.; Fazio, D.D.; Ma, T.; Luo, B.; Wang, J.; Ott, A.K.; Yoon, D.; et al. Broadband, electrically tunable third-harmonic generation in graphene. *Nat. Nanotechnol.* **2018**, *13*, 583–588. [CrossRef] [PubMed]
64. Jiang, T.; Huang, D.; Cheng, J.; Fan, X.; Zhang, Z.; Shan, Y.; Yi, Y.; Dai, Y.; Shi, L.; Liu, K.; et al. Gate-tunable third-order nonlinear optical response of massless Dirac fermions in graphene. *Nat. Nanotechnol.* **2018**, *12*, 430–436.
65. Hendry, E.; Hale, P.J.; Moger, J.; Savchenko, A.K. Coherent Nonlinear Optical Response of Graphene. *Phys. Rev. Lett.* **2010**, *105*, 091401. [CrossRef] [PubMed]
66. Virga, A.; Ferrante, C.; Batingnani, G.; Fazio, D.D.; Nunn, A.D.G.; Ferrari, A.C.; Cerullo, G.; Scopigno, T. Coherent anti-Stokes Raman Spectroscopy of single and multi-layer graphene. *Nat. Commun.* **2019**, *10*, 3658. [CrossRef] [PubMed]
67. Khurgin, J.B. Graphene—A rather ordinary nonlinear optical material. *Appl. Phys. Lett.* **2014**, *104*, 16116. [CrossRef]
68. Zhang, H.; Virally, S.; Bao, Q.; Ping, L.K.; Massar, S.; Godbout, N.; Kockaert, P. Z-scan measurement of the nonlinear refractive index of graphene. *Opt. Lett.* **2012**, *37*, 1856–1858. [CrossRef]
69. Cheng, J.L.; Vermeulen, N.; Sipe, J.E. Third order optical nonlinearity of graphene. *New J. Phys.* **2014**, *16*, 053014. [CrossRef]
70. Gu, T.; Petrone, N.; McMillian, J.F.; Zande, A.V.D.; Yu, M.; Lo, G.Q.; Kwong, D.L.; Hone, J.; Wong, C.W. Regenerative oscillation and four-wave mixing in graphene optoelectronics. *Nat. Photonics* **2012**, *6*, 554–559. [CrossRef]
71. Johnson, S.G.; Joannopoulos, J.D. Block-iterative frequency-domain methods for Maxwell's equations in a planewave basis. *Opt. Express* **2001**, *8*, 173–190. [CrossRef]
72. Berman, O.L.; Boyko, V.S.; Kezerashvili, R.Y.; Kolesnikov, A.A.; Lozovik, Y.E. Graphene-based photonic crystal. *Phys. Lett. A* **2010**, *374*, 4784–4786. [CrossRef]
73. Saleh, B.E.A.; Teich, M.C. *Fundamentals of Photonics*; John Wiley & Sons: Hoboken, NJ, USA, 1991; Chapter 19.
74. Sullivan, D.M. *Electromagnetic Simulation Using the FDTD Method*; John Wiley & Sons: Hoboken, NJ, USA, 2013.
75. Daghooghi, T.; Soroosh, M.; Ansari-Asl, K. Ultra-fast all-optical decoder based on nonlinear photonic crystal ring resonators. *Appl. Opt.* **2018**, *57*, 2250–2257. [CrossRef] [PubMed]

Article

Photonic Crystal Cavity-Based Intensity Modulation for Integrated Optical Frequency Comb Generation

Henry Francis [1,*], Si Chen [1], Kai-Jun Che [2], Mark Hopkinson [1] and Chaoyuan Jin [1,3,*]

[1] Department of Electronic and Electrical Engineering, University of Sheffield, Sheffield S3 7HQ, UK; schen62@sheffield.ac.uk (S.C.); m.hopkinson@sheffield.ac.uk (M.H.)

[2] Department of Electronic Engineering, Xiamen University, Fujian 361005, China; chekaijun@xmu.edu.cn

[3] College of Information Science and Electronic Engineering, Zhejiang University, Hangzhou 310007, China

* Correspondence: hfrancis1@sheffield.ac.uk (H.F.); jincy@zju.edu.cn (C.J.)

Received: 30 August 2019; Accepted: 23 September 2019; Published: 25 September 2019

Abstract: A simple scheme to generate an integrated, nanoscale optical frequency comb (OFC) is numerically studied. In this study, all optical intensity modulators based on photonic crystal (PhC) cavities are cascaded both in series and parallel. By adjusting the modulation parameters, such as the repetition rate, phase, and coupling efficiency of the modulating wave, it is possible to produce combs with a variety of different characteristics. Unique to PhC intensity modulators, in comparison with standard lithium niobate modulators, is the ability to control the amplitude of the light via a cavity rather than controlling the phase through one arm of a Mach–Zehnder interferometer. This opens up modulation-based OFC generation to new possibilities in both nanoscale operation and cavity-based schemes.

Keywords: photonic crystals; microwave photonics; optical frequency combs

1. Introduction

In recent years, optical frequency combs (OFC) have become an increasing popular research area [1–3]. The rapid adoption of frequency combs into many different research topics has meant the need for diverse functionality and integration [4]. There are many schemes for generating an OFC, which include mode locked lasers [5], electro-optic (EO) modulators [6], and micro toroidal cavities [7]. Mode locked lasers have the ability to generate combs with a fixed phase relationship and stable operation. However, the frequency space between comb lines is governed by the device length due to fundamental operation principles and hence limit their abilities and flexibility to be integrated onto photonic circuits. Another well-established area of OFC generation is in micro-toroidal cavities [8]. This technique relies on nondegenerate four wave mixing within the cavity to generate a broad band of intensity spikes in the frequency domain. Using this technique, on chip integration of an OFC generator has been achieved [9]. This provided a major step forward within the research community. Currently, OFC generation by this method is receiving a lot of attention, a comprehensive review is given in ref. [1]. While the work outlined in ref. [1] shows OFCs with large spectral range and on-chip integration, the work presented here is in parallel and looks towards new technologies for generating OFCs. By generating an OFC from photonic crystal (PhC) cavities and waveguides, the work looks towards improving spectral homogeneity, reducing the device volume and exploring the potential for nanoscale photonic integration of OFCs.

A common method of OFC generation is the use of cascaded EO $LiNbO_3$ intensity modulators (IM) [10]. In this method, light is split into two arms and the phase of the light through one of arms is modulated via a radio frequency (RF) electrical signal. The RF signal induces a refractive index change of the waveguide material, hence changing the phase of the light. The recombination of the two carrier signals with a controllable phase difference causes modulation of the carrier lights intensity. A DC bias is also present through each arm so that a constant phase shift can be applied. The voltage of the RF signal used to modulate the light will determine the modulation depth and the voltage of the DC bias will determine a constant phase shift. To generate an OFC, the light is modulated using a sinusoidal electrical signal. This generates sidebands in the frequency response of the modulated light, and the position of these sidebands, relative to the initial frequency of the carrier light, has a direct relation to the frequency of the electrical signal. The amplitude of the sidebands is determined by the modulation depth and phase shift. By fine control of these parameters, an intensity modulator can generate multiple comb lines in the frequency response of the modulated light [11]. A second IM leads from the output of the first IM, where the voltage of the RF and DC bias are kept the same but the modulation frequency is a fifth of the first IM. This generates a large number of homogeneous comb lines centered around the initial frequency of the input light. This technique provides a simple solution to generating a multiple wavelength source, however this method cannot be implemented with chip-scale components due to the need for relatively large waveguides to achieve deep enough modulation. In the work presented here, the operating principle outlined above is adapted and built upon for the implementation in PhC waveguides and cavities.

The standard modulation process of $LiNbO_3$ modulator is fundamentally very different to the process undergone in photonic crystal cavity modulators. Here, a mismatch between the cavity resonance and the propagating wave will either cause transmission or reflection, depending on the implemented scheme. In this paper, two intensity modulation schemes are analyzed; direct and side coupling, as shown in Figure 1. In the direct coupling scheme, the transmitted power is proportional to the energy inside the cavity, this will decrease the achievable contrast ratio. This scheme will also lead to the transmitted power having a similar pulse shape to the control light. However, in the side coupled scheme, interference between the decaying wave from the cavity and the propagating wave in the waveguide mean a higher contrast ratio and variable pulse shape. In both schemes, an increase in the optical excitation power leads to a larger cavity resonance shift, this has a large effect on the pulse shape of the carrier wave and the sidebands produced from the modulation process. By arranging two or more PhC IMs in either a series or parallel formation, it is possible to increase the number of generated sidebands and hence observe an OFC. In this paper, detailed analysis of the different modulation schemes will be undertaken. This analysis will confirm that an OFC can be produced from a PhC device, thus broadening the application of PhC devices into the mature and prosperous research field of OFC generation.

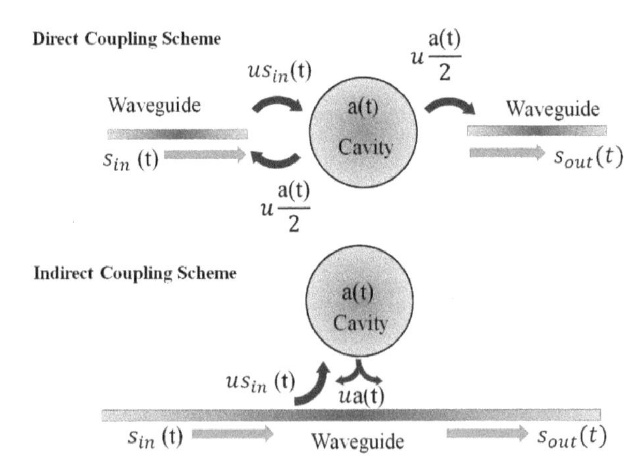

Figure 1. The two all-optical intensity modulator schemes studied in this paper. $a(t)$ is the cavity field amplitude, u is the coupling factor, and s_{in} (t) and s_{out} (t) represent the amplitude of the propagating wave in the input and output waveguide, respectively.

2. Materials and Methods

The scheme proposed here is based on photonic crystal waveguides and cavities to construct all-optical intensity modulators, as shown in Figure 1. This method of all-optical modulation has been studied in great detail by different research groups [12–15], which has led to a rich and common understanding of the fundamental principles behind PhC switches. PhC cavities have a very high quality factor (Q) while maintaining a low mode volume (V). This means that nonlinearity within the cavity can be greatly enhanced, given that the field intensity scales with Q/V. A shift in the cavity resonant wavelength, due to a refractive index change, is therefore possible with a relatively low-power pump light. Intensity modulation of a carrier light is therefore possible when a pump light is used to change the transmission wavelength. The device itself is made up of holes in triangular lattice formation fabricated in an InP membrane with embedded InAs quantum dots (QDs), with broad band emission around 1550 nm. The waveguides are made up of a line of missing holes and the cavity consists of a three hole defect, known as an L3 cavity, as shown in Figure 2. To ensure the cavity has a resonant frequency around 1550 nm the filling factor and lattice constant have been carefully calculated. The lattice constant is calculated to be 480 nm and the filling factor at 0.29. The thickness of the membrane will be 220 nm. The parameters given were calculated in COMSOL by finding the eigenfrequency via the finite element method (FEM). A scanning electron microscope (SEM) image of an L3 cavity along with the FEM simulation results for an L3 cavity with a fundamental mode at 1550 nm is shown in Figure 2.

Figure 2. (**a**) An scanning electron microscope (SEM) image of a photonic crystal membrane with an L3 cavity in the center. (**b**) Finite element method (FEM) calculation of the fundamental mode at 1550 nm.

Coupled mode theory (CMT), a standard PhC analysis tool, is used in this paper to simulate the device described above. The equations employed in this paper are based on an extension of CMT models that have been presented previously [13,16,17]. In the simulated device, it is assumed that an active region is present in the form of InAs QDs. This will lead to a dominating third order nonlinearity process based on the saturable absorption (SA) of QDs. Therefore, the refractive index change will be dominated by the Kramers–Kronig relation to the absorption coefficient. This has a direct relation to the field amplitude within the photonic crystal cavity and hence will be the main dynamic variable used throughout. A pump and carrier light are injected into the input waveguide, denoted by $s_{in}^{p,c}$ in Figure 1. The dynamical equations for the cavity modes is written as:

$$\frac{da_{p,c}}{dt} = (-i(\omega_0 + \Delta\omega_p(t) - \omega_{p,c}) - \gamma_{total})a_{p,c}(t) + \sqrt{2\gamma_c}s_{in}^{p,c}(t) \quad (1)$$

Here, $a_{p,c}$ is the amplitude of the cavity mode which is excited by the pump and carrier light, respectively and the energy inside the cavity is represented by $|a_{p,c}(t)|^2$. The field of the cavity is represented by $A_{in}^{p,c}(t) = a_{in}^{p,c}(t)\,exp(-i\omega_{p,c}t)$ and the input light through the waveguide is given by $S_{in}^{p,c}(t) = s_{in}^{p,c}(t)\,exp(-i\omega_{p,c}t)$, the power of the light is represented by $|s_{in}^{p,c}(t)|^2$. It is assumed that a_p is far higher than a_c, such that only a_p will have an effect on the cavity resonance shift. $\omega_{p,c}$ is the angular frequency of the input pulses, for either the pump or carrier light. In order for the cavity resonance to change, the amplitude of the pump light must increase. The change is then dependent on the energy in the cavity due to the pump light and a third order non-linearity, denoted as the Kerr constant. This is given below:

$$\Delta\omega_p(t) = K_{kerr}|a(t)|^2 \tag{2}$$

K_{kerr} represents a simplified value for the nonlinear dynamics in the system [12,17], given by:

$$K_{kerr} = \frac{\omega_0 c n_2}{n_{eff} V_{kerr}} \tag{3}$$

n_2 is the enhanced Kerr coefficient, due to the SA of QDs in the membrane structure and the average rate of SA within the cavity is calculated and given by V_{kerr}. These values are greatly enhanced by the PhC cavity due to the small mode volume and high Q attainable in PhC structures. n_{eff} is the effective refractive index, this due to the membrane thickness, material properties and resonant wavelength of the cavity. The total loss rate in the cavity is split into two parts; the intrinsic loss, γ_{int}, and the coupling loss, γ_c. The intrinsic loss due to coupling into non-relevant modes, predominantly by vertical emission, is dependent on the Q factor of the cavity, such that $\gamma_{int} = \omega_0/2Q$. The coupling loss is due to the coupling coefficient between the cavity and the waveguide. The light in the L3 cavity will reflect off each side of the cavity in the longitudinal direction which leads to light coupling to the waveguide at each end, hence $2\gamma_c$.

In order to induce a change in the transmission through the device, a change in the refractive index is needed. The amount that the refractive index must shift by is dependent on the linewidth of the cavity mode. In the device simulated here, a resonance shift of 1 nm will cause the carrier light to either reflect or transmit through the device. To obtain a shift of 1 nm, it is calculated that a refractive index shift of 0.005 is needed. The relationship between refractive index change and cavity resonance change is defined as $\frac{\Delta\omega}{\omega} = -\frac{\Delta n}{n}$. Therefore, the profile of the refractive index change is in relation to the profile of the pump light, as given in Equation (1) and shown by the dashed line in Figure 3. Achieving a refractive index change of 0.005 using QD saturable absorption as the refractive index non-linearity is common among nanophotonics research community [18,19].

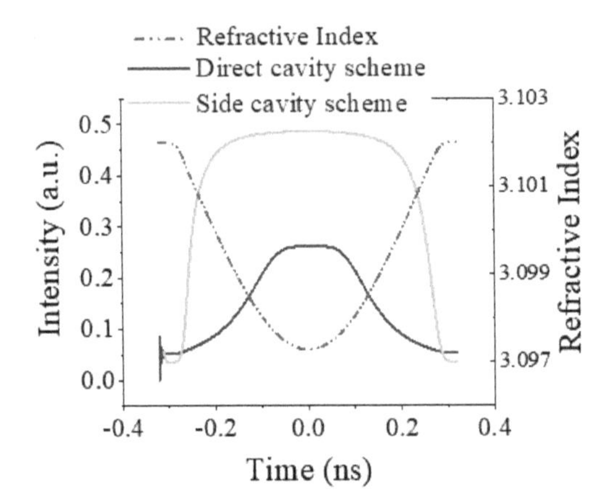

Figure 3. The power of the carrier wave at the output of the device is given by the solid lines. The refractive index change inside the cavity is given by the dotted line.

Using the parameters stated in Table 1, Equations (1)–(7) are solved numerically. This is done using standard routines available for solving differential equations via a self-written computing program.

Table 1. Physical parameters used in coupled mode theory (CMT) calculations.

Parameter	Symbol	Value	Source
Coupling loss	γ_c	1×10^{11}	Estimated
Cavity quality factor	Q	10,000	FEM calculation
Effective refractive index	n_{eff}	3.1	[13]
Intrinsic loss	γ_{int}	1.1×10^{11}	Calculated
Kerr coefficient	n_2	1.5×10^{-14} m^2/W	[20]
Kerr mode volume	V_{Kerr}	0.21 μm^3	Calculated [16]
Resonant wavelength	ω_0	1.550 μm	Calculated

3. Results and Discussion

As discussed earlier, two common cavity-waveguide all-optical modulator configurations have been considered in this paper: side and direct cavity coupling; these are schematically represented in Figure 1. Given that the waveguide is unbroken in the side coupling scheme, the output will be dependent on both the cavity energy and the energy in the waveguide, giving rise to Equation (4) and the green line in Figure 3.

$$s_{out}^{c}(t) = u\frac{a_c(t)}{2} + s_{in}^{c}(t) \tag{4}$$

The coupling factor, u, is defined as $u = \sqrt{2\gamma_c}$ [21]. In the direct coupling scheme the output will only rely on the energy in the cavity due to the carrier pump, such that:

$$s_{out}^{c}(t) = u\frac{a_c(t)}{2} \tag{5}$$

The output of this is shown by the solid purple line in Figure 3. The control light has a sinusoidal wave form with a repetition frequency in the GHz range, as shown in Figure 3. This leads to different output dynamics for the carrier light, depending on which scheme is used. In this paper, the PhC-based IMs are cascaded in both series and parallel in order to generate an OFC, as shown in Figure 4. When two IMs are in parallel with each other, it is assumed that the carrier light is split evenly into each arm. Each arm is then coupled to a cavity via the side coupling scheme, as shown in Figure 4 and when they recombine, the output is given by:

$$s_{out}^{c}(t) = u\frac{a_{c1}(t)}{2} + s_{in}^{c1}(t) - (u\frac{a_{c2}(t)}{2} + s_{in}^{c2}(t)) \tag{6}$$

where $s_{in}^{c1,c2}(t)$ represents the input to the bottom or top cavity and $a_{in}^{c1,c2}(t)$ represents the amplitude of the carrier signal in the bottom or top cavity. When two IMs are put in parallel with each other, as shown in Figure 4a, and the phase of the modulating wave in each arm can be controlled, it displays properties similar to that of a conventional IM. The PhC design, outlined in Figure 4a, consists of a waveguide leading into a Y junction symmetric splitter [22,23], where the pump and carrier light split evenly into each arm. An L3 cavity is coupled to each arm of the device to induce intensity modulation of the split carrier light. The two intensity modulated carrier lights then recombine via a Y junction to produce a dual modulated carrier signal. This technique of waveguide splitting and recombining in PhC waveguides is common when implementing PhC-based Mach–Zehnder interferometers [24,25]. Using this process, the shape of the carrier light will be determined by a number of parameters unique to this type of modulation, these include the phase shift of the pump light in each arm, the coupling efficiency of the light into the cavity and the Q factor of the cavity.

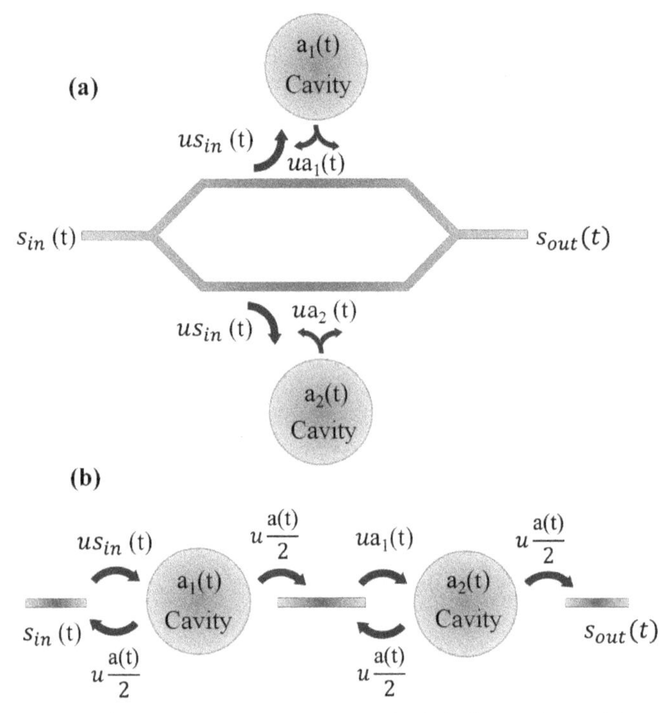

Figure 4. Schematics for the proposed photonic crystal (PhC) all-optical optical frequency combs (OFC) generators. (**a**) Two side coupled cavity modulators in parallel. (**b**) Two direct coupled cavity modulators in series.

This method of modulation can generate 3 comb lines with a maximum intensity difference of 1 dB on each side of the central carrier frequency, hence 7 comb lines all together as shown in Figure 5b. Additional comb lines exist at much weaker intensities further away from the initial carrier frequency. The number of generated intensity spikes in the frequency domain is limited by the attainable modulation profile of the carrier light in the time domain. As can be seen, the carrier wavelength has a much stronger intensity than the sidebands produced. Nevertheless, the homogeneity in intensity between the sidebands remains very uniform. The homogeneity of these sidebands is dependent on the parameters stated above and can therefore be optimized to generate a comb, as shown in Figure 5b. During experimental implementation of the proposed scheme, there will be disparity between the optimised parameters given here and the fabricated device. Although this will have an effect on the OFC quality, it is expected that a 7 line OFC can be observed for a broad range of parameter values. Figure 5a shows that for Q factors between 10,000 and 25,000 the side band intensity fluctuates to a maximum of 1.5 dB. This shows that a wide range of Q factor values will still achieve a 7 line OFC. The extinction ratio of the carrier signal is dominated by the coupling factor between the cavity and the waveguide. This will have an effect on the intensity of the generated comb lines relative to the intensity of the central carrier frequency. However, 3 sidebands on either side of the initial carrier frequency can still be generated for a broad range of coupling factors.

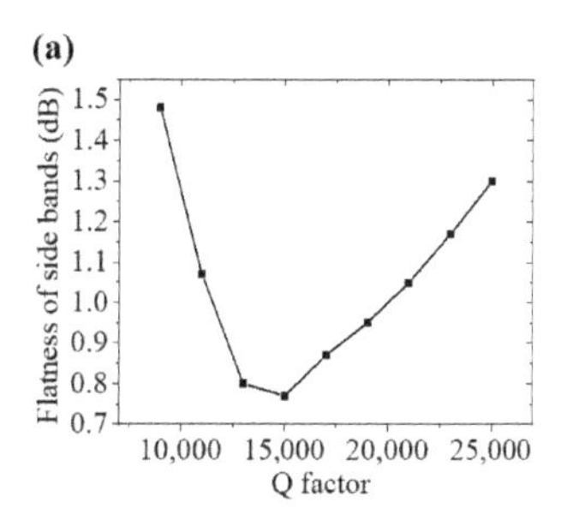

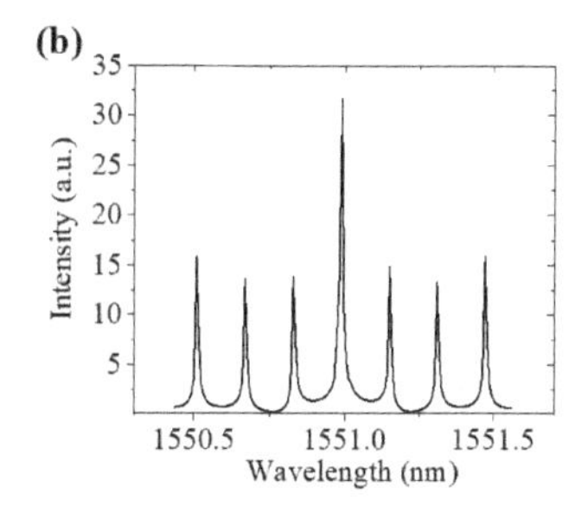

Figure 5. (**a**) Homogeneity between the sidebands against the cavity Q factor. (**b**) The generated OFC.

Figure 5a shows the effect the Q factor has on the uniformity with an optimum Q factor of 13,000. This is due to the shape of the carrier light through each arm when it recombines to generate the intensity modulated signal. Cavity Q factors in PhC L3 cavities can be greatly enhanced by altering the position of the holes surrounding the L3 cavity [26]. It is therefore possible to design the L3 cavity with the desired Q factor via fine tuning of the hole positions. The 7 comb lines observed have with a maximum fluctuation in intensity of less than 1 dB if the carrier frequency is not taken into consideration. In previous work, the intensity of the initial carrier frequency has also posed a problem in modulation-based combs [27] and micro-cavity-based combs, due to the high energy needed to induce nonlinearity within the micro-cavity [28]. A method to overcome this is to include a cavity-based notch filter. To obtain the required suppression of the carrier signal the coupling coefficient, size and loss coefficient of the cavity can be determined via calculation. The filter will induce an overall loss to the carrier wave signal of around 5 dB, but the flatness of the 7-line OFC will be greatly enhanced.

When two IMs are in series with each other, as shown in Figure 4b, the output from the first IM will be the input for the second. The equation for the energy inside the second cavity is given by:

$$\frac{da_{c2}}{dt} = (-i(\omega_0 + \Delta\omega_{p2}(t) - \omega_c) - \gamma_{total})a_{c2}(t) + \sqrt{2\gamma_c}a_{c1}(t) \tag{7}$$

The output of the second cavity is equivalent to the output of the single cavity device, as given in Equation (5). In the parallel scheme outlined above, side coupled cavities were used to generate an OFC because a broad pulse was able to be produced from a sinusoidal control light. The high extinction ratio and pulse shape gained from this scheme meant a short, high intensity pulse could be produced and generate an OFC. However, in order to produce an OFC from two IMs in parallel, the sinusoidal pulse shape at the output of the IM is more desirable. Sidebands are produced in the frequency domain from the initial carrier light by a single IM with a sinusoidal pump light profile [29]. This is shown in Figure 6a where a sideband either side of the initial carrier frequency has been generated. The sideband intensity is within 5 dB of the carrier intensity and their FSR is in direct relation to the modulation frequency. The sidebands produced from the first IM then act as carrier waves leading into the second IM. The second IM has a modulation frequency one third that of the first, meaning that the generated sidebands will have an FSR one third that of the original FSR. Two extra comb lines are generated on each side of the carrier frequency; they are the sum of the generated sidebands from both the initial carrier frequency and the initial sidebands generated from a single IM.

This scheme then generates multiple sidebands that are of equal distance from each other in the frequency domain and of similar intensity. In this scheme, 7 comb lines have been generated, Figure 6b shows the generated OFC, while Figure 6a shows sideband generation from a single IM, in which the FSR is directly related to the repetition rate of the modulating signal. In this case, this is within the microwave range, at around 10 GHz. A microwave signal is then generated from the beat note between the carrier and the sideband, this is a common tool in microwave photonics [30].

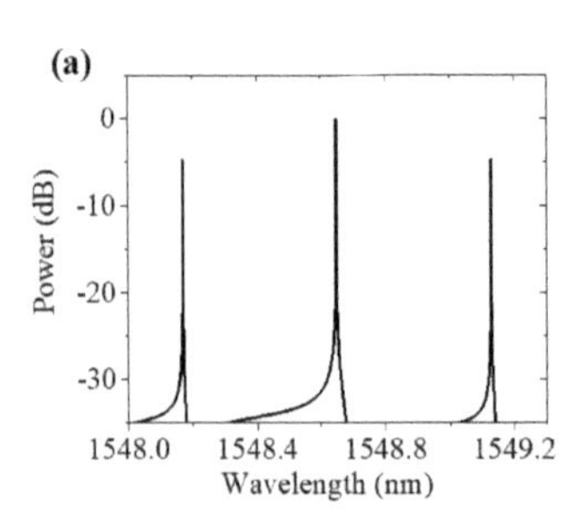

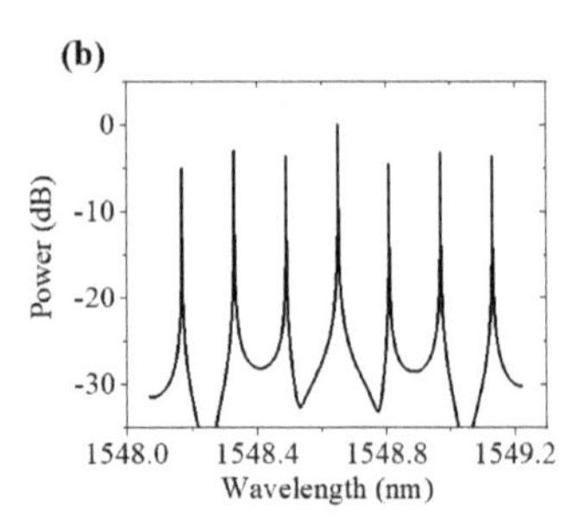

Figure 6. (**a**) The frequency response of the carrier wave after a single direct coupled intensity modulators (IM). (**b**) The generated OFC from two all-optical PhC direct coupled modulators in series.

4. Conclusions

To conclude, a new and novel approach to OFC generation using PhC structures is proposed. By integrating multiple PhC cavities, it is possible to observe OFC generation on the nanoscale. A model based on CMT is built upon to simulate the modulation of a carrier light through a PhC modulator using a pump light with a repetition rate in the RF range. By cascading multiple cavities in either series or parallel, sidebands in the frequency response of the carrier wave can be observed. By careful manipulation of the device parameters, these sidebands can be optimised to give a flat spectrum across 7 comb lines. Using the parallel-based scheme, the Q factor of both cavities can be specified to give a flat comb over the generated sidebands. However, the initial carried frequency will have a far greater intensity, which could limit the OFC operation. To overcome this, it is possible to use a specifically designed notch filter. Using a scheme with two cavities in series, it is possible to generate a flatter comb across all lines without the need for a notch filter. In this scheme, two separate RF lines are needed to drive each cavity separately, which will cause complexity within the system.

PhC devices are a mature enough research field that fabrication techniques are commonplace to many institutes; therefore, implementation of this technique experimentally is very promising. PhC devices based on cavities and waveguides are a well-established and rich research field which have the potential to enable on chip photonic networks. To this end, the device proposed here will contribute towards both the field of PhCs as well as expanding OFC generation techniques into a new domain.

Author Contributions: Conceptualization, H.F. and C.J.; methodology, K.C. and S.C.; software, H.F., K.C. and S.C.; validation, C.J., M.H. and K.C.; investigation, H.F.; resources, C.J.; writing–original draft preparation, H.F.; writing–review and editing, C.J. and M.H.; visualization, H.F and C.J.; supervision, C.J. and M.H.

Funding: This research was funded by EPSRC First Grant, the Royal Society Research Grant (UK) and NSFC Grant No. 61574138 and No. 61974131 (China).

Conflicts of Interest: The authors declare no conflict of interest.

Abbreviations

The following abbreviations are used in this manuscript:

PhC	Photonic Crystal
OFC	Optical Frequency Comb
FEM	Finite Element Method
SEM	Scanning Electron Microscope
FSR	Free Spectral Range
CMT	Coupled Mode Theory
QD	Quantum Dot

References

1. Gaeta, A.L.; Lipson, M.; Kippenberg, T.J. Photonic-chip-based frequency combs. *Nat. Photonics* **2019**, *13*. [CrossRef]
2. Pasquazi, A.; Peccianti, M.; Razzari, L.; Moss, D.J.; Coen, S.; Erkintalo, M.; Chembo, Y.K.; Hansson, T.; Wabnitz, S.; Del'Haye, P.; et al. Micro-combs: A novel generation of optical sources. *Phys. Rep.* **2017**, *729*, 1–81. [CrossRef]
3. Demirtzioglou, I.; Lacava, C.; Bottrill, K.R.H.; Thomson, D.J.; Reed, G.T.; Richardson, D.J.; Petropoulos, P. Frequency comb generation in a silicon ring resonator modulator. *Opt. Express* **2018**, *26*, 790. [CrossRef] [PubMed]
4. Diddams, S.A. The evolving optical frequency comb. *J. Opt. Soc. Am.* **2010**, *27*, B51–B62. [CrossRef]
5. Udem, T.; Holzwarth, R.; Hansch, T.W. Optical frequency metrology. *Nature* **2002**, *416*, 233–237. [CrossRef] [PubMed]
6. Torres-Company, V.; Lancis, J.; Andres, P. Lossless equalization of frequency combs. *Opt. Lett.* **2008**, *33*, 1822–1824. [CrossRef] [PubMed]
7. Spencer, D.T.; Drake, T.; Briles, T.C.; Stone, J.; Sinclair, L.C.; Fredrick, C.; Li, Q.; Westly, D.; Ilic, B.R.; Bluestone, A.; et al. An optical-frequency synthesizer using integrated photonics. *Nature* **2018**, *557*, 81–85. [CrossRef] [PubMed]
8. Chembo, Y.K. Kerr optical frequency combs: Theory, applications and perspectives. *Nanophotonics* **2016**, *5*, 214–230. [CrossRef]
9. Razzari, L.; Duchesne, D.; Ferrera, M.; Morandotti, R.; Chu, S.; Little, B.E.; Moss, D.J. CMOS-compatible integrated optical hyper-parametric oscillator. *Nat. Photonics* **2010**, *4*, 41–45. [CrossRef]
10. Shang, L.; Wen, A.; Lin, G.; Gao, Y. A fl at and broadband optical frequency comb with tunable bandwidth and frequency spacing. *Opt. Commun.* **2014**, *331*, 262–266. [CrossRef]
11. Healy, T.; Gunning, F.C.G.; Ellis, A.D.; Bull, J.D. Multi-wavelength source using low drive-voltage amplitude modulators for optical communications. *Opt. Express* **2007**, *15*, 2981–2986. [PubMed]
12. Nozaki, K.; Tanabe, T.; Shinya, A.; Matsuo, S.; Sato, T.; Taniyama, H.; Notomi, M. Sub-femtojoule all-optical switching using a photonic-crystal nanocavity. *Nat. Photonics* **2010**, *4*, 477–483. [CrossRef]
13. Yu, Y.; Palushani, E.; Heuck, M.; Kuznetsova, N.; Kristensen, P.T.; Ek, S.; Vukovic, D.; Peucheret, C.; Oxenløwe, L.K.; Combrié, S.; et al. Switching characteristics of an InP photonic crystal nanocavity: Experiment and theory. *Opt. Express* **2013**, *21*, 31047. [CrossRef] [PubMed]
14. Notomi, M.; Shinya, A.; Mitsugi, S.; Kira, G.; Kuramochi, E.; Tanabe, T. Optical bistable switching action of Si high-Q photonic-crystal nanocavities. *Opt. Express* **2005**, *13*, 2678. [CrossRef] [PubMed]
15. Jin, C.Y.; Johne, R.; Swinkels, M.Y.; Hoang, T.B.; Midolo, L.; Van Veldhoven, P.J.; Fiore, A. Ultrafast non-local control of spontaneous emission. *Nat. Nanotechnol.* **2014**, *9*, 886–890. [CrossRef] [PubMed]
16. Uesugi, T.; Song, B.S.; Asano, T.; Noda, S. Investigation of optical nonlinearities in an ultra-high-Q Si nanocavity in a two-dimensional photonic crystal slab. *Opt. Express* **2006**, *14*, 377–386. [CrossRef]
17. De Rossi, A.; Lauritano, M.; Combrié, S.; Tran, Q.V.; Husko, C. Interplay of plasma-induced and fast thermal nonlinearities in a GaAs-based photonic crystal nanocavity. *Phys. Rev. At. Mol. Opt. Phys.* **2009**, *79*, 1–9. [CrossRef]
18. Jin, C.Y.; Wada, O. Photonic switching devices based on semiconductor nano-structures. *J. Phys. Appl. Phys.* **2014**, *47*. [CrossRef]
19. Prasanth, R.; Haverkort, J.E.; Deepthy, A.; Bogaart, E.W.; Van Der Tol, J.J.; Patent, E.A.; Zhao, G.; Gong, Q.; Van Veldhoven, P.J.; Nötzel, R.; et al. All-optical switching due to state filling in quantum dots. *Appl. Phys. Lett.* **2004**, *84*, 4059–4061. [CrossRef]
20. Sauvage, S.; Boucaud, P.; Glotin, F.; Prazeres, R.; Ortega, J.M.; Lemaître, A.; Gérard, J.M.; Thierry-Mieg, V. Third-harmonic generation in InAs/GaAs self-assembled quantum dots. *Phys. Rev. Condens. Matter Mater. Phys.* **1999**, *59*, 9830–9833.
21. Joannopoulos, J.; Johnson, S.; Winn, J.; Meade, R. *Photonic Crystals: Molding the Flow of Light*, 2nd ed.; Princeton University Press: Princeton, NJ, USA, 2008.
22. Boscolo, S.; Midrio, M. Y junctions in photonic crystal channel waveguides: high transmission and impedance matching. *Opt. Lett.* **2002**, *27*, 1001–1003. [PubMed]

23. Turner, J.S.; Lourtioz, J.M.; Berger, V.; Gérard, J.M.; Tchelnokov, A. *Photonic Crystals*, 2nd ed.; Springer: Berlin/Heidelberg, Germany, 2008.
24. Martinez, A.; Sanchis, P.; Marti, J. Mach – Zehnder interferometers in photonic crystals. *Opt. Quantum Electron.* **2005**, *37*, 77–93. [CrossRef]
25. Nakamura, H.; Sugimoto, Y.; Kanamoto, K.; Ikeda, N.; Tanaka, Y.; Nakamura, Y.; Ohkouchi, S.; Watanabe, Y.; Inoue, K.; Ishikawa, H.; et al. Ultra-fast photonic crystal/quantum dot all- optical switch for future photonic networks. *Opt. Express* **2004**, *12*, 6606–6614.
26. Chalcraft, A.R.; Lam, S.; O'Brien, D.; Krauss, T.F.; Sahin, M.; Szymanski, D.; Sanvitto, D.; Oulton, R.; Skolnick, M.S.; Fox, A.M.; et al. Mode structure of the L3 photonic crystal cavity. *Appl. Phys. Lett.* **2007**, *90*, 1–4. [CrossRef]
27. Nagarjun, K.P.; Vadivukarassi, J.; Kumar, S.; Supradeepa, S. Generation of tunable, high repetition rate frequency combs with equalized spectra using carrier injection-based silicon modulators. *Opt. Express* **2018**, *26*, 975218. [CrossRef]
28. Kippenberg, T.J.; Holzwarth, R.; Diddams, S.A. Microresonator-Based Optical Frequency Combs. *Science* **2011**, *332*, 555–560. [CrossRef] [PubMed]
29. Francis, H.; Chen, S.; Ho, C.H.; Che, K.J.; Wang, Y.R.; Hopkinson, M.; Jin, C.Y. Generation of optical frequency combs using a photonic crystal cavity. *IET Optoelectron.* **2018**, *13*, 23–26. [CrossRef]
30. Yao, J. Microwave Photonics. *J. Light. Technol.* **2014**, *27*, 314–335. [CrossRef]

Article

Quantitative Analysis of Photon Density of States for One-Dimensional Photonic Crystals in a Rectangular Waveguide

Ruei-Fu Jao [1] and Ming-Chieh Lin [2,*]

[1] School of Information Technology, Guangdong Industry Polytechnic, Guangdong 510300, China; 2014108048@gditc.edu.cn

[2] Multidisciplinary Computational Laboratory, Department of Electrical and Biomedical Engineering, Hanyang University, Seoul 04763, Korea

* Correspondence: mclin@hanyang.ac.kr; Tel.: +82-2-2220-0358

Received: 1 August 2019; Accepted: 30 October 2019; Published: 4 November 2019

Abstract: Light propagation in one-dimensional (1D) photonic crystals (PCs) enclosed in a rectangular waveguide is investigated in order to achieve a complete photonic band gap (PBG) while avoiding the difficulty in fabricating 3D PCs. This work complements our two previous articles (Phys. Rev. E) that quantitatively analyzed omnidirectional light propagation in 1D and 2D PCs, respectively, both showing that a complete PBG cannot exist if an evanescent wave propagation is involved. Here, we present a quantitative analysis of the transmission functions, the band structures, and the photon density of states (PDOS) for both the transverse electric (TE) and transverse magnetic (TM) polarization modes of the periodic multilayer heterostructure confined in a rectangular waveguide. The PDOS of the quasi-1D photonic crystal for both the TE and TM modes are obtained, respectively. It is demonstrated that a "complete PBG" can be obtained for some frequency ranges and categorized into three types: (1) below the cutoff frequency of the fundamental TE mode, (2) within the PBG of the fundamental TE mode but below the cutoff frequency of the next higher order mode, and (3) within an overlap of the PBGs of either TE modes, TM modes, or both. These results are of general importance and relevance to the dipole radiation or spontaneous emission by an atom in quasi-1D periodic structures and may have applications in future photonic quantum technologies.

Keywords: photonic crystals; photonic band gap; waveguide; complete PBG; PDOS; TE; TM

---

1. Introduction

Photonic crystals (PCs), also known as artificial materials, have attracted much attention in the past three decades due to the tremendous needs of gaining complete control over light propagation and emission [1,2]. PCs, according to the dimension of the periodicity, are divided into three categories: one- (1D), two- (2D), and three-dimensional (3D) crystals. Due to the periodicity, the stop band and the pass band are formed, according to the Bloch theorem [3]. Therefore, periodic dielectric materials are characterized by photonic band gaps (PBGs). As an analogy to the electronic band gaps in solid state materials [4], a PBG in PCs can prohibit the propagation of electromagnetic (EM) waves whose frequencies fall within the band gap region. The continuing success of the semiconductor industry in controlling electric properties of materials from the last century has encouraged us to manipulate the flow of light in PCs and control their optical properties. These PC-based materials are expected to have many applications in optoelectronics, optical communications, and photonic quantum technologies in the next decades [5]. In optical range, PCs have been extensively studied. It was proposed that the emission of EM radiation can be modified by the environment [6,7]. Several environments such as metallic cavities [8], dielectric cavities [9], superlattices [10–15], and 2D PCs [16] have been studied.

The environmental effects have been described by the photon density of states (PDOS), which is related to the transition rate of the Fermi golden rule. In principle, a PC with a complete PBG can be best realized in a 3D system to prohibit the propagation of electromagnetic waves of any polarization traveling in any direction from any source [1]. However, the difficulty in fabricating such 3D crystals with PBGs in the optical regime prohibits the progress of many applications.

On the other hand, there has been a lot of interest in microwave and millimeter wave applications of PBG, such as the significant progress in the design of filters [17,18], microstrip antennas [19], and slow wave structures [20,21], and so on. However, the design of PBG in this frequency range is still difficult due to complexities of the modeling. There are too many parameters affecting the PBG properties, such as the number of lattice periods [22], lattice shapes [23,24], lattice spacing [25], and relative volume fraction [26–30]. Since the actual fabrication of 3D PCs remains difficult, another simpler choice is periodic dielectric or PC waveguides, which have only a one-dimensional periodic pattern [1]. The rigorous study of the PC waveguide can be traced back to the 1970s, where a more detailed review can be found [31,32]. Recently, it was demonstrated that, by considering a quantum dot spin coupled to a PC waveguide mode, the light–matter interaction can be asymmetric, leading to unidirectional emission and a deterministic entangled photon source, which might have application in future optical quantum devices [5,33]. One interesting feature of electromagnetism in dielectric media is that there is no fundamental length scale, namely the scaling properties of Maxwell's equations, i.e., the solution of problem at one length scale determines the solutions at all other length scale [1]. In a previous work [34], a multilayer dielectric window in a rectangular waveguide had been studied to achieve a wide-bandwidth transmission of a millimeter wave. A transfer matrix approach was successfully employed to discretize the dielectric function profile and the transmission functions could be calculated efficiently. In principle, the approach can be extended to study a quasi-1D PC, a PC confined in a waveguide. However, the transmission method is limited to study radiation modes in a finite-length system. In order to study the PBG phenomena such as the suppression of spontaneous emission [35] in a quasi-1D PC, the calculation of the dispersion relations or band structures (BS) and the PDOS are needed. Metallic waveguides and cavities are widely used to control microwave propagation. One of the main concerns is visible light energy is quickly dissipated within the metallic components, which makes this method of optical control almost impossible to generalize. Recently, an unconventional superconductivity in magic-angle graphene superlattices had been discovered and studied [36]. The superconductivity might help realize the metallic waveguide confinement of optics in the near future.

In this work, light propagation in 1D PCs enclosed in a rectangular waveguide or quasi-1D PCs is investigated in order to achieve a complete PBG while avoiding the difficulty in fabricating 3D PCs. This work complements two previous articles [15,16] that quantitatively analyzed omnidirectional light propagation in 1D and 2D PCs, respectively, both showing that a complete PBG cannot exist if an evanescent wave propagation is involved. The transfer matrix method is extended to study the transmittance of the quasi-1D PCs for both TE and TM polarization modes [34]. The corresponding BS are obtained by solving the eigenvalue equations with proper periodic boundary conditions following the Bloch theorem [3,4]. The formulas for evaluating the PDOS of the quasi-1D PCs for TE and TM modes are derived, respectively, for determining the PBGs. The contributions of the PDOS from each modes can be distinguished. The model is formulated in Section 2. The calculated results and discussion are presented in Section 3. The conclusions are given in Section 4.

2. Formulations

A transfer matrix approach is employed to discretize the dielectric function profile of the dielectric multilayer heterostructures and the transmission functions are calculated by matching the boundary conditions at each interfaces. In order to solve the PDOS, it is necessary to calculate the dispersion relation, and the corresponding band structures are obtained by solving the eigenvalue equations with proper periodic boundary conditions.

2.1. Transfer Matrix Method

Let us consider a waveguide with its rectangular cross section of sides a and b, and the enclosed multilayer dielectric slab with thickness, $(t_1, t_2, t_1, t_2, ...)$ and dielectric function, $(\varepsilon_1, \varepsilon_2, \varepsilon_1, \varepsilon_2, ...)$, as shown in Figure 1. The TE mode (H mode) and TM mode (E mode) in the rectangular waveguide are characterized by the z components of the magnetic field and the electric field, H_z and E_z, respectively. By definition, these components are never absent in the corresponding modes. The z components of the Helmholtz's equations for the inhomogeneous media are

$$\left\{\varepsilon(z)\vec{\nabla} \times \left[\frac{1}{\varepsilon(z)}\vec{\nabla} \times \vec{H}\right]\right\}_z + \omega^2\varepsilon(z)\mu(z)H_z = 0 \tag{1}$$

and

$$\left\{\mu(z)\vec{\nabla} \times \left[\frac{1}{\mu(z)}\vec{\nabla} \times \vec{E}\right]\right\}_z + \omega^2\varepsilon(z)\mu(z)E_z = 0. \tag{2}$$

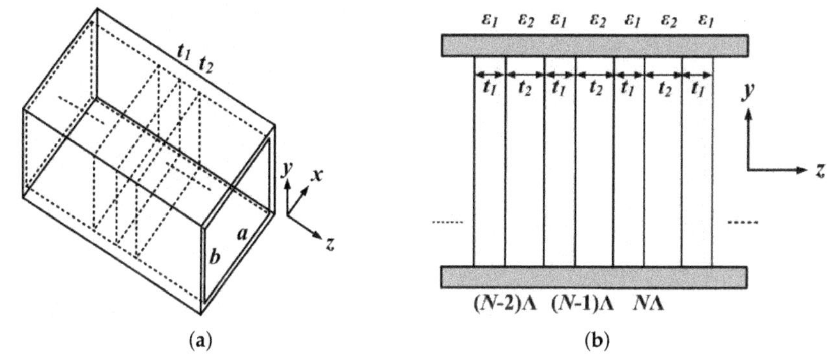

Figure 1. (**a**) 3D schematic of periodic multilayer heterostructure along the z-direction confined in a rectangular waveguide with a width a and a height b and (**b**) the corresponding dielectric function profile $\varepsilon(z)$ in the cross-sectional view.

In these cases, the effect of losses of the medium inside the waveguide is characterized by the complex permittivity $\varepsilon(z)$ and permeability $\mu(z)$. Thus, Equations (1) and (2) can be rearranged as:

$$\left[\frac{\partial^2}{\partial x^2} + \frac{\partial^2}{\partial y^2} + \varepsilon(z)\frac{\partial}{\partial z}\frac{1}{\varepsilon(z)}\frac{\partial}{\partial z}\right] H_z + \omega^2\varepsilon(z)\mu(z)H_z = 0 \tag{3}$$

and

$$\left[\frac{\partial^2}{\partial x^2} + \frac{\partial^2}{\partial y^2} + \mu(z)\frac{\partial}{\partial z}\frac{1}{\mu(z)}\frac{\partial}{\partial z}\right] E_z + \omega^2\varepsilon(z)\mu(z)E_z = 0. \tag{4}$$

By symmetry, using separation of variables, Equations (3) and (4) can be split into transverse and longitudinal parts, and the problem can be simplified as solving the one-dimensional Helmholtz's equation along the z direction, the longitudinal parts,

$$\varepsilon(z)\frac{\partial}{\partial z}\frac{1}{\varepsilon(z)}\frac{\partial}{\partial z}\psi(z)_{TE} + \left[\omega^2\varepsilon(z)\mu(z) - k_c^2\right]\psi(z)_{TE} = 0 \tag{5}$$

and

$$\mu(z)\frac{\partial}{\partial z}\frac{1}{\mu(z)}\frac{\partial}{\partial z}\psi(z)_{TM} + \left[\omega^2\varepsilon(z)\mu(z) - k_c^2\right]\psi(z)_{TM} = 0, \tag{6}$$

with eigenvalues k_c which are determined by the following eigenvalue equations, the transverse parts,

$$\left(\frac{\partial^2}{\partial x^2} + \frac{\partial^2}{\partial y^2} + k_c^2\right) \phi(x,y)_{TE} = 0 \tag{7}$$

and

$$\left(\frac{\partial^2}{\partial x^2} + \frac{\partial^2}{\partial y^2} + k_c^2\right) \phi(x,y)_{TM} = 0. \tag{8}$$

The corresponding boundary conditions for $\phi(x,y)_{TE}$ and $\phi(x,y)_{TM}$ are

$$\tfrac{\partial}{\partial x}\phi(x,y)_{TE}|_{x=0,a} = 0 \quad and \quad \tfrac{\partial}{\partial y}\phi(x,y)_{TE}|_{y=0,b} = 0, \tag{9}$$

and

$$\phi(x,y)_{TM}|_{x=0,a} = 0 \quad and \quad \phi(x,y)_{TM}|_{y=0,b} = 0. \tag{10}$$

It then follows that

$$k_c = \frac{2\pi}{\lambda_c} = \sqrt{\left(\frac{m\pi}{a}\right)^2 + \left(\frac{n\pi}{b}\right)^2}. \tag{11}$$

Applying the eigenvalue equations, Equations (7) and (8), and the boundary conditions, Equations (9) and (10), to the general solutions of Equations (1) and (2), the following particular solutions can be found by separation of variables,

$$H_z(x,y,z) = H_0 \cos\left(\frac{m\pi x}{a}\right) \cos\left(\frac{n\pi y}{b}\right) \psi(z)_{TE} \tag{12}$$

and

$$E_z(x,y,z) = E_0 \sin\left(\frac{m\pi x}{a}\right) \sin\left(\frac{n\pi y}{b}\right) \psi(z)_{TM}, \tag{13}$$

where H_0 and E_0 are determined by the energy of electromagnetic waves propagating inside the waveguide, and m and n are integers. The function $\psi(z)$ is chosen for a particular solution since it represents propagating waves in the z-direction. A transfer matrix approach is employed to discretize the dielectric function profile of the heterostructure. For an N-layer dielectric-filled waveguide, $\varepsilon(z)$ and $\mu(z)$ can be divided into $p = 1, 2, ..., (N+2)$ layers with a piecewise constant permittivity ε_p and constant permeability μ_p, respectively. The discretized one-dimensional Helmholtz's equation, for the pth region with constant permittivity ε_p and constant permeability μ_p can be written as

$$\frac{d^2}{dz^2}\psi_p(z) + k_p^2\psi_p(z) = 0 \text{ for } z_{p-1} \le z \le z_p \tag{14}$$

with

$$k_p = \sqrt{\omega^2\varepsilon_p\mu_p - k_c^2} = \frac{2\pi}{\lambda}\sqrt{\varepsilon_{rp}\mu_{rp} - \left(\frac{\lambda}{\lambda_c}\right)^2}, \tag{15}$$

where $\psi_p(z)$ represents the wave function in the pth layer, and k_p defines the complex wavevector in the same layer along the z-direction, λ represents the wavelength in free space at the operating angular frequency ω, ε_{rp} and μ_{rp} are the relative dielectric constant and permeability of the medium, respectively, and λ_c is the cutoff wavelength. The solutions of Equations (12) and (13) can be written as a superposition of the forward and backward traveling wave functions:

$$\psi_p = a_p \exp(-jk_pz) + b_p \exp(jk_pz) \quad for \quad z_{p-1} \le z \le z_p. \tag{16}$$

The boundary conditions for $\psi(z)$ at the interface between layers p and $(p+1)$ at position $z = z_p$ where $p = 1, 2, ..., (N+1)$ are (for TE modes)

$$\mu_p \psi_p(z_p) = \mu_{p+1} \psi_{p+1}(z_p) \quad and \quad \tfrac{d}{dz} \psi_p(z_p) = \tfrac{d}{dz} \psi_{p+1}(z_p), \tag{17}$$

and (for TM modes)

$$\varepsilon_p \psi_p(z_p) = \varepsilon_{p+1} \psi_{p+1}(z_p) \quad and \quad \tfrac{d}{dz} \psi_p(z_p) = \tfrac{d}{dz} \psi_{p+1}(z_p). \tag{18}$$

By matching the boundary conditions at each discontinuity, we arrive at

$$\begin{pmatrix} a_{N+2} \\ b_{N+2} \end{pmatrix} = M_{N+1} \cdots M_p \cdots M_1 \begin{pmatrix} a_1 \\ b_1 \end{pmatrix} = \begin{pmatrix} M_{11} & M_{12} \\ M_{21} & M_{12} \end{pmatrix} \begin{pmatrix} a_1 \\ b_1 \end{pmatrix}, \tag{19}$$

where for TE modes:

$$M_p = \frac{1}{2} \begin{bmatrix} \exp(jk_{p+1} z_p) & 0 \\ 0 & \exp(-jk_{p+1} z_p) \end{bmatrix} \cdot \begin{pmatrix} \frac{\mu_p}{\mu_{p+1}} + \frac{k_p}{k_{p+1}} & \frac{\mu_p}{\mu_{p+1}} - \frac{k_p}{k_{p+1}} \\ \frac{\mu_p}{\mu_{p+1}} - \frac{k_p}{k_{p+1}} & \frac{\mu_p}{\mu_{p+1}} + \frac{k_p}{k_{p+1}} \end{pmatrix} \cdot \begin{bmatrix} \exp(-jk_{p+1} z_p) & 0 \\ 0 & \exp(jk_{p+1} z_p) \end{bmatrix}, \tag{20}$$

and for TM modes:

$$M_p = \frac{1}{2} \begin{bmatrix} \exp(jk_{p+1} z_p) & 0 \\ 0 & \exp(-jk_{p+1} z_p) \end{bmatrix} \cdot \begin{pmatrix} \frac{\varepsilon_p}{\varepsilon_{p+1}} + \frac{k_p}{k_{p+1}} & \frac{\varepsilon_p}{\varepsilon_{p+1}} - \frac{k_p}{k_{p+1}} \\ \frac{\varepsilon_p}{\varepsilon_{p+1}} - \frac{k_p}{k_{p+1}} & \frac{\varepsilon_p}{\varepsilon_{p+1}} + \frac{k_p}{k_{p+1}} \end{pmatrix} \cdot \begin{bmatrix} \exp(-jk_{p+1} z_p) & 0 \\ 0 & \exp(jk_{p+1} z_p) \end{bmatrix}. \tag{21}$$

Using Equation (19) with $a_1 = 1$, $b_1 = r$, $a_{N+2} = t$, and $b_{N+2} = 0$, the reflection and transmission amplitudes, r and t, can be obtained, respectively, by

$$r = -\frac{M_{21}}{M_{22}} \tag{22}$$

and

$$t = \frac{M_{11} \cdot M_{22} - M_{12} \cdot M_{21}}{M_{22}}. \tag{23}$$

The reflection and transmission coefficients, R and T, can be implicitly represented by S-parameters, $S_{11}(dB)$ and $S_{12}(dB)$, as a function of the operating frequency, , for TE modes

$$S_{11}(dB) = 10log\left|\tfrac{M_{21}}{M_{22}}\right|^2 \quad and \quad S_{12}(dB) = 10log\tfrac{\mu_1 k_{N+2}}{\mu_{N+2} k_1}\left|\tfrac{M_{11}\cdot M_{22} - M_{12}\cdot M_{21}}{M_{22}}\right|^2, \tag{24}$$

and for TM modes

$$S_{11}(dB) = 10log\left|\tfrac{M_{21}}{M_{22}}\right|^2 \quad and \quad S_{12}(dB) = 10log\tfrac{\varepsilon_1 k_{N+2}}{\varepsilon_{N+2} k_1}\left|\tfrac{M_{11}\cdot M_{22} - M_{12}\cdot M_{21}}{M_{22}}\right|^2. \tag{25}$$

2.2. Band Structures

In this model, E_z and H_z both are periodic with period Λ. According to the Bloch theorem, the electric and magnetic fields in a periodic layered medium are $E_z(z) = E_z(z+\Lambda)$ and $H_z(z) = H_z(z+\Lambda)$, respectively. The column vector of the Bloch wave satisfies the following eigenvalue equation for consistency

$$\begin{pmatrix} a_3 \\ b_3 \end{pmatrix} = \frac{1}{4} \begin{pmatrix} A_{TE/TM} & B_{TE/TM} \\ C_{TE/TM} & D_{TE/TM} \end{pmatrix} \cdot \begin{pmatrix} a_1 \\ b_1 \end{pmatrix} = e^{jk_B\Lambda} \begin{pmatrix} a_1 \\ b_1 \end{pmatrix}. \quad (26)$$

For the TE mode:

$$A_{TE} = e^{j(k_1t_1-k_2t_2)} \left(\frac{\mu_1}{\mu_2} - \frac{k_1}{k_2}\right)\left(\frac{\mu_2}{\mu_1} - \frac{k_2}{k_1}\right) + e^{-j(k_1t_1+k_2t_2)} \left(\frac{\mu_1}{\mu_2} + \frac{k_1}{k_2}\right)\left(\frac{\mu_2}{\mu_1} + \frac{k_2}{k_1}\right), \quad (27)$$

$$B_{TE} = e^{-j[k_1t_1-k_2(2t_1+t_2)]} \left(\frac{\mu_1}{\mu_2} + \frac{k_1}{k_2}\right)\left(\frac{\mu_2}{\mu_1} - \frac{k_2}{k_1}\right) + e^{j[k_1t_1+k_2(2t_1+t_2)]} \left(\frac{\mu_1}{\mu_2} - \frac{k_1}{k_2}\right)\left(\frac{\mu_2}{\mu_1} + \frac{k_2}{k_1}\right), \quad (28)$$

$$C_{TE} = B^*_{TE}, \quad (29)$$

and

$$D_{TE} = A^*_{TE}. \quad (30)$$

For the TM mode:

$$A_{TM} = e^{j(k_1t_1-k_2t_2)} \left(\frac{\varepsilon_1}{\varepsilon_2} - \frac{k_1}{k_2}\right)\left(\frac{\varepsilon_2}{\varepsilon_1} - \frac{k_2}{k_1}\right) + e^{-j(k_1t_1+k_2t_2)} \left(\frac{\varepsilon_1}{\varepsilon_2} + \frac{k_1}{k_2}\right)\left(\frac{\varepsilon_2}{\varepsilon_1} + \frac{k_2}{k_1}\right), \quad (31)$$

$$B_{TM} = e^{-j[k_1t_1-k_2(2t_1+t_2)]} \left(\frac{\varepsilon_1}{\varepsilon_2} + \frac{k_1}{k_2}\right)\left(\frac{\varepsilon_2}{\varepsilon_1} - \frac{k_2}{k_1}\right) + e^{j[k_1t_1+k_2(2t_1+t_2)]} \left(\frac{\varepsilon_1}{\varepsilon_2} - \frac{k_1}{k_2}\right)\left(\frac{\varepsilon_2}{\varepsilon_1} + \frac{k_2}{k_1}\right), \quad (32)$$

$$C_{TM} = B^*_{TM}, \quad (33)$$

and

$$D_{TM} = A^*_{TM}. \quad (34)$$

The phase factor $e^{jk_B\Lambda}$ is thus the eigenvalue and satisfies the secular equation

$$\begin{vmatrix} A_{TE/TM} - e^{jk_B\Lambda} & B_{TE/TM} \\ C_{TE/TM} & D_{TE/TM} - e^{jk_B\Lambda} \end{vmatrix} = 0. \quad (35)$$

Finally, the dispersion relation for the Bloch wave function is

$$k_B(k_z, \omega) = \frac{1}{t_1+t_2} \cos^{-1}[\frac{1}{2}(A_{TE/TM} + D_{TE/TM})]. \quad (36)$$

2.3. Photon Density of States

The quasi-one-dimensional photonic crystal has been confined in the xy-plane, so the wave vectors k_x and k_y are determined according to the guiding modes. To perform the PDOS calculation, it is required to use the formal definition which is the number of available photon modes per unit frequency range. Then, we construct two frequencies, namely, $\omega(k_B) = \omega$ and $\omega(k_B) = \omega + \Delta\omega$, where $\Delta\omega$ is a small increment. We calculate the line therein, and divide it by the line segment occupied by a single mode. The differential line element in $\boldsymbol{K}$ space within the frequency range, is given by $\Delta L_k = \Delta k_B$. Now, Δk_B is defined as

$$\Delta k_B = \frac{\Delta\omega}{|\nabla_k \omega|}. \tag{37}$$

Integrating over the frequency increment, we have that the total phase space line segment contributing to the frequency range $(\omega, \omega + d\omega)$ is

$$\int_{\omega_k} dL_k = \int_{\omega_k} \frac{1}{|\nabla_k \omega|} d\omega, \tag{38}$$

where we take the limit of infinitesimal increments. The number of modes within the range $(\omega, \omega + d\omega)$ is obtained by dividing the length calculated in Equation (38) by the line segment corresponding to one mode, $2\pi/\Lambda$ in the phase space. This yields

$$dN(\omega) = \frac{\Lambda}{2\pi} \int_{\omega_k} \frac{1}{|\nabla_k \omega|} d\omega \equiv D(\omega) d\omega. \tag{39}$$

Because ω is a function of k , we can write

$$\nabla_k \omega = \frac{d\omega}{dk_B} \hat{\mathbf{z}}. \tag{40}$$

For the TE mode:

$$\nabla_k \omega_{TE} = -\frac{(t_1 + t_2) \sin[0.5 k_B (t_1 + t_2)]}{\alpha_1 + \alpha_2 + \alpha_3 + \alpha_4 + \alpha_5 + \alpha_6} \hat{\mathbf{z}} \tag{41}$$

and for the TM mode:

$$\nabla_k \omega_{TM} = -\frac{(t_1 + t_2) \sin[0.5 k_B (t_1 + t_2)]}{\beta_1 + \beta_2 + \beta_3 + \beta_4 + \beta_5 + \beta_6} \hat{\mathbf{z}}, \tag{42}$$

where the functions, α_i and β_i, with $i = 1, 2, ..., 6$, are listed in the Appendix A. Finally, the formula for evaluating the PDOS can be expressed as

$$D(\omega)_{TE/TM} = \frac{\Lambda}{2\pi} \int_{\omega_k} \frac{1}{|\nabla_k \omega_{TE/TM}|} d\omega. \tag{43}$$

3. Results and Discussion

All of macroscopic electromagnetism, including the propagation of light in a photonic crystal, is governed by the four macroscopic Maxwell's equations with no free charges or currents. One interesting feature of electromagnetism in PCs is that there is no fundamental length scale other than the assumption that the system is macroscopic [1]. Therefore, to study physical phenomena in PCs, one may scale a system from the optical frequency range to the microwave one and vice versa if suitable conditions are fulfilled. For the purposes of demonstration and easier verification by experimentalists, a WR28 (7.11 mm $\times$ 3.555 mm) rectangular waveguide, usually used for Ka-band millimeter waves, is chosen. The periodic dielectric heterostructure is arranged along the longitudinal (z) direction to form the quasi-1D PCs [34]. In the microwave or millimeter wave frequency ranges, the PC experiments are very popular and more affordable compared to those in the optical frequency ones. Nevertheless, in order to extend the results for general use, the data are normalized for solutions at all length scale. Consider the quasi-1D PCs with the arrangement shown in Figure 1, 15 double-layer stacks (n = 30) have been included in the waveguide for investigation. The transmittance is calculated using the transfer matrix approach mentioned above. In order to demonstrate the stop band and pass band, the transmittances, expressed as the S-parameters, S_{12}, in dB, of the lowest TE and TM modes for three quasi-1D PCs have been calculated and plotted in Figure 2. For the three cases, the dielectric constants used are the same as $\varepsilon_1 = 3.8$ (quartz) and $\varepsilon_2 = 1.0$ (air), while the thicknesses are varied as (t_1, t_2) = (1.00, 3.30), (1.00, 3.60), and (1.00, 3.90) mm, corresponding to the filling ratios $t_1/\Lambda = 23.26\%$, 21.74%, and 20.41% for the periods of the stacks Λ = 4.3, 4.6, and 4.9 mm, respectively. As one can see, the central frequencies of the PBGs for both the TE_{10} and TM_{11} modes, as shown in Figure 2a,b respectively, are shifted to lower values as the filling ratio decreases. As mentioned above, the frequency axes in the plots are normalized to the cutoff frequency of TE_{10} for general use.

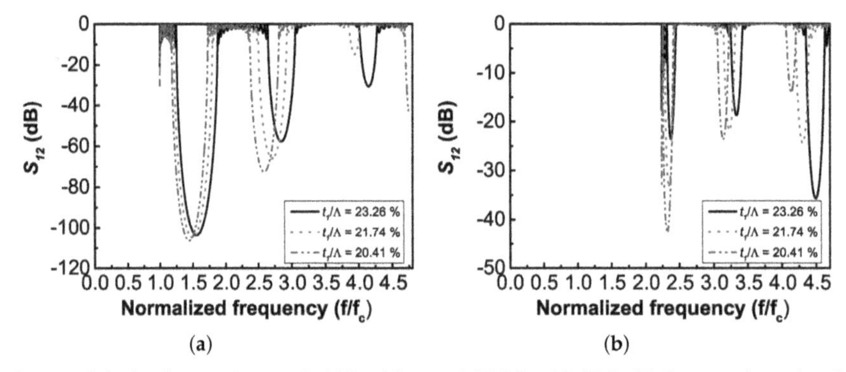

Figure 2. Calculated transmittance, S_{12}(dB), of the quasi-1D PCs with 15 double-layer stacks enclosed in the rectangular waveguide for (**a**) TE_{10} and (**b**) TM_{11} modes of light propagation. The dielectric constants used are $\varepsilon_1 = 3.8$ and $\varepsilon_2 = 1.0$, and the corresponding filling ratios are $t_1/\Lambda = 23.26\%$, 21.74%, and 20.41% in the three cases. The frequency axes are normalized to the cutoff frequency of TE_{10}.

In order to look for a complete PBG for all possible modes in the quasi-1D PCs, one may want to enlarge a PBG for a specific mode of choice. For the following four cases, the dielectric constants used are (ε_1, ε_2) = (2.3, 1.0), (3.8, 1.0), (4.9, 1.0), and (11.4, 1.0), while the thicknesses are varied as (t_1, t_2) = (2.15, 2.15), (1.00, 3.30), (0.72, 3.58), and (0.27, 4.03) mm, corresponding to the filling ratios $t_1/\Lambda = 50.00\%$, 23.26%, 16.74%, and 6.28% for the fixed period of the stacks Λ = 4.3 mm, respectively. The dielectric constants of 2.3, 3.8, 4.9 and 11.4 used in the chosen microwave frequency range correspond to the dielectric materials, polyethylene, quartz, phenolic resin, and barium sulfate, respectively. As one can see, the calculated transmittances, S_{12}(dB), of the four quasi-1D PCs with different 15 double-layer stacks for the TE_{10} and TM_{11} modes are plotted in Figure 3a,b, respectively. The width of PBGs for both the TE_{10} and TM_{11} modes in the quasi-1D PCs is widened to larger sizes as decreasing the filling ratio while increasing the dielectric contrast between ε_1 and ε_2. One may notice that the configurations of these quasi-1D PCs have been specially arranged so that the central frequencies of the PBGs of the TE_{10} mode are kept the same while tuning the PBG sizes. However, those of the PBGs of the TM_{11} mode are not centralized. This indicates that a tremendous computational effort is still needed to find a complete PBG in spite of the approach with a closed form developed is very efficient for the quasi-1D PCs.

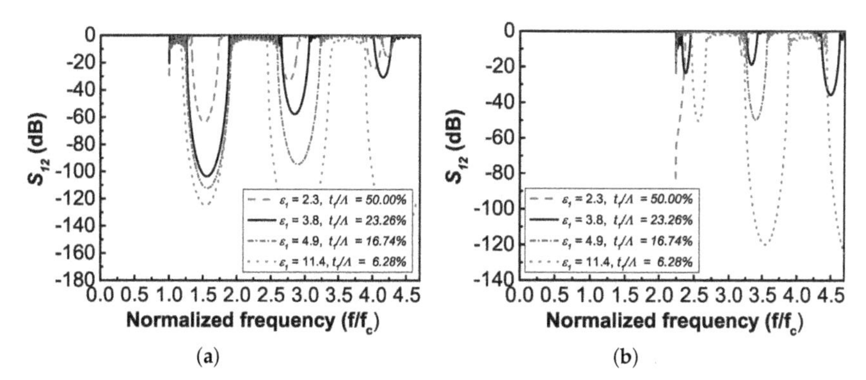

(a) (b)

Figure 3. Calculated transmittance, S_{12}(dB), of the quasi-1D PCs with 15 double-layer stacks enclosed in the rectangular waveguide for (**a**) TE_{10} and (**b**) TM_{11} modes of light propagation. The dielectric constants used are ε_1 = 2.3, 3.8, 4.9, and 11.4 and $\varepsilon_2 = 1$, and the corresponding filling ratios are $t_1/\Lambda = 50.00\%$, 23.26%, 16.74%, and 6.28% in the four cases. The frequency axes are normalized to the cutoff frequency of TE_{10}.

A better way to identify PBGs more accurately and efficiently is to perform the BS calculation and further compute the PDOS [15,16]. Figure 4 shows the band structures of (a) TE_{10} and (b) TM_{11} modes for the quasi-1D PC with the dielectric constants $\varepsilon_1 = 3.8$, $\varepsilon_2 = 1$ and the thicknesses $t_1 = 1.00$ mm, $t_2 = 3.30$ mm, corresponding to the case presented in Figures 2 and 3 in a black solid line with a filling ratio of 23.26%. The PBGs of the TE_{10} mode show no overlap with those of TM_{11} mode, all marked in gray stripes in the figure. However, from Figure 4, one may notice that there are no photon states in the frequency ranges below the cutoff frequency of the TE_{10} mode nor within the first PBG of TE_{10} which is under the cutoff frequency of the TM_{11} mode. Note that only the lowest TE amd TM modes are plotted in Figure 4. As there might be other modes involved within the frequency range of the PBG of interest, it is easier to identify a complete PBG from the PDOS plots compared to the BS ones. Figure 5 shows the PDOS of TE_{10}, TE_{01}, TE_{11}, and TM_{11} modes for the same quasi-1D PC. As one can see, the PDOS of the TE_{10} and TM_{11} modes are consistent with the BS calculations shown in Figure 4. The PDOS contributed from the TE_{01} and TE_{11} modes tend to fill up the first PBG of the TE_{10} mode. Other higher order modes are not considered as their cutoff frequencies are too high to contribute any PDOS in the frequency range of the PBG. Finally, the combined PDOS of TE_{10}, TE_{01}, TE_{11}, and TM_{11} modes shows no photon states in some frequency ranges. The first one is below the cutoff frequency of the TE_{10} mode, (0–0.77)f_c, the second one is within the first PBG of TE_{10} but under the cutoff frequency of the TE_{01} mode, (1.26–1.48)f_c, and the third one is the overlap of the PBGs of TE_{10}, TE_{01}, and TE_{11} modes, (1.79–1.87)f_c. Therefore, a "complete PBG" can be obtained for some frequency ranges and categorized into three types: (1) below the cutoff frequency of the fundamental TE mode, (2) within the PBG of the fundamental TE mode but below the cutoff frequency of the next higher order mode, and (3) within an overlap of the PBGs of either TE modes, TM modes, or both.

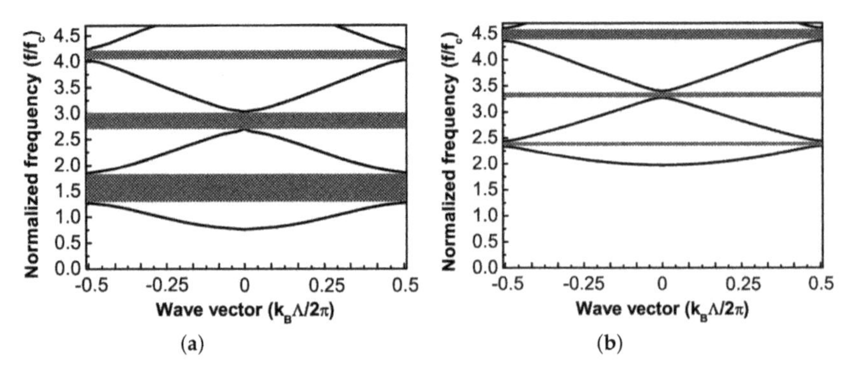

Figure 4. Band structures of (**a**) TE_{10} and (**b**) TM_{11} modes for the quasi-1D PC with the dielectric constants $\varepsilon_1 = 3.8$, $\varepsilon_2 = 1$ and the thicknesses $t_1 = 1.00$ mm, $t_2 = 3.30$ mm, corresponding to a filling ratio of 23.26%. The PBGs of TE_{10} mode show no overlap with those of TM_{11} mode.

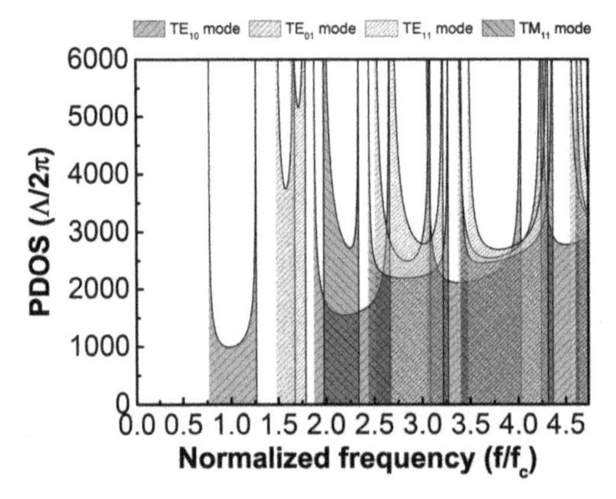

Figure 5. PDOS of TE_{10}, TE_{01}, TE_{11}, and TM_{11} modes for the quasi-1D PC with the dielectric constants $\varepsilon_1 = 3.8$, $\varepsilon_2 = 1$ and the thicknesses $t_1 = 1.00$ mm, $t_2 = 3.30$ mm, corresponding to a filling ratio of 23.26%. The combined PDOS of TE_{10}, TE_{01}, TE_{11}, and TM_{11} modes shows no photon states in some frequency ranges.

4. Conclusions

In summary, light propagation in quasi-1D PCs has been investigated quantitatively. The transmittances for both the TE and TM modes through a periodic multilayer heterostructure in a rectangular waveguide are calculated using the transfer matrix method. The corresponding band structures are obtained by solving the eigenvalue equations with proper periodic boundary conditions following the Bloch theorem. The formulas for determining the PDOS have been obtained to facilitate identifying the photonic band gaps for all the modes residing in the system. With our approach, the quantitative determination of PDOS can be performed very accurately and efficiently. A complete PBG can exist in these quasi-1D PCs, but the determination must be carefully conducted and verified. It is demonstrated that three types of "complete PBG" can be found and categorized as follows. The first type is the frequency range within which the TE and TM modes are both cutoff, the second type is for which the fundamental TE mode has a PBG while other higher order modes are cutoff, and the last type is an overlap of the PBGs of either TE modes, TM modes, or both. The model might be easier for an experimental validation in a millimeter wave frequency range while the optical counterpart might be possibly pursued as well. We believe these results are of general importance and relevance to the dipole radiation or spontaneous emission by an atom in quasi-1D periodic structures and may have applications in future photonic quantum technologies.

Author Contributions: Conceptualization, M.-C.L.; methodology, R.-F.J. and M.-C.L.; software, R.-F.J. and M.-C.L.; validation, R.-F.J. and M.-C.L.; formal analysis, R.-F.J.; investigation, R.-F.J. and M.-C.L.; resources, R.-F.J. and M.-C.L.; data curation, R.-F.J.; writing—original draft preparation, R.-F.J. and M.-C.L.; writing—review and editing, R.-F.J. and M.-C.L.; visualization, R.-F.J.; supervision, M.-C.L.; project administration, M.-C.L.; funding acquisition, R.-F.J. and M.-C.L.

Funding: This work was partially supported by Guangdong Industry Polytechnic, P. R. China, under Grant Nos. KYRC2018-001 and RC2016-005, the research fund of Hanyang University (HY- 201400000002393), National Research Foundation of South Korea (2015R1D1A1A01061017), and the Alexander von Humboldt Foundation of Germany.

Acknowledgments: The authors would like to thank the late B. Y. Gu at the Institute of Physics, CAS and C. T. Chan at the Department of Physics, HKUST for the helpful comments, discussions, and encouragement.

Conflicts of Interest: The authors declare no conflict of interest. The funders had no role in the design of the study; in the collection, analyses, or interpretation of data; in the writing of the manuscript, or in the decision to publish the results.

Abbreviations

The following abbreviations are used in this manuscript:

1D	one-dimensional
2D	two-dimensional
3D	three-dimensional
BS	band structure
EM	electromagnetic
PBG	photonic band gap
PC	photonic crystal
PDOS	photon density of states
TE	transverse electric
TM	transverse magnetic

Appendix A. FORMULAS

Here, we give the formulas of the functions employed in Equations (41) and (42).

$$\alpha_1 = \exp[-j(k_1t_1 + k_2t_2)]\,[\exp(j2k_1t_1) + \exp(j2k_2t_2)] \left[\frac{\omega}{k_1}\left(\frac{\varepsilon_1\mu_1k_2}{k_1^2} - \frac{\varepsilon_2\mu_2}{k_2}\right)\right]\left(-\frac{k_1}{k_2} + \frac{\mu_1}{\mu_2}\right), \tag{A1}$$

$$\alpha_2 = \exp[-j(k_1t_1 + k_2t_2)]\{1 + \exp[j2(k_1t_1 + k_2t_2)]\} \left[\frac{\omega}{k_1}\left(-\frac{\varepsilon_1\mu_1k_2}{k_1^2} + \frac{\varepsilon_2\mu_2}{k_2}\right)\right]\left(\frac{k_1}{k_2} + \frac{\mu_1}{\mu_2}\right), \tag{A2}$$

$$\alpha_3 = \exp[-j(k_1t_1 + k_2t_2)]\,[\exp(j2k_1t_1) + \exp(j2k_2t_2)] \left[\frac{\omega}{k_2}\left(-\frac{\varepsilon_1\mu_1}{k_1} + \frac{\varepsilon_2\mu_2k_1}{k_2^2}\right)\right]\left(-\frac{k_2}{k_1} + \frac{\mu_2}{\mu_1}\right), \tag{A3}$$

$$\alpha_4 = -j\exp[-j(k_1t_1 + k_2t_2)]\,[\exp(j2k_1t_1) - \exp(j2k_2t_2)] \left[\omega\left(-\frac{\varepsilon_1\mu_1t_1}{k_1} + \frac{\varepsilon_2\mu_2t_2}{k_2}\right)\right]\left(-\frac{k_1}{k_2} + \frac{\mu_1}{\mu_2}\right)\left(\frac{k_2}{k_1} + \frac{\mu_2}{\mu_1}\right), \tag{A4}$$

$$\alpha_5 = \exp[-j(k_1t_1 + k_2t_2)]\{1 + \exp[j2(k_1t_1 + k_2t_2)]\}$$
$$\left[\frac{\omega}{k_2}\left(\frac{\varepsilon_1\mu_1}{k_1} - \frac{\varepsilon_2\mu_2k_1}{k_2^2}\right)\right]\left(\frac{k_2}{k_1} + \frac{\mu_2}{\mu_1}\right), \tag{A5}$$

$$\alpha_6 = -j\exp[-j(k_1t_1 + k_2t_2)]\{-1 + \exp[j2(k_1t_1 + k_2t_2)]\}$$
$$\left[\omega\left(-\frac{\varepsilon_1\mu_1t_1}{k_1} - \frac{\varepsilon_2\mu_2t_2}{k_2}\right)\right]\left(\frac{k_1}{k_2} + \frac{\mu_1}{\mu_2}\right)\left(\frac{k_2}{k_1} + \frac{\mu_2}{\mu_1}\right), \tag{A6}$$

$$\beta_1 = \exp[-j(k_1t_1 + k_2t_2)]\,[\exp(j2k_1t_1) + \exp(j2k_2t_2)]$$
$$\left[\frac{\omega}{k_1}\left(\frac{\varepsilon_1\mu_1k_2}{k_1^2} - \frac{\varepsilon_2\mu_2}{k_2}\right)\right]\left(-\frac{k_1}{k_2} + \frac{\varepsilon_1}{\varepsilon_2}\right), \tag{A7}$$

$$\beta_2 = \exp[-j(k_1t_1 + k_2t_2)]\{1 + \exp[j2(k_1t_1 + k_2t_2)]\}$$
$$\left[\frac{\omega}{k_1}\left(-\frac{\varepsilon_1\mu_1k_2}{k_1^2} + \frac{\varepsilon_2\mu_2}{k_2}\right)\right]\left(\frac{k_1}{k_2} + \frac{\varepsilon_1}{\varepsilon_2}\right), \tag{A8}$$

$$\beta_3 = \exp[-j(k_1t_1 + k_2t_2)]\,[exp(j2k_1t_1) + exp(j2k_2t_2)]$$
$$\left[\frac{\omega}{k_2}\left(-\frac{\varepsilon_1\mu_1}{k_1} + \frac{\varepsilon_2\mu_2k_1}{k_2^2}\right)\right]\left(-\frac{k_2}{k_1} + \frac{\varepsilon_2}{\varepsilon_1}\right), \tag{A9}$$

$$\beta_4 = -j\exp[-j(k_1t_1 + k_2t_2)]\,[\exp(j2k_1t_1) - \exp(j2k_2t_2)]$$
$$\left[\omega\left(-\frac{\varepsilon_1\mu_1t_1}{k_1} + \frac{\varepsilon_2\mu_2t_2}{k_2}\right)\right]\left(-\frac{k_1}{k_2} + \frac{\varepsilon_1}{\varepsilon_2}\right)\left(\frac{k_2}{k_1} + \frac{\varepsilon_2}{\varepsilon_1}\right), \tag{A10}$$

$$\beta_5 = \exp[-j(k_1t_1 + k_2t_2)]\{1 + \exp[j2(k_1t_1 + k_2t_2)]\}$$
$$\left[\frac{\omega}{k_2}\left(\frac{\varepsilon_1\mu_1}{k_1} - \frac{\varepsilon_2\mu_2k_1}{k_2^2}\right)\right]\left(\frac{k_2}{k_1} + \frac{\varepsilon_2}{\varepsilon_1}\right), \tag{A11}$$

and

$$\beta_6 = -j\exp[-j(k_1t_1 + k_2t_2)]\{-1 + \exp[j2(k_1t_1 + k_2t_2)]\}$$
$$\left[\omega\left(-\frac{\varepsilon_1\mu_1t_1}{k_1} - \frac{\varepsilon_2\mu_2t_2}{k_2}\right)\right]\left(\frac{k_1}{k_2} + \frac{\varepsilon_1}{\varepsilon_2}\right)\left(\frac{k_2}{k_1} + \frac{\varepsilon_2}{\varepsilon_1}\right). \tag{A12}$$

References

1. Joannopoulos, J.D.; Johnson, S.G.; Winn, J.N.; Meade, R.D. *Photonic Crystals: Molding the Flow of Light*; Princeton University Press: Princeton, NJ, USA, 2008.
2. Joannopoulos, J.D.; Villeneuve, P.R.; Fan, S. Photonic crystals: putting a new twist on light. *Nature* **1997**, *386*, 143–149. [CrossRef]
3. Yariv, A.; Yeh, P. *Optical Waves in Crystals*; Wiley: New York, NY, USA, 1984.
4. Kittel, C. *Introduction to Solid State Physics*; Wiley: New York, NY, USA, 1976.
5. O'Brien, J.L.; Furusawa, A.; Vučković, J. Photonic quantum technologies. *Nat. Photonics* **2009**, *3*, 687–691. [CrossRef]
6. Purcell, E.M. Spontaneous emission probabilities at radio frequencies. *Phys. Rev.* **1946**, *69*, 681.
7. Kleppner, D. Inhibited Spontaneous Emission. *Phys. Rev. Lett.* **1981**, *47*, 233–236. [CrossRef]

8. Barut, A.O.; Dowling, J.P. Quantum electrodynamics based on self-energy: Spontaneous emission in cavities. *Phys. Rev. A* **1987**, *36*, 649–654. [CrossRef] [PubMed]
9. Rigneault, H.; Monneret, S. Modal analysis of spontaneous emission in a planar microcavity. *Phys. Rev. A* **1996**, *54*, 2356–2368. [CrossRef]
10. Dowlin, J.P.; Bowden, C.M. Atomic emission rates in inhomogeneous media with applications to photonic band structures. *Phys. Rev. A* **1992**, *46*, 612–622. [CrossRef]
11. Suzuki, T.; Yu, P.K.L. Emission power of an electric dipole in the photonic band structure of the fcc lattice. *Opt. Soc. Am. B* **1995**, *12*, 570–582. [CrossRef]
12. Kamli, A.; Babiker, M.; Al-Hajry, A.; Enfati, N. Dipole relaxation in dispersive photonic band-gap structures. *Phys. Rev. A* **1997**, *55*, 1454–1461. [CrossRef]
13. Sánchez, A.S.; Halevi, P. Spontaneous emission in one-dimensional photonic crystals. *Phys. Rev. E* **2005**, *72*, 056609. [CrossRef]
14. Halevi, P.; Sánchez, A.S. Spontaneous emission in a high-contrast one-dimensional photonic crystal. *Opt. Commun.* **2005**, *251*, 109–114. [CrossRef]
15. Lin, M.C.; Jao, R.F. Quantitative analysis of photon density of states for a realistic superlattice with omnidirectional light propagation. *Phys. Rev. E* **2006**, *74*, 046613. [CrossRef] [PubMed]
16. Jao, R.F.; Lin, M.C. Efficient and quantitative analysis of photon density of states for two-dimensional photonic crystals with omnidirectional light propagation. *Phys. Rev. E* **2018**, *98*, 053306. [CrossRef]
17. Rumsey, I.; Melinda, P.M.; Kelly, P.K. Photonic bandgap structures used as filters in microstrip circuits. *IEEE Microw. Guided Wave Lett.* **1998**, *8*, 336–338. [CrossRef]
18. Kim, T.; Seo, C. A novel photonic bandgap structure for low-pass filter of wide stopband. *IEEE Microw. Guided Wave Lett.* **2000**, *10*, 13–15.
19. Iluz, Z.; Shavit, R.; Bauer, R. Microstrip antenna phased array with electromagnetic bandgap substrate. *IEEE Trans. Antennas Prop.* **2004**, *52*, 1446–1453. [CrossRef]
20. Yang, F.R.; Qian, Y.; Coccioli, R.; Itoh, T. A novel low slow-wave microstrip structure. *IEEE Microw. Guided Wave Lett.* **1998**, *8*, 372–374. [CrossRef]
21. Xue, Q.; Shum, K.; Chan, C. Novel 1D microstrip PBG cells. *IEEE Microw. Wirel. Compon. Lett.* **2000**, *10*, 403–405.
22. Yang, H.W.; Zhong, W.X.; Sui, Y.K. Analysis of dielectric layer PBG structure using precise integration. In Proceedings of the International Symposium on Computational Mechanics, Beijing, China, 30 July–1 August 2007; pp. 1239–1245.
23. Kshetrimayum, R.S.; Zhu, L. Guided-wave characteristics of waveguide based periodic structures loaded with various FSS strip layers. *IEEE Trans. Antennas Prop.* **2005**, *53*, 120–124. [CrossRef]
24. Marini, S.; Covers, Á.; Boria, V.E.; Gimeno, B. Full-wave modal analysis of slow-wave periodic structures loaded with elliptical waveguides. *IEEE Trans. Electron Devices* **2010**, *57*, 516–524. [CrossRef]
25. Razavizadeh, S.M.; Sadeghzadeh, R.; Navarri-Cía, M.; Ghattan, Z. Compact THz waveguide filter based on periodic dielectric-gold rings. In Proceedings of the 5th International Conference on Millimeter-Wave and Terahertz Technologies, Tehran, Iran, 18–20 December 2018; pp. 42–44.
26. Pelster, R.; Gasparian, V.; Nimtz, G. Propagation of plane waves and of waveguide modes in quasiperiodic dielectric heterostructures. *Phys. Rev. E* **1997**, *55*, 7645–7655. [CrossRef]
27. Amari, S.; Vahldieck, R.; Bornemann, J.; Leuchtmann, P. Propagation in a circular waveguide periodically loaded with dielectric disks. *IEEE MTT-S Digest.* **1998**, *3*, 1535–1538.
28. Kesari, V.; Basu, B.N. Analysis of some periodic structures of microwave tubes: part II: Analysis of disc-loaded fast-wave circular waveguide structures for gyro-travelling-wave tubes. *J. Electomagn. Waves Appl.* **2018**, *32*, 1–36. [CrossRef]
29. Christie, L.; Erabati, G.; Jana, M. Analysis of propagation characteristics of circular waveguide loaded with dielectric disks using coupled integral equation technique. In Proceedings of the 5th International Conference on Advances in Computing and Communications, Kochi, India, 2–4 September 2015; pp. 231–234.
30. Miller, R.D.; Jones, T.B. On the effective dielectric constant of columns or layers of dielectric spheres. *J. Phys. D Appl. Phys.* **1988**, *21*, 527–532. [CrossRef]
31. Peng, S.T.; Tamir, T.; Bertoni, H.L. Theory of periodic dielectric waveguides. *IEEE Trans. Microw. Theory Technol.* **1975**, *MTT-23*, 123–133. [CrossRef]

32. Villeneuve, P.R.; Fan, S.; Johnson, S.G.; Joannopoulos, J.D. Three-dimensional photonic confinement in photonic crystals of low-dimensional periodicity. *IEE Proc. Optoelectron.* **1998**, *145*, 384–390. [CrossRef]
33. Young, A.B.; Thijssen, A.C.T.; Beggs, D.M.; Androvitsaneas, P.; Kuipers, L.; Rarity, J.G.; Hughes, S.; Oulton, R. Polarization engineering in photonic crystal waveguides for spin-photon entanlers. *Phys. Rev. Lett.* **2015**, *115*, 153901. [CrossRef] [PubMed]
34. Lin, M.C. A multilayer waveguide window for wide-bandwidth millimeter wave tubes. *Int. J. Infrared Millim. Waves* **2007**, *28*, 355–362. [CrossRef]
35. Yablonovitch, E. Inhibited spontaneous emission in solid-state physics and electronics. *Phys. Rev. Lett.* **1987**, *58*, 2059–2062. [CrossRef]
36. Cao, Y.; Fatemi, V.; Fang, S.; Watanabe, K.; Taniguchi, T.; Kaxiras, E.; Jarillo-Herrero, P. Unconventional superconductivity in magic-angle graphene superlattices. *Nature* **2018**, *556*, 43–50. [CrossRef]

Article

Spatial Beam Filtering with Autocloned Photonic Crystals

Pei-Yu Wang, Yi-Chen Lai and Yu-Chieh Cheng *

Department of Electro-Optical Engineering, National Taipei University of Technology, No. 1, Sec. 3, Chung-Hsiao E. Rd, Taipei 10608, Taiwan; peggy831110pp@gmail.com (P.-Y.W.); 26789jimmy@gmail.com (Y.-C.L.)

* Correspondence: yu-chieh.cheng@mail.ntut.edu.tw

Received: 18 October 2019; Accepted: 6 November 2019; Published: 8 November 2019

Abstract: We have been numerically demonstrated the mechanism of spatial beam filtering with autocloned photonic crystals. The spatial filtering through different configurations of the multilayered structures based on a harmonically modulated substrate profile is considered. The paper demonstrates a series of parameter studies to look for the best spatial beam filtering performance. The optimization results show that a beam spectral width of 39.2° can be reduced to that of 5.92°, leading to high potential applications for integrated optical microsystems.

Keywords: photonic crystal; beam shaping; angular filtering; autocloning; multilayered structures

1. Introduction

Angular/spatial filtering devices based on photonic crystals (PhCs) [1,2] provide diffraction of the angular components of an incident beam. The effect of a PhCs-based spatial/angular filtering device that works on a spatial frequency spectrum relies on an angular band-gap [3–7]. For spatial filtering, a range of angular components of a beam can be removed due to the angular band-gaps, that is, the waves can be reflected in a backward direction [3–5] or deflected at large angles in a forward one [6,7].

Furthermore, double-periodic photonic structures enable manipulation of the zero diffraction order of a transmitted beam [8]. For example, some angular components of an incident light source diffract from the zero diffraction order to the other orders at resonance conditions. On the other hand, some angular components, out of resonance, directly propagate through the PhCs. In this way, low-angle-pass or high-angle-pass filtering devices are achievable through a proper interplay among the grating characteristics.

In particular, the PhCs filtering has been already implemented for intracavity angular filtering in an integrated platform such as microchip lasers [9]. Such a PhCs-based confocal filtering device presents an alternative method for replacing conventional filtering devices [10], but has a critical disadvantage that is the presence of an optical axis [11]. Therefore, the transmitted axisymmetric concentric ring structures result in the limitation of angular filter merely for on-axis incident light.

It is noted that there have been other approaches to spatial filtering such as passive [12] or light-induced [13] Bragg gratings and pulse-induced population density gratings [14]. However, these alternative methods require not only more sophisticated schemes but additional optical components, leading to limited applications in the compact micro-systems. Therefore, more compact PhCs-based angular filters are desirable solutions for the use in the microlasers, e.g., autocloned PhCs, which preserves the initial modulation during the deposition of multilayers [15]. For example, a multilayered photonic microstructure based on a sinusoidal or braze profile was demonstrated experimentally as one of the most compact PhCs-based angular filters [16]. However, this proposed filter presents the weak filtering performance, so further investigation is required for practical use.

For example, a compact filter with a transverse invariance performs both the narrow angular bandwidth and the high efficiency. The filtering for the application of such a compact filter toward a Gaussian beam has not been well studied in the prior study [16]. Although the fabrication of the autocloned PhCs has been demonstrated experimentally [17,18], it is still unavoidable that the variation of the amplitude (Amp) of the harmonic modulation of the autoclaved PhCs increases with the number of layers [19]. Further studies for feasible parameters in fabrication such as less number of layers are a concern.

In this paper, we provide a numerical study of an angular/spatial filtering based on multilayered PhCs gratings with a harmonically modulated substrate profile. The multilayered gratings are all-dielectric and periodic, where the variation of the periods or wavelength results in a low- or high-angle-pass filter. A spectral width (SW), defined as the full spectral width at half maximum (FWHM), is calculated by using the finite-difference time-domain (FDTD) method [20]. To enhance the spatial filtering, we focus on narrowing filtering angular distributions by the design of the low-angle-pass filter and the results are compared with the study in [16].

2. Numerical Far-Field Simulations for Autocloned PhCs

The configuration of the proposed multilayered PhCs grating for spatial filtering is schematically shown in Figure 1. The multilayers consisting of alternating high-refractive-index (N_H) and low-refractive-index (N_L) layers with the number of layers (N) on top of a grating substrate (N_{sub}) that results in a sinusoidal profile with a peak-to-peak Amp. Such a multilayered grating with a sinusoidal profile provides a transversal and a longitudinal modulations of the period (Dx) and an alternating thicknesses (Dz) of the multilayers. The thicknesses of the high- and low-refractive-index layers are equal to Dz/2. The wavelength of the incident light is equal to λ = 582 nm and a transverse electric (TE) polarization is considered. A more detailed calculation procedure is further described in the following section, Method.

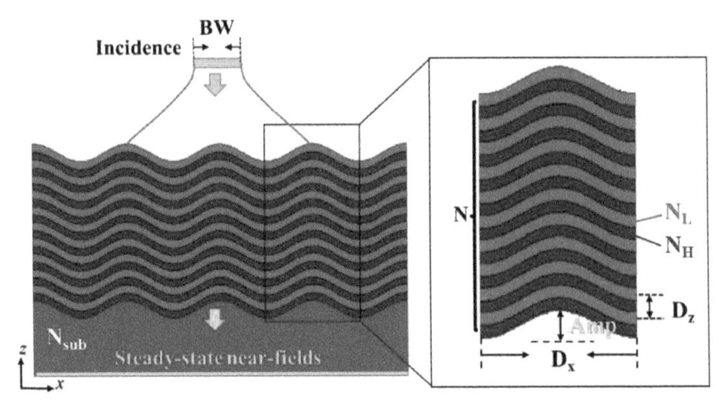

Figure 1. Illustration of the proposed autocloning photonic crystals (PhCs) with a harmonic modulation. The inset shows the parameters of the wavy structure. Dx and Dz represent the horizontal and longitudinal periods, respectively. N is the number of layers. Low- and high-refractive-index materials are shown in red and blue colors, respectively. The parameter Amp indicates the peak-to-peak value of the multilayer structure, also regarded as the amplitude of the harmonic modulation. The beam width (BW) represents the full spectral width at half maximum (FWHM) of a launched beam. The steady-state near-field plane is defined as the output plane of the simulation domain.

We first analyze diffraction patterns of the periodic multilayered structure by using a rigorous coupled-wave analysis (RCWA) method [21] and its 0th-order transmission is as plotted in Figure 2a. The parameters of the studied structure are identical as those in [16]: N = 33 layers, Dx = 1.67 µm, Dz = 0.24 µm, N_{Sub} = 1.5, N_H = 1.42, and N_L = 1.3. Since the parameter of Amp was not mentioned in [16], we found the narrowest SW by scanning Amp up to 0.5 μm. The narrowest SW of 9.21° is obtained for Amp = 0.27 μm. Furthermore, the angular dependence of the 0th-order diffraction efficiency (DE_{t0}) (Appendix A Method) presents a low-pass filtering design where the angular components at around 10° are removed or coupled out to the others, as shown in Figure 2b. Thus, the far-field distribution of a Gaussian beam, regarded as different plane-wave components at different angles of incidence, can be narrowed down. Figure 2c,d show that the filtered far-field distributions for two incident beams with different beam widths (BW1= 1 λ and BW2 = 2 λ). Their filtered SWs, SW1′ (9.21°) and SW2′ (9.42°), are narrower than SW1 (39.2°) and SW2 (20.99°), respectively.

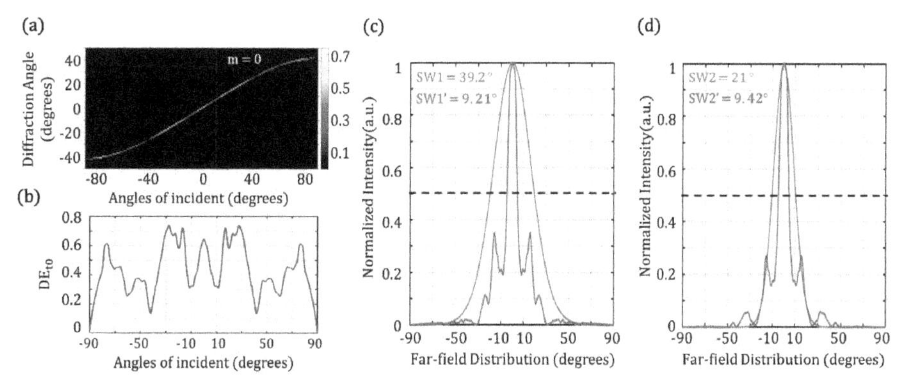

Figure 2. Numerical far-field simulation of autocloning-mode-based PhCs: (**a**) the diffraction map of 0th order versus angles of incidence; (**b**) diffraction efficiency of 0th-order transmission (DE_{t0}) with respect to angles of incidence; far-field distributions for two different incident Gaussian beam widths: (c) BW1 = λ; (**d**) BW2 = 2λ. The parameters of the structure in (**a**) and (**b**) are as following: N = 33 layers, Amp = 0.27 μm, Dz = 0.24 μm, and Dx = 1.67 μm. These two spectral width SW1 and SW2 mean the FWHM of the spectral width of the Gaussian beam for BW1 = λ and BW2 = 2λ, respectively. SW1 and SW1′ in (**c**) represent the normalized far-field distributions of the Gaussian beam passing without PhCs in blue and with the PhCs in red for BW1 = λ. SW2 and SW2′ in (**d**) represent the normalized far-field distributions of the Gaussian beam passing without PhCs in blue and with the PhCs in red for BW2 = 2λ.

3. Minimum SWs for Different Configurations of Autocloned PhCs

In this section, a series of numerical calculations is executed for the best filtering effect. We focus on scanning structural parameters such as Amp and Dz for obtaining a narrow FWHM of SW. First, three configurations of the wavy structures are considered with different layer numbers (N = 40, 30, and 20). Three related maps of their SWs are calculated by scanning two parameters Amp from 0.005 to 0.5 μm and Dz from 0.2 to 0.4 μm, as shown in Figure 3a. It is noted that their transversal periods are identical, referred to [15].

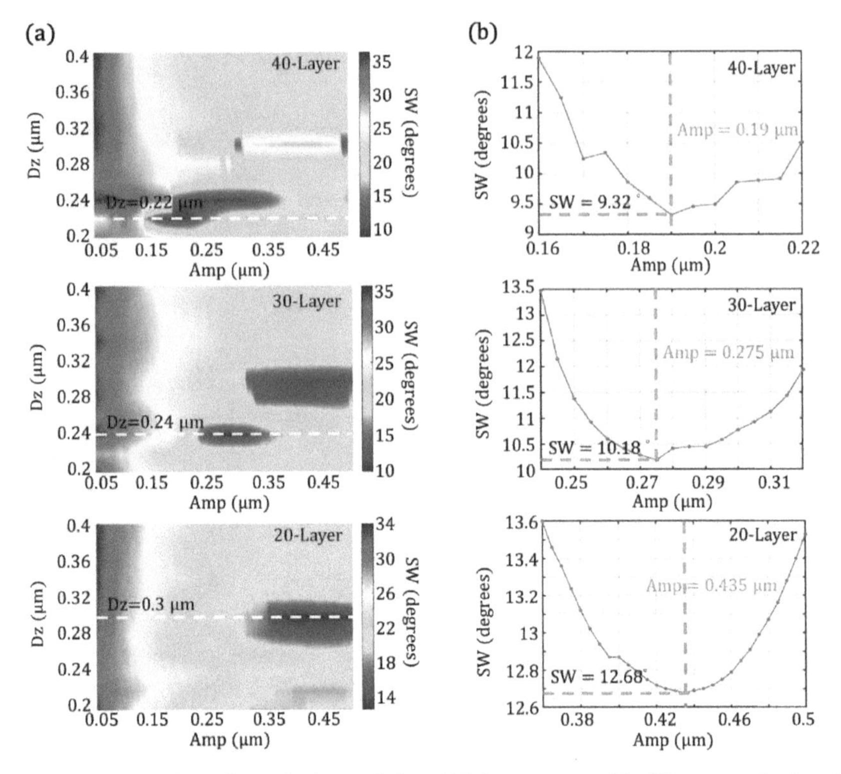

Figure 3. (**a**) The SWs of autocloning-mode-based PhCs structures with different amplitudes of harmonic modulation and longitudinal periods Dz for different layer structures. The smallest SW can be obtained at Dz indicated by the dashed white lines. (**b**) Variation of the SW as a function of Amp at Dz indicated by the dashed white lines in (**a**).

The smallest SWs for the 40-, 30-, and 20-layer configurations are found at the longitudinal periods Dz of 0.22, 0.24, and 0.3 μm, respectively. The corresponding variances of the SWs at the specific Dz, highlighted by white dashed lines in Figure 3a, are shown in Figure 3b. The optimal Amps of these three harmonic structures with N = 40, 30, and 20 are 0.19, 0.275, and 0.435 μm, respectively. As a result, in the three configurations with the different layer numbers, the 40-layer structure presents the smallest SW of 9.32° where BW of 1 λ is considered.

Furthermore, the diffraction efficiency for the center of the filtered beam is also considered as depicted in Figure 4. The variations of the SW and DE_{t0} with respect to Dz are studied for the three different configurations of the wavy structures. The SWs of 9.32°, 10.18°, and 12.68° and DE_{t0} values of 42%, 60%, and 58% are achievable for the 40-, 30-, and 20-layer structures, respectively. Although the narrowest SW of 9.32° of a filtering beam is achievable by the configuration with 40 layers, it brings in the lowest diffraction efficiency.

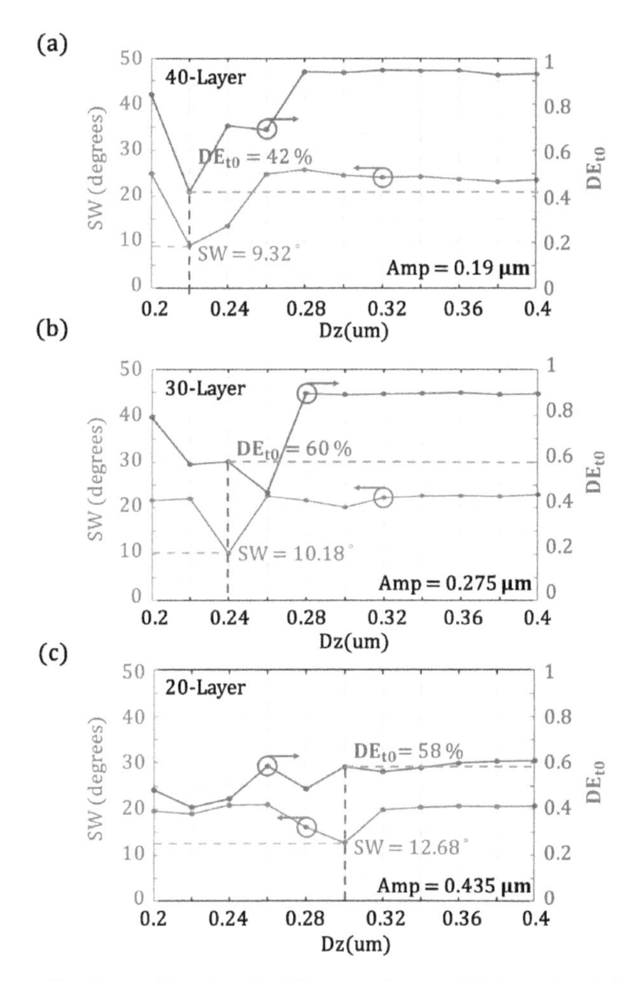

Figure 4. The smallest SWs and the 0th-order diffraction efficiency (DE_{t0}) varied with the longitudinal period Dz for three autocloning-mode-based PhCs with N = 40 layers (**a**), 30 layers (**b**), and 20 layers (**c**). The minimum of the SW of 9.32° can be obtained for the 40-layer structure at Dz = 0.22 μm and Amp = 0.19 μm. The minimum of the SW of 10.18° can be obtained for the 30-layer structure at Dz = 0.24 μm and Amp = 0.275 μm. The minimum of the SW of 12.68° can be obtained for the 20-layer structure at Dz = 0.3 μm and Amp = 0.435 μm. The DE_{t0} values of the 40-, 30-, and 20-layered structure are 42%, 60%, and 58%, respectively. The transversal period D*x* is constant, i.e., 1.67 μm.

For all the above simulations, the transversal period Dx of 1.67 μm is considered for the low-pass filtering design. As generally known, the transversal period of PhCs plays an important role in diffraction patterns. As the transversal period Dx is changed, the far-field distribution changes dramatically, leading to the low- and high-pass filtering effects. Hence, Figure 5 shows the variation of far-field distributions by varying Dx for three configurations with 40-, 30-, and 20-layer structures. The best low-angle-pass filtering performance (SW = 8.89° and DE_{t0} = 60%) is found at the 30-layer structure with Amp = 0.275 μm, Dz = 0.24 μm, and Dx = 1.7 μm.

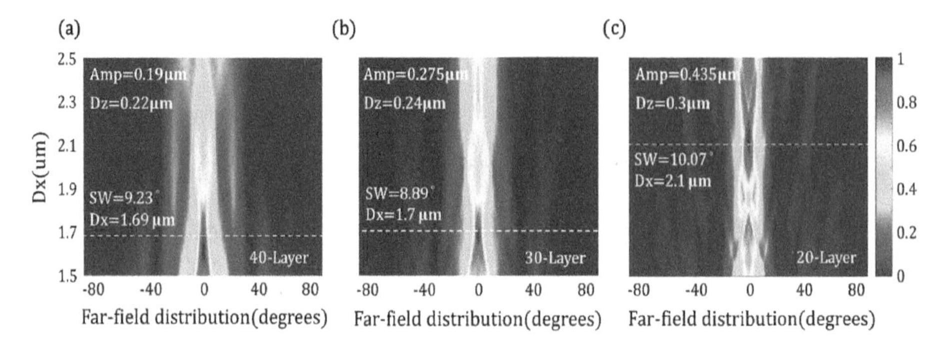

Figure 5. 2-D maps of normalized far-field distributions with respect to the angle of incidence and transversal period Dx for the multilayered structures with 40 layers (**a**), 30 layers (**b**), and 20 layers (**c**). In the 40-layer structure, the minimum FWHM is 9.23° at Dx = 1.69 μm with the DE_{t0} of 39%. In the 30-layer structure, the minimum FWHM is 8.89° at Dx = 1.7 μm with the DE_{t0} of 60%. In the 20-layer structure, the minimum FWHM is 10.07° at Dx = 2.1 μm with the DE_{t0} of 49%.

A more optimal parameter Dx of 1.7 μm is found after scanning the far-field distribution with respect to Dz, as shown in Figure 4. Even though the longitudinal period Dx of 1.67 μm is considered for the previously studied structure, it has shown that the procedure of the search of optimal parameters such as Amp, layers, and Dz for the low-pass angular filtering is beneficial. For example, as shown in Figures 3 and 4, we obtain SWs of 12.68° and 9.32° and DE_{t0} values of 58% and 42% for the two configurations with 20 layers and 40 layers, respectively. Although the 40-layer structure presents stronger spatially filtering, it requires more layers to achieve the narrower SW; however, the DE_{t0} is lower.

Furthermore, the number of layers is also a concern for the fabrication of a wavy-like multilayered structure because the experimental modulation of the wavy structure may be reduced with the increased number of layers [19]. Even if the fabrication of the 20-layer structure may be controllable, a weaker filtering effect is obtained. Obviously, the fabricating feasibility concerning the number of layers should be also considered at the stage in the design process. Therefore, a series of the analysis of structural parameters is helpful to define the optimization range and consider fabrication feasibility.

In our work, the optimization of this wavy structure by using the simplex method has been demonstrated, as shown in Figure 6. All parameters of the multilayered structure are optimized simultaneously and the target of the optimization is to realize a low-pass filter design that achieves the narrowest SW for a Gaussian beam. To ensure fabrication feasibility, the optimizing range of the number of the layer should not be more than 40 layers. The optimizing ranges of Amp and the longitudinal period Dz should be less than 0.5 μm and 0.4 μm, respectively, as referred to the previous study. As a result, the final optimal parameters of the structure are obtained as follows: N = 27, Amp = 0.36 μm, Dz = 0.26 μm, and Dx = 2 μm. The far-field distributions in Figure 6b,c show the narrowest SW1′ = 5.92° and SW2′ = 6.58° for two Gaussian beams with BW1 = λ and BW2 = 2λ, respectively. The DE_{t0} of the optimal result is 64%. The narrower SW and the higher diffraction efficiency are achieved by the optimized structure with the less number of layer, compared with that of the previous study. Therefore, the optimizing results not only present the best spatial filtering to narrow the SW, but also demonstrate the use of the structures with fewer layers is a more feasible approach to a manufacturing process.

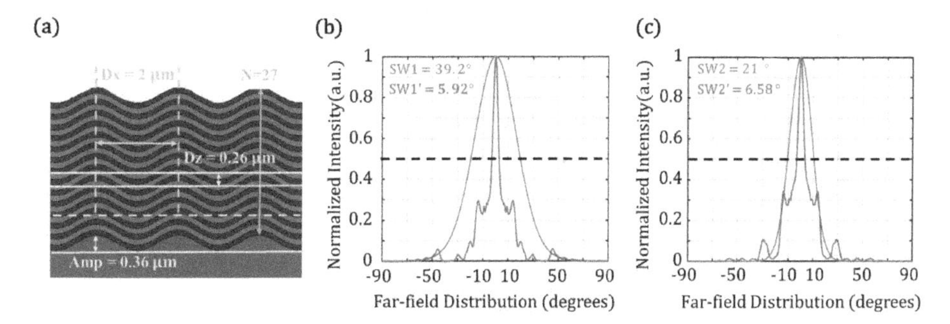

Figure 6. The optimization results of the wavy structure by using the simplex method. (**a**) The illustration of the optimal structures with parameters: N = 27, Amp = 0.36 μm, Dz = 0.26 μm, and Dx = 2 μm. Far-field distributions for two different incident Gaussian beam widths BW1 = λ (**b**) and BW2 = 2λ (**c**), where λ is the operating wavelength. The narrowest FWHM of a Gaussian beam with BW = λ is 5.92°.

4. Conclusions

This paper has demonstrated the PhCs filtering effect for a light beam. The autocloned PhCs present transversal and longitudinal periods, the key element to modulate spatial spectra. The paper has studied different configurations of the multilayered structures based on the harmonic modulation. The narrowest SW of 8.89° and the diffraction efficiency of 60% are obtained for a 30-layer structure after a series of the scanning procedure in the structural parameters.

Considering feasible fabrication, the optimization has been further studied for practical use by using the simplex method. The number of layers N for our optimal structure is reduced to 27 and a stringer spatial filtering performance provides an SW of 5.92° and the diffraction efficiency of 0th transmission is 64%. For an autocloned PhC with several layers, another approach, such as an autocloned blazed modulation, could be more applicable for fabrication, although the fabrication of a tilted blazed profile is not as simple as that of the sinusoidal modulation.

Author Contributions: Conceptualization, Y.-C.C.; methodology, software and validation, Y.-C.C., Y.-C.L. and P.-Y.W.; investigation, writing—original draft preparation, writing—review and writing—editing, Y.-C.C. and P.-Y.W.; visualization, Y.-C.C., P.-Y.W. and Y.-C.L.; supervision, Y.-C.C.; funding acquisition, Y.-C.C.

Funding: This work was financially supported by the Young Scholar Fellowship Program by the Ministry of Science and Technology (MOST) in Taiwan, under grant number MOST108-2636-M-027-001.

Acknowledgments: The authors would like to thank the Headquarters of University Advancement at National Cheng Kung University (NCKU) for the funding support of the Higher Education Sprout Project of Ministry of Education (MOE).

Conflicts of Interest: The authors declare no conflict of interest. The funders had no role in the design of the study; in the collection, analyses, or interpretation of data; in the writing of the manuscript, or in the decision to publish the results.

Appendix A. Method

The diffraction efficiency of zero-order transmission is calculated by using the RCWA method, which solves full vectorial versions of Maxwell's equations. It covers a wide range of scattering problems on the wavy structures with horizontally periodic boundary conditions and vertically perfectly matched layer (PML) boundary. The incident and transmitted fields are defined at the simulation domain along the z-direction. A plane wave incidence is used in the RCWA simulation.

The far-field distribution is obtained by using two methods. First, the FDTD method provides steady-state near-field distributions at an output plane while an incident Gaussian beam with a BW considered. Second, we use the discrete Fourier transform (DFT) to convert the steady-state near-field distribution to the far-field distribution. The SW is defined as the FWHM of the far-field distribution. The Gaussian beam is launched at the position of 1 μm above the wavy structure. The horizontal simulation domain is equal to 50 periods of the proposed wavy structure and the vertical boundary condition is the PML.

References

1. Yablonovitch, E. Inhibited Spontaneous Emission in Solid-State Physics and Electronics. *Phys. Rev. Lett.* **1987**, *58*, 2059–2062. [CrossRef] [PubMed]
2. Sajeev, J. Strong localization of photons in certain disordered dielectric superlattices. *Phys. Rev. Lett.* **1987**, *58*, 2468–2489.
3. Luo, Z.; Tang, Z.; Xiang, Y.; Luo, H.; Wen, S. Polarization-independent low-pass spatial filters based on one-dimensional photonic crystals containing negative-index materials. *Appl. Phys. B* **2009**, *94*, 641–646. [CrossRef]
4. Colak, E.; Cakmak, A.O.; Serebryannikov, A.E.; Ozbay, E. Spatial filtering using dielectric photonic crystals at beam-type excitation. *J. Appl. Phys.* **2010**, *108*, 113106. [CrossRef]
5. Hamam, R.E.; Celanovic, I.; Soljačić, M. Angular photonic band gap. *Phys. Rev. A* **2011**, *83*, 035806. [CrossRef]
6. Maigyte, L.; Gertus, T.; Peckus, M.; Trull, J.; Cojocaru, C.; Sirutkaitis, V.; Staliunas, K. Signatures of light-beam spatial filtering in a three-dimensional photonic crystal. *Phys. Rev. A* **2010**, *82*, 043819. [CrossRef]
7. Maigyte, L.; Staliunas, K. Spatial filtering with photonic crystals. *Appl. Phys. Rev.* **2010**, *2*, 011102. [CrossRef]
8. Purlys, V.; Maigyte, L.; Gailevicius, D.; Peckus, M.; Malinauskas, M.; Staliunas, K. Spatial filtering by chirped photonic crystals. *Phys. Rev. A* **2013**, *87*, 033805. [CrossRef]
9. Gailevicius, D.; Koliadenko, V.; Purlys, V.; Peckus, M.; Taranenko, V.; Staliunas, K. Photonic Crystal Microchip Laser. *Sci Rep.* **2016**, *6*, 34173. [CrossRef] [PubMed]
10. Chen, G.; Leger, J.R.; Gopinath, A. Angular filtering of spatial modes in a vertical-cavity surface-emitting laser by a Fabry–Perot étalon. *Appl. Phys. Lett.* **1999**, *74*, 1069–1071. [CrossRef]
11. Koch, B.J.; Leger, J.R.; Gopinath, A.; Wang, Z.; Morgan, R.A. Single-mode vertical cavity surface emitting laser by graded-index lens spatial filtering. *Appl. Phys. Lett.* **1997**, *70*, 2359–2361. [CrossRef]
12. Svyakhovskiy, S.E.; Skorynin, A.A.; Bushuev, V.A.; Chekalin, S.V.; Kompanets, V.O.; Maydykovskiy, A.I.; Murzina, T.V.; Mantsyzov, B.I. Experimental demonstration of selective compression of femtosecond pulses in the Laue scheme of the dynamical Bragg diffraction in 1D photonic crystals. *Opt. Express.* **2014**, *22*, 31002–31007. [CrossRef] [PubMed]
13. Brown, A.W.; Xiao, M. All-optical switching and routing based on an electromagnetically induced absorption grating. *Opt. Lett.* **2005**, *30*, 699–701. [CrossRef] [PubMed]
14. Arkhipov, R.M.; Pakhomov, A.V.; Arkhipov, M.V.; Babushkin, I.; Demircan, A.; Morgner, U.U.; Rosanov, N.N. Population density gratings induced by few-cycle optical pulses in a resonant medium. *Sci. Rep.* **2017**, *7*, 12467. [CrossRef] [PubMed]
15. Kawakami, S. Fabrication of submicrometre 3D periodic structures composed of Si/SiO/sub 2/. *Electron. Lett.* **1997**, *33*, 1260–1261. [CrossRef]
16. Grineviciute, L.; Babayigit, C.; Gailevičius, D.; Bor, E.; Turduev, M.; Purlys, V.; Tolenis, T.; Kurt, H.; Staliunas, K. Angular filtering by Bragg photonic microstructures fabricated by physical vapour deposition. *Appl. Surf. Sci.* **2019**, *481*, 353–359. [CrossRef]
17. Ohtera, Y.; Kawashima, T.; Sakai, Y.; Sato, T.; Yokohama, I.; Ozawa, A.; Kawakami, S. Photonic crystal waveguides utilizing a modulated lattice structure. *Opt. Lett.* **2002**, *27*, 2158–2160. [CrossRef] [PubMed]
18. Notomi, M.; Shinya, A.; Kuramochi, E.; Yokohama, I.; Tamamura, T.; Takahashi, J.; Takahashi, C.; Kawashima, T.; Kawakami, S. Si-Based Composite-Dimensional Photonic Crystals towards Si Photonics, Technical Digest. In Proceedings of the CLEO/Pacific Rim 2001 4th Pacific Rim Conference on Lasers and Electro-Optics, Chiba, Japan, 15–19 July 2001.
19. Huang, C.Y.; Ku, H.M.; Chao, S. Surface profile control of the autocloned photonic crystal by ion-beam-sputter deposition with radio-frequency-bias etching. *Appl. Opt.* **2009**, *48*, 69–73. [CrossRef] [PubMed]

20. Taflove, A. *Computational Electrodynamics: The Finite-Difference Time-Domain Method*; Artech House: Norwood, MA, USA, 1995; pp. 93–105.
21. Moharam, M.G.; Grann, E.B.; Pommet, D.A.; Gaylord, T.K. Formulation for stable and efficient implementation of the rigorous coupled-wave analysis of binary gratings. *J. Opt. Soc. Am. A* **1995**, *12*, 1068–1076. [CrossRef]

 *crystals*

Article

Dual Photonic–Phononic Crystal Slot Nanobeam with Gradient Cavity for Liquid Sensing

Nan-Nong Huang [1], Yi-Cheng Chung [1], Hsiao-Ting Chiu [1], Jin-Chen Hsu [2,*], Yu-Feng Lin [1], Chien-Ting Kuo [1], Yu-Wen Chang [1], Chun-Yu Chen [1] and Tzy-Rong Lin [1,3,4,*]

1 Department of Mechanical and Mechatronic Engineering, National Taiwan Ocean University, Keelung 20224, Taiwan; b0120@mail.ntou.edu.tw (N.-N.H.); whitesheep29@gmail.com (Y.-C.C.); 00672026@email.ntou.edu.tw (H.-T.C.); cblinyubee@gmail.com (Y.-F.L.); a1998011431@gmail.com (C.-T.K.); 00672032@email.ntou.edu.tw (Y.-W.C.); jerry9902023@hotmail.com (C.-Y.C.)
2 Department of Mechanical Engineering, National Yunlin University of Science and Technology, Yunlin 64002, Taiwan
3 Institute of Optoelectronic Sciences, National Taiwan Ocean University, Keelung 20224, Taiwan
4 Center of Excellence for Ocean Engineering, National Taiwan Ocean University, Keelung 20224, Taiwan
* Correspondence: hsujc@yuntech.edu.tw (J.-C.H.); trlin@ntou.edu.tw (T.-R.L.)

Received: 14 April 2020; Accepted: 22 May 2020; Published: 25 May 2020

Abstract: A dual photonic–phononic crystal slot nanobeam with a gradient cavity for liquid sensing is proposed and analyzed using the finite-element method. Based on the photonic and phononic crystals with mode bandgaps, both optical and acoustic waves can be confined within the slot and holes to enhance interactions between sound/light and analyte solution. The incorporation of a gradient cavity can further concentrate energy in the cavity and reduce energy loss by avoiding abrupt changes in lattices. The newly designed sensor is aimed at determining both the refractive index and sound velocity of the analyte solution by utilizing optical and acoustic waves. The effect of the cavity gradient on the optical sensing performance of the nanobeam is thoroughly examined. By optimizing the design of the gradient cavity, the photonic–phononic sensor has significant sensing performances on the test of glucose solutions. The currently proposed device provides both optical and acoustic detections. The analyte can be cross-examined, which consequently will reduce the sample sensing uncertainty and increase the sensing precision.

Keywords: photonic crystal; phononic crystal; sensor; sensitivity; figure of merit

1. Introduction

Photonic crystals (PTCs) are periodically arranged structures composed of materials with different refractive indices (RIs). These periodic structures have photonic band gaps (PTBGs) that can block light propagation for a wave with a specific range of frequencies, which, thereby, affects optical wave behaviors in the artificial crystal [1]. PTCs can be designed for measuring RI. Due to their high sensing performances for high sensitivity, real-time and label-free features, PTCs have been developed into sensing applications [2–6], including biomedical sensing and environmental detection. In the same manner as PTCs, phononic crystals (PNCs) consist of periodic materials with different mass densities and elastic coefficients. Besides, PNC structures have phononic band gaps (PNBGs) as well. Recently, PNC sensors have received increasing attention [7,8]. More recently, photonic–phononic crystal sensors with dual PTBGs and PNBGs have also been investigated [9–11]. Researchers have created the defect, i.e., cavity, in a periodic crystal so that light/sound is confined in the cavity to produce a localized defect state for performing both photonic and phononic sensing.

Single nanobeams and dual nanobeams have been studied intensively in recent years because of their high-quality factor and small mode volume. Dual nanobeam consists of two beams with a slot between the beams. Each beam is patterned with a one-dimensional line of holes. Wang et al. [12] verified experimentally that the dual nanobeam exhibits high sensibility and quality factors in refractive index sensing for glucose solutions. At the same time, a gradual lattice variation in the defect has been introduced to enhance the energy confinement in cavities [13,14]. Researchers have shown that the gradient design can reduce the mode mismatch at the interface between the cavity space region and the PTC mirrors, which can lead to lower energy loss and higher quality factor for the cavity.

In this study, a dual crystal nanobeam incorporated with a gradient cavity is proposed as the photonic–phononic sensor. The optical and acoustic waves were utilized to detect the physical parameters RI and sound velocity of glucose solutions. The gradient cavity was characterized by gradual variations in the diameter and spacing of holes. By introducing the gradient cavity, the optical and acoustic waves can further be confined in the slot and holes, and avoid energy leakage from the edges of the nanobeam sensor. It can be expected to have a stronger interaction between light/sound and analyte solution, and thus, yield a higher sensing performance. In addition, the current simulation showed that the gradient cavity is indispensable for exciting localized acoustic resonances.

The usage of dual photonic–phononic sensing can increase the selectivity of the sample analysis. A "single sensing" based on either optical sensing or acoustic sensing alone may have trouble in assessing unequivocally the analyte that could be misplaced or contaminated. The simultaneous detection of the optical and acoustic properties from a single device can cross-examine the analyte, which consequently reduces the sensing uncertainty and increases the sensing precision.

This paper is organized as follows. The design of the crystal sensor and theories are presented in Section 2, which includes the geometry of the nanobeam sensor, material properties of the device and glucose solutions, and fundamental theories for the optical and acoustic modal analyses. In Section 3, the numerical results of both the optical and acoustic analyses and the sensing performance for photonic–phononic sensors are given. Pertinent discussions on the results are also included in this section. Section 3.1 focuses on the optical sensing of RI of the glucose solutions. The effect of the cavity gradient on the sensor performance is examined in detail. Meanwhile, the acoustic sensing for the sound velocity of the solutions is shown in Section 3.2. Finally, Section 4 concludes this study.

2. Structure Design and Theory

The current photonic analysis was based on the electromagnetic waves theory. Meanwhile, the phononic analysis was based on the wave propagation theory of elastic solids and fluids. The finite element software COMSOL Multiphysics [15] was employed to analyze the acoustic and optical characteristics of the photonic–phononic sensor, which was immersed in the glucose solution. In the current numerical analysis, to simulate the infinite solution domain, the perfect matching layers (PML) were placed on the exteriors of a rectangular parallelepiped that contains the nanobeam.

2.1. Device Design

The structures of coupled nanobeams with slot waveguide can increase the sensitivity of optical sensing due to the enhanced overlap between the localized optical field and the analyte in the slot region [16]. The PTC slot nanobeam slow-light waveguides for RI sensing was investigated by Wang et al. [12]. On the basis of their study, a gradient cavity version of the device for both the optical and acoustic sensing applications is proposed. The schematic three-dimensional view and top view of the device are shown in Figures 1a and 1b, respectively. The enlarged area of the gradient cavity is also depicted in Figure 1b. The lattice constants a_i and hole diameters d_i, as indicated in the figure, vary according to the following formulations:

$$\begin{aligned} a_i &= a \times [0.98 - (i-1) \times N]\,,\ d_i = d \times [1 - (i-1) \times N]\,,\ (i = 1 \sim 9) \\ a &= 490\text{nm}\,,\ d = 310\text{nm}\,,\ N = 0.01, 0.02, 0.03, 0.04 \end{aligned} \tag{1}$$

where a and d are periodic lattice constant and regular hole diameter, respectively. Variable N, referred to as the hole tuning parameter, controls the cavity gradient. A larger value of N represents a larger variation in hole diameter and spacing. In this study, the effect of N, on the sensing performance of the device was examined. The design of this gradient cavity structure was classified as a heterogeneous-type resonant cavity [17], which is aimed at making the optical energy more tightly confined in the slot and holes. It is noted from Figure 1b that the length of the slot was also dependent on N. To prevent energy loss from two ends, mirrors with a lattice constant of 490 nm and hole diameter of 400 nm were added [12], as indicated in Figure 1b. The top and cross-section views of the unit cell of the corresponding perfect crystal structure are presented in Figure 1c, which shows that the width of a single nanobeam was 450 nm, the width of the slot was 200 nm, and the beam thickness was 220 nm.

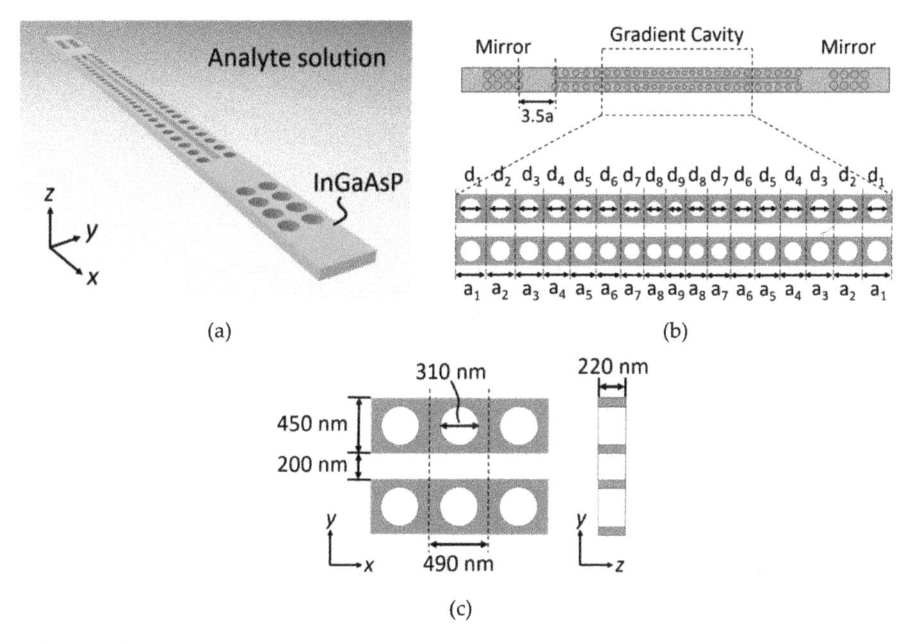

Figure 1. Geometry of the crystal slot nanobeam with a gradient cavity: (**a**) Schematic three-dimensional view of the device; (**b**) Top view of the device and the enlarged area of gradient cavity; (**c**) The top and cross-section views of the unit cell (marked by two dashed lines) of the corresponding perfect structure.

The structure was assumed to be made of indium gallium arsenide phosphide (InGaAsP). From the point of optical view, InGaAsP behaves as an isotropic medium. The RI was taken as $n = 3.2395$ [18]. The elastic constants of InGaAsP, a cubic material, were $C_{11} = 1.0081 \times 10^{11}$ N/m^2, $C_{12} = 5.4468 \times 10^{10}$ N/m^2, and $C_{44} = 4.6392 \times 10^{10}$ N/m^2 [19]. Meanwhile, the mass density of InGaAsP was $\rho = 4936.9$ kg/m^3 [19]. The analyte to be sensed was glucose solution with percent concentrations (by mass) of 0%, 20%, 40% and 60% (w/w), separately. The corresponding RIs of the solutions with various concentrations were $n = 1.3330, 1.3639, 1.3999$ and 1.4419, respectively [20]. The corresponding mass densities were $\rho = 998, 1084, 1180$ and 1273 kg/m^3, respectively [21]. The sound speeds in the solutions were taken as 1490, 1558, 1671 and 1781 m/s, respectively [21].

2.2. Theories

The time-harmonic governing equations for the optical and acoustic modes are provided in Sections 2.2.1 and 2.2.2, respectively. All equations presented hereafter are in phasor notations.

The formulations for calculating the sensitivity, quality factor, and figure of merit for the photonic–phononic device are listed in Section 2.2.3.

2.2.1. Optical Modes

The magnetic field **H** of time-harmonic optical waves is governed by the Helmholtz equation [22]:

$$\nabla\times\left(\frac{1}{\varepsilon_r}\nabla\times\mathbf{H}\right) = \frac{\omega^2}{c^2}\mathbf{H} \tag{2}$$

and subjected to the transverse condition:

$$\nabla\cdot\mathbf{H} = 0 \tag{3}$$

where ε_r, ω, and c are the relative permittivity, angular frequency, and the light speed in a vacuum, respectively. Once the magnetic field was solved, the electric field could be obtained as

$$\mathbf{E} = \frac{i}{\omega\varepsilon}\nabla\times\mathbf{H} \tag{4}$$

where ε is the permittivity.

2.2.2. Acoustic Modes

The sensor was immersed in the glucose solution. Therefore, the acoustic analysis involved the solid–fluid interaction. The relevant equations include not only the governing equations for both the solid and fluid but also the interface coupling conditions between two media. The pertinent equations in the phasor forms are presented.

The propagation of time-harmonic waves in an elastic body was governed by the Cauchy equation of motion. Without considering body force, the governing equation is [23]:

$$\nabla\cdot\mathbf{T} + \rho_s\omega^2\mathbf{u}_s = 0 \tag{5}$$

where **T**, ρ_s, and $\mathbf{u}_s$ are the stress, mass density, and displacement of the solid, respectively. For an anisotropic elastic material, the constitutive equation connecting stress **T** and strain **S** is

$$\mathbf{T} = \mathbf{C} : S \tag{6}$$

where **C** is the stiffness tensor. The strain **S** caused by the displacement field $\mathbf{u_s}$ can be expressed as:

$$S = \frac{1}{2}\left[\nabla\mathbf{u}_s + (\nabla\mathbf{u}_s)^T\right] \tag{7}$$

The time-harmonic wave propagation in the fluid domain was governed by the Helmholtz equation for sound pressure [24]:

$$\nabla\cdot\left(\frac{1}{\rho_f}\nabla p\right) + \frac{\omega^2}{\rho_f c_f^2}p = 0 \tag{8}$$

where p, ρ_f, and c_f are the pressure, mass density, and sound speed of the fluid, respectively.

On the interface, the force balance and displacement continuity are required [24]. The force balance condition, or traction condition, is:

$$\mathbf{T}\cdot\mathbf{n} = -p\mathbf{n} \tag{9}$$

where **n** is the unit outward normal vector to the boundary of the solid domain. Meanwhile, when the body force is neglected, the displacement continuity condition can be expressed as:

$$\rho_f \omega^2 \mathbf{u}_s \cdot \mathbf{n} = \frac{\partial p}{\partial n} \quad (10)$$

2.2.3. Sensitivity and Figure of Merit

The sensing parameters, namely sensitivity, quality factor, and figure of merit, were introduced to evaluate the optical and acoustic sensing performances of the nanobeam device. The sensitivities of photonic and phononic sensors are defined separately as [2]

$$S_{photonic} = \frac{\Delta\lambda_r}{\Delta n} \text{ (nm/RIU) }; \ S_{photonic} = \frac{\Delta f_r}{\Delta c} \left(\text{kHz/ms}^{-1}\right) \quad (11)$$

In the above expression, $\Delta\lambda_r$ and Δn are the resonant wavelength shift and RI difference of two solutions when performing optical sensing. Parameters Δf_r and Δc are the resonant frequency shift and sound-velocity difference of two solutions when performing acoustic sensing. Unit RIU stands for the Refractive Index Unit.

Figure of merit, also widely used in photonic and phononic sensors to characterize the sensing capability, is defined as [2]

$$FOM_{photonic} = \frac{S_{photonic} \times Q_o}{\lambda_r} \left(\text{RIU}^{-1}\right); \ FOM_{photonic} = \frac{S_{photonic} \times Q_a}{f_r} \text{ (m/s)}^{-1} \quad (12)$$

where λ_r and f_r are the resonant wavelength shift and frequency shift for optical sensing and acoustic sensing, respectively. Parameters Q_o and Q_a are the quality factors for photonic and phononic sensors, which are defined as [2]:

$$Q_o = \frac{\lambda_r}{FWHM_{,\lambda}}; \ Q_a = \frac{f_r}{FWHM_{,f}} \quad (13)$$

where $FWHM_\lambda$ and $FWHM_f$ are the full width at half maximum of the resonant wavelength and frequency, respectively.

3. Results and Discussions

This section presents and discusses the calculated results of the optical and acoustic behaviors and the sensing performance of the photonic–phononic device. The results based on the optical and acoustic analyses are given in Sections 3.1 and 3.2, respectively.

3.1. Optical Behaviors

The results of the optical behavior analysis are presented in this subsection. First, the band diagram and eigenmodes of the perfect crystal structure immersed in the distilled water were studied. Second, the defect modes of the finite-length nanobeam with $N = 0.04$ were found. Third, the resonant spectra and wavelength shifts for glucose solutions with various concentrations are presented. The sensing performance for each excited mode was evaluated by examining the evaluation parameters. Finally, the effect of the cavity gradient on the sensing performance was investigated.

3.1.1. Perfect Modes of Photonic Crystal

The time-harmonic optical wave equation (2) was solved numerically based on the unit-cell model of the perfect crystal slot nanobeam. The perfect crystal waveguide under consideration was assumed to be immersed in pure water. The calculated dispersion curves below the light cone are shown in Figure 2a. The four band curves corresponded to four photonic eigenmodes that can propagate in the perfect crystal structure. At the Brillouin zone boundary, their corresponding wavelengths were 1338.6, 1407.9, 1630.9, and 1707.1 nm, respectively. It is noted that there was a PTBG between Modes B and C.

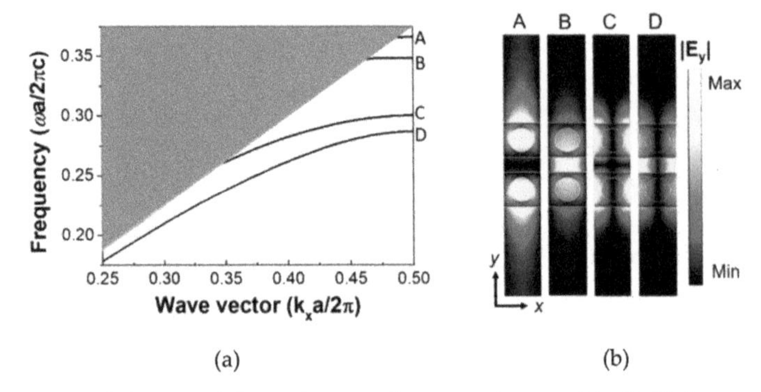

Figure 2. (**a**) Band diagram of the perfect crystal structure immersed in distilled water; (**b**) The electric-field intensity ($|\mathbf{E}_y|$) profiles of the four band-edge modes.

The electric-field intensity patterns associated with these four band-edge Modes A–D are shown in Figure 2b. For all those modes, the electric field was primarily polarized on the xy plane and dictated by E_y; meanwhile, the magnetic field was predominately in the transverse (z) direction. The highest-frequency Mode A was the closest mode to the light line; its mode profile revealed that the electric field extended into the surrounding water. Meanwhile, the lowest-frequency Mode D was far away from the light line; its mode pattern suggested that this mode had more optical energy concentrated in the slot and InGaAsP.

To understand the modal characteristics of the grooved nanobeam, the sensing performance of the perfect lattice modes of the PTC was further analyzed. The surrounding pure water was replaced by a glucose solution with a concentration of 60%. The wavelength shift ratios and sensitivity values of these four band-edge modes are reported in Table 1. It is noted that the high-frequency Modes A and B had higher wavelength shift ratios and sensitivity values than the low-frequency Modes C and D. These results were consistent with the electric field distributions shown in Figure 2b, which indicates that there is a stronger interaction between light and fluid for the high-frequency modes. For the low-frequency modes, Mode D had higher values of wavelength shift and sensitivity than Mode C. It is because the optical energy of Mode C was primarily located in InGaAsP, while Mode D had more optical energy confined within the slot infiltrated with solution. From the above results, it is further confirmed that an increase in the active region between optical resonance and analyte will generally improve sensitivity.

Table 1. The wavelength shifts and sensitivity values of perfect crystal modes for the 60% glucose solution.

Mode	$\Delta\lambda_{60\%}/\lambda_r$	$S_{60\%}$ (nm/RIU)
A	5.94%	730
B	5.58%	721
C	3.20%	479
D	3.59%	563

3.1.2. Defect Modes of Photonic Crystal (with $N = 0.04$)

For the finite-length crystal slot nanobeam (with $N = 0.04$) immersed in the pure water, nine localized defect modes were found and are indicated by dashed lines in Figure 3a. The bands of the perfect crystal structure are also repeatedly shown in Figure 3a. The six higher-frequency defect modes 1–6 were inside the band gap, and their wavelengths were 1445.0, 1472.2, 1526.6, 1572.0, 1576.2, and 1620.0 nm, respectively. The distributions of their electric field are shown in Figure 3b. Meanwhile, three lower-frequency modes 7–9 were below the frequency band gap with wavelengths 1640.9, 1699.5, and 1746.7 nm, respectively. The electric field profiles are shown in Figure 3c, which indicate that the optical energy was also largely confined. In general, the high-frequency modes had more optical energy concentrated in the central part of the nanobeam. Meanwhile, the optical energy of low-frequency modes spread to both ends but did not leak to the surrounding water due to the existence of mirrors and gradient cavity. The energy of Modes 1, 3, 5, and 6 was primarily concentrated in the hole cavities only. However, the energy of Modes 2, 4, 7, 8, and 9 was concentrated in both holes and the slot.

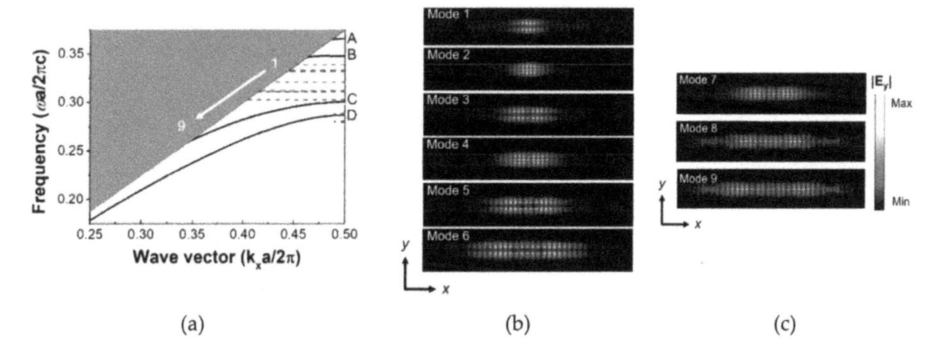

Figure 3. (**a**) Frequencies of the defect Modes 1–9 for the nanobeam (with $N = 0.04$) immersed in the pure water are indicated on the band diagram of the perfect crystal; (**b**) Electric strength profiles of Modes 1–6 that were inside the band gap; (**c**) Electric strength profiles of Modes 7–9 that were below the band gap.3.1.3. Sensing Performance for Photonic Device (with $N = 0.04$)

To analyze the sensing performance for these cavity modes of the photonic device, external excitations are applied to the cavity structure (with $N = 0.04$) immersed in a sensing fluid to obtain the resonance spectrum. By changing the surrounding fluids with different RIs, the offsets of the resonance wavelengths in the spectrum are observed and used for the RI sensing. The resonance spectrum of the nanobeam sensor with infiltration of distilled water is shown in Figure 4a. Five peaks were observed for wavelengths ranging from 1400 nm to 1770 nm. These five excited modes corresponded to defect crystal Modes 2, 4, 7, 8, and 9 with resonance wavelengths 1472.1, 1572.0, 1640.9, 1701.8, and 1746.5 nm, respectively. The electric field patterns of the excited modes are illustrated in Figure 4b. It is seen that both the resonance wavelengths and excited mode profiles were consistent with frequencies and eigenmodes based on the eigen-analysis. It is also noted from resonance spectra that Modes 4 and 7 had higher peaks, which implies that these two modes had more optical energy stored in the structure than the other modes. However, those modes associated with higher peaks did not necessarily have a better sensing performance. The quality of the sensing performance depended on the interaction between the optical mode profile and the object to be tested, and on the energy storage capacity in the cavity.

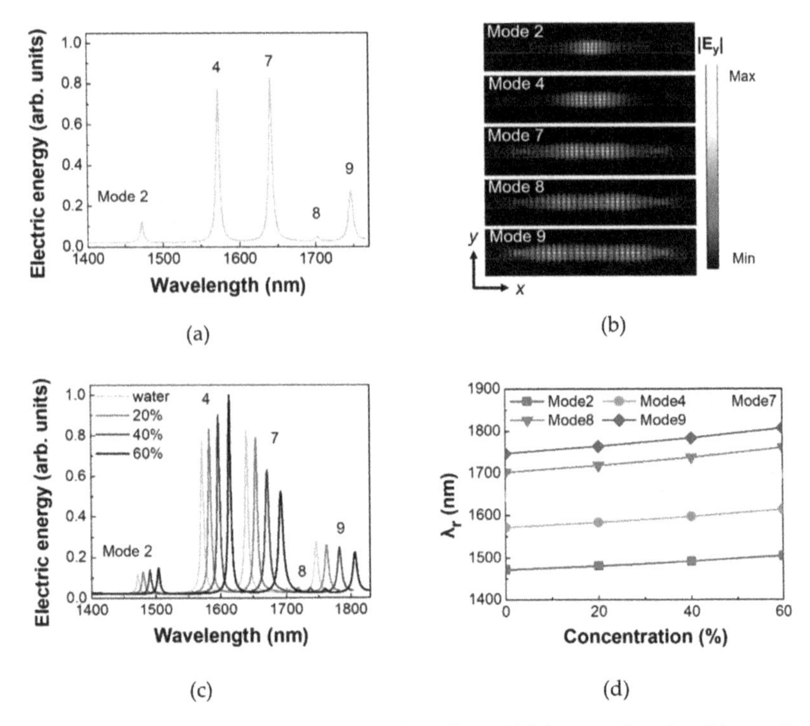

Figure 4. Resonance spectra, resonance wavelengths, and excited defect crystal modes of the nanobeam sensor device with $N = 0.04$: (**a**) Resonance spectrum of the nanobeam immersed in distilled water; (**b**) Electric field distributions of the excited defect crystal modes of the nanobeam immersed in distilled water; (**c**) Resonance spectra and wavelength shifts of the nanobeam immersed in the glucose solutions with various concentrations for various modes; (**d**) The resonance wavelengths of various modes are plotted against concentrations of glucose solutions.

To analyze the sensing performance of these resonance modes, the surrounding distilled water was replaced by glucose solutions with concentrations of 20%, 40%, and 60%, separately. The resulting resonance spectra are shown in Figure 4c. It is observed that a larger redshift occurred for a solution with a higher concentration. The redshift phenomenon can be justified by the following perturbation formula for frequency shift [25]:

$$\frac{\omega - \omega_0}{\omega} = \frac{-\iiint_V dV[(\Delta\mu\mathbf{H})\cdot\mathbf{H}^* + (\Delta\varepsilon\mathbf{E})\cdot\mathbf{E}^*]}{\iiint_V dV[\mu\mathbf{H}\cdot\mathbf{H}^* + \varepsilon\mathbf{E}\cdot\mathbf{E}^*]} \tag{14}$$

where ω_0 and ω are the resonance frequencies of the structure immersed in pure water and in glucose solution, respectively. Parameter $\Delta\mu$ is the difference in permeability between solution and water, which can be neglected since both permeability values are taken to be unity. Meanwhile, $\Delta\varepsilon$ is the difference in dielectric constants between glucose solution and water. A glucose solution with higher concentration will have a larger RI and consequently will have a higher value in dielectric constant. Therefore, it can be concluded from Equation (14) that replacing water with glucose solution will cause a redshift in resonance wavelength. The resonance wavelengths of the above-mentioned five excitation modes are plotted against solution concentrations in Figure 4d. It is seen from the figure that the relation between resonant wavelength and solution concentration was nearly linear up to a concentration of 60% for every mode, which posts a great advantage in the detection of an analyte with various concentrations.

The resonance wavelengths and the full width at half maximum (FWHM) for various excited defect modes for a solution with 60% concentration are listed in Table 2. The resonance wavelength shifts corresponding to distilled water and 60% solution are also included in Table 2. The performance evaluation parameters Q, S, and FOM were calculated; the results based on the 60% solution are tabulated in the table. It is noted that, in general, higher-frequency modes had higher Q values than the lower-frequency modes. This observation can be confirmed by examining the mode profiles illustrated in Figure 4b, which shows the optical energy distributions of high-frequency modes were more concentrated and confined in the center of the nanobeam. One exception was low-frequency Mode 8, whose energy distribution spread all over the holes and slot, but it had the highest Q value. This may be due to the mirrors on both ends, which can effectively reflect light back to the structure. It is further noted from Table 2 that the lower-frequency modes consistently had a larger sensitivity S than the higher-frequency modes. This is because those low-frequency modes had more extensive interaction between light resonance and sensing fluids. As a final observation of Table 2, it is no surprise that Mode 8 had a much higher FOM than the other modes since FOM incorporated both S and Q, and Mode 8 had nearly the largest values in both parameters.

Table 2. The resonance wavelengths, full width at half maximum (FWHM) and evaluation parameters for various excited defect modes corresponding to the 60% glucose solution and $N = 0.04$.

Mode	$\lambda_{0\%}$ (nm)	$\lambda_{60\%}$ (nm)	FWHM(nm)	$\Delta\lambda_{60\%}$ (nm)	$Q_{60\%}$	$S_{60\%}$ (nm/RIU)	$FOM_{60\%}$ (1/RIU)
2	1472.1	1504.3	4.64	32.2	324	296	64
4	1572.0	1614.6	4.63	42.6	349	391	84
7	1640.9	1692.1	7.87	51.2	215	470	60
8	1701.8	1760.1	4.30	58.3	409	535	124
9	1746.5	1806.3	7.71	59.8	234	549	71

3.1.3. Gradient Cavity Optimization

The design of the gradient cavity with varying hole diameters and heterogeneous lattices is aimed at creating a cavity that can produce more localized modes. In this subsubsection, the effect of gradient cavity, controlled by the hole tuning parameter N on the sensing performance of the nanobeam, is studied in detail. The glucose solution with a 60% concentration was considered. The values of N were 0, 0.01, 0.02, 0.03, and 0.04, respectively. According to Equation (1), the diameters and spacing of holes decreased when the value of N increased.

The electric-field intensity patterns corresponding to the highest-frequency Mode 2 and the low-frequency Mode 8 for various values of N are illustrated in Figure 5a,b. It is noted that from both figures that modes with larger N had a more localized optical energy distribution. For Mode 2, a large cavity gradient (i.e., large N) yielded an extremely localized distribution; meanwhile, this mode had a more extended pattern when the cavity gradient was smaller. For low-frequency Mode 8, it is observed that optical energy leaks from both ends of the nanobeam when gradient defect was absent ($N = 0$). On the other hand, the leaky mode could be tuned into a more localized mode by creating a larger gradient cavity. It is worth mentioning that the lowest-frequency resonant Mode 9 could not be excited when the nanobeam lacked a cavity gradient.

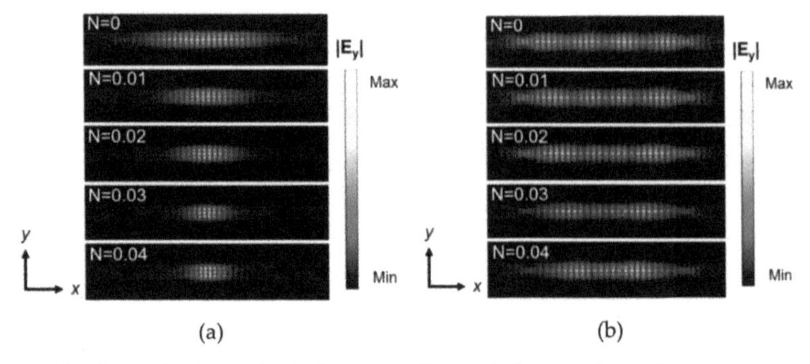

Figure 5. The electric-field intensity ($|\mathbf{E}_y|$) patterns of the 60% glucose solution for various degrees of cavity gradients (N = 0, 0.01, 0.02, 0.03, 0.04): (**a**) Profiles of high-frequency Mode 2, (**b**) Profiles of low-frequency Mode 8.

The evaluation parameters Q, S, and FOM are plotted against the hole tuning parameter N in Figure 6a–c, separately for the 60% glucose solution. It is noted from Figure 6a that while quality factors of the high-frequency modes were not much affected by the cavity gradient, the quality factor Q of low-frequency Modes 8 and 9 improved as the cavity gradient increases. This is because a dispersive pattern can be transformed into a more concentrated mode by increasing N, as illustrated in Figure 5b. Figure 6b reveals that sensitivity S decreased as cavity gradient parameter N increased for all the modes considered, especially for high-frequency Modes 2 and 4. As indicated in Figure 5a, Mode 2 became more concentrated, and the volume of mode was greatly reduced as N increased. The reduction in the active region between optical resonance and analyte caused a decrease in S. Figure 6c shows the effect of hole tuning parameter N on the FOM, which incorporated both S and Q. It shows that the design of gradient cavity was more beneficial to the low-frequency modes. The FOM of Mode 9 improved steadily as N increased. Mode 8 had an overall highest FOM when N = 0.03.

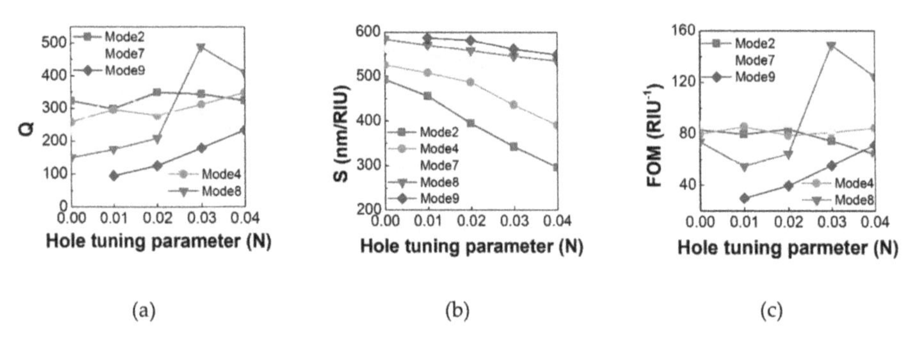

Figure 6. The evaluation parameters Q, S, and FOM are plotted against the hole tuning parameter N in (**a–c**), separately for the 60% glucose solution.

3.2. Acoustic Behaviors

The nanobeams, coupled with gradient cavity design, can work not only as optical sensors but also as acoustic sensors, which is demonstrated in this subsection. The results of the phononic analysis are presented hereafter. First, the band diagram and eigenmodes of the perfect crystal structure immersed in distilled water are analyzed. Among various eigenmodes, those modes associated with volume dilation of cavities are singled out because they possess a larger solid–fluid interaction. Then, the defect modes of the finite-length nanobeam with N = 0.04 were investigated. Finally, the acoustic resonant spectra for glucose solutions with various concentrations are presented. The acoustic sensing performance for each excited mode was evaluated by examining the evaluation parameters.

3.2.1. Perfect Modes of Phononic Crystal

From the sound analysis of the perfect crystal dual nanobeam immersed in water, twelve acoustic dispersion curves were found, as shown in Figure 7a. The corresponding displacement and sound pressure fields for these modes are depicted in Figure 7b. The deformation patterns of the unit cell associated with these modes are also illustrated in Figure 7b. It is noted that the solid black line in Figure 7a denotes the sound line in the water. In light of the results of the optical part, it can be shown that when the mode distributions penetrated the water domain, their interaction with the water solution could be stronger. However, when sound waves propagated in a structure surrounded by water, acoustic energy may radiate freely into the water. Here, the crystal dual nanobeam was used to confine the sound pressure field in the slot and holes to sense changes in the liquid domain. Confining sound fields requires the frequency to be below the sound cone of water. Since the perfect crystal structure is composed of two identical beams separated by a gap, there exist similar vibration patterns in these two beams. The vibrations of the beams are either in-phase or out-of-phase, as illustrated by the deformation patterns shown in Figure 7b. Therefore, there were six pairs of modes among the twelve modes. In Figure 7a, the modes, in pairs, in the dispersion curves are labeled with the same band number; the in-phase modes are distinguished from the out-of-phase modes with a superscript prime (′). The in-phase and out-of-phase motions of the dual nanobeams in the water will yield distinct displacement and pressure fields, but with very close eigen-frequencies.

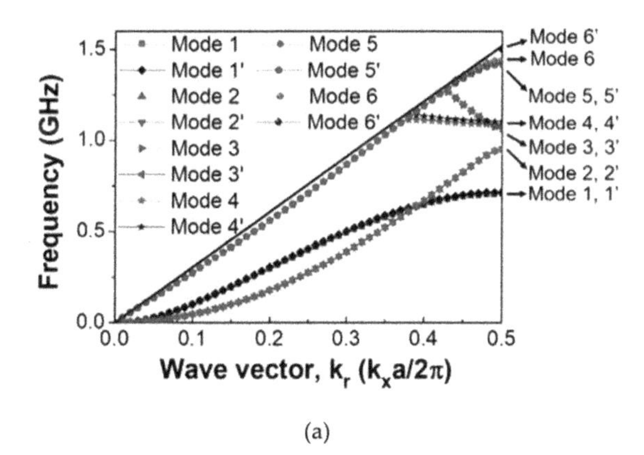

(a)

Figure 7. *Cont.*

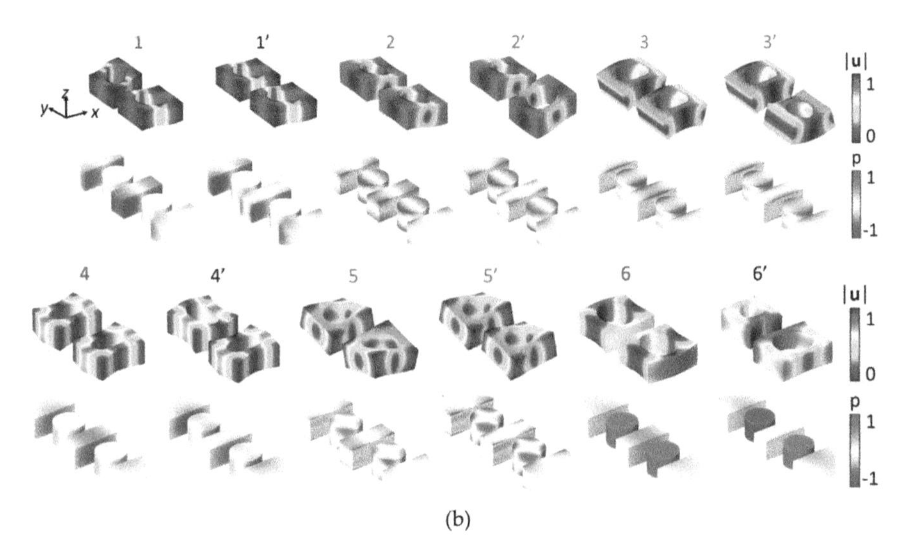

Figure 7. (**a**) Acoustic dispersion curves of the perfect crystal dual beams immersed in water; (**b**) The displacement and sound pressure fields in the unit cell, respectively.

Figure 7b shows the displacement and sound pressure fields in the crystal dual nanobeams and water, respectively. The modes can be classified into three types based on their displacement fields. The modes of the first type had a dominated displacement polarization in the z-direction, and this type of the mode can be divided into the bending and twisting forms. It is noted from the three-dimensional deformation figures that Modes 2, 2′, 3, 3′, 5, and 5′ belong to this type. Because of their polarization direction, they were less likely to be influenced by the deformation of the gap and holes, so the solid–fluid coupling and the resulting sound pressure field were relatively weaker. The polarizations of the second and third types were both on the x-y plane. The second type is called the **dilatation mode**, which is characterized by the volume dilatation (expansion or contraction) in both the gap and holes. Modes 1, 4, and 6 were seen to be the dilatation modes. The dilatation was primarily caused by the out-of-phase motions in the y-direction of the dual nanobeams. The opposite movement of the dual nanobeams creates a symmetric deformation pattern with the center line of the gap as the line of reflection symmetry, which consequently will result in a dilatation in the gap. The dilatation modes were more likely to produce a stronger coupling between the beam vibration and the sound pressure. As an example, Mode 1 showed a concentrated sound pressure field in the gap region, which was induced by the expansion and contraction of the gap due to the opposite movements of the two beams. In addition, for this mode, the opposite motion deformed the circular holes into oval cavities, which generated a quadrupole sound pressure pattern in both holes. On the other hand, the third-type mode lacked a significant volume change in the gap, which is primarily attributed to the same y-directional movement of the dual nanobeams. This type of motion causes a shear/bend pattern of the gap; the volume change in the gap due to this type of deformation appears to be trivial. The remaining modes 1′, 4′, and 6′ fell into this category. Due to the fact that the dilatation modes are more likely to cause the volume changes in the gap and holes, they are expected to enhance the solid–fluid coupling, and thus, yields a better sensing performance. Therefore, the attention of acoustic sensing is focused on the dilatation modes associated with the dispersion curves 1, 4, and 6.

3.2.2. Defect Modes of Phononic Crystal (with $N = 0.04$)

The identical slot dual nanobeam with a gradient cavity used for optical sensing was employed for acoustic sensing. The previous photonic analysis has shown that an extended mode can be tuned into a more concentrated mode by increasing the cavity gradient. The current phononic analysis shows that the incorporation of a gradient cavity is even more critical for producing localized defect modes. In this study, no defect modes could be excited in case that nanobeams lack a gradient cavity. Hereafter, only crystal nanobeams with a large cavity gradient ($N = 0.04$) are considered.

The band diagram of perfect crystal dual nanobeams with dispersion curves associated with dilatation modes 1, 4, and 6 are drawn in Figure 8a. There were two acoustic frequency gaps for the dilatation modes, whose frequencies ranged from 0.709 to 1.087 GHz and from 1.105 to 1.331 GHz, respectively. For the sensor with gradient defect $N = 0.04$, one defect mode in the upper-frequency gap and two defect modes in the lower frequency gap were found. The displacement and sound pressure fields of these three defect modes in the gradient cavity area are shown in Figure 8b. It is seen that the displacements of the dual nanobeams were reflection-symmetric with respect to the centerline of the slot, which results in the width variation of the slot. The dilatation pressed the liquid in the slot and holes and subsequently induced a high sound pressure field in the cavity. Accordingly, these confined cavity modes can be used to sense the change in the acoustic properties of the glucose solutions. Moreover, it is noted from modal patterns shown in Figure 8b that Mode C was more dispersive than Modes A and B, which suggests that Mode C had a larger wavelength, that is, a lower resonance frequency than Modes A and B.

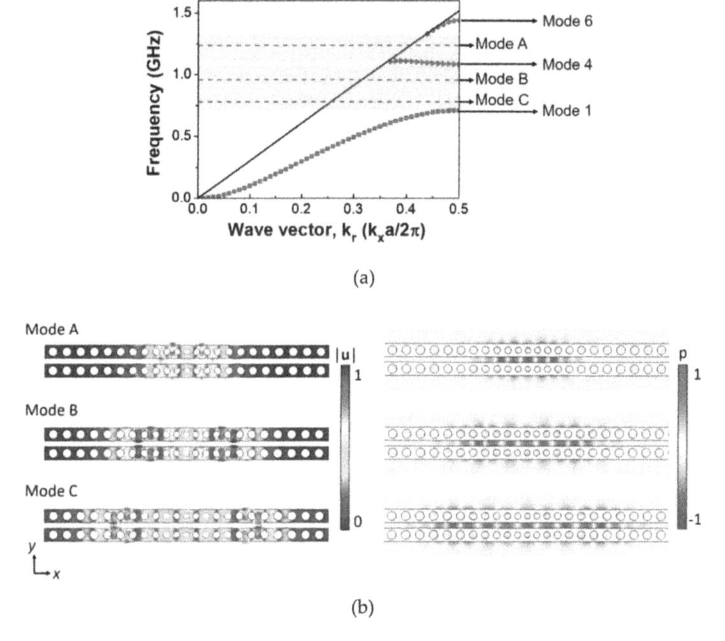

Figure 8. (**a**) Acoustic dispersion curves associated with dilatation modes (1, 4, and 6) of perfect crystal nanobeams immersed in pure water are drawn. Frequencies of the defect modes (A, B, and C) with $N = 0.04$ are indicated in the band diagram; (**b**) The displacement and sound pressure fields of these three defect modes are shown in the gradient cavity area of the dual nanobeam device.

3.2.3. Sensing Performance for Phononic Device

To analyze the acoustic sensing performance of the nanobeam with a gradient cavity (with $N = 0.04$) using the resonant cavity, an external excitation method was applied to obtain the acoustic cavity mode spectra for the 20%, 40%, and 60% glucose solutions. The calculated results of the resonance spectra of the three acoustic cavity modes are shown in Figure 9a–c. It is found for all these three modes that a system involving a higher-concentration solution had a lower system resonant frequency. As the glucose concentration increased, so did the density of the surrounding fluid. In this situation, the overall system was less likely to be pushed around to produce oscillations, so the resonance frequency of the system reduced. It is observed from Figure 9a–c that all the resonance spectra had very narrow peaks and small FWHMs, especially in Modes B and C, which yielded very large values of quality factor. This apparently unrealistic result is largely thanks to the idealized numerical simulation adopted in the current study, without considering material losses, viscous damping, etc. The resonance frequencies of these three modes are plotted against various solution concentrations in Figure 9d. It is noted that their relationship was nearly linear for concentrations up to 60%, and the acoustic resonance frequency offset reached about 13 MHz.

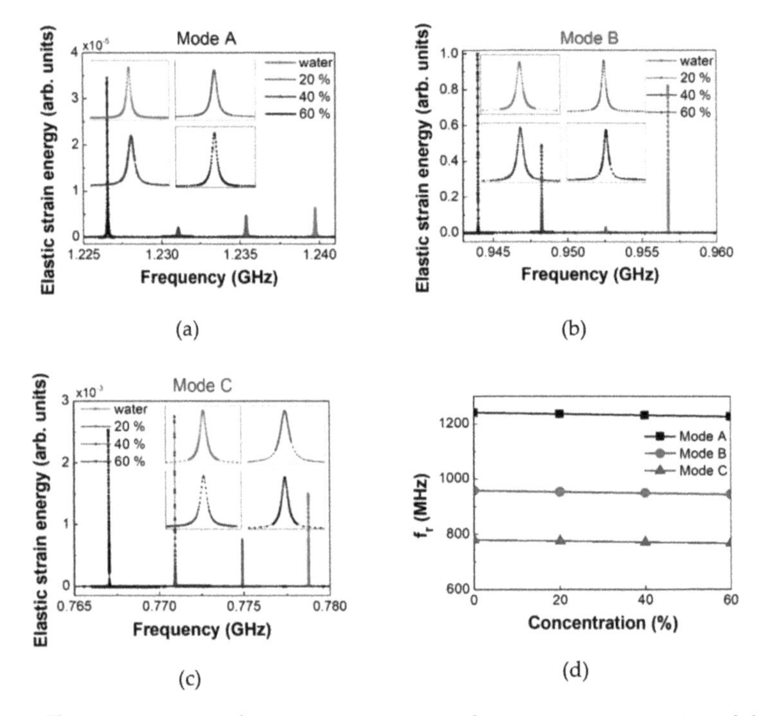

Figure 9. The acoustic resonance frequency spectra corresponding to various concentrations of glucose solutions for Modes A, B, and C are plotted in (**a**–**c**), separately; (**d**) The resonance frequencies of various modes are plotted against the concentrations of glucose solutions.

Table 3 lists the acoustic sensing evaluation parameters when the liquid was changed from water to a 60% glucose solution. It is noted that the sensitivity of Mode A was slightly higher than Modes B and C. Meanwhile, Mode B had the highest value of quality factor, more than tenfold that of the other two modes. Consequently, Mode B possessed a much better value of figure of merit than Modes A and C. Figure 8b shows that Mode A had the best concentrated field. Nonetheless, it can be seen from the strain energy intensity spectrum in Figure 9b that Mode B had the highest energy intensity. This is because the displacement of Mode B can induce a stronger solid–fluid coupling, which results in higher quality factor and figure of merit. As a final remark, it is worthwhile to repeatedly mention that the values of quality factor and figure of merit listed in Table 3 were the consequence of the "highly" idealized analysis. This study neglects many factors that could lead to energy losses. The readers should treat the current numerical results with caution when the proposed design is to be realized into a sensor.

Table 3. Acoustic evaluation parameters for various excited defect modes corresponding to the 60% glucose solution.

Mode	Frequency$f_{r,60\%}$ (GHz)	$S_{60\%}$ (kHz/ms^{-1})	$Q_{60\%}$	$FOM_{60\%}$ (m/s)$^{-1}$
A	1.240	45.3	2.31×10^4	1.98
B	0.9567	44.0	8.08×10^6	411
C	0.7787	40.3	3.32×10^5	22.6

4. Conclusions

A dual photonic-phononic crystal slot nanobeam incorporated with a gradient cavity was proposed for liquid sensing in this study. The device was able to confine both optical and acoustic waves with mode bandgaps to enhance the interactions between light/sound and analyte solution. The introduction of the gradient cavity can further confine energy and reduce energy loss by avoiding the abrupt changes in lattices. The calculated results for photonic devices showed that the high-frequency defect modes had more optical energy concentrated in the central part of the nanobeam. Meanwhile, the optical energy of low-frequency defect modes spread to both ends but did not leak to the surrounding liquid due to the existence of mirrors and gradient cavity. The effect of the cavity gradient on the optical sensing performance for each defect mode was thoroughly examined. It is concluded that the gradient cavity design is more beneficial to low-frequency optical modes. The low-frequency defect modes tend to have a more dispersive pattern. A dispersive mode can be tuned into a localized mode by increasing the degree of cavity gradient, and the energy leakage can thus be reduced. Therefore, low-frequency optical modes can have more extensive interaction between light and analyte solution.

The integration of the gradient cavity also makes it feasible for the nanobeam to be used for acoustic sensing. The acoustic defect modes can be excited when the gradient cavity is present. The phononic analysis revealed that the acoustic dilatation modes, associated with the volume changes in the slot and holes, are more likely to produce a stronger coupling between the nanobeam vibration and sound pressure, and thus, yields a high acoustic sensing performance. The current photonic–phononic crystal sensors with gradient cavity have a noteworthy sensing performance, which is attributed to the strong interaction between light/sound and analyte solution. The dual optical/acoustic sensing mechanism provides opportunities for various applications of photonic–phononic crystals in sensing and detection.

Author Contributions: N.-N.H., Y.-C.C., J.-C.H., and T.-R.L. initiated the work. Y.-C.C., H.-T.C., Y.-F.L., C.-T.K., Y.-W.C., and C.-Y.C. performed the numerical simulation and modeling supervised by N.-N.H., J.-C.H., and T.-R.L. The manuscript was written by N.-N.H., Y.-C.C., J.-C.H., and T.-R.L. The funding was acquired by T.-R.L. and J.-C.H. All authors have discussed the results and commented on the manuscript. All authors have read and agreed to the published version of the manuscript.

Funding: This work was supported by Minister of Science and Technology (MOST) in Taiwan under Contracts MOST 105-2221-E-019-049-MY3, MOST 108-2221-E-019-055-MY3, MOST 108-2221-E-019-057-MY3, and MOST 106-2628-E-224-001-MY3.

Conflicts of Interest: The authors declare no conflict of interest.

References

1. Yablonovitch, E. Photonic crystals: Semiconductors of light. *Sci. Am.* **2001**, *285*, 47–55. [CrossRef] [PubMed]
2. Cheng, P.-J.; Huang, Z.-T.; Li, J.-H.; Chou, B.-T.; Chou, Y.-H.; Lo, W.-C.; Chen, K.-P.; Lu, T.-C.; Lin, T.-R. High-performance plasmonic nanolasers with a nanotrench defect cavity for sensing applications. *ACS Photonics* **2018**, *5*, 2638–2644. [CrossRef]
3. Song, Y.; Bai, J.; Zhang, R.; Wu, E.; Wang, J.; Li, S.; Ning, B.; Wang, M.; Gao, Z.; Peng, Y. LSPR-enhanced photonic crystal allows ultrasensitive and label-free detection of hazardous chemicals. *Sens. Actuators B* **2020**, *310*, 127671. [CrossRef]

4. Li, T.; Gao, D.; Zhang, D.; Cassan, E. High-Q and high-sensitivity one-dimensional photonic crystal slot nanobeam cavity sensors. *IEEE Photonics Technol. Lett.* **2016**, *28*, 689–692. [CrossRef]
5. Watanabe, T.; Saijo, Y.; Hasegawa, Y.; Watanabe, K.; Nishijima, Y.; Baba, T. Ion-sensitive photonic-crystal nanolaser sensors. *Opt. Express* **2017**, *25*, 24469–24479. [CrossRef] [PubMed]
6. Rodriguez, G.A.; Markov, P.; Cartwright, A.P.; Choudhury, M.H.; Afzal, F.O.; Cao, T.; Halimi, S.I.; Retterer, S.T.; Kravchenko, I.I.; Weiss, S.M. Photonic crystal nanobeam biosensors based on porous silicon. *Opt. Express* **2019**, *27*, 9536–9549. [CrossRef] [PubMed]
7. Lucklum, R.; Ke, M.; Zubtsov, M. Two-dimensional phononic crystal sensor based on a cavity mode. *Sens. Actuators B* **2012**, *171–172*, 271–277. [CrossRef]
8. Oseev, A.; Mukhin, N.; Lucklum, R.; Zubtsov, M.; Schmidt, M.-P.; Steinmann, U.; Fomin, A.; Kozyrev, A.; Hirsch, S. Study of liquid resonances in solid-liquid composite periodic structures (phononic crystals)—Theoretical investigations and practical application for in-line analysis of conventional petroleum products. *Sens. Actuators B* **2018**, *257*, 469–477. [CrossRef]
9. Amoudache, S.; Pennec, Y.; Rouhani, B.D.; Khater, A.; Lucklum, R.; Tigrine, R. Simultaneous sensing of light and sound velocities of fluids in a two-dimensional phoXonic crystal with defects. *J. Appl. Phys.* **2014**, *115*, 134503. [CrossRef]
10. Amoudache, S.; Moiseyenko, R.; Pennec, Y.; Rouhani, B.D.; Khater, A.; Lucklum, R.; Tigrine, R. Optical and acoustic sensing using Fano-like resonances in dual phononic and photonic crystal plate. *J. Appl. Phys.* **2016**, *119*, 114502. [CrossRef]
11. Ma, T.-X.; Wang, Y.-S.; Zhang, C.; Su, X.-X. Theoretical research on a two-dimensional phoxonic crystal liquid sensor by utilizing surface optical and acoustic waves. *Sens. Actuator A Phys.* **2016**, *242*, 123–131. [CrossRef]
12. Wang, B.; Dündar, M.A.; Nötzel, R.; Karouta, F.; He, S.; van der Heijden, R.W. Photonic crystal slot nanobeam slow light waveguides for refractive index sensing. *Appl. Phys. Lett.* **2010**, *97*, 151105. [CrossRef]
13. Zain, A.R.M.; Johnson, N.P.; Sorel, M.; De la Rue, R.M. Ultra high quality factor one dimensional photonic crystal/photonic wire micro-cavities in silicon-on-insulator (SOI). *Opt. Express* **2008**, *16*, 12084–12089. [CrossRef] [PubMed]
14. Lin, T.-R.; Lin, C.-H.; Hsu, J.-C. Strong optomechanical interaction in hybrid plasmonic-photonic crystal nanocavities with surface acoustic waves. *Sci. Rep.* **2015**, *5*, 13782. [CrossRef] [PubMed]
15. COMSOL Multiphysics®®Modeling Software. Available online: https://www.comsol.com/ (accessed on 13 April 2020).
16. Almeida, V.R.; Xu, Q.; Barios, C.A.; Lipson, M. Guiding and confining light in void nanostructure. *Opt. Lett.* **2004**, *29*, 1209–1211. [CrossRef] [PubMed]
17. Zhang, Y.; Zhao, Y.; Zhou, T.; Wu, Q. Applications and developments of on-chip biochemical sensors based on optofluidic photonic crystal cavities. *Lab Chip* **2018**, *18*, 57–74. [CrossRef] [PubMed]
18. Adachi, S. Optical dispersion relation for GaP, GaAs, GaSb, InP, InAs, InSb, $Al_xGa_{1-x}As$ and $In_{1-x}Ga_xAs_yP_{1-y}$. *J. Appl. Phys.* **1989**, *66*, 6030–6040. [CrossRef]
19. Adachi, S. *Physical Properties of III-V Semiconductor Compounds*; Wiley: New York, NY, USA, 1992; pp. 11–23.
20. Rosenbruch, K.J.; Emmerich, A. The refractometric determination of aqueous sugar solutions. *Sugar Technol. Rev.* **1988**, *14*, 137–205.
21. Saggin, R.; Coupland, J.N. Concentration measurement by acoustic reflectance. *J. Food Sci.* **2001**, *66*, 681–685. [CrossRef]
22. Joannopoulos, J.D.; Johnson, S.G.; Winn, J.N.; Meade, R.D. *Photonic Crystal*, 2nd ed.; Princeton University Press: New Jersey, NJ, USA, 2008; pp. 9–10.
23. Royer, D.; Dieulesaint, E. *Elastic Waves in Solids I: Free and Guided Propagation*, 1st ed.; Springer: New York, NY, USA, 2000; p. 128.
24. Monkola, S. Numerical Simulation of Fluid-Structure Interaction between Acoustic and Elastic Waves. Ph.D. Dissertation, University of Jyvaskyla, Jyvaskyla, Finland, 2011.
25. Lin, T.-R.; Chang, S.-W.; Chuang, S.L.; Zhang, Z.; Schuck, P.J. Coating effect on optical resonance of plasmonic nanobowtie antenna. *Appl. Phys. Lett.* **2010**, *97*, 063106. [CrossRef]

MDPI
St. Alban-Anlage 66
4052 Basel
Switzerland
Tel. +41 61 683 77 34
Fax +41 61 302 89 18
www.mdpi.com

Crystals Editorial Office
E-mail: crystals@mdpi.com
www.mdpi.com/journal/crystals

CPSIA information can be obtained
at www.ICGtesting.com
Printed in the USA
BVHW021037270321
603557BV00023B/99